Construction Technology

SUBIR K. SARKAR
Former Visiting Faculty
Department of Construction Engineering
Jadavpur University

SUBHAJIT SARASWATI
Former Head
Department of Construction Engineering
Jadavpur University

OXFORD
UNIVERSITY PRESS

OXFORD
UNIVERSITY PRESS

YMCA Library Building, Jai Singh Road, New Delhi 110001

Oxford University Press is a department of the University of Oxford.
It furthers the University's objective of excellence in research, scholarship,
and education by publishing worldwide in

Oxford   New York
Auckland   Cape Town   Dar es Salaam   Hong Kong   Karachi
Kuala Lumpur   Madrid   Melbourne   Mexico City   Nairobi
New Delhi   Shanghai   Taipei   Toronto

With offices in
Argentina   Austria   Brazil   Chile   Czech Republic   France   Greece
Guatemala   Hungary   Italy   Japan   Poland   Portugal   Singapore
South Korea   Switzerland   Thailand   Turkey   Ukraine   Vietnam

Oxford is a registered trade mark of Oxford University Press
in the UK and in certain other countries.

Published in India
by Oxford University Press

© Oxford University Press 2008

The moral rights of the author/s have been asserted.

Database right Oxford University Press (maker)

First published 2008

All rights reserved. No part of this publication may be reproduced,
stored in a retrieval system, or transmitted, in any form or by any means,
without the prior permission in writing of Oxford University Press,
or as expressly permitted by law, or under terms agreed with the appropriate
reprographics rights organization. Enquiries concerning reproduction
outside the scope of the above should be sent to the Rights Department,
Oxford University Press, at the address above.

You must not circulate this book in any other binding or cover
and you must impose this same condition on any acquirer.

Third-party website addresses mentioned in this book are provided
by Oxford University Press in good faith and for information only.
Oxford University Press disclaims any responsibility for the material contained therein.

ISBN-13: 978-0-19-569483-3
ISBN-10: 0-19-569483-X

Typeset in Times Roman
by Text-o-Graphics, Noida 201301
Printed in India by Ram Book Binding House, New Delhi 110020
and published by Oxford University Press
YMCA Library Building, Jai Singh Road, New Delhi 110001

To
Archana, my wife,
who was always by my side during
the preparation of the manuscript

                Subir K. Sarkar

To
my revered parents
Atoshi and late Rajat Kanti

                Subhajit Saraswati

Preface

In India, construction is a labour-intensive process, while in developed countries, the construction process is mostly mechanized. The degree of mechanization, however, depends on the nature of the construction work involved and the time schedule. Construction activities are now being increasingly mechanized all over the world, including India, to ensure timely completion of projects. High-capacity machines with better output and greater efficiency have made the construction process less stressful. Nevertheless, the construction process in India still involves a huge labour force because of the easy availability of cheap labour of all categories.

Infrastructural development is the need of the hour to ensure overall growth. Building infrastructures in India has now become a formidable challenge which can only be met by adopting innovative construction technology. Earlier, development projects, mostly in the public sector, were plagued with financial constraints. The scenario has now changed; there is no dearth of funds anymore. The projects are now being financed by the Central and State governments, private enterprises, and foreign direct investment. However, timely completion of projects is still a problem. This is where construction technology becomes relevant. We can no more rely on manual efforts as the same is beset with the problem of fatigue. Therefore, it becomes necessary to switch to more reliable technology to attain desired results.

Employing high-capacity equipment is the only way to expedite the construction process, thereby maintaining quality and enhancing productivity. For example, earthmoving is a very arduous and time-consuming process if done by manual efforts, whereas the time taken can be drastically reduced if heavy earthmoving equipment and machineries are judiciously selected and deployed. Technology can turn a giant project with tight construction schedules into a manageable one.

ABOUT THE BOOK

This book provides comprehensive coverage of all aspects of construction activities, starting from the initial survey to the completion of a project. It also covers the latest relevant developments in construction technology, such as mechanized concreting and excavation by controlled blasting. Each chapter begins with an introduction outlining the topics discussed in the chapter and gradually discusses each topic at length.

The book contains a detailed description of various topics, such as earthmoving in excavation/filling, rock blasting, concrete technology, welding technology, fabrication and erection of structural steel, transportation and handling, finishing work, and erection of precast concrete elements and bridges. The fundamentals of mechanization and automation have been explained along with the latest relevant developments in construction technology. The theory is presented in a simple and lucid style with a number of illustrative graphics and plenty of review questions at the end of each chapter. The extensive coverage of topics makes this book a perfect companion of the students of civil/mining/mechanical engineering and construction technology. The book is also expected to find favour among professionals and project management consultants involved in the field of construction.

CONTENTS AND COVERAGE

The book is divided into 15 chapters. The first three chapters discuss the basic issues. The last two chapters deal with quality and safety measures. The rest of the chapters present the different types of construction activities. A brief outline of the chapter-wise coverage is presented here.

Chapter 1 describes the role of 'technology' in transforming 'construction' from primitive to a highly sophisticated process. It also serves as an introduction to the fundamental issues related to construction, including quality and safety. *Chapter 2* deals with preparatory work which are the prerequisites for starting construction work. *Chapter 3* discusses transportation, handling, and storage of materials.

Chapter 4 presents earthmoving in excavation/filling in detail. Starting with a peek into the soil mechanics, it goes on to discuss groundwater problems, no-dig excavation, embankment formation, and dredging. *Chapter 5* deals with excavation in hard rock by blasting. This chapter contains the theory of rock blasting, drilling, and explosives in detail. *Chapter 6* covers classification of piles, pile driving methods, and quality control tests.

Chapter 7 contains a detailed discussion on concrete technology. It presents the properties of both plastic and hardened concrete giving an understanding of their ingredients. The details of various admixtures which may modify concrete properties to suit specific requirements are also discussed. Readymixed concrete, light/heavyweight concrete, high-performance concrete, self compacting concrete, extreme-weather concrete, polymer-modified concrete, fibre-reinforced concrete, and prestressed concrete are discussed in detail. *Chapter 8* covers fabrication and erection of structural steel. It includes welding technology and quality assurance in welding work. Methods of erection of steel structures as well as precast/prestressed concrete elements and bridges are discussed.

Chapter 9 is devoted to cladding along the periphery and partition walls with masonry work. *Chapter 10* talks about different types of roofing over the building structures. *Chapter 11* deals with the finishing work, such as flooring, plastering, glazing, painting, etc. *Chapter 12* discusses construction of roads and highways, including earthwork as well as rock excavation by controlled blasting. Road construction involves extensive drainage work, including construction of culverts, bridges, etc. *Chapter 13* deals with the fundamentals of mechanized construction work involved in earthwork, transportation, movement and handling, concrete production and placing, and scaffolding. It contains the fundamentals of mechanization, and detailed descriptions of all kinds of construction equipment.

Chapter 14 explains the globally accepted ISO 9000 Quality Systems. *Chapter 15* gives an insight into the safety and health measures required to be strictly followed at the project site. This chapter also deals with environmental pollution.

We encourage the readers to point out slip-ups, if any, which may have crept in inadvertently. Any suggestions for improvement of this book are also welcome.

ACKNOWLEDGEMENTS

Help and guidance from the following sources during the development of this book is gratefully acknowledged:

- Authors of the books and articles referred to in the Bibliography
- Mr Subir Dasgupta, an executive of Development Consultants Pvt Ltd, Kolkata, who stressed the necessity of writing a book on construction technology
- Dr Prasenjit Sarkar of R&D, IBM, Almaden, USA, who extended help on the electronic version of the manuscript
- A team of distinguished engineers led by M/s Dipak K Sarkar, Dipta Sundar Mallick, Sampuran Dhar, Pranab K Bhaduri, Shyamal Ganguli, Shyamadas Mukherjee, Achinta K Biswas, Sunirmal Neogi, Chandan Bhattacharya, and Anuj K Banerjee who extended all kinds of help whenever approached
- Prof. Santanu K Karmakar, PhD, Secretary, Global Alumni Association of Bengal Engineering and Science University (GAABESU) and HOD (Mechanical Engineering), BESU, who allowed us to use the University's library whenever required
- A team of veteran engineers led by Mr Moloy Roy, Jadav Bhattacharjee, and Gautamlal Basu of Bridge and Roof Co. (India) Ltd
- Mr Milan Mukherjee, a senior executive of Sika India Pvt Ltd, Kolkata
- Prof. Shailaja Venkatasubramanyan, PhD, for providing information on the latest developments on some of the subjects included in the book
- Mr Debashis Roy (Jadavpur University) for extending secretarial help whenever required
- The staff of Jadavpur University library (Salt Lake campus)
- The entire editorial team of Oxford University Press India who guided us at every step during the development of this book.

SUBIR K. SARKAR
SUBHAJIT SARASWATI

Contents

Preface   v

1. Fundamentals of Construction Technology   1
    Introduction *1*
    1.1   Definitions and Discussion *1*
    1.2   Construction Activities *2*
    1.3   Construction Processes *3*
    1.4   Construction Workers *4*
    1.5   Construction Estimating *5*
    1.6   Construction Schedule *5*
    1.7   Productivity and Mechanized Construction *9*
    1.8   Construction Documents *9*
    1.9   Construction Records *10*
    1.10   Quality *11*
    1.11   Safety *12*
    1.12   Codes and Regulations *13*
    Summary *13*
    Review Questions *14*

2. Preparatory Work and Implementation   15
    Introduction *15*
    2.1   Site Layout *15*
    2.2   Infrastructure Development *18*
    2.3   Construction Methods *26*
    2.4   Construction Materials *27*
    2.5   Deployment of Construction Equipment *29*
    2.6   Prefabrication in Construction *30*
    2.7   Falsework and Temporary Works *30*
    Summary *32*
    Review Questions *32*

3. Transportation and Handling   34
    Introduction *34*
    3.1   Basic Principles *34*
    3.2   Road Transportation *36*

3.3   Railway Transportation   *37*
   3.4   Waterway Transportation   *38*
   3.5   Airways Transportation   *39*
   3.6   Hauling and Handling by Construction Equipment   *39*
   3.7   Loading and Unloading Operations   *41*
   3.8   Storage and Preservation   *41*
   Summary   *43*
   Review Questions   *43*

4. **Earthwork**   44
   Introduction   *44*
   4.1   Classification of Soils   *44*
   4.2   Project Site Development   *50*
   4.3   Setting Out   *51*
   4.4   Mechanized Excavation   *53*
   4.5   Groundwater Control   *56*
   4.6   Trenchless (No-dig) Technology   *81*
   4.7   Grading   *85*
   4.8   Dredging   *86*
   Summary   *101*
   Review Questions   *101*

5. **Excavation by Blasting**   103
   Introduction   *103*
   5.1   Rock Excavation   *103*
   5.2   Basic Mechanics of Breakage   *104*
   5.3   Blasting Theory   *105*
   5.4   Drillability of Rocks   *109*
   5.5   Kinds of Drilling   *111*
   5.6   Selection of the Drilling Method and Equipment   *117*
   5.7   Explosives   *118*
   5.8   Blasting Patterns and Firing Sequence   *128*
   5.9   Smooth Blasting   *130*
   5.10   Environmental Effect of Blasting   *135*
   Summary   *137*
   Review Questions   *137*

6. **Piling**   139
   Introduction   *139*
   6.1   Basic Concept   *139*
   6.2   Classification of Piles   *140*
   6.3   Pile Driving Methods   *157*
   6.4   Load Tests and Quality Control   *167*
   Summary   *170*
   Review Questions   *171*

7. Concrete and Concreting — 172

Introduction *172*
- 7.1 Definition of Concrete *172*
- 7.2 Important Properties of Concrete *173*
- 7.3 Composition and Fineness of Cement *194*
- 7.4 Quality of Fine and Coarse Aggregates *197*
- 7.5 Quality of Water *199*
- 7.6 Use of Admixtures *200*
- 7.7 Formwork Including Enabling Work *206*
- 7.8 Reinforcing Steel *214*
- 7.9 Shotcrete *217*
- 7.10 Lightweight and Heavyweight Concrete *219*
- 7.11 Ready-mixed Concrete *226*
- 7.12 High Performance Concrete *227*
- 7.13 Self-compacting Concrete *229*
- 7.14 Extreme Weather Concreting *232*
- 7.15 Fibre-reinforced Concrete *237*
- 7.16 Prestressed Concrete *239*
- 7.17 Underwater Concreting *241*
- 7.18 Polymers in Concrete *243*
- 7.19 Stripping of Forms *247*
- 7.20 Curing of Concrete *248*
- 7.21 Inspection and Acceptance of Finished Concrete *249*
- 7.22 Mechanization of Concreting *250*
- 7.23 Laboratory Testing Facilities at a Site *250*
- 7.24 Non-destructive Testing of Hardened Concrete *253*

Summary 257
Review Questions 258

8. Fabrication and Erection Work — 260

Introduction *260*
- 8.1 Fabrication of Structural Steel at Shops and Sites *261*
- 8.2 Welding Technology *263*
- 8.3 Qualification of Welders *284*
- 8.4 Supervision of Welding Work and Approval *287*
- 8.5 Handling and Transportation of Units to be Erected *289*
- 8.6 Erection of Fabricated Steel Structures *290*
- 8.7 Erection of Precast Reinforced Concrete Structures *293*
- 8.8 Erection of Bridges *296*
- 8.9 Grouting of Joints of Precast Reinforced Concrete Structures *302*
- 8.10 Anti-corrosive Painting *302*

Summary 306
Review Questions 307

9. Cladding and Wall 309
Introduction *309*
- 9.1 Masonry Materials *309*
- 9.2 Masonry Bonding *311*
- 9.3 Stone Masonry *312*
- 9.4 Solid Brickwork *316*
- 9.5 Refractory Masonry *318*
- 9.6 Enabling Work for Cladding *318*
- 9.7 Supervision and Approval of Executed Cladding and Wall *319*

Summary 320
Review Questions 321

10. Roof and Roofing 322
Introduction *322*
- 10.1 Cast-in-situ Reinforced Concrete Roofs *322*
- 10.2 Precast Reinforced Concrete Roofs *324*
- 10.3 Roofs Covered with Sheets *325*
- 10.4 Thermal Insulation Over Roofs *326*
- 10.5 Waterproofing Over Roofs *327*
- 10.6 Shell Roofs *329*

Summary 336
Review Questions 336

11. Finishing Work 338
Introduction *338*
- 11.1 Plastering *338*
- 11.2 Facing *341*
- 11.3 Glazing *343*
- 11.4 Flooring *345*
- 11.5 Painting *347*

Summary 350
Review Questions 350

12. External Work 351
Introduction *351*
- 12.1 Roads *351*
- 12.2 Drainage *375*
- 12.3 Construction—Accommodation of Services and Impact *381*

Summary 382
Review Questions 383

13. Mechanized Construction 384
Introduction *384*
- 13.1 General Considerations *384*

13.2 Fundamentals of Mechanization *388*
13.3 Plants and Tools *401*
13.4 Plants for Eearthwork *405*
13.5 Plants for Transportation, Movement, and Handling *446*
13.6 Concrete Mixers and Pumps *469*
13.7 Scaffolding *494*
Summary 499
Review Questions 500

14. Quality Control and Assurance   501
Introduction *501*
14.1 Definitions *501*
14.2 ISO 9000 Quality System *509*
Summary 526
Review Questions 527

15. Safety   528
Introduction *528*
15.1 Basic Principles on Safety *528*
15.2 Housekeeping *532*
15.3 Personal Safety *533*
15.4 Fire Protection *534*
15.5 Electrical Safety *535*
15.6 Mechanical Handling *537*
15.7 Transportation *539*
15.8 Welding and Flame Cutting *540*
15.9 Scaffolds and Ladders *542*
15.10 Fabrication and Erection *544*
15.11 Excavation *545*
15.12 Blasting *547*
15.13 Formwork *548*
15.14 Concreting *549*
15.15 Floors *550*
15.16 Environment at Site *551*
15.17 First Aid *553*
15.18 Accidents *554*
Summary 556
Review Questions 557

Bibliography   558
Index   563

Fundamentals of Construction Technology

INTRODUCTION

Construction work at its rudimentary stage was based totally on manual efforts. The scenario is different now, and technology has changed it for the better. Construction work involves different activities carried out by diverse processes by skilled and unskilled workers, and needs to be completed within planned time schedule. To maintain such time schedule, construction processes need to be mechanized by deploying construction equipment capable of large production of satisfactory quality. Mechanized construction would require assignment of qualified and skilled workers for carrying out the planned production safely maintaining quality conforming to the specifications, standards, and codes. Not only that, the client should be satisfied with the quality of work produced. This chapter deals with the basics of the construction technology.

1.1 DEFINITIONS AND DISCUSSION

Technology is rapidly shaping and reshaping the world. What appears to be impossible today could be made possible by technology tomorrow. What is this technology that is causing such phenomenal changes and reshaping the world? The dictionary meaning of technology is "The practice, description, and terminology of any or all of the applied sciences which have practical value and/or industrial use" or "systematic application of knowledge to practical tasks in industry." Here knowledge means "ascertained and tested" knowledge.

Decades ago, the editors of a journal, Engineering News Record, asked 32 departmental heads of universities in the USA to define civil engineering. On the basis of the responses received, the editors defined civil engineering as follows:

"Engineering is the application of laws of science, mathematics and economics for the production of things. And civil engineering is the principal branch of engineering concerned with things constructed as opposed to things manufactured, mined, grown, or generated."

Prof. K. A. Padmanabhan of IIT, Kanpur mentioned in an article that "But science is preoccupied with understanding and explaining, while engineering is concerned with doing, realizing and implementing. Thus, the aim of future of engineering education should be the integration of knowledge, skills, understanding and experience." He further added, "An engineering design integrates mathematics, basic sciences and complimentary studies in developing elements, systems and processes to meet specific needs. It is a creative, iterative, open-ended process subject to constraints, which may be governed by standards or legislation to varying degree depending upon the discipline. These constraints may relate to economic, health, safety, environmental, social or other pertinent factors. Thus, the neat and rigorous' solution obtainable in pure science is mostly unattainable in engineering."

Chancellor T. R. Anantharaman of Ashram Atmadeep, New Delhi clarified in an article that "However, there is a basic difference in approach since the technologist is concerned with the application of science to satisfy or fulfil one or more human or social needs or aspiration, while the scientist pursues knowledge for its own sake ... The technology developed in a laboratory has to satisfy many special requirements and pass through one or more intermediate developmental stages before it enters the realm of engineering ... Scientists are concerned with concepts, theories, proofs and explanations, while technologists emphasize tangible processes, products and results. Engineers worry about designs, costs, productivity, regulatory decisions and patent protection ... To put in a nutshell, it can be said that the main base of science is original thinking; that of technology, innovative thinking; and that of engineering, practical thinking."

Construction is thus execution of mostly civil engineering work. And construction technology is application of applied sciences in order to enhance productivity and quality. Production evaluated in time scale is productivity.

The end product of construction engineering can be a completed building or a utility or an industry or an infrastructure – something useful and good for a country and its people.

1.2 CONSTRUCTION ACTIVITIES

Construction work comprises many construction activities performed by a few or a great many number of construction workers. Deployment of manpower can be reduced drastically by deploying construction equipment of high capacities. For that, the requirements are: (i) executing agencies must be familiar with the construction technology involved in the work to be implemented, (ii) whether construction equipment necessary for adopting the appropriate technology is/are available or not.

General construction work involves all civil engineering work starting from substructures to reinforced concrete and structural steel superstructures, highways including bridges, airports, silos, dams, etc. In some countries, superstructure frames of buildings are also made of timber. General construction work comprises the bulk of all construction activities.

Specialized construction work involves all mechanical and electrical erection work, sanitary and plumbing work, roofing/insulation work, and other similar work of specialized nature. In rare occasions, specialized construction work may comprise bulk of the construction activities and general construction work may be of minor nature.

Auxiliary construction work involves preparatory and enabling work that would facilitate general and specialised construction work. Auxiliary work is meant to be of temporary nature. But if any auxiliary work is retained as a necessity, it should be deemed as a part of the general construction work. Sheet piles, for example, are not removed on many occasions after completion of construction work to ensure stability of what they are retaining in place.

Construction activities cannot be continued indefinitely. Therefore, all construction activities need to be planned sequentially and executed in minimum possible time. Men, machines and materials need to be mobilized for timely execution of all the construction work.

1.3 CONSTRUCTION PROCESSES

The nature of the construction activities involved, the place where construction work is to be carried out and the time available for construction work are the three factors that determine the effective construction process. The sequence of different activities depend on availability of vacant spaces that can be earmarked for allotment to different executing agencies to mobilize men, material and machine for timely implementation of contracted work. Each executing agency has to build its own infrastructures to produce planned output. The owners make provisions for basic facilities (water, communication, access to power, overall security, etc.) so as to enable the construction agencies to perform effectively and efficiently.

Construction process varies worldwide. In India, construction is a very labour intensive process, whereas, it is highly mechanized in western countries. Degree of mechanization, however, depends on the nature of construction work. As time schedule is becoming an important factor in completion of construction activities, the construction process is gradually being mechanized all over the world including India. Machines of very high capacity and output are now available. Nevertheless, construction process still involves a huge labour force in India even now because of availability of cheap labour of all categories.

Construction process can be simple as well as complicated. In a simple process, a worker can execute the entire work involved using a simple tool or machine, for instance, lifting materials onto a platform with the help of a mobile/stationary crane. Such lifting would be complicated if the lifting would depend on another simple process; for instance, lifting a truss connecting two columns. In such a case, more workers would be needed on the columns to place the truss at the precise level and position maintaining correct orientation. If the connection of the truss with the columns is taken up simultaneously, the process would become complex as the columns are to be stiffened laterally by connecting them with adjacent columns.

Executing agencies, in general, are well-experienced, and they decide the construction process according to their convenience, unless the owners have reasons to influence such decision because of unavoidable reasons.

1.4 CONSTRUCTION WORKERS

Human element is an important factor in construction activities. The progress of construction work conforming to the time schedule depends, to a large extent, on the quality and efficiency of workers deployed in actual execution.

The workers deployed at construction sites are generally classified into three categories—unskilled, semiskilled and skilled. They are deployed in different construction activities on the basis of their skill, efficiency, and experience. A worker endowed with only muscle power falls under unskilled category. A smart unskilled worker, who acquires skill after a long tenure at construction sites, falls under semiskilled category. A deserving semiskilled worker may be promoted to skilled category. A worker with appropriate education and training who is deployed in different trades such as welding, gas cutting, carpentry and other specialized work falls under skilled category. Highly skilled workers operate construction equipment at optimum efficiency. Semiskilled workers are often deployed as helpers of skilled workers.

There are many institutes in India where aspiring skilled workers can acquire both theoretical knowledge and practical training to qualify as skilled workers. These trained personnel would have to work at construction sites under strict supervision so as to perform satisfactorily. What they learn at the institutes would be the basis of their employment and assignment at construction sites, and how they actually perform during their assignment would be the basis of their continuity in service and promotion.

Training is important to ensure that the quality of work is maintained at satisfactory level. Construction workers are trained periodically to perform more efficiently. On-job training can be arranged on continuous basis. Apart from that, workers can also be trained at specialized training centres or educational institutions, wherever possible.

Qualified personnel deployed for supervision of construction have to ensure that construction work is carried out as per the drawings and specifications maintaining satisfactory workmanship. Visual supervision may not be good enough to approve executed work. Destructive and/or non-destructive tests are carried out to check quality and integrity of work. Thus the performance of workers would depend on the quality of supervision they would be working under.

Apart from aforementioned supervision, performance of workers would depend considerably on the appreciation of their efforts by the management in the form of remuneration and incentives. A happy worker definitely performs better than an unhappy one.

Agencies responsible for execution of work know the productivity of their workers. They work out the number of different categories of workers to be deployed trade-wise on the basis of quantum of work involved and the time allowed for execution. For occasional high volume work, manpower can be hired from outside.

1.5 CONSTRUCTION ESTIMATING

To turn a concept into a construction project, investment is to be made. For that, construction estimating is necessary.

In the initial stage, the owner has to be sure about the investment for which the potential cost is to be assessed. Subsequently, more reliable estimate is made on the basis of available information relating to the project. A Detailed Project Report (DPR) contains adequate information to initiate construction estimating. As a project of sizable investment involves work of all disciplines of engineering, the estimating process becomes complex.

In practice, the entire construction work is divided into several discipline-wise packages so that specialized executing agencies may be hired to complete their assigned work within scheduled time without incurring extra expenditure. However, construction estimating being a complex process, unforeseen extra work escalates cost as well as results in time overrun. For proper estimating, the scope of the work involved should be looked into keeping the following in view:

- Technology involved in the construction and installation activities
- Availability of the required materials and construction equipment for planned output
- Critical milestones in the construction schedule
- Manpower requirement for efficient implementation

It takes a lot to transform a design-on-paper into a functioning facility or an infrastructure or a utility or an industry. It involves utilization/consumption of multitude of resources of the following categories:

- Finance
- Materials
- Construction equipment
- Manpower
- Time

The process of cost estimating remains the same as before, but the tools used in the process have changed in recent times. The computer has become a great help for the personnel engaged in construction estimating.

Direct cost is related to the cost of installed equipment, materials and labour involved in actual or physical construction. Indirect cost is related to the costs that are required for the orderly completion of all the construction work.

1.6 CONSTRUCTION SCHEDULE

A project is an undertaking with a defined starting point and a defined completion point. A project has also defined objectives, and the project is regarded as complete only when the objectives are fulfilled. A series of tasks and activities in a project are to be completed with defined and limited resources. All these tasks and activities (distinct and identifiable operations within a project) are to be grouped into a number of packages. Tasks and activities of different packages are inter-related, and they need to be accomplished in proper logical sequence within project completion schedule taking existing constraints and available resources into consideration. The construction schedule lists all the pertinent activities and indicates the duration of each activity sequentially. The total duration of the project should be equal to the sum of the duration of the individual activities in sequence. Each project is a unique undertaking as no two projects can be exactly similar. Scheduling of a construction project involves determining the time required for each activity within the overall time span of the project. Time span or duration of each activity is determined using the following equation:

$$t = \frac{t_0 + 4t_m + t_p}{6}$$

Where t_0 is the optimistic time, t_p is the pessimistic time, t_m is the most likely time, and t is the time to be used in construction schedule.

Once the activities in a package are determined, they must be arranged into an executing plan in the network logic diagram. Starting from the initial activity in a package, all the remaining activities in a package must fall into one of the three categories:

- Must precede the activity in question
- Must succeed the activity in question
- Can be performed concurrently with the activity in question

Each activity has an early start date and early finish date. Early start is possible only if all preceding activities are already over. The early finish date is the date arrived at by adding duration of an activity with the early start date. The late start date is the latest date by which an activity may be started without delaying the project. The late finish date is the date arrived at by adding duration of the activity with the late start date. Free float is the period by which an activity may be delayed without affecting the early start of the succeeding activity. Total float is the period by which an activity may be delayed from its early start without delaying the project completion date. Independent float is related to a particular activity considering the float between the late finish date of the preceding event and the early start of the succeeding event.

In construction schedule, duration of activities is represented in network form: PERT or CPM. PERT is the abbreviation of Programme Evaluation Review Technique. PERT is more suitable for projects wherein completion periods of various activities may vary considerably because of uncertainties as in research and development projects. CPM is the abbreviation of Critical Path Method and it is extensively used in construction project scheduling. A network diagram (Fig. 1.1) shows periods of different activities. As the activities are inter-related, periods of execution are shown logically and sequentially which ultimately end up as a network comprising lines, arrows, and nodes. Such a network would form a great number of paths leading to completion date. Each path should have key dates on achieving milestones. The paths, which do not have any

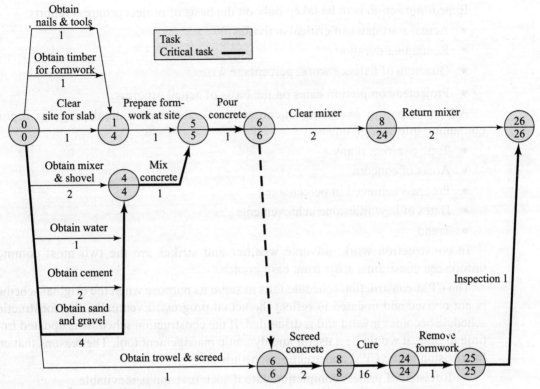

Fig. 1.1 Typical network diagram

float, are the critical paths and would require special attention so that the completion date does not slip. All activities in critical paths may be expedited by crashing, that is, by employing more resources such as men, materials, machines, money, and moments (overtime). Crashing is possible only at extra cost.

$$\text{Daily crashing cost} = \frac{\text{Cost increase due to crashing}}{\text{Time decrease due to crashing}} = \frac{\text{Crash cost} - \text{Normal cost}}{\text{Normal time} - \text{Crash time}}$$

A realistic construction schedule, which is agreed upon by the owner and the executing agency, should be the yardstick for progress measurement. Such agreed construction schedule should not have any negative floats. In case the finish dates are delayed, possible measures to be taken are:

- Overlapping activities in critical paths, wherever possible, irrespective of preceding and succeeding activities
- Working overtime
- Mobilizing more resources by deploying more men and construction equipment
- Re-allocating resources from non-critical areas to critical areas.

Expediting action is to be taken only on the basis of project progress reports:
- Actual start dates of critical activities
- Remaining duration
- Quantum of balance work, percentage wise
- Projected completion dates on the basis of actual progress.

All participating executing agencies must submit periodical progress reports containing information on:
- Time overrun, if any
- Areas of concern
- Progress achieved in percentage
- Dates of key/milestone achievements
- Trend

In construction work, adverse weather and strikes are the two most common unforeseen constraints apart from cash crunch.

The CPM construction schedule fails to serve its purpose when the original schedule is not revised and updated to reflect the actual progress. Eventually, the construction schedule becomes invalid and is discarded. If the construction schedule is updated from time to time, it would be a useful and dynamic management tool. The reasons that call for revision of the CPM construction schedule are:
- Revision of project completion date if such revision is inevitable
- Changes in project plans, specifications, or site conditions
- Activity durations not equal to planned durations
- Construction delays due to weather, delayed delivery, subcontractors' lapses, labour problems, natural disasters, owners' indecision

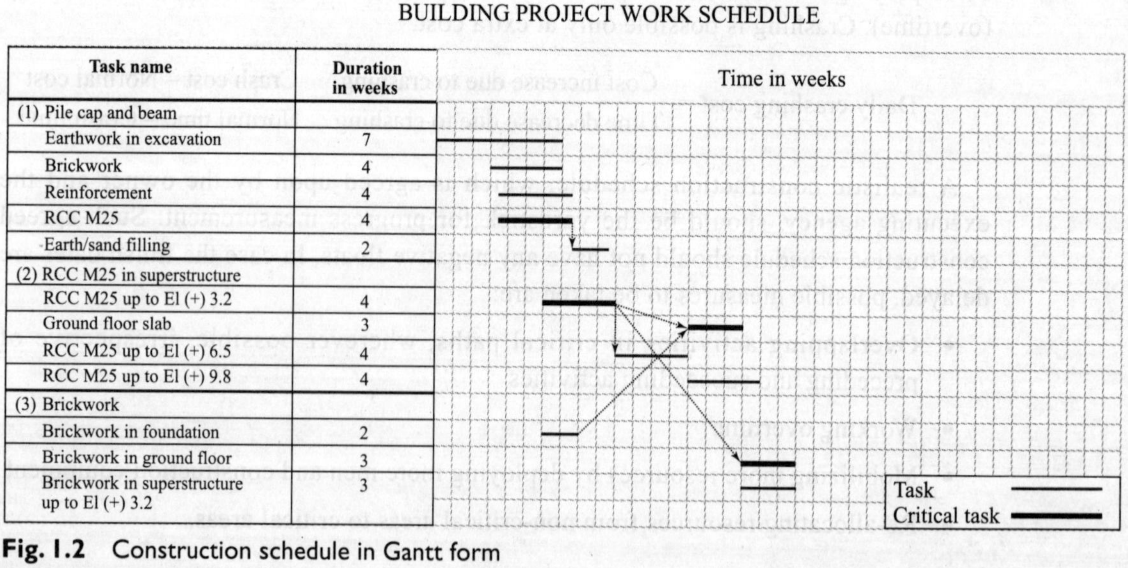

BUILDING PROJECT WORK SCHEDULE

Task name	Duration in weeks	Time in weeks
(1) Pile cap and beam		
Earthwork in excavation	7	
Brickwork	4	
Reinforcement	4	
RCC M25	4	
Earth/sand filling	2	
(2) RCC M25 in superstructure		
RCC M25 up to El (+) 3.2	4	
Ground floor slab	3	
RCC M25 up to El (+) 6.5	4	
RCC M25 up to El (+) 9.8	4	
(3) Brickwork		
Brickwork in foundation	2	
Brickwork in ground floor	3	
Brickwork in superstructure up to El (+) 3.2	3	

Fig. 1.2 Construction schedule in Gantt form

In practice, construction schedule is generally prepared package wise in bar-chart form. Arrows connect inter-related activities. Critical tasks are shown in thicker lines as shown in Fig. 1.2.

1.7 PRODUCTIVITY AND MECHANIZED CONSTRUCTION

The productivity of any construction equipment is a term that indicates how many units of output the equipment produces in an hour depending on the job conditions and management as well as the operator's skill, persistence and co-ordination with other construction forces.

Productivity signifies the rate at which things are produced. Technology, competitive design, external constraints, human elements, managerial efficiency; and most importantly, uninterrupted financing contribute to productivity. In simple terms, productivity refers to the ratio of output versus input (output/input). Output here has to be of specified quality produced within scheduled time. Input are the resources deployed for achieving the desired output.

Construction schedule shows the time allotted for each and every construction activity. For speedy execution of construction activities, larger targets are planned. Construction equipment of high capacity are mobilized and pressed into service to achieve such large targets consistently. In voluminous construction work where large targets are planned, the only factor that would necessitate the use of construction equipment is human fatigue. As humans get easily tired, deployment of construction equipment becomes necessary to leap ahead and achieve targeted output and maintain scheduled dates. Other problems concerning manual efforts are related to job dissatisfaction, militancy, and slackness. Deployment of construction equipment eliminates such uncertainties in production.

Mechanization means deployment of machines, or getting the work done by machines rather than human beings. Since lever, pulley, screw are all simple machines, a bit of mechanization would always be there. However, what we really mean by mechanization is the large-scale deployment of construction equipment to significantly increase the rate of output of construction activities.

1.8 CONSTRUCTION DOCUMENTS

Once a new undertaking is conceived, it is necessary to check its feasibility. Owners' engineers can do this. If owners' engineers do not have the necessary knowledge and experience, then specialists or consultants can do the job on their behalf. On approval of the *Feasibility Reports*, Project Reports are prepared. The Feasibility Report, depending on its contents, is sometimes called *Pre-feasibility Reports*. The Project Report, also depending on its scope and contents, is sometimes called Detailed Project Report. Thus the owners may have two, three or four reports to suit their requirement.

What is conceived (the ideas on paper) becomes a project when personnel are assigned and investment is made for implementation. The documents required by engineers for supervising the construction work are:

- Project reports/detailed project reports
- Contract documents
- Specifications
- Design and detailed drawings
- Erection manuals
- Relevant Indian and international standards
- Documents containing applicable statutory rules and regulations
- Quality assurance documents
- Safety system documents
- Copies of statutory rules and regulations

1.9 CONSTRUCTION RECORDS

On completion of project work, the owners must have the following records at the time of handover. Records are documents, which are to be preserved for future reference as evidence of conformance to the contractual provisions, specifications, codes and national/international standards. The records as mentioned below are documents, which cannot be revised or modified. A document, which is not a record, can be revised or modified if required. Records would be different depending on the nature of projects. Compared to an industrial project, a housing project generates very few records. A utility project like building a power station generates very many records vis-à-vis an industrial project. An infrastructure project like construction of highways stretching on shore and off shore may generate very many or few records depending on the route lengths.

- Updated contract documents
- Updated specifications
- Soil investigation reports/ground water data
- As-built drawings and sketches
- Updated erection manuals
- Updated operation and maintenance manuals
- Approved contractors' logs
- Procurement documents of bought-out items
- Material qualification records
- Skilled workers' qualification records
- Meteorological data
- Inspection and test records
- Quality system records
- Safety records including accident records, if any
- Statutory clearance records
- Commissioning check lists and protocols

- Performance test records
- Handing over protocols
- Storage and preservation records
- Equipment data
- Important correspondences

1.10 QUALITY

Quality should be the most important consideration in all construction activities. Reliability, durability, and safety of constructed work depend mostly on quality. Reliability is the probability that a product, system, or service will perform its intended function satisfactorily over a stipulated period of time under specified conditions.

According to traditional definition, "Quality is the totality of features and characteristics of a product or service that bear on its ability to satisfy stated or implied needs."

Different authors have defined quality in different ways. Quality does not mean quality of the highest level, but it means a predictable degree of uniformity and reliability. The definition would be more meaningful if quality is defined in measurable terms. Thus, quality should mean conformance to drawings, specifications, codes and statutory regulations. Degree of such conformance could be measured, and it should be satisfactory to the owners/customers.

According to traditional definition, quality assurance (QA) is defined traditionally as 'all those planned and systematic actions necessary to provide adequate confidence that a structure, a system or a component will render or perform safe and trouble-free services and satisfy specified requirements.' QA is a management tool. The owners or their representatives are responsible for quality assurance.

Quality control (QC) is traditionally defined as 'Those QA actions required to keep control and regulate factors to attain predetermined qualitative characteristics related to materials, processes, and services.' Executing agencies use operational techniques and activities to fulfill the requirements for quality. As rectification of unsatisfactory work would escalate cost and delay the progress of construction activities, executing agencies need to pay utmost attention to quality control so as to avoid subsequent rectification/reworking. QC is a production tool.

The basic purpose of QA is not only to assure quality, but also to ensure continual improvement by rigorous control and training of personnel engaged in construction activities (vide Chapter 14).

Training programmes should be structured for different categories of employees to ensure that they are adequately qualified for the work they do. They must have appropriate education, experience, and training. A procedure is to be worked out to identify the training requirements and then regular training is to be arranged on-job, in-house or outside to inculcate a sense of excellence among the employees.

1.11 SAFETY

Risk is an inherent part of all construction activities. Risk is traditionally defined as 'a combination of probability of an abnormal event or failure and the consequence of that event or failure to a system's operators, users, or its environment.' Event is traditionally defined as "an internal or external occurrence involving equipment performance or human action that causes a system upset." Risk is likely to affect safety, health, and environment (vide Chapter 15).

Safety depends on human attitude. Construction activities expose both men and material to risks. Faulty attitude is inherent in inaction, lack of interest, worry, and impulsiveness. Faulty attitude may cause serious accidents. A judicious combination of application of behavioural science and appropriate technology is essential for preventing such accidents. Modern concept of safety relies more on involving construction workers directly in safety efforts apart from their adhering to prescriptive approach. If construction workers identify hazardous situation ahead and take necessary corrective action, probability of accidents at construction sites would be drastically reduced.

Safety is to be viewed not as add-on expenditure but as another consideration just like operability, security, maintainability etc. Personnel to be deployed to assure safety should be properly selected. They must ensure that workers are well-trained and not overworked.

Training is a mode of learning that changes behaviour and attitude. If attitude has a bearing on safety, then training is essential for correcting faulty attitude. Training also motivates the workers to excel in performance.

A number of agencies are involved in construction activities. These agencies must state their safety policies and adopt construction methods inline with their stated safety policies. Since these agencies execute inter-dependent activities, it would not be possible for any agency to formulate its safety policy independently. The owners or customers would have to conscientiously frame comprehensive safety policies and form Safety Committees at sites, thereby allowing all agencies to interact and resolve interdependent constraints. Overlapping of responsibilities should be avoided. The owners' representatives should head such safety committees. The safety status of work of all agencies is to be reviewed in periodic meetings and follow-up action is to be taken to resolve all kinds of problems and setbacks.

As accidents resulting from risks involved in construction activities may cause bodily injury/death, delay in project implementation as well as cost escalation, the onus of safe execution of construction activities rests with the owners/customers. Executing agencies would follow the lead taken by the owners on safety. The causes of accidents at construction sites are listed below:

- Fall from height—persons or materials
- Slip or fall at the same level
- Struck by falling, speeding, or moving objects
- Injury due to projected reinforcing steel bars

- Electric shock—contact with electric current
- Injury due to welding, loose scaffolding, machinery, tunnelling, poisonous gas, toxic material, impact noise and excavators
- Cave in
- Caught in or between objects or machinery
- Striking against objects
- Contact with high temperature
- Exposure to or contact with potentially harmful substances
- Environmental problems—noise, dust, radiation, toxic materials, heat and cold
- Drowning
- Fall into pits
- Overexertion

As risk cannot be totally eliminated, accidents may take place without warning. Both owners and executing agencies should, therefore, make provisions for first aid and subsequent medical treatment at sites or avail services of nearby hospitals/nursing homes for which ambulances should be available.

Post-accident investigation reports should be comprehensive so that preventive measures can be taken in subsequent construction activities. The present trend is to assure even off-the-job safety of all personnel involved in construction work.

1.12 CODES AND REGULATIONS

For design and construction of projects, conformance to various codes, regulations, statutes, laws, and guidelines is mandatory. The owners' or their consultant's engineers are required by law to carry out design work complying with the applicable laws and regulations, which vary from one jurisdiction to another. If relevant, sophisticated codes are not available under any jurisdiction, then recognized national or international codes should be followed. This should be agreed upon by all concerned at the onset of a project so that everyone understands the rules that would govern the design and construction.

SUMMARY

Project implementation involves multifarious activities that could be executed only by experienced and knowledgeable personnel following processes of diverse nature. Estimating, planning, mobilizing (men, material and plants) are essential preparatory work on which successful implementation could be based. However, actual execution would depend on various documents such as contracts, specifications, drawings, and standards. All these documents must be preserved as records for future reference. The execution processes could be hazardous, and accidents at construction site cannot be forecast. Therefore, safety deserves utmost attention. As for quality, it can be controlled progressively by carrying out destructive and non-destructive tests. The owner must feel assured that the project as a whole has been completed conforming to the contractual provisions on quality.

REVIEW QUESTIONS

1. How is technology related to science and engineering?
2. How would you differentiate between general, specialized, and auxiliary construction activities?
3. Why is construction process being increasingly mechanized in India?
4. Who could be designated as a skilled worker?
5. What has made project estimating much easier now?
6. How does crashing help in expediting all the activities in critical paths?
7. What does mechanization truly mean?
8. List the documents required for actual project implementation.
9. What is the difference between a record and a document?
10. List the documents the owners must have at the time of handover on completion of the project work.

2
Preparatory Work and Implementation

INTRODUCTION

Construction work cannot be started without adequate preparation, such as acquisition of land, collection of comprehensive information/data required for design work, and preparation of specifications and contract documents. Enough land should be acquired for project work as well as for building infrastructures for executing agencies to complete their contractual obligations safely maintaining quality without time overrun. Construction methods should be adopted keeping all these in view to ensure assured quality output.

In project implementation, different kinds of materials conforming to the detailed specifications are required. Execution of project work should be planned so as to allow parallel execution of different activities, thereby reducing execution time. Project execution work quite often requires falsework and temporary structures that do not form part of the permanent features. These extra expenditures should be included in the budgeted cost.

2.1 SITE LAYOUT

A decision is the alternative chosen by the management from various different conceived ideas for setting up an undertaking. A concept is turned into a project when investment is made for implementation after assignment of personnel to form an organization for the purpose.

Acquisition of land is a prerequisite for starting construction activities. The acquired land must be sufficient to meet the requirements of both the project layout as well as the construction agencies. A construction agency would need sufficient land for constructing their temporary offices, covered stores, open storage yard, space for pre-assembly or site-fabrication work if needed, storage areas for spare parts and mountings, movement and garaging of construction equipment, workshop facilities and space near the location of construction activities for working/manoeuvring. It is necessary to place the construction equipment in advantageous positions so that general circulation at site does not affect the performance of such equipment.

Before initiating the implementation process, the owners need to estimate the land requirement for their projects, land requirement of executing agencies as well as land required for developing infrastructures to facilitate smooth execution of project work. Land is also required for accommodating construction workers, who would be engaged in construction activities. Land required only for the duration of implementation process may be leased instead of outright purchase. Acquisition of land is a difficult process in India and calls for judicious thinking and utmost attention. If expansion of the proposed undertaking is envisaged in future, outright purchase of more land initially could be advantageous in many respects.

For initiating preparation of project layout by owners (or their consultants) and working out schedule of items for competitive bidding by interested construction agencies, the following information should be gathered by visiting the actual location where construction work is to be executed.

- Site location—(i) how to reach from the nearest town/city, airport, railhead, or harbour (ii) distance involved and cost of travel (iii) availability of public transport (iv) load and dimension restriction on road bridges, if any, on the way (v) economic mode of transportation of materials and equipment (vi) if improvement of existing road to the site is necessary
- Accommodation—availability of hotels/guest houses/restaurants/residential accommodation, and expenses involved to avail such facilities
- Services—availability of facilities like banks, post office, telephone services, water supply, power supply, markets, schools/colleges, hospitals/nursing homes/ medical facilities, clubs, fire station, testing laboratories, petrol/diesel filling stations, police station and security provider—if any of such services is located at the site of construction, whether relocation of the same would be necessary
- Records of seasonal temperature, humidity, rainfall, visibility, seismic damages, tidal water, flooding risk, drainage pattern and environment
- Records of soil investigation including information on groundwater, water table, and contour map
- Manpower—availability of local unskilled, semi-skilled, and skilled workers, their wage structure, and record of industrial relation at and around the project site
- Materials—availability of raw materials, finished products, and spare parts locally
- Sub-contractors—availability of local sub-contractors including their backgrounds and track records
- Applicable local laws, taxes, and service charges

The owners or their consultants finalize project layouts on the basis of the technological and/or other requirements based on the site survey and data/information collected during the site visits. Project layouts are for giving shape to whatever is planned and finalized for implementation. As regards the construction agencies, they would prepare their temporary site layout plans showing their land requirement for executing the work contracted to them. A sample layout of a construction agency's site facilities is shown in Fig. 2.1. Exact nature of the layout would, however, depend on the spare space available at the construction site, and availability of time and fund with the construction agency to build infrastructural facilities.

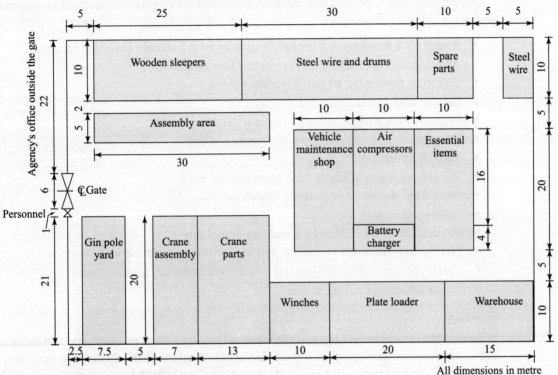

Fig. 2.1 A construction agency's site layout

Site layouts of construction agencies should be prepared keeping in view economy, efficiency, and security. Avoiding double handling, easy accessibility, and short movement/hauling would economize construction activities. Judicious selection of sequence of operation so as to accelerate progress of work would boost efficiency. Availability of materials and equipment as and when required is essential for both economy and efficiency. Security deserves prime consideration for avoiding loss by theft or vandalism. This can best be done by fencing along the boundary of the allotted space and posting security guards.

The layout of the project should be developed keeping in view the extent of mechanization contemplated. If an owner desires to induct mechanization at a later stage to expedite execution work, it would not be possible to do so if the layout is prepared without giving much attention to future requirements.

2.2 INFRASTRUCTURE DEVELOPMENT

Once the Detailed Project Report (DPR) has been prepared and sanctioned by the appropriate authority, necessary fund should be made available for building up infrastructures to facilitate project implementation. Assignment of site personnel is the first step that should be taken towards initiating implementation process. Acquisition of land must be completed before assignment of site personnel, mainly because delays on any account whatsoever would only escalate the project cost.

Speedy and timely project implementation would not be possible without having the following infrastructures ready:

- Boundary wall along the project boundary with lockable gates for security
- Arrangement for office and residential accomodation
- Electrical power for all participating agencies
- Potable and construction water
- Project roads with necessary temporary linking roads
- Drainage facilities to avert water-logging at site
- Yard illumination
- Warehouse, cement sheds, and open storage yard
- Workshop, garage, and parking facilities
- Testing laboratory facilities
- Medical facilities including a standby ambulance
- Firefighting facilities including parking a fire-tender at the site
- Communication facilities—postal, courier, electronic, internet, etc.
- Diesel and petrol filling station

Security

Security at the site is to be very strict to ward off loss by pilferage, theft, and vandalism. Elaborate security arrangement would necessitate construction of masonry boundary walls along the periphery of a site with strong steel gates at the crossing of roads and rail-tracks. Security guards should be posted to check entry and exit of personnel, vehicles, wagons as well as all kinds of construction equipment and containers.

Office and Residence

Arrangement for office and residential accommodation would depend on the exact site location and existing accommodation available nearby. If the site is located in a remote and isolated area, new accommodation would have to be built. Office accommodation can be built in two phases. In the first phase, temporary offices may be built with tubular structures with cladding by masonry work. In the next phase, permanent office would be built where project personnel could be accommodated till the project is completed. A cafeteria should cater hygienic food, beverages, and water to all people at the project site. Temporary structures are normally dismantled when vacated. Construction agencies build all buildings or sheds on temporary basis. Both owners and executing agencies provide rented residential accommodation to the managerial and supervisory personnel even at far off places, as vehicles are also available on rental basis. Temporary accommodation is built near the site for junior staff and workers. Workers should not be allowed to reside in the project area within the boundary wall, as it becomes a difficult task to dislodge them from the site after completion of work.

Human resource is vital for project implementation. So, providing them with just residential accommodation would not be enough. There should be basic amenities like marketing, education, recreation, healthcare, sanitation, transport, etc.

Power

The DPR contains an estimated requirement of unit-wise as well as total electrical power (load) for a project. This load comprises static loads in kW plus motor loads plus additional load for starting motors. Similarly, load demand during implementation process is also worked out. On the basis of such estimated load, an application is addressed to the Government agency or private enterprise responsible for distribution of metered power at the project site. Cable laying for power supply during implementation stage should be done in such a way that the same cables, if required, could also be used in the post-implementation stage. An example of load estimate during implementation stage is shown below for a part of the project under implementation. As disruption of contracted power cannot be totally ruled out and as cost of power disruption is substantial, installation of on-site diesel generator sets offer protection from power outage. To assure generation of minimum power to continue uninterrupted progress, more than one generator set should be installed so that disruption of power supply from a generator set would not grind the progress of construction activities to halt. All power connections need to be switched off before starting the diesel generator sets on, and then selectively restoring power to continue most essential work so as not to hamper the overall progress. In case of a power project, construction power would be necessary for construction of foundations, powerhouse buildings, cooling towers, water intake structures, water treatment plant, erection of turbo generators, boilers, coal handling plant, ash handling plant, both high and low pressure piping, etc. Construction power requirement for constructing a small powerhouse building would be as follows:

Powerhouse building	Power requirement
Builder's office/workshop	25 kVA
Welding sets	180 kVA
Winch	20 kVA
Pumps	50 kVA
Batching plant	100 kVA
Concrete pumps	40 kVA
Compressor	75 kVA
Tower crane	160 kVA
Total	650 kVA power factor = 0.85 (1 HP = 0.7455 kW) 552.5 kW

Similarly, total construction power requirement of the above project including the power requirement of (i) erection of turbo generator, boiler(s), coal handling plant, ash handling plant, and water treatment plant, (ii) fabrication and laying of both high- and low-pressure pipes, (iii) construction of cooling towers including erection of fans, etc., (iv) construction of water intake pump house including erection of pumps, etc., (v) construction of water treatment plant civil work, and (vi) inside/outside illumination, can be worked out for all activities in the same way as done above. The total construction power requirement for a small power project may add up to 3000 kVA. However, power would not be required continuously for all the equipment, machinery, and tools at all times. Considering a diversity factor of 0.6, the total construction power requirement would be about 1800 kVA. Similarly, requirement of back-up power generated by standby diesel generators can be worked out taking into account the essential construction activities that cannot be interrupted.

Potable and construction water

Like electrical power, demand of both potable and construction water requirement could be worked out. Water would be required for (a) construction activities, and (b) human consumption. In construction activities, water is required for (i) concrete production, (ii) masonry work, (iii) curing of concrete, and (iv) soil compaction. In case of human consumption, water is mainly required for drinking and sanitation. Maximum demand for water is in concrete production, and such demand would depend on the concrete mix design for different grades of concrete. A sample calculation is shown below for production of 750 m^3 of concrete per day:

Peak daily production considered	750 m^3
Cement in concrete as per mix design	350 kg/m^3
Water:cement ratio as per mix design	0.35
Water requirement	$350 \times 0.35 = 122.5$ kg/m^3 = 122.5 litres/m^3
Total water requirement for concrete production	$750 \times 122.5 = 91875$ litres $\approx$ 100000 litres

Considering 1500 personnel at site during the peak period of construction, water requirement for human consumption would be as follows:

Personnel present at site	1500
Water requirement of an individual per day	40 litres
Total water requirement	$1500 \times 40 = 60000$ litres

Total water requirement for concrete production and human consumption adds up to 160000 litres. Provision should be there for an equal volume of water for other activities, such as testing, loss, and emergency use like firefighting. Loss results mainly from negligence. Provision of water should be made for $160000 + 160000 = 320000 \approx$ 0.4 million litres so that supply of construction water can be continued uninterrupted. Water should be stored in storage tanks and distributed through feeder pipelines. Source of water could be rivers, canals, or groundwater obtained from tube wells. Whatever be the source, there should always be a back-up source so that non-availability of water does not impede the planned progress.

Linking roads

Movement of heavy transport vehicles and construction equipment would need firm base. While track-mounted crawler construction equipment can move on difficult uneven terrain, tyre-mounted construction equipment would need road surfaces for movement. Even track-mounted equipment would get stuck on soft soil. Construction of roads envisaged for a project should be taken up right at the beginning, unless the surface existing at a project site is good enough for movement of both tyre-mounted and track-mounted construction equipment. If not, then construction of project roads should be taken up and completed at the beginning, leaving top surfacing layers for completion after construction activities are over. For the construction period, roads should be 9.0 to 10.5 metres wide to allow movement of construction equipment and carrier vehicles both ways. Road layout and curves at junctions should allow easy movement, negotiation, and maneuverability. Temporary roads, linking project roads or linking a particular area, are also constructed to facilitate construction activities. Quality of these linking roads would depend on the purpose for which these are constructed. Availability of roads is important to achieve planned progress.

Sometimes, the project layout could be compact requiring minimum road length. Such compact layout would involve interdependent construction sequence that could impede the progress of project execution. This problem may be overcome by construction of more temporary roads if such temporary construction is possible within the available space.

Apart from the road carriers, materials and equipment could be transported by rail if there is provision of railway siding for a project. If the provision for a railway siding is there, the same should be made ready for utilization at the time of implementation. A railway consignment can bring in construction materials in bulk as well as very heavy construction equipment. For coastal or riverbank projects, transportation by sea or inland waterways would be possible if unloading facilities are available at the site. Both feasibility as well as the cost of transportation should be taken into consideration in deciding the mode of transportation.

Drainage

Water-logging at the site may disrupt the progress of construction activities to a great extent. Arrangement for drainage of both rainwater as well as wastewater should be made from the beginning. Where the project site is generally flat, such water is collected in excavated pits or low-lying areas and then pumped out. Collected water may also be disposed off by connecting pits or low-lying areas with any existing drainage system/river/canal using precast pipes or digging ditches up to the disposal area. Where the project site is sloped or having uneven gradient, then the existing slope or gradient could be used to drain the water out of the areas where construction activities are planned. Even in such project sites, rain or wastewater may be collected in pits or low-lying areas and then pumped out or disposed off through precast pipes or dug ditches.

Yard illumination

Yard illumination is necessary for (i) extending working period beyond daylight hours, (ii) security, and (iii) safety measures. The work needs to be continued uninterrupted to adhere to the planned progress schedule or to accelerate delayed activities. In such cases, executing agencies arrange for illumination at work locations, but owners must arrange for yard illumination. The level of illumination at which a worker can safely carry out the assigned task is to be taken into consideration while yard illumination is designed. The SI unit of illumination is lux, which is equal to one lumen per square metre. One lumen is the amount of light that falls on unit area when the surface area is at unit distance from a source of one candela. Illumination level should be 10 lux for movement and material handling. For working areas, illumination level should be 15 lux. Illumination level should be 200 for fine craftwork, 100 for plastering work, and 50 for reinforced concrete work. The requirement of high level of illumination is localized, and executing agencies should arrange for that. Illumination level can be checked by hand-held light meter. Required illumination level would vary depending on the project site's geographical location, weather condition, and season of the year.

Yard illumination also helps avoidance of loss by theft and vandalism. In the cover of darkness, while personnel engaged in the construction activities may indulge in pilfering, thieves and vandals would be more daring. Such losses would not only escalate the cost but also upset the planned progress. Security guards would be able to perform more effectively if the project site is illuminated up to the required level.

Yard illumination level varies according to the work involved. Illumination levels of different areas of work are indicated below:

Area of activity	Illumination level in lux
Building/Civil work	
Movement only	5
Concrete casting	10–20
Moving plant	50
Detailed work	50–100
Quarries	
General/movement	5
Excavating	20
Special working area	50
Road works	
Low hazard	5–10
Moderate hazard	10–50
Extreme hazard	50–100
Site offices	
Office work	200
Drafting work	300

Lux = Lumens per m^2

Safety of personnel engaged at construction sites is jeopardized if yard illumination is not arranged. Site personnel may fall in excavated pits and underground construction areas. Those who are engaged in construction activities at heights may slip and fall in darkness. Apart from providing lamps on poles along roadsides, searchlights are also provided to illuminate a site to the required level. Construction agencies arrange for local illumination where construction work is in progress.

Loading and unloading as well as handling of materials and equipment need to be carried out on arrival of carriers at the site. As such activities should not be carried out

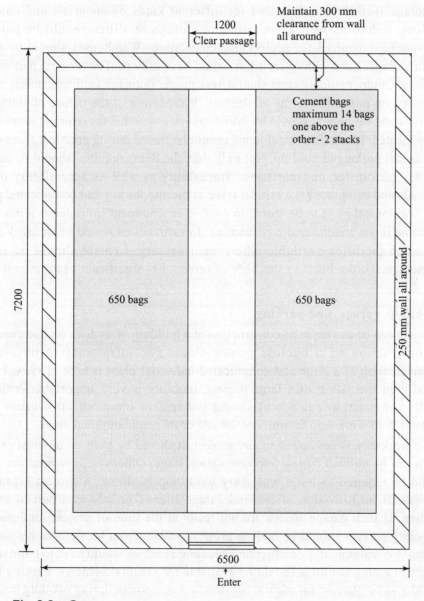

Fig. 2.2 Cement store

in darkness or semi-darkness, sufficient illumination should be arranged as holding up carriers could result in congestion at the site. Such congestion may delay planned progress.

Storage

All kinds of goods ranging from heavy to light and sophisticated equipment for a project, steel structures fabricated at shops, raw materials for both cast-in-situ and precast concrete work as well as raw steel sections/plates for structural steel fabrication at site, and various other raw or finished materials are received at a site. Different types of storage facilities are required for different kinds of materials and equipment. Outdoor, indoor as well as semi-indoor storage facilities should be part of the infrastructure required for project implementation. Warehouses should be built for indoor storage. Semi-indoor facilities would have only roof cover with part or no side cladding. Outdoor storage areas should have proper drainage facilities, unless materials are stored on raised platforms or sleepers. Irrespective of the nature of storage, both materials and equipment should be stored in such a way that the required items could be easily located. If certain critical items cannot be traced during erection, there could be serious and prolonged hold-up. Not only that, the items required should be accessible for easy pickup for transportation. Traceability as well as accessibility of stored materials and equipment is a critical issue at the site for big and complicated projects. Some equipment need to be stored in passive environment (nitrogen is a passive gas) and some in air-conditioned environment. Infrastructures would be deemed complete with such facilities available wherever necessary. Cement should be stored in chronological order because shelf life of cement has significant bearing on its quality (Fig. 2.2).

Workshop, garage, and parking

A project can be as simple as construction of a building or as difficult and complicated as construction of a nuclear power plant. For infrastructure development, implementation of a large and complicated industrial plant is kept in view. For time-bound implementation of a large project, mobility is very important. Vehicles and mobile construction equipment should not remain immobilized because of non-availability of workshop facilities at the site or its neighbourhood areas.

If a workshop is envisaged in any project, it should be built on topmost priority so that it can be utilized during implementation stage. Otherwise, construction agencies should be required to build temporary workshop facilities. A project layout would always call for provision of covered garage spaces, mostly as a part of the office building. If such garage spaces are not ready at the time of project implementation, temporary garages should be built to protect vehicles from hostile weather conditions and dusty environment. Construction/executing agencies would have to build their own garages. Parking facilities should be available for visitors' vehicles. Parking facilities should also be planned for mobile construction equipment during off-duty hours.

Testing facilities

Quality is defined in measurable terms as conformance to drawings, specifications, codes, and statutory regulations. For quality assurance, conformance to specifications and applicable codes need to be checked and tested by taking samples of raw materials or products and testing the same in recognized laboratories. If such facilities are not available near the project site, then a laboratory should be set up at the site with all the necessary testing facilities. The size and type of the laboratory would depend upon tests to be carried out. For civil work, all raw materials (aggregates, water, bricks, etc.) need to be tested to ensure conformance to the specifications. Cement and soil samples are sent to outside laboratories of repute for testing. Executing agencies often set up some testing facilities to check quality of materials being supplied by their suppliers and also quality of what they themselves are producing. Erectors responsible for mechanical erection and piping work set up sophisticated facilities for testing quality of welding being carried out at the site. Hydraulic tests and leak tests are carried out in-situ. For the construction of roads and highways, contractors are required to set up full-fledged laboratories in remote areas.

Medical care

If a medical care centre is not located within a reasonable distance from the project site, then at least a room should be earmarked at the site for providing first aid to an injured person or a casualty of an accident. Time interval between an accident and a casualty receiving medical treatment is very critical as the question of life and death hangs on it. First aid is the first help that can be provided at the site at all times before a victim or casualty is shifted to a place where medical facilities are available. Access to first-aid room should be easy as first medical help is to be provided at the earliest possible time. First-aid box for providing first aid should contain all kinds of bandages, adhesive tapes, antiseptic lotion/cream, absorbent cotton, calamine lotion, safety pins, blunt-end scissors, pair of tweezers/forceps, wooden splints, eye bath, torch, and a book containing instructions on first aid. Contents of first-aid box would depend on what medical experts would advise. If none among the personnel engaged at a project site is trained on first aid, then either a few of them should be trained for the purpose or trained personnel should be separately engaged. A trained person would know not only what to do but also what not to do. A well-equipped ambulance should remain parked at the site round the clock for use in emergency. For more information on first aid, refer to Chapter 15, Section 15.17.

Firefighting facilities

If a Government Fire Station is located nearby, then firefighting planning should be done in consultation with them. Insurance agency should also be a party in this planning. Ravages of fire can be fought if necessary infrastructures exist at the site. Adequate water should always be available at the project site round the clock. Also a fire tender with sufficient fire hoses should be available at the site so that water can be used for firefighting. Portable fire extinguishers should be installed at all vulnerable points. Also buckets filled with sand or water should also be kept at vulnerable points so that those who cannot use fire extinguishers can use the contents of buckets for fire mitigation. Vulnerable points can be located keeping in view that fuel, heat, and oxygen are the three constituents which, when in contact, can cause a fire.

Communication facilities

Quick and efficient communication facilities should be available at the project site so that there is no communication gap between the top management and the project site management. Nanotechnology has created revolutionary changes in the way people communicate. Cell phones are extensively used these days. Internet is used for communication around the world. Drawings and details can be e-mailed to remote sites if Internet connectivity is arranged at the site. However, if electronic medium is not accessible, drawings and documents can be sent by courier or speed post in the least possible time.

Fuel filling station

Vehicles are extensively used at the project site for movement of both people and materials. There are areas where mechanization pays rich dividends. Construction equipment and machineries are, therefore, deployed more and more to cut down on project implementation period as much as possible. For standby electrical power, diesel generators are an absolute necessity. In case diesel- and petrol-filling stations are not located in the proximity of a project site, then it would be better to have filling station at the project site itself.

2.3 CONSTRUCTION METHODS

The basic purpose of construction work is to execute a job at a reasonable cost within the scheduled time frame with assured quality. Construction method should be adopted keeping all these in view.

Construction methods as followed in India are dependent largely on the labour force. Both timely implementation and quality assurance calls for sustained efforts by the labour force engaged in the project implementation work. However, it is not easy to maintain sustained effort because human beings have physical limitations, and they give away to fatigue. In evolving construction methods, the following should be kept in view:

- Quality of construction is to be assured—quality assurance creates a bright working environment conducive for the best of human efforts
- Project implementation is to be completed within the scheduled time—timely execution is cost effective

Excellent advancement in technology and engineering is changing the scenario in India. There is a visible shift towards deployment of construction equipment instead of adopting labour-intensive construction methods. Quality can be totally assured even where large outputs on continuous basis are planned if quality remains inherent in the construction methods. Both input and output of construction equipment can be programmed by putting relevant data into its software, and the performance of construction equipment would be in tune with what is programmed. A compaction roller fitted with 'Compaction Meter System' guarantees compaction to set level—appropriate signal would indicate achievement of specified compaction. Similarly, if parameters are set in a modern batching plant, it would keep on producing concrete of uniform quality. Pumpable concrete has to be homogeneous and consistent in quality, otherwise pumping would not be possible due to clogging. Quality apart, modern construction equipment and machineries come in all sizes—micro-mini to mega sizes with the production matching the sizes. Target outputs, therefore, could be very large. Mechanized construction is more reliable method for timely completion of project work.

Executing agencies operate in an atmosphere of cut-throat competition. Cost of construction equipment could amount to as high as 20% of the total cost of the job to be executed. If cost of construction equipment is to be added to the production expenses, then it would be impossible to be competitive. So, repeated and diversified use of all construction equipment is to be planned and only planned equipment and machineries are procured. Construction equipment can also be hired or taken on lease. Cost involved on hiring or leasing could be considerable. Maximum utilization of hired or leased equipment should be planned in advance so that hiring or leasing could be cost efficient.

Construction equipment of high capacity may be deployed to ensure very large output and consistent performance. Earthmoving is very arduous and time-consuming process for manual efforts, whereas heavy earthmoving equipment of very high capacity could turn a massive job into a manageable one. In concrete work, mechanization could push the output sky-high and cut the execution time drastically. The question of manual efforts does not arise as time is the essence of all contracts.

2.4 CONSTRUCTION MATERIALS

Various types of construction materials are discussed below.

Reinforced cement concrete

Concrete is the most widely used construction material in the world. Plain concrete is a mixture of cement, water, and inert aggregates (vide Chapter 7). An admixture is an optional ingredient mixed as and when required to produce special-purpose concrete or to enhance the quality of concrete. Concrete is generally produced at the construction sites although ready-mixed concrete (RMC) is available from the central batching plants, which are located in areas where there is regular demand for the RMC.

Portland cement, which is generally used for concrete production, unless otherwise specified, is a manufactured item and can be procured directly from the producers. Cement is marketed under different 'brand' names. The details of different kinds of cement including blended cement are furnished in Chapter 7, Section 7.3.

Aggregates comprise the bulk of the ingredients of concrete. Size-wise, aggregates are categorized into fine and coarse aggregates. The aggregates are obtained from rock materials. The natural fine (passing through 4.75 mm mesh) aggregates are picked up from the riverbanks or river beds. If natural fine aggregates are not available, then rocks are crushed to obtain the same. Similarly, coarse (retained by 4.75 mm mesh) aggregates are also available from the natural sources or obtained from the rock-crushing plants.

The water used for preparation of concrete should be of potable quality. Apart from the cement, aggregates, and water, whatever material is added to the concrete is an admixture. Different kinds of admixtures are available to produce different kinds of concrete to serve different requirements. Admixtures used to modify concrete properties are discussed in Chapter 7, Section 7.6.

Steel is another widely used construction material. Steel which is used in the construction of reinforced cement concrete (RCC) is low-to-medium carbon mild steel. Steel is used in concrete to reinforce it against tension. Reinforcing steel is produced in the form of mild steel plain and deformed bars. High-tensile steel is used in prestressed RCC members. Wire fabric is used in construction of roads, highways, and floor slabs.

Steel structures

Structural steel is produced in the form of plates, sections, and shapes. Steel structures are fabricated out of steel sections and shapes by gas cutting and welding. Acetylene and oxygen gases are used for gas cutting. Electrodes are bought-out items and made according to various specifications to suit divergent requirements. High-tensile bolts are used to connect structural steel members wherever the drawings show bolted connections instead of welded ones. Rivets are still used to manufacture wagons and containers for the railways.

Metal products

Cast iron and ductile iron pipes and fittings are extensively used in plumbing and drainage work. Large-diameter steel pipes are used in conveying or drainage of water/wastewater. Steel is beset with the problem of corrosion. Therefore, diameter of pipes should be large enough so that the problem of corrosion can be taken care of as and when required. Measures against corrosion are painting, galvanizing, treatment with phosphate solutions, and cathodic protection.

Aluminium and aluminium alloys are used in buildings in various forms. Pure aluminium is soft, weak, ductile, and corrosion resistant.

Structural aluminium sections and shapes are all extruded—angles, channels, I-sections and plates. These sections and shapes are used for installation of false ceilings, glazed air-conditioned enclosures, door frames, window frames, etc. Connections are either welded, bolted or even riveted.

Timber

Timber is mostly used for temporary work. Bulk of the timber materials are used as formwork for concreting operation. Because of the plastic state, all green concrete at the time of construction requires some kind of formwork to mould concrete to the required shape. As and when the green concrete is hardened and matured sufficiently, the forms are stripped. Forms are made of timber planks or plywood/particle boards stiffened by square or rectangular timber pieces and propped by timber poles. The same arrangement is followed in shoring of vertical sides of excavation. Timber is used in large scale in building structures in many countries.

Coal derivatives

Asphalt and coal tar are bituminous substances and used for roof waterproofing. These substances are heated to high temperature and applied on the roof. Waterproofing felts are laid over the hot bituminous coating to form an impervious layer. The process is repeated by applying hot bitumen over the felt and laying another layer of felt over it. While built-up waterproofing comprises one or more plies of felts, damp-proofing generally means application of bituminous coating only. Asphalts are also used for building pavements.

Masonary work

Multistoried residential or industrial buildings are built of RCC or steel structures. The gaps between the columns, beams, and floors are filled up by masonry cladding/partitions comprising brick masonry or stone masonry. Bricks are made of clay or fly ash. Cladding is also done using concrete bricks or blocks. Fly ash bricks are also available for masonry work.

Soils

If the construction site is located in a low lying area, then the site has to be developed by transporting large volumes of soils from outside. River beds or even seabeds are dredged to obtain soils for land reclamation.

2.5 DEPLOYMENT OF CONSTRUCTION EQUIPMENT

Selection of construction equipment to be deployed at any particular project would depend on the type of job involved, time schedule, location, infrastructures, climate, environment, economic viability, and management attitude. Project layout should have sufficient space available for movement, maneuverability, and parking of mobile construction equipment.

Earthwork calls for heavy construction equipment, which comprise of power units with different mountings for earthmoving, excavating, loading, and hauling. Since mechanization has to be cost efficient, efforts should be made to use the same power unit for multipurpose application, using different mountings/attachments. Bulldozers, scrapers, shovels, backhoes, draglines, trenchers, dredgers, boring machines, dump trucks, etc. are used in earthwork after careful selection. Among all these equipment, shovels in combination with bulldozers and dump trucks are commonly used. Rollers and rammers are used for compaction of earth filling. In rock blasting and mucking (disposal of spoils), manual efforts cannot measure up to the task; and hence, construction equipment like compressors, jack hammers, wagon drills, loaders, dump trucks are very commonly used. Pumps are extensively used for dewatering purpose.

Mixers and batching plants are used for concrete production; truck mixers, dumpers, pumps, cranes, and conveyors are used for transporting and placing concrete; and vibrators as well as rollers are used for compacting green concrete. If aggregates for concrete are not readily available, then crushing plants are installed for aggregate production. Slip-forms with hydraulic jacks are used for construction of chimney stacks and high-rise building columns and walls.

Deployment of construction equipment and machineries facilitate almost all operations in construction work. Some of the commonly used construction equipment and machineries are: dump trucks, tractor trailers, hoists, derricks, cranes, etc. Cranes can be of various types, and they are used extensively. Tower cranes, if deployed, serve useful purpose and accelerate the overall construction process.

In road construction, the commonly used construction equipment and machineries are: bulldozers, scrapers, graders, backhoes, road rollers, impact compactors, slip-form pavers, etc. Extensive deployment of concrete slip-form paving equipment is the need of the hour in India. For this, construction equipment should also be deployed for recycling old road materials. Most of all, deployment of heavy-duty compactors is of prime necessity to ensure proper compaction of earth filling and also of sub-base above earth filling.

There are other versatile construction equipment and machineries available in the global market which would enhance the productivity manifold, assure specified quality, reduce execution time, and thus make the execution cost efficient. Such equipment could be deployed in a project if mobilization of the same is economically viable.

2.6 PREFABRICATION IN CONSTRUCTION

Prefabrication in construction allows parallel execution of different activities whereby execution time could be significantly reduced. In building construction, steel superstructures are designed over reinforced cement concrete (RCC) foundations. Roofs are also designed as precast RCC concrete slabs to be lifted and placed on steel beams or trusses. All these could be produced simultaneously, and erected progressively in sequence. Cast-in-situ reinforced concrete foundations and sub-structures are constructed at or below the ground level, while fabrication of steel structures is carried out at the shops. When the foundations or bases are ready, fabricated structures could be erected over the completed foundations. Simultaneously, RCC roof elements are precast at the sites. RCC slabs are also precast at the project sites to cover RCC or masonry trenches constructed for (i) laying cables, pipes, etc., and (ii) drainage purpose.

Instead of structural steel superstructures, RCC superstructures may also be prefabricated at the shops. But such prefabrication of RCC structural members is not favoured in India, as steel is readily available. Besides, transportation, handling, and erection of steel structures are relatively less problematic. But in case of bridge construction, pre-stressed RCC girders and deck segments are prefabricated on-shores. However, many fabrication shops are there only for prefabrication of precast, pre-stressed RCC sleepers for the railways, pipes for conveying fluids, and poles for power distribution.

There are large shops for fabrication of structural steel. Engineering design of steel structures is done, keeping shop fabrication of the steel structures in mind. Fabricators prepare fabrication drawings on the basis of design drawings. Fabrication is done on the basis of fabrication drawings approved by the designers. Shop fabrication calls for transportation of fabricated steel to the sites. Fabrication detailing is done, keeping possible transportation problems in mind. Capacity of transports would determine the dimensions and weight of fabricated steel to be transported. Fabricated members could easily be spliced at the sites. Splicing may be done by welding or by using high-strength bolts. Riveting is used for fabrication of wagons, etc., for the railways, but not for splicing or joining anymore.

2.7 FALSEWORK AND TEMPORARY WORKS

Falsework may be defined as temporary structure that is intended to support a permanent structure while it is not self-supporting. And temporary work covers a wide range of processes like exclusion of groundwater, dewatering, piling, excavation, shoring, scaffolding, pipe/box jacking, bridging, and so on. Falsework is also temporary work and does not form part of the permanent feature of any project, except where the cost of removal would be prohibitive.

Because concrete takes time to harden, forms are needed to mould concrete into design shapes. As forms are required temporarily to cast concrete on the basis of design drawings furnished by the owners or their consultants, executing agencies design forms so that the same can withstand load of plastic concrete without bulging or yielding. When design of forms in special cases is different from normal formwork, the owners' engineers or consultants review and approve such design. The same principle is followed on building temporary platforms and access ladders with steel tubes or timbers. In short, focused attention is needed in providing supporting scaffolding, planks, ladders, guardrails, barriers, and slinging devices for putting forms in position; fixing reinforcement in place; and pouring concrete in-situ.

Scaffolding is temporarily assembled timber or steel tube platforms or working areas. Such platforms or working areas are required for working at heights. Steel tubes are coupled and braced together. Similarly, timber poles are bolted or tied and braced together. As it is easier to make steel scaffoldings, the trend now is to use steel tubes as much as possible. Scaffoldings can be assembled independent of permanent structures or, if necessary, can be tied to the permanent structures also. Steel scaffoldings can be fitted with wheels at the bottom for mobility and manoeuvring. Scaffoldings can also be supported or hung from permanent structures.

High water table may pose serious problem for construction of foundations or sub-structures. In such cases, water may be pumped out of sump dug below excavation pit. The other alternatives are lowering of water table by well-point system or sinking shallow-bored well (vide Chapter 4, Section 4.5). Sheet pile cofferdams isolate both water and soil that exist outside such cofferdams. Once the cofferdam is in place, the construction work can be carried out inside after digging soil to the required level and then sealing the bottom. As retrieving the cofferdam steel sheet piles completely is difficult and costly, such retrieving is abandoned on many occasions especially when cost of retrieving becomes prohibitive and time-consuming.

Cofferdams are very important for offshore construction. Offshore structures are connected with the shore by placing precast slabs or timber planks on steel girders. Temporary piling is often required for such construction activities. Even temporary bridging may be required, as access along such bridge would facilitate offshore construction.

Shoring is done to retain the soil temporarily in position so as to complete the construction work. Shoring is also done to retain part(s) of buildings. Both timber and steel materials are used for shoring and propping.

For construction of underpass or drain under existing roads or railways, it would be necessary to excavate the embankments on which the railways or roads exist resulting in disruption of services. Pipe or box jacking is now a widely used method of boring through embankments without disrupting any traffic or work on the embankments. Concrete or steel pipe of adequate diameter or box section of adequate size is jacked through the embankment without causing any disruption of railways/road traffic. Pipe or box that is jacked through should be designed to support the load it would be subjected to. Soil inside such pipe or box is scooped or dug out. A number of pipes or boxes can be jacked side by side depending on the requirement of underpass or drainage. For more details, refer to Chapter 4, Section 4.6.

Executing/construction agencies are responsible for design of falsework and temporary work. However, owners or their consultants should pay attention to the quality and conditions of the materials that are being used for falsework and temporary work, examining particularly how they are loaded, laced, braced, and tied. Special attention needs to be focused on the bases, foundations, or structures on which the falsework and temporary work are supported. Similar attention is needed while checking where the temporary vertical members are laterally tied or jointed.

SUMMARY

Timely project implementation maintaining quality within the budgeted cost requires completion of preparatory work before implementation is initiated. This chapter contains detailed information on the required preparatory work and throws some light on the method of working out the requirement vital for such preparatory work. Site layout should be prepared keeping provision of sufficient spaces to cater to the requirement of the executing agencies and movement/parking of construction equipment. Infrastructures are of vital importance for uninterrupted continuation of construction work. Equally important is the construction methods based on deployment of sophisticated construction equipment. Construction materials in general comprise ingredients for concrete production, steel bars for reinforcing concrete and steel shapes/sections/plates for structural fabrication. Besides, cast and ductile iron pipes, aluminium, timber and bituminous products are also used extensively. Prefabrication in construction allows parallel execution of different activities whereby execution time could be significantly reduced. False and temporary works do not form part of permanent feature of any project except where cost of removal would be prohibitive or removal could jeopardize safety.

REVIEW QUESTIONS

1. How should requirement-wise total areas be worked out during land acquisition?
2. What is the first step towards the implementation of a project work?
3. What is the diversity factor that is used in estimating the power requirement for project implementation?

4. How would you work out the total water requirement for a project during implementation stage?
5. What should be the illumination level in lux where moving equipment and plants are deployed for construction work?
6. What kind of storage facilities should be available at a construction site?
7. If there is provision for a workshop in a project, why should it be built on priority?
8. How does setting up of laboratory testing facilities help a project?
9. How can firefighting be arranged at construction sites?
10. Why is it required to plan repeated and diversified use of all construction equipment?
11. What is the role of cast iron and ductile iron pipes and fittings in construction work?
12. How can the scheduled progress be hastened by prefabrication work?
13. Explain why temporary and falseworks are unavoidable in many cases.
14. How do temporary and falseworks escalate the project cost?

3
Transportation and Handling

INTRODUCTION

Transportation of all kinds of building materials, construction equipment, plants and machineries is to be meticulously planned well before the commencement of project implementation. Contractors responsible for execution of civil, mechanical, electrical, and other work arrange transportation of both raw materials and construction equipment required for contracted work. Owners have the option of engaging separate agencies for transportation including handling of all items at both stores/shops and construction sites.

Modes of transportation vary from project to project. Bulk materials are transported by surface transport, such as *railways* using freight wagons or by *roads/highways* using trucks or wheel mounted tractor-trailer combination. By sea, transportation is done by cargo ships; whereas in case of inland waterways, transportation is done using barges towed by launches. Aeroplanes, helicopters, or special cargo planes are used to transport goods by air. Movement of materials and plants at the project site can be horizontal, vertical, or a combination of both.

Materials can be of any form: solid, liquid, or gaseous. The simplest way to move fluids is to use pumps. However, containerized fluids would be like any other packaged goods hauled by trucks or tractor-trailer combination. A lot of hauling is carried out by winch using manpower or mechanical/electrical drives. Dumpers are most versatile haulers. Transported materials/plants are to be handled, stored, and preserved in line with the specifications furnished by the suppliers/manufacturers.

3.1 BASIC PRINCIPLES

A project can be the extension of an existing building or utility or industry, in which case there would be much less problem of infrastructures and other related facilities. If such extension is of stand-alone nature, all existing infrastructures may not be available for extension. If a project is set up in an open area like a green field, there will be nothing there to start with. All kinds of building materials, construction equipment, plant and machineries are to be transported to the site. Transportation cost in case of a green field project would be substantial.

Executing agencies responsible for construction of civil engineering work arrange transportation of both raw material and construction equipment required for contracted work. Suppliers depending on the nature of contract deliver materials at the site. Erection contractors depending also on the nature of contract may be responsible for all the transportation involved. If so, then they have to pick up plant and machineries including structures from manufacturing shops and deliver at the the construction site. Owners have the option of engaging separate agencies for transportation including handling of all items at both shops and construction sites. Involving more and more agencies would call for more coordination by the owners.

When India was dependent on imported plant and machinery, transportation by sea was substantial. Plant and machineries for Tarapur Atomic Power Project reached Indian shore in 1960s by sea. Urgently required items were even airlifted. As the country developed and advanced financially, transportation generally meant surface transportation. Now, again with the increasing trend of direct foreign investments in new projects, transportation by both sea and air is being taken into account.

Sometimes, there may be dislocation of items while the goods are despatched or loaded at the shops for transportation to a project site. Such dislocation may be disastrous for a project as dislocated items may cause total stoppage of work where dislocated items are of vital nature. To circumvent such hazards, cautious owners assign experienced personnel to escort important consignment(s). Where the dislocated items cannot be traced at all, there would be no option but to re-fabricate or remanufacture the lost items. This would escalate cost and delay the progress, thus upsetting both time schedule and budget.

Weight and dimensions of individual items to be transported are very important. Be it by surface or by sea/river or by air, the carrier should be capable of carrying the item dimension-wise and/or weight-wise to destinations. Any problem in transportation would mean additional work at extra cost.

Surface transport means transportation by road or by railways. Bulk materials are transported by railways using freight wagons. By road, transportation is done by trucks or wheel-mounted tractor-trailer combination. By sea, transportation is done by cargo ships whereas in case of inland waterways, transportation is done by barge towed by launch. Some barges have in-built power units for hauling. By air means transportation by aeroplanes, helicopters, or special cargo planes.

Apart from transportation up to the construction site boundary, there is the question of transportation inside the boundary of any project site. If roads are ready within the project boundary, then transportation could reach the unloading areas. In case of road transportation, continuation up to unloading areas would not be any problem. But in case of railway transportation, there would be problem if there is no railway siding at the project site. Unloading from rail wagons outside would involve transportation again up to unloading area(s) of the project site.

Loads comprising small pieces are put into containers or packaged properly for handling for transportation.

3.2 ROAD TRANSPORTATION

Road transportation is indispensable for movement of materials, plant, and machinery from outside locations/terminals to project sites. Transportation is an economic function and is a part of the construction process. Increase in efficiency of transportation would reduce the cost of construction. Road transportation is experiencing booming growth because of its simplicity, independence of operation, capability of negotiating high gradient, manoeuvrability in difficult conditions, and door-to-door service. Despite such growth, capacity of road transportation is still not enough to meet the high demand. Road freight traffic has increased from 6 billion tonne kilometres in 1951 to 800 billion tonne kilometres in 1999.

For economic transportation, roads should be fit for vehicle movement. It should be checked whether or not condition of roads to be used are fit for transportation of heavy and oversized freight. While checking condition of roads, condition of bridges and culverts should also be checked.

Condition of roads in India, in general, is not good. Maintenance and improvement of roads is hampered by fund crunch and lack of proper maintenance. The country's road network generates large revenues out of which only about one-third or so is spent on maintenance and improvement whereas almost the entire earning is spent for the same purpose in advanced countries. Because of the bad condition of roads in India, survey of the road(s) is necessary before freight loaded transport start plying on such roads. If necessary, roads should be improved and made strong to be able to bear the loads of heavy freight. Both bridges and culverts may have side railings/guard-walls, which may foul with oversized freight. If it is not possible to make any road freight-worthy, then alternative to such road, if any, is selected even if it means longer distance to travel. If such detour is not possible, then the probable solution could be rail or other modes of transportation.

In a fertilizer project, the ammonia converter having 28.2 m length and 3.9 m diameter weighed 410 tonnes. Because of the non-availability of any alternative like rail transportation, a lot of strengthening of bridges and road was done to carry it on a 58 metre long special trailer weighing 150 tonnes having 32 axles and 256 wheels with hydraulic suspension. An engine of 350 HP pulled the trailer and another engine of the same power was attached at the rear to regulate the movement. The movement of the trailer was under surveillance round-the-clock.

Union and state governments build and maintain roadways for development of the country. Indian laws now permit levying of toll for the use of roads. This would encourage private participation on building better roads and make the road transportation more efficient.

Private operators control the bulk of road transportation. Private operation is mostly unorganized and individual operators run the business. Road transportation, therefore, can be availed easily and efficiently.

Once freight-laden transport reaches the project site, project roads should be ready for movement of heavy transport. Project site roads may either be temporary roads built for heavy traffic and rough use or permanent roads built without the top-wearing course. As permanent project roads are meant for double lane two-way traffic, there should be proper curves at road crossings so that heavy transport vehicles or tractor-trailer combination could manoeuvre and negotiate all turning points. In case of temporary roads, width and turning radii should be sufficient for accommodating heavy trucks or tractor-trailer combination. Temporary roads are built along the routes of permanent roads as shown in the project layout.

Road construction and paving are discussed in Chapter 12. And for transportation, movement and handling at the construction sites, vide Section 3.6 below and Chapter 13, Section 13.5.

3.3 RAILWAY TRANSPORTATION

Compared to road transportation, railway freight (Fig. 3.1) traffic is much slower. As such, railway freight traffic is coming down. The rail transportation came down from 89% in 1951 to 40% in 1995.

Railways have different gauges of tracks—broad gauge (1.676 m), metre gauge (1.0 m), and narrow gauge (0.762 m/0.610 m). Laying of metre and narrow gauge tracks has been stopped. Not only that, conversion of the metre and narrow gauge tracks to broad gauge tracks is going on. For hauling of rail freight traffic, different types of locomotives are used. Locomotives may be powered by steam, diesel, or electricity. Steam-powered locomotives are being phased out. Diesel locomotives are widely used now. There were 17 diesel locomotives in 1950–52. This figure went up to 4313 in 1995–96. Diesel locomotives consume less fuel, require no preparation for start-up, and operate economically as well as efficiently. Electric locomotives are most environment friendly haulers, but their use is restricted to such places having specified power (25 kV, 60 cycles). 220 V power is to be stepped up to 25 kV using transformer and separate overhead transmission line is to be strung for energizing electric locomotives. The

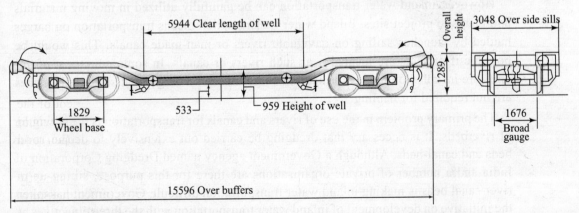

Fig. 3.1 Open type freight carrier on rails

number of locomotives in use in India has increased from 72 in 1950–51 to 2387 in 1995–96.

Another serious problem of rail transportation is the existence of single track and double track broad gauge lines. The records of broad gauge tracks for the year 1995-96 are available. The records show single track of 25,516 km length and that of double tracks is 15,064 km. Single track would obviously cause slowing down of freight movement.

Possibility of dislocation of freight is more in rail transportation than in case of road transportation. Where timely transportation is of vital importance, project management must assign personnel to escort freight booked for transportation by the railways.

3.4 WATERWAY TRANSPORTATION

India has a long coastline and vast hinterland. There were close to 200 ports including 11 major ports at one time. More ports are being built in both public and private sectors to facilitate trade with other countries. Some ports would be on private ownership basis for exclusive use. World trade is dominated by shipping or sea transportation, as there is hardly any economically viable alternative mode of transport to it. (Name of the 11 major ports – Kolkata, Haldia, Paradeep, Mumbai, Chennai, Cochin, Tuticorin, Jawaharlal Nehru Port at Nava Sheva, Kandla, Vishakhapatnam, New Mangalore, Marmugao.) Nearly 95% of India's foreign cargo (by volume) moves by sea and, therefore, ports and their development assume an important place in policy making. Development and maintenance of India's major ports are the responsibility of the Central Government, while other ports are in the concurrent list.

As construction-related projects are not dependent on imports except for such projects as are tied up with foreign direct investment, transport by sea is not a common feature now in construction activities in project execution. This is good as sea transportation is costly, and there is uncertainty of delivery in time because of too much of loading and unloading operations at ports as well as places of origin and destination. While 30% time in sea transportation is spent on voyage, 70% time is used up at the ports.

However, inland water transportation can be gainfully utilized in moving materials and plants to project sites. Inland water transportation means transportation on barges hauled by launches sailing on navigable rivers or man-made canals. This would be possible if projects are located near such rivers or canals. In some countries, power units are installed on barges for hauling cargo loaded on barges and separate launches are not required for hauling.

The primary problem in the use of rivers and canals for transportation is the silting up of riverbeds. It is necessary that dredging be carried out extensively to deepen river beds and canal-beds. Although a Government agency named Dredging Corporation of India and a number of private organisations are there for this purpose, silting up of river/canal beds is making inland water transportation difficult. Government has taken the initiative on development of inland water transportation with the three objectives as follows:

- Development of inland water transportation in the regions having natural advantages
- Modernization of barges and launches to suit local conditions
- Improvement in the productivity of assets—the Inland Waterway Authority has been set up which is a big step forward and should help in the accelerated development of inland water transportation

Waterway transportation in India is confined to Eastern and North-Eastern states and Goa.

3.5 AIRWAYS TRANSPORTATION

Airways transportation is the quickest mode of transportation. However, it is expensive and unsuitable for heavy and bulky cargo.

Both domestic and international trade is growing briskly with India's increasing trade with other countries. But for project construction work, air transportation is not very relevant except for high-value light goods and items required on emergency basis. Both aeroplanes and helicopters are used in air transportation.

3.6 HAULING AND HANDLING BY CONSTRUCTION EQUIPMENT

Movement of plants and materials at project sites can be horizontal, vertical, as well as combination of both. Materials can be of any form—solid, liquid, or gaseous. The simplest way to move a fluid is to pump it to any distance and height according to a pump's capacity. However, containerized or bottled fluids would be like any other solid or packaged goods hauled by trucks or tractor-trailer combination. Tractor details are given in Chapter 13, Section 13.4.

The most common movement is hauling by self-propelled transport vehicle. And, the most commonly used transport for hauling plants and materials is truck as its movement is much simpler with more flexibility and manoeuvrability in operation apart from their high speeds when operating on suitable roads at relatively low hauling costs. Either temporary or permanent roads with sufficiently firm and smooth surface is necessary for truck mobility and movement. Road grades should not be excessively steep. Mobility is related to time factor. Tractor is used for both on-road and off-road hauling.

Trucks used for hauling are powered by diesel engines to reduce cost. A vast range of trucks is manufactured to cater to different requirements of a project work. Trucks are fitted with hydraulic power transfer system for tipping, hoisting, and loading for versatile use. The trucks may be provided with a body to transport different goods, special purpose body to transport loose and viscous materials, and tank body to transport liquids. Trucks are classified according to their load carrying capacity and cross-country capability. Payload capacity of a truck can be 200 tonnes or more. Movement of such a heavily laden truck would depend on whether or not the condition of roads, bridges, and culverts are designed to sustain its maximum wheel load.

Trucks may be classified according to several factors including the following:

- Size and type of engine—petrol, diesel, butane, propane
- Number of gears
- Kind of drive—two-wheel, four-wheel, six-wheel, etc.
- Number of axles and wheels and arrangement of driving wheels
- Method of dumping the load—rear dump, side dump
- Method of rear dumping—hydraulic or cable-operated
- Class of materials hauled—earth, rock, aggregates, bricks, plants, etc.
- Capacity in tonnes and cubic metres

The *struck capacity* of a truck is the volume of materials that it would haul when it is filled to the top of the sides, with no materials above the sides. The *heaped capacity* is the volume of materials that it would haul when the load is heaped above the sides. The capacity should be expressed in cubic metres. While the struck capacity remains fixed for a particular unit, the heaped capacity would vary with the height to which material may extend above the sides and with the length and width of the body. Moist earth or sandy clay heaped with a slope of 1:1(H:V) may be hauled, but hauling in case of dry sand and gravel would be possible if heaped only up to a slope of 3:1. For working out the heaped capacity of a unit, it is necessary to know the following details:

- Struck capacity—length × width × height of the unit
- Slope at which the material would remain stable while the unit is moving

Smooth paved roads would allow larger heaped capacity than rough haul roads.

Truck mixers are readily available trucks used for hauling concrete ingredients minus water, plastic concrete, and even mortar. In case of dry ingredients, water is added at the location of concreting, keeping in view initial and final setting time of concrete mix.

Efficient and economic use of a truck depends on its condition and age, type and volume of freight, weight of goods to be hauled, distance involved, road conditions, and the management responsible for operation and maintenance of trucks.

A *tractor* is a wheel-mounted or track-mounted (crawler) self-propelled vehicle used as a power unit for moving goods or towing trailers, vide Chapter 13, Fig. 13.3. Engines in both trucks and tractors convert thermal energy into mechanical energy. Drive line comprising clutch, flexible coupling, gearbox, and rear axle transmit torque developed by the engine to the driving wheels or tracks and change the driving torque both in magnitude and direction to haul the goods.

A *full trailer* is a truck without engine whose wheels fully support the vertical load due to its own weight plus weight of goods it carries. A semi-trailer is a trailer whose front end is supported by the tractor in such a way that a part of the vertical load is transmitted to the rear wheels of the tractor. Trailers may be of two-wheeled, four-wheeled or multi-wheeled type; thus, they may have one, two, or more axles. Trailer(s) is/are coupled with suitable tractor on the basis of size, shape, and weight of materials or plants to be hauled. Semi-trailers may be equipped with jockey wheels to support their front-ends whenever they are detached from their tractors.

Dumpers are most versatile haulers used extensively at construction projects right from carrying raw materials to concrete or fabricated/manufactured items that are ready for erection/installation. Dumpers of various design and capacities are available to be used for hauling – two/four wheel drive, hydraulic and gravity operated container, side or high level discharge and self-loading facilities.

A lot of hauling is carried out by winch at the project sites using manpower or mechanical/electrical power. Wire ropes are extensively used for hauling and handling. Steel wires are twisted together to form strands, and strands are twisted together to form wire ropes, vide Chapter 13, Fig. 13.18. Wire rope is used in a very simple way in winches. Materials or plants are placed on pipes or rails and tied with wire ropes uncoiled from winch drum so as to facilitate hauling. This, however, is a relatively slow process.

3.7 LOADING AND UNLOADING OPERATIONS

Self-propelled loaders are equipment used extensively in construction work for their mobility and versatility. Loaders are mounted on tracks or pneumatic wheels.

A *derrick*, vide Chapter 13, Fig. 13.20 and Fig. 13.21, is a lattice mast held in position by five or more guy wire ropes, which are anchored at ground level. In a simple system, a wire rope can be used over a pulley hooked at the mast top for lifting materials. This mast may be converted into a static crane by incorporating a jib of height less than that of the mast. The jib is to be of low-pivot type. Shorter jib would allow its rotation all around.

Hoists (Fig. 3.2) are meant for vertical movement of materials, simple plants, and personnel within its capacity. If a hoist is designed for movement of materials only, then personnel should not be allowed to use the same. A stationary hoist comprises a mast or tower with cantilever platform and is rigidly tied to a permanent structure or a strong scaffold. The platform can also be hung from the top, with guide rails on two or three sides. A mobile hoist is a self-supporting structure moving on rails with similar platform. There should be platforms for unloading and exit at desired levels.

3.8 STORAGE AND PRESERVATION

Some guidelines are mentioned below for storage of civil engineering construction materials:

Bricks – Must be stacked in regular tiers, and not dumped at the site—there should not be any problem of picking up the right type of brick when required

Stones – Dressed stones, particularly, should be stored carefully so as to avoid defacement and damaged edges—stones should be free of damp and rust stains

Lime – Must be stored in covered sheds

Cement – Must be stored in covered and weatherproof stores above the ground level in such a manner so as to facilitate 'first in, first out' principle—vide Chapter 2, Fig. 2.2.

Aggregate – Both fine and coarse aggregates should be stored size-wise on brick soling

42 Construction Technology

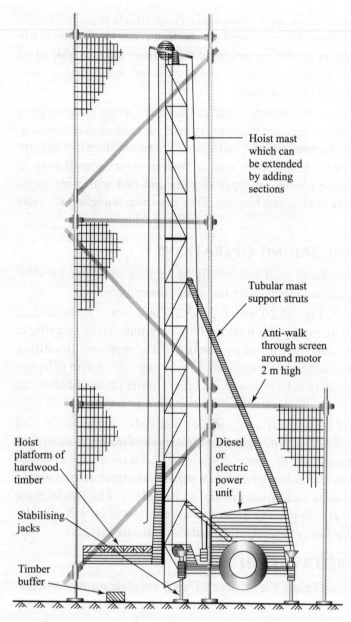

Fig. 3.2 Hoist

or appropriate platform—the materials must be protected from dirt, clay, vegetation, and other deleterious substances

Reinforcing steel – Should be stored off the ground preferably under shed—steel bars must be protected against corrosion

Some guidelines are mentioned below for equipment and plants:

Open storage

- It should be 400 mm above the firm ground. Periodic painting is recommended wherever required
- Milled/machined ends of columns are to be covered with plastic covers/wooden covering
- HSFG (High Strength Friction Grip) bolts are to be stored in hermetically sealed packages
- Loose components are to be stored under covered sheds—turn buckles, nuts, suspension rods, springs, constant load hangers
- Mill drive assembly to be stored under tarpaulin cover

Covered sheds

- Fans
- Boiler feed pumps
- Vertical motors to be stored vertically

Semi-closed shed

- Feet of suction chamber and diffuser
- Connecting pipes

Checkpoints during storage

- Turbo-generators—fully closed, properly ventilated, dust/moist free premises—to ensure that inside temperature is 5°C above the ambient, especially during monsoon and damp climatic conditions

Air-conditioned stores

- Measuring instruments and sophisticated items

Anti-corrosion

- Nitrogen blanketing is done against possible corrosion as in transformer tank

IS: 4082—Recommendation on stacking and storage of construction materials at site; and IS: 7969—Safety code for handling and storage of building materials are safety related codes issued by the Bureau of Indian standards.

SUMMARY

Transportation and handling of materials as well as plants and machineries are vital for successful project implementation, as construction activities could only be continued without interruption if materials/plants/machineries are available when required. Transportation includes: (i) surface transportation by road or railways, (ii) waterway transportation by sea or inland waterways, (iii) air transportation by aeroplanes, helicopters, or special cargo planes. Surface transportation is vital nowadays. As India is surging ahead economically, there is less dependence on imported materials and plants/machineries. Surface transportation is, therefore, becoming an essential feature of project implementation so much so that highways and railways are being increasingly upgraded and modernized. Government is planning special freight corridors. The entire country is being linked by multilane highways. Not only that, use of inland waterways is receiving utmost attention. These trends are encouraging as these would result in squeezing of project implementation time schedules, thereby increasing the cost-effeciency of projects.

Handling involves movement of materials and plants at project sites horizontally, vertically, and both vertically as well as horizontally. Materials handled can be of any form – solid, liquid, or gaseous. Of these materials, fluids can be pumped to any reasonable distance or height. However, containerized or bottled fluids would be like any other solid or packaged goods. A lot of hauling is carried out at the project sites by winch or dumpers. Modern trend, however, is to use stationary or mobile tower cranes as tower cranes turn handling very easy.

REVIEW QUESTIONS

1. What is meant by surface transport?
2. Why are experienced personnel assigned to escort important consignments?
3. Why is road transportation experiencing booming growth in India?
4. For over-sized freight, road survey is carried out prior to actual transportation. What is checked during such road survey?
5. Why are diesel locomotives preferred for hauling railway freight traffic?
6. What should be the specification for electrical power for energizing electric locomotives?
7. What are the problems that beset sea transportation?
8. How can inland water transportation be gainfully utilized?
9. Why is dredging necessary for inland water transportation to remain viable?
10. Why is air transportation not availed although it is the quickest mode of transportation?
11. Self-propelled transport vehicles are the most common mode of hauling at site. Why?
12. Define struck and heaped capacities of trucks.
13. What is the power source of tractor for hauling trailers?
14. Describe how a derrick can be used for loading and unloading operations.
15. List the protective measures required for open storage of equipment and plants.

4
Earthwork

INTRODUCTION

Earthwork is earthmoving, either in excavation or filling. Soils in earthwork may be broadly classified into three categories: (i) coarse grained or non-cohesive, (ii) fine grained or cohesive, and (iii) organic.

The topsoil is unsuitable for supporting foundations or substructures as the same comprises mineral matter, organic matter, air, and water. Soils existing at a project site should be classified before planning any earthwork.

Project work should be started with site development including: (i) replacement of topsoil wherever marshy land exists, (ii) removal of water and slush from waterlogged areas, (iii) uprooting and clearing of trees, bushes, and vegetation, (iv) dismantling and removal of existing buildings/structures, (v) removal of loose/partly buried rocky or waste materials, and (vi) leveling of uneven areas.

Setting out can be started only on developed sites so that survey grid pillars can be established without any problem. Buildings or structures are laid out on the basis of grid pillars and reference level.

Earthwork in most projects is voluminous and deployment of earthmoving equipment is now very much a part of the construction in major projects. Earthwork in excavation is beset with more problems compared to backfilling or embankment formation.

Groundwater poses serious problem in executing excavation work under the water table. There are both permanent and temporary solutions to this problem. Trenchless technology is adopted for underground utility services in urban or developed areas underneath existing buildings, streets, roads, railways, embankments, installations, or wherever open excavation or trenching is not possible. Earthwork in grading is necessary for proper and accurate finishes.

Dredging is the process of excavating underwater soil or rock (hard rock by blasting) for deepening river/sea beds or land reclamation.

4.1 CLASSIFICATION OF SOILS

Soil is the result of disintegration and decomposition of the earth's rocky crust. Disintegration of earth's crust could be caused by various factors—frost, rain, temperature, gravity, wind, ice, and water. Factors like frost, rain, and temperature variation disrupt the solid rock of earth's crust; gravity disturbs balance of the earth's rocky mass; and wind, water as well as ice have inherent energy to transport disintegrated rock particles all over. Some soil mass is mainly of organic origin.

The mineral particles differ in size and chemical composition, depending upon the original rock mass or the nature of physical disintegration or chemical decomposition the rock mass was subjected to. The organic matter may be dead or decaying plant and animal remains including macro-organism like earthworms and micro-organisms like bacteria. The pores in mineral particles include atmospheric gases plus gases produced during biological and chemical activities.

Soils may be broadly classified into the following three categories:
- Coarse grained or non-cohesive
- Fine grained or cohesive
- Organic

Coarse grained or non-cohesive

Very coarse-grained soils like boulders and gravel are composed of rock fragments. Boulders are over 200 mm in size. Particle sizes of cobbles lie between 60–200 mm. Each fragment of gravel may be composed of one or more minerals. Quartz could be predominant mineral in gravels when fragments are rounded in shape. Grain sizes of gravels lie between 2–60 mm.

Coarse-grained materials like sands predominantly have quartz in their composition. Sand particles that are not rounded may contain one or more minerals other than quartz. The particles are more angular in fine sands than in coarse variety. Grain sizes of sand lie between 0.06–2 mm. Depending on the higher percentage of grain sizes, sands are further classified as coarse or medium or fine sands. Sand is gritty to touch and the individual grains or particles can be seen with naked eye. The soil in which sand particles are abundant is classified, logically enough, as a sand-textured soil or simply *sandy soil*. The term 'texture' refers to the degree of fineness and uniformity of soil. Sandy soils are coarse in texture and have high permeability and low compressibility with low void ratio.

Principal soil type	Further classification	Particle size (mm)
Gravel	Coarse	20–60
	Medium	6–20
	Fine	2–6
Sand	Coarse	0.6–2
	Medium	0.2–0.6
	Fine	0.06–0.2

Fine grained or cohesive

Fine-grained soils like silt and clay are composed of fine particles each of which contain only one mineral. Fine particles are not rounded, but are more angular and flake-shaped. Silt is a type of soil, intermediate between fine sand and clay. Grain sizes of silt lies between 0.002–0.06 mm. Silt is smooth and slippery to touch when wet, and the individual particles are much smaller than those of sand. These individual particles can only be seen with the aid of a microscope. Silt-textured or silty soils contain relatively large amounts of silt.

Clays are very fine-grained soils possessing plasticity. Grain size of clay particles is less than 0.002 mm. Clay is sticky and plastic-like to handle when wet. Clay-textured or clayey soils are rich in clay and fine in texture. Fine-grained soils have low permeability and high compressibility due to high void ratio. A soil sample generally consists of materials of all the three phases—solid, liquid, and gas. Liquids and gases occupy the void spaces in solid soil ingredients. More water content in clay means more volume but less strength. With less water content, clay contracts but gains in strength. The force that sticks the particles of clay together is called cohesion; and unless there is any significant movement of pore water, cohesion would remain the predominant shear strength of the fine soil.

Soils in general consist of a mixture of two or more different grain size particles. On the basis of predominant grain-size composition, soils are designated as *sandy silt*, *sandy clay*, *silty clay* and so on. *Silty sand*, for example, indicates predominance of sand in soil with a small amount of silt. The predominant component is used as a noun and the lesser component is used as a modifying adjective.

Topsoil and organic matter

Soil at the top is a mixture of mineral matter, organic matter, air, and water. Topsoil is thus capable of supporting plant growth but unsuitable for supporting foundations or substructures. The organic matter owes its existence to decomposed vegetable matter and humus, which is a dark brown colloidal material derived from the decomposition of vegetable and animal matter. The topsoil is removed for construction work. *Peat* is accumulated decomposed plant material in varying degrees of alteration and is considered as coal in its early stage of formation. Peat is used as fuel. Peat is highly compressible and, therefore, it is unsuitable for civil engineering work.

Soil Investigation

Soil can be classified as outlined above or in more details on the basis of soil investigation carried out at project sites. There are agencies that specialize on soil investigation work. Soil investigation is carried out for obtaining data on the properties and characteristics of soil by collecting disturbed and undisturbed samples at project sites and testing the same at approved laboratories. Soil samples are obtained as follows:

- Trial pits
- Exploratory boreholes

Trial pits are dug manually for visual inspection as well as collection of samples from excavated pits. Only disturbed samples may be collected from trial pits. However, undisturbed samples may be collected from the sides of excavated pits for laboratory tests. Trial pits would be dug in accordance with the specification furnished for this purpose.

Trial pits serve the useful purpose of examining the quality of weathered rocks for shallow foundations. Trial pits extended to trenches provide means for investigation of filled ground on the following three aspects:

- The engineering characteristics like compressibility, permeability, and existence of buried obstructions, if any
- The effect of organic and mineral substances on the durability of foundations and services
- Whether contaminants that may be present in the fill materials or seeping water are environment friendly or not

Boreholes are drilled for collection of both disturbed and undisturbed samples from every stratum encountered in the boring process. The most common methods of borehole exploration are wash borings and drilling using augers.

In wash borings, a casing (about 65 mm diameter) is driven into the soil with the weight suspended from a tripod. A 25 mm wash pipe is inserted in the casing and water is pumped through the wash pipe. Wash pipe is churned up and down to cut the bottom soil loose. Soil sample in water is collected in a tub at the top, as the water comes out through the annular space between casing and wash pipe. As the colour of water changes, experienced driller stops pumping and takes the spoon sample of soil for laboratory testing. Spoon samples are also taken at regular intervals of particular driving depth like 1.5 metres or so.

Commonly used augers are of 100 mm diameter. If soil particles are of larger size, then 200 or 250 mm augers may be used. An auger is bored into the soil and then withdrawn to collect samples. The auger is re-inserted for further boring. If side collapses in the process, the auger must be lined with casing. Soil is disturbed due to the driving process, and spoon samples need to be collected for testing. Drilling in rock is done by using a diamond drill bit as discussed in Chapter 5.

On completion of a bore/drill hole, it is necessary to record the groundwater levels over a period of time. The upper surface of gravitational water that occurs at some depth below the ground level is known as the water table.

The *standard penetration test* is a simple method of obtaining some information on compactness of soil. 'N' represents the number of blows required for driving the sampling spoon by 1 ft (305 mm) with a fall of 2.5 ft (762 mm) by a weight of 140 lbs (63.6 kg). 'N' is regarded as penetration resistance.

Soil comprises three components: solid particles, water, and air. Specific gravity of soil, expressed as percentage, is the ratio of the weight of solid per unit volume to the weight of water per unit volume.

Porosity is the ratio of the volume of voids to the total volume. The term 'volume of voids' means that portion of the volume not occupied by mineral grains. Porosity has a powerful influence on the physical characteristics of materials, especially water-absorbing capacity, water permeability, strength, and so on.

Void ratio is the ratio of the volume of the voids to the volume of solids. The pores are minute interstices in solid rock filled with water or air, whereas the voids are air-filled openings formed among the fragments of bulk materials. Unit weight or density is the ratio of the total weight to the total volume.

Moisture content

Moisture content is the ratio of the weight of water to the weight of solids in a given volume. The moisture content is a measure of water saturation of soil. If a soil sample is oven dry, its unit weight is called *dry density*. The unit weight with water content intact is called *bulk density*.

$$\text{Total volume} = \text{Volume of (air + water + solids)}$$

$$\text{Degree of saturation} = \frac{\text{Volume of water}}{\text{Volume of (air + water)}} \times 100\%$$

$$\text{Void ratio} = \frac{\text{Volume of (air + water)}}{\text{Volume of solids}}$$

$$\text{Porosity} = \frac{\text{Volume of (air + water)}}{\text{Total volume}}$$

The term 'bulking of sand' is used to describe the increase in volume that takes place when water is added to dry sand. Water content of 6–8 % in sand often causes increase in its volume by as much as 20–30%. With addition of more water, however, the volume reduces again. The low density due to bulking that is obtained at low water contents is due to capillary forces resisting rearrangement of sand grains.

The *angle of repose* is the angle with the horizontal level at which a heap of dry loose soil like sand will stand without support and without either impact or vibration.

Permeable material contains continuous voids. Because all kinds of soils including stiffest clays contain such voids, all soils are permeable to some extent. Water exerts pressure on the porous material through which it percolates. This pressure, known as *seepage pressure*, is independent of the rate of percolation and, therefore, may be of great magnitude even in a soil of very low permeability.

To resolve the following problems, it is necessary to be aware of the laws governing the interaction between soils and percolating water:

(i) Estimation of volume of water that would be collected by percolation in a construction pit or volume of water that would be lost by percolation through a dam or its subsoil,

(ii) Effect of seepage pressure on the stability of slopes or on foundations, and

(iii) Influence of the permeability on the rate at which the excess water drains from loaded clay strata.

Compressibility and consolidation

Deformation occurs when the soil particles are wedged closer together by compressive pressure by the expulsion of water and/or air and rearrangement of soil particles. This soil property is known as *compressibility*. When the compression is effected by mechanical means like rolling or tamping, it is called *compaction*. When such compression is effected by steady pressure exerted by the weight of the structure built on the soil or earth filling carried out over the soil, it is called *consolidation*. The two relevant factors on compressibility are: (i) rate of compression taking place, and (ii) total amount of compression under full load. In case of non-cohesive soils like sands and gravels, the rate of compression would keep pace with the completion of project. There would be no further settlement later. In cohesive soil, pores are saturated with water. Expulsion of water from voids would be a slow process. With expulsion of water, there would be settlement. Uneven settlement of foundations and substructures could be very damaging.

Shearing resistance

The shearing resistance of soil, s, may be represented by the following equation,

$$s = c + p.\tan \phi$$

In this equation, p is the normal stress on the surface of sliding, and c is cohesion. ϕ is the angle of internal friction or the angle of shearing resistance for which a soil subjected to normal pressure exhibits resistance against shearing. Cohesion is a property of soil that imparts some resistance in soil against sliding even if there is no normal pressure between the soil particles.

Every soil mass exerts lateral pressure on walls retaining it. If the wall moves away from the soil, the pressure exerted is called the 'active earth pressure'. And if the wall moves towards the soil, the pressure exerted is called the 'passive earth pressure'.

Consistency limits

Plasticity of soil is its property to remain deformed without elastic rebound or rupture after the load has been removed. This property enables soil to undergo considerable deformation without cracking or crumbling.

Fine-grained soils have been formed by the gradual settlement of solid particles that are in suspension in water followed by consolidation and drying out in stages: (i) suspension of soil particles in water (liquid), (ii) viscous liquid, (iii) plastic solid (iv) semi-plastic solid, (v) solid. Important changes in physical properties accompany change from each stage to the next. The moisture contents at which the soil passes from one stage to the next are known as consistency limits. These limits are better known as *Atterberg limits*. The consistency limits are expressed as percentage moisture content reckoned on the dry weight. Moisture content studies are commonly made in soil improvement studies.

The term 'liquid limit' (w_L) is the minimum moisture content at which the soil would flow under its own weight—the moisture content at which the soil stops acting as liquid and starts acting as plastic solid.

The term 'plastic limit' (w_P) is defined as the lowest moisture content at which a thread of 3 mm diameter could be rolled out of the soil sample without breaking. As more moisture is driven from the soil, it becomes possible for the soil to resist large shearing stresses. Eventually, the soil exhibits no permanent deformation and simply fractures with no plastic deformation—it acts as brittle solid.

The moisture content at which further loss of moisture does not cause a decrease in the volume of soil is the 'shrinkage limit'. Below the shrinkage limit, the volume of the soil remains constant with further drying, but the weight of the soil decreases until the soil is fully dried.

A measure of the range of moisture content over which a soil is plastic is the *plasticity index*, which is the difference between the liquid and plastic limits, $I_P = w_L - w_P$. Soils having no cohesion have no plastic stage, that is, the liquid and plastic limits may be said to coincide—plasticity index is zero. The finer the soil, the grater is its plasticity index. The plasticity index is commonly used in strength correlations; the liquid limit is also used primarily for consolidation estimates.

4.2 PROJECT SITE DEVELOPMENT

All preparatory work should be completed to make a project site ready for all executing agencies to start construction of their site offices and mobilise resources to start construction work assigned to them. How a project site should be made ready for executing agencies to step in would depend on the existing conditions there. Possible site development work may include:

- If existing ground is marshy, it may be necessary that the entire site topsoil be replaced with approved soil. If any site is partly marshy, replacement may be made with available soil from other areas of the site if found suitable for the purpose.

- If there are any waterlogged areas, water should be pumped out beyond the site's boundary or drained out by digging narrow ditches. Slush in waterlogged areas need to be replaced.

- Trees, bushes, and vegetation should be uprooted and cleared to make a project site ready for starting construction work. In case of big trees, stumps may have to be removed by deploying construction equipment or by blasting with explosives if manual efforts turn out to be ineffective or time consuming. If not cleared, stumps may create problem during excavation for foundations or sub-structures. Sometimes, special attachments are fitted to the power units of construction equipment for effective and quick uprooting/clearing. If land is available, waste materials should be buried at the site. The other option, apart from moving them from the site, is to dispose them by burning. It is better to burn machine-cleared vegetation immediately after being piled up.

- In case of existing buildings or structures in any project site, the same should be pulled down. One or more existing buildings may be used temporarily if project implementation is not hampered by such use.

- Site should be cleared of loose stones or boulders, if any, lying on project implementation areas.
- A site may be strewn with big loose or partially buried boulders. They should be cleared by using earthmoving machines. If necessary, big boulders should be blasted before removal.
- If a project site is uneven in levels, the site is to be levelled manually or by deploying construction equipment depending upon the quantum of work involved.
- If uneven site is rocky, a lot of blasting using explosives should be carried out. Levelling of rocky site could be a time-consuming activity. To reduce blasting, different levels may be formed by benching to the extent possible in accordance with the project layouts.
- All the spoils and unusable rubbish should be burned or removed and dumped in the areas earmarked for the purpose.

Manual labour is used extensively in land clearing as a part of site development. However, if the job involved in site development is large and/or difficult, construction equipment like bulldozers, scrapers, shovels and backhoes are deployed.

4.3 SETTING OUT

Project site layout is prepared according to the technological or design requirement, strictly following the statutory rules and regulations. Orientation of the buildings and/or units is shown with reference to the true North line (magnetic North line as seen in theodolites is used at project sites because the difference between magnetic and true North lines is almost negligible). Locations are shown with reference to the project site's boundary lines or any landmark existing within the boundary lines. A project site's total area is divided into small squares by drawing reference North-South and East-West grid lines. The reference North line could be at an angle with the true North line. Site gridlines normally form squares of 30, 50, or 100 metre sides. This 30/50/100-metre side could be more or even less depending upon the total area of a project site. The setting out is done with respect to reference baseline by the use of a theodolite and steel or fibreglass tapes. Permanent grid pillars (Fig. 4.1) are solidly built outside the project layout area by means of closed traverse survey, where the survey ends at the starting point, so as to allow executing agencies to execute their contracted work at correct locations by constructing temporary grid pillars with reference to the permanent grid pillars. All grid pillars, permanent or temporary, should be built outside the limits of excavation. Centre lines on grid pillars are marked on galvanized steel or just steel plates. Once a unit or a building is raised above plinth level, then other units and buildings are built with reference to the already built unit or building. The relative distances and level differences between different buildings and units are maintained correctly and meticulously.

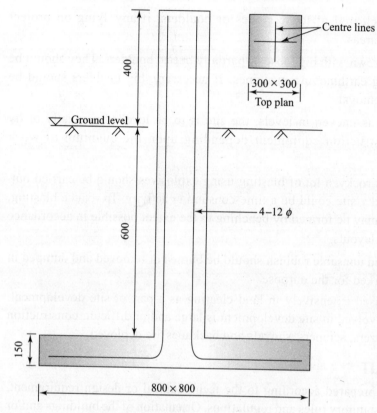

Fig. 4.1 Permanent grid pillar

Mean Sea Level (MSL) is the datum for elevation. MSL is determined by tidal observations at some selected places on the seacoast of a country. Elevations, therefore, indicate relative heights with respect to the MSL. The Survey of India during survey and mapping of the country construct Benchmarks (BM) of different types. These BMs are solidly built with concrete base. Inscriptions that these are GTS (Great Trigonometrical Survey) benchmarks are engraved on them generally on zinc or copper plates showing their elevations from the MSL. The Indian Railways also construct BMs for laying railway tracks and leave them for their own and others' future reference and use. Engineers assigned for project implementation, therefore, would have to locate either GTS or Railways BMs and transfer the level to a site to establish BMs at different site locations to facilitate project implementation. In absence of the GTS and the Railways BMs, the PWD or the Irrigation Departments' BMs could be used. A contour map of a project site is prepared on the basis of the BM established at the site. Both grid pillars and BMs should be guarded with fencing to protect them from physical damage. Modern trend for site survey is to utilize orbiting satellites by using 'total station' (vide Chapter 12, Section 12.1).

Executing/construction agencies prepare excavation plans on the basis of design drawings showing all dimensions of pits/trenches with appropriate side slopes with/without benches, elevations, access to pits/trenches, berm (horizontal ledge on the side of cutting to intercept earth rolling down slopes), dewatering sumps and drains. Surfaces surrounding excavated pits/trenches should be graded away from pits/trenches. Arrangement of dewatering of collected water or rainwater should also be shown. If shoring is to be done because of non-availability of sufficient vacant area, arrangement of shoring for vertical-sides' protection is to be shown. Surface water should not be allowed to flow down excavated slopes. Executing agencies would start excavation work on approval of their excavation plans by the owners' engineers.

Executing agencies would be responsible for setting out different components of a project under their scope of work with reference to the grid lines and the BM established at a project site. This is to be done as a part of the contractual obligation. The accuracy of setting out is to be checked by the owner's engineers.

Centrelines of excavation pits/trenches need to be marked at the ground level by driving pegs along the excavation lines. Outlines can be marked with lime powder or pegs. Areas earmarked for dumping excavated soil and possible borrow pit areas earmarked for backfilling are also marked likewise with lime powder or pegs.

Whether or not the survey instruments like theodolites, levels, tapes, etc. are in good condition should be checked before commencement of setting out and levelling work. Properly qualified personnel should be assigned to carry out all surveys including the levelling work. Setting out of earthwork in embankment formation is discussed in Chapter 12.

4.4 MECHANIZED EXCAVATION

The activity of earthwork in excavation for foundations and substructures is of temporary nature. Once foundations and substructures are constructed, the adjacent areas are backfilled and levelled. But earthwork in excavation for tanks or reservoirs or canals and in dredging is of permanent nature. Earthwork in formation of embankments for roads, railways, and so on is also of permanent nature.

Excavation areas are marked with lime powder or pegs. Existing elevations of such areas should be checked and recorded for determining the volume of excavation involved. The executing agencies should jointly carry out checking of ground levels with personnel assigned by the owners so as to avoid later disputes on the quantum of excavation work done.

Earthwork in excavation in most projects is large in volume. Bigger a project, more and more is the earthwork involved. Depending on the existing levels of a project site and design grade elevations, backfilling could be equally voluminous. Earthwork, therefore, could be very strenuous and labour-intensive process. Large labour forces were used to be deployed in India when maximum number of projects was in the public sector. As timely completion of projects has become a very important factor because of large-scale participation of the private sector in projects, deployment of construction equipment for earthwork is taken into account in almost all projects now. However, if piling is involved in foundation work, then deployment of construction equipment can be limited to (i) site development prior to starting of piling work and (ii) subsequent backfilling up to design grade levels.

General excavation, manually or by deploying construction equipment, should be done up to a maximum depth of 150 mm above the final level. The trimming of 150 mm depth should be done with special care by engaging experienced manual labour or by deploying grader. If soft or unsuitable spots are found at the final levels, the same should be removed and replaced with good soil, concrete, or whatever is approved by the concerned engineers.

Like the trimming of the final layers of soil as mentioned above, excavation work need to be carried out with utmost care as the excavation reaches the final stages in conformity with required dimensions, lines, and grades. Possibility of over-excavation should be avoided. Where there is apprehension of upheaval, slips, and similar deterioration; the final excavation should be done just before commencing the actual concrete work in foundations/substructures. Causes of such deterioration may be the quality of soil being excavated, rainfall and heavy movement of loaded vehicles or heavy construction equipment by the side of excavation.

Earthwork in excavation involves the following operations:
- Loosening of existing earth mass/materials to put it in workable state
- Excavating earth mass/materials to start moving the spoil from the original location
- Hauling of the excavated spoil from the original location to the dumping point
- Dumping the spoil at its place of deposit
- Working on the spoil to put it into specified condition at its place of use

Considering different methods of excavation, the types of excavation/filling can be as follows:
- Confined excavation
- Sloped excavation
- Open or bulk excavation
- Excavation in rock
- Embankment formation

Confined Excavation

Confined excavation is related to excavation carried out for construction of individual foundations and trenches. As already mentioned in Section 4.3, outline of the foundations are marked with wooden pegs or lime powder before construction equipment are deployed. Slewing (360°) type backhoe excavators are suitable for confined excavation. Excavated spoil is dumped alongside for later backfilling or

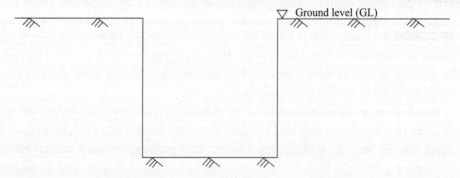

Fig. 4.2 Confined excavation

loaded on transport vehicles for disposal at areas earmarked for the purpose. Number of transport vehicles to be deployed per excavator would depend on the cycle time of a transport vehicle. This cycle time would include time for loading the transport, time for travel both ways, and time for dumping. Travel time would depend on how far away the disposal area is located. Number of excavators and transports is calculated on the basis of cycle time and volume of excavation involved.

Sloped Excavation

Sloped excavation is involved in case of relatively deeper excavation where the sides of excavated pits or trenches need to be in slope to be stable. Side slopes would vary depending on the type of soil in which earthwork in excavation is to be executed. The advantage of this type of excavation is that it takes away the worry of supporting the sides with sheet piles or props. And also, there is unrestricted access to the excavation level via a ramp on one side. Compared to this advantage, the minus point is the volume of excavation would be more covering more land. However, carrying out excavation, if possible, in two stages could reduce the volume of excavation. Sloped excavation is to be carried out up to a level wherefrom confined excavation could be done. If sloped excavation is to be carried out in existing sloped ground, then setting out is to be done after working out the dimensions of excavated sides and heights of excavation around. After initial rough cutting, sides are to be trimmed carefully. Backhoe excavators may also be deployed in combination with transport vehicles.

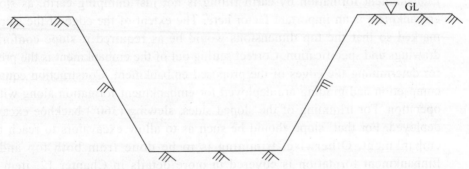

Fig. 4.3 Sloped excavation

Bulk Excavation

Bulk excavation can be carried out where quantum of excavation involved is voluminous. Big versatile construction equipment can be deployed in such cases, as there would be larger areas for manoeuvring plus massive earthmoving operation involved. A number of scrapers may be deployed for earthmoving, a grader to maintain haul roads and a dozer to supplement the efforts of both scrapers and graders. A 360° slewing excavator is sometimes deployed for trimming the slopes and spreading topsoil. If the required slope is steep, trimming is done either from the top or the bottom. Since an excavator's hydraulic arm can reach a limited height, bulk excavation needs to be interrupted when an excavator is deployed at the bottom. Even in case of bulk excavation, the excavation is carried out in two stages as shown in the sketch below.

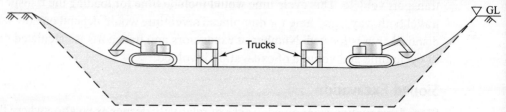

Fig. 4.4 Bulk excavation

Excavation in Rock

Excavation in rock by deploying construction equipment is possible if the rock is weak. However, rock excavation in confined spaces is not possible by fitting rippers with earthmoving equipment. In open areas, weak rocks like siltstone or mudstone can be excavated with common earthmoving equipment. Relatively hard rocks like slate and sandstone can be excavated using rippers or pneumatic breakers. Very hard rocks like limestone or granite are required to be blasted for excavation purposes. Rock blasting is covered in Chapter 5, and ripping is covered in Chapter 13, Section 13.4.

Embankment Formation

Embankment formation by earth filling is not just dumping earth, as slope of the embankment is an important factor here. The extent of the edge of the slope is to be marked so that the top dimensions would be as required if slope conforms to the drawings and specification. Correct setting out of the embankment is the pre-requisite for determining the edges of the proposed embankment. Construction equipment for compaction and grading are deployed for embankment formation along with hauling operation. For trimming of the sloped sides, slewing (360°) backhoe excavators are deployed. For that, slope should be such as to allow excavators to reach the top of embankment. Otherwise, trimming is to be done from both top and bottom. Embankment formation is covered in more details in Chapter 12, item 12.1 and Chapter 13, item 13.4.

4.5 GROUNDWATER CONTROL

The soil is naturally full of water up to the water table. This is groundwater. The soil above the water table also holds some water because of the capillary action and surface tension of the voids. Water both below and above the water table as well as the water collected above the ground level or in hollows and depressions may cause problems in execution of underground work and need to be controlled effectively. There are two ways of controlling water as follows:

- Temporary exclusion of water to facilitate construction work
- Permanent exclusion of groundwater by lowering water table

In order to design methods of exclusion of water in construction projects, the quality of soil that exists at the exact location should be taken into consideration. Such particular locations receive special attention during soil investigation work. Cohesive soil like clay is impermeable and does not pose any problem during excavation except for taking care of retaining the excavation sides in place. But over a long period of time, water may permeate out of or into the clay. There is also the question of heaving of excavated bottom due to compressible nature of clayey soil – this is a feature related to design of foundations on clay or clayey soil. As regards strength, the more the water content of clay, more would be its volume but strength would be reduced.

But non-cohesive soil like sand or gravel is permeable in nature. There is no question of water permeating over a period of time. Water content does not pose any problem and strength is also not affected. When excavation is carried out in non-cohesive soil, fine materials may be washed out due to permeability, thus causing problem. Gravels need to be contained to retain its strength. If non-cohesive soil is allowed to remain in slope in excavation, it is called 'open cast' excavation. Both dry sand and gravels would fall in slope at the angles of repose.

Permeability is relevant for groundwater control. There are different ways of controlling both temporary and permanent exclusion of water as mentioned above. The principles on which water exclusion is based are:

- Dewatering by pumping to the required level
- Construction of shield/barrier to keep water level below the excavation level

The different ways of controlling groundwater are given in the following table.

Temporary exclusion	Permanent exclusion
• Sump pumping	• Drainage/sand drains
• Sheet piling	• Sheet piling
• Cofferdams	• Diaphragm walls
• Well point system	• Contiguous piling
• Deep bored wells	• Secant piled walls
• Horizontal drains	• Slurry trench cut-off
• Electro-osmosis	• Thin grouted membrane
• Ground freezing	• Pressure grouting
	• Caissons

Temporary Exclusion

Temporary exclusion can be achived by the following: (a) Sump pumping, (b) Sheet piling, (c) Cofferdams, (d) Well point system, (e) Deep bored wells, (f) Horizontal drains, (g) Electro-osmosis, and (h) Ground freezing.

Sump pumping

Sump pumping is a temporary solution where excavation is carried out to a level below the water table or where water gets into the excavated pit. The sump is excavated at one or more corners of the excavated pit below the required formation level so that the excavated pit remains free of water. Small ditches filled with gravels may be dug along the perimeter of the excavation pit leading to sumps. Open sump pumping is limited by the suction lift of the pump. The disadvantage of sump pumping is related to the stability of side slopes and is only suitable in soils whose stability is not affected by the action of seepage, for example, stiff clay or soil having high percentage of gravel or boulders. Seepage velocity and pressure would be maximum at the bottom as a result of which there would be softening and sloughing at the lower levels. If the soil strata through which water is seeping contain fine sand or coarse silt, there would be concurrent removal of water and soil from the bottom and the side slopes. If this happens in fine-grained granular soils like fine sands and silts, the bottom of the excavation may become 'quick' and the side slopes unstable because of the seepage force imparted to the soil grains by the water seeping into the excavation pit. This may result in collapse of the sides. The sides can be protected against collapse by sheet piling.

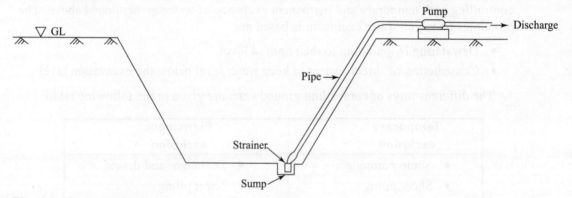

Fig. 4.5 Sump pumping

Sheet piling

Sheet piling is very common water exclusion as well as earth-retaining process during and after excavation for both temporary and permanent solution. The soil properties, which are relevant for design of earth-retaining structures, are:

- Unit weights of soils in both natural as well as in submerged conditions
- Angle of internal friction in case of non-cohesive soils
- Shear strength or cohesion in case of cohesive soils

Steel sheet piles are commonly used for sheet piling. Sheet pile driving is discussed under 'permanent exclusion' below. Timber is rarely used. Temporary sheet piling would not be cost-efficient if used sheet piles are not extracted for reuse. Pile extractors are used to pull out such piles. The advantages of using steel sheet piles are:

- High strength—resistant to the high driving stresses developed in hard soil and rocky strata because the steel sheet pile is strong along its length in bending

- Relatively light weight
- A pile can be used and reused several times
- Less driving time
- Pile details are available in tabulated forms
- Steel sheet piles may be used for both temporary and permanent work
- Steel sheet piles can be extended by welding or bolting
- Extra lengths can be cut off using flame
- Long piles can be driven with heavy pile hammers
- Special piles are available for interlocking corners, junctions, etc.
- It has long service life, either above or below water, if it is provided with protective coating

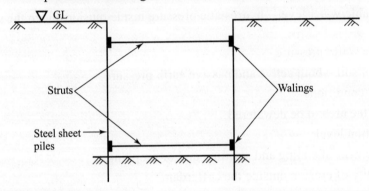

Fig. 4.6 Sheet piling

Cofferdams

A cofferdam is a water-tight dam that is built for temporary exclusion of soil and/or water so that construction, both off-shore and on-shore, can be carried out in the 'dry' below the water table or water level. Although interlocking steel sheet piling is the most common method of constructing cofferdams, there are other types of cofferdams also as follows:

- *Earth embankments*—ideally impervious clay core with sand fill on either side. Fit where water flow is slow and water height is not high as soil offers little resistance to erosion from moving water or wave action. Blanketing the sides with rocky materials, tarpaulins, and geotextiles would protect the same against wave action. Well-points need to be installed to dewater both the embankments and the excavation, as and when required.
- *Rock-fill embankments*—rock-fill embankments are similar to earth embankments with steeper slopes. Clay or fly ash core could remove the disadvantage of this type of pervious embankments. The weight of rock gives stability to the dam.

- *Sandbag embankments*—instead of directly using sands to form embankments, the same are filled in bags and placed in layers to form embankments.
- *Timber sheet piling*—in single or double walls sheet piling where timber is readily available and where the cofferdam head would be low.
- *Bored cast-in-situ piling*—where sheet piling is not possible because of headroom limitation or where vibrations due to piling is to be avoided or sheet piling in ground containing boulders may cause splitting sheet piles and damage interlocks. A cofferdam may be formed by a row of bored cast-in-situ piles cast in contact with each other.
- *Precast concrete block-work*—precast concrete blocks may be used to form gravity cofferdams. For this, the blocks should be cast to fine tolerances with joggled or dovetailed joints to give water tightness and stability against sliding.

Design of a cofferdam would depend on:
- Water table/water level—hydrostatic pressure inside/outside due to the weight of water
- Variable water pressure
- Types of soil—both active and passive earth pressure
- Subsoil conditions
- Size of the area to be dewatered
- Excavation level
- Arrangement of waling and strutting and/or propping
- Possibility of erosion outside the cofferdam
- Surcharge
- Seepage
- Effect of wind/wave
- Effect of earthquake
- Cofferdam—temporary (need not be dismantled) or permanent

The following requirements are to be considered in design of cofferdam:
- Heaving of soil at the bottom and the consequences thereof
- Inward yielding of sheet piling and the consequent failure
- Failure from piping or boiling resulting from heavy upward seepage of groundwater into the cofferdam especially where the sheet piles are not driven to the impervious strata—severe piping can result in loss of soil outside the cofferdam and its eventual collapse
- Scour that may destroy an offshore cofferdam
- Obstruction to water flow during working and monsoon seasons—hydrostatic pressure
- Earth pressure outside cofferdam
- Temporary cofferdams may fail during removal/dismantling

Steel sheet piles are used widely because of:
- Structural strength—high strength in bending and shear
- High interlock strength
- Water-tightness possible because of interlocking joints
- Ease of driving and removal
- Driving to full penetration is possible in all types of soils
- High salvage and reuse value
- Can be spliced by full penetration butt weld

Steel sheet piles need to be adequately braced and strutted to hold them in position or anchored with tie rods. The struts are essentially columns, which need to be supported in both horizontal and vertical planes. Intersecting struts in the other direction provides lateral support making the struts behave as short columns. Circular-shaped cofferdams should be braced adequately without strutting. Modern development is the use of hydraulically pressurized modular aluminium horizontal frames in place of conventional waling and strutting. Seepage through the interlocking joints should not be much and should be pumped out. Otherwise, the interlocking joints should be caulked with sealing compounds. Sheet piles should be procured with handling holes at one end for ease in lifting and pulling. These holes should be plugged later by welding.

Steel sheet pile cofferdams may be built in two ways:
- Single skin
- Double skin

Single skin cofferdam

A single skin cofferdam comprises a single enclosure built of sheet piles and should be good for loading up to a depth of about 15 m below existing soil or water level. This type of sheet piling needs some form of bracing or strutting against soil and water pressure as well as surcharge, if any, and internal bracings provide the most economical solution. In tidal water, single skin cofferdams should be fitted with sluice gates for pressure equalization or for neutralizing unanticipated pressure. Or, the top portions of some sheet piles should be so fabricated as to allow raising them to create opening in the cofferdam.

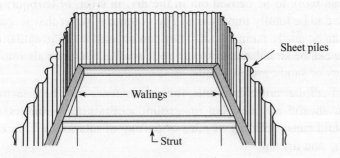

Fig. 4.7 Single skin cofferdam

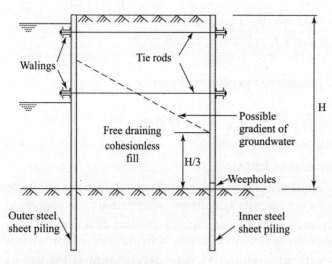

Fig. 4.8 Double skin cofferdam

Double skin cofferdam

A double skin cofferdam comprises two parallel rows of sheet piles (equidistant two rows in case of curved cofferdams). The gap between the two rows is filled up with sand, gravel, crushed rock, or broken bricks. A double skin cofferdam is a self-supporting and freestanding gravity structure where cross-bracing would be uneconomically long. While the inner wall is designed as ordinary sheet pile retaining wall, the outer wall is intended to be an anchor wall for the tie rods. Tie rods tie the two rows of piles together. The piles are restrained at the bottom by the passive pressure of the soil. The inner wall is provided with weep holes to drain out water so as to relieve the filling materials of high water pressure.

A self-supporting cellular cofferdam made of straight web sheet piles does not need any walings, struts, or tie rods; and used primarily as water-retaining structures. High web strength and specially designed interlocking joint give straight web sheet pile strength to resist high-tensile stresses. The interlocking joint has adequate angular manoeuvring arrangement to construct cellular cofferdam of any shape. The stability of the cellular cofferdams depends on:

- Tensile strength of sheet piling
- Size and shape of the cells
- Types of filling materials
- Types of soil at foundation level

Sand, gravel, crushed stone, and broken bricks are suitable materials for filling. The ratio of the average width of the cofferdam to the retained height should not be less than 0.8 to 0.9 against sliding or tilting or lateral distortion. Cellular cofferdams are required for large construction work to be carried out in the dry, in river, or harbour; although they are not supposed to be totally impervious. The volume of water that seeps through should be such as can be easily pumped out. They can be used on an irregular rock bed as sheet pile lengths can be so tailored as to suit rock profile. They are also suitable for founding on stiff clay or sandy/gravelly soils.

If the interlock of cellular cofferdam fails, the consequence may be disastrous. The cellular cofferdam should not be used on ground containing boulders or other obstructions that could cause splitting of piles or parting of interlocks. Utmost care is necessary in pitching and driving.

Cellular cofferdams

Cellular cofferdams are of three basic types as follows:
- Circular—with smaller connecting partial cells
- Diaphragm—made of a series of circular arcs connected by cross walls
- Cloverleaf—assembled from sheet piling shapes and fabricated connections

Circular cellular cofferdams are commonly constructed using circular cells and connecting smaller partial cells. The cloverleaf type can be effectively used as a corner or anchor cell in conjunction with circular cells. Cellular cofferdams may be used for structures such as breakwaters, and retaining walls. They may be used for mooring barges etc and to function as piers for loading and unloading cargoes.

Excavation of soil inside cofferdams is carried out by deploying cranes (dredging using clamshell) or by hydro-mechanical means. Soil is turned into pulp in case of hydro-mechanical means and then pumped out. Water, both inside and outside the cofferdam, is maintained at the same level thereby the bottom of the excavation remains hydraulically stable. On completion of excavation the excavation is plugged with concrete placed using tremie. The thickness of the plug should be sufficient to withstand hydrostatic uplift pressure as and when the inside of the cofferdam is dewatered. Leakage through sheet-pile interlocks and/or concrete plug is to be taken care of by proper application of appropriate remedial measures to minimize leakage to negligible level.

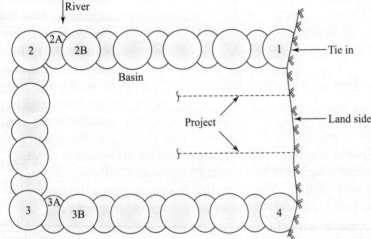

Fig. 4.9 Cellular cofferdam

Well point system

Well point system is the most effective method of lowering groundwater table to a point below the required level of excavation by pumping out water through a group of well points (riser pipes) which are connected to an interconnecting header pipe (about 150 mm in diameter) which in turn is connected to a self-priming centrifugal pump. Well points are smaller-diameter pipes (75 mm or less in diameter) that are perforated at the water bearing strata so that water can easily be pumped out. Wire mesh or strainer covers the perforations. Well points are installed at relatively close spacing, usually in the range of 500 mm to 2.5 m, to adequately dewater fine-grained or stratified soils. The purpose of the well point system is to produce a cone of depression in the water table so that excavation can be executed in relatively dry condition. There should be one or

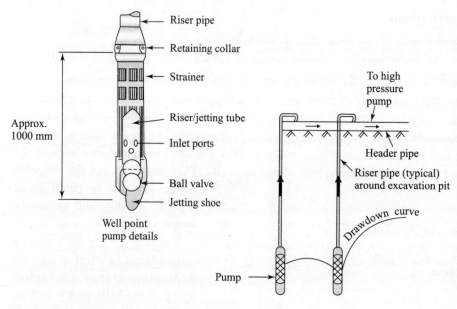

Fig. 4.10 Well point system

more standby suction pumps to take care of possible outage of the main suction pump due to breakdown/failure, and there should also be arrangement for standby power supply arrangement. This is necessary to avoid collapse of the excavation pit due to well point not functioning. The standby pump can also be operated if the inflow from the well points is more. The pumps are connected to the header by means of flexible suction hoses. Pumping has to be a continuous process. Well point pumps are designed to handle both water and air. Water entering the pump passes through an air separation tank from which air is separated by a supplementary vacuum pump.

The header pipe is designed as a ring main around the perimeter of the excavation area at a distance of at least 500 mm from the shoulder of the slope of excavation. A well point is placed at the desired location for installation. The water pressure is then turned on. As a result, the jetting action causes the pipe to penetrate the soil by its own weight. The well point head is very important equipment. Disposable well point head is available now. A well point head ends in a cutter tip that houses ball and ring valves. Water pumped into the well point head forces back the ball valve, thus letting water washing the soil below. Jetting is continued till the required depth is reached. This procedure is repeated until all the water points are jetted in. On completion of work, the voids left are filled up with sand. Well points may also be installed by driving or drilling.

Well points can lower water up to 5 m to 6 m in non-cohesive or granular soil in single stage pumping. For dewatering of excavation area beyond this depth, two or multi-stage installation would be required.

Deep bored wells

An alternative to multi-stage well point system is resorting to deep bored wells. Deep bored wells are effective in excavation depths in excess of 6 m right up to 100 m or more. The hole is bored using conventional rotary boring machines and a temporary outer steel lining tube of 300 mm to 600 mm diameter is left in place into the ground to the required depth based on the sub-soil conditions. Another perforated steel tube called inner liner (150–300 mm in diameter) is inserted inside the already installed liner

Earthwork 65

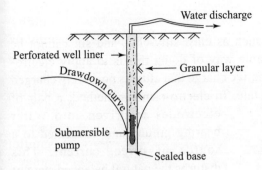

Fig. 4.11 Deep bored wells

and the bottom is sealed by plugging to have stable base. The annular space between the liners is filled with permeable materials like sand and gravels. The outer liner tube is then removed. This outer liner may, however, be retained if so required by the soil conditions. Before the installation of the submersible pump, water is forced through the inner liner to purge the well. The water under pressure is forced through the perforations and granular filler materials. A submersible pump is then installed.

When pumping out of water is continued, water table is lowered. The drawdown curve forms conical depression. The radius of the drawdown curve as well as spacing of deep bored wells would depend on the permeability of soil. The excavation should be carried out above the drawdown curve. Horizontal ditches may be dug below the water table radially from the borewell to make this system more productive.

Horizontal Drains

As an alternative to the vertical well points and borewells, horizontal system of dewatering may be used. A PVC pipe is horizontally installed underground using a special machine that excavates a narrow trench, lays the pipe and backfills the excavation in one operation. The trench could be as deep as 5 m. The perforated PVC suction pipe is covered with a nylon filter sleeve to stop infiltration of fine particles. A single pump of the special machine under average conditions can pump up to a length of about 230 m of pipeline. Two consecutive pipe lengths must have an overlap of about 4 m.

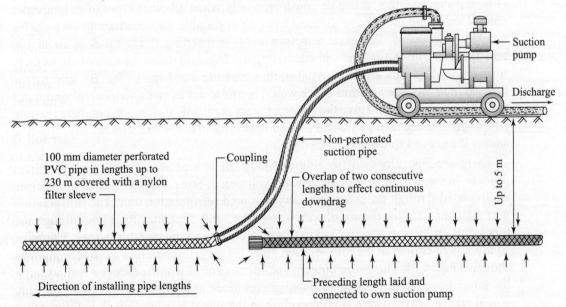

Fig. 4.12 Horizontal drains

Electro-osmosis

Fine-grained low strength cohesive soils such as clay, silty-clay, and silt cannot be dewatered and stabilized easily. One possible way of doing this is to apply electro-osmosis method. All soil particles carry negative charge that attracts positively charged water particles, thus creating a balanced state. In electro-osmosis method, a pair of electrodes is driven into water bearing ground and connected to a source of direct current. This disturbs the natural balanced state of the ground and actuates flow of water from anode to cathode. The positive electrode (anode) can be steel rods or sheet piles or steel pipes, and a well point is installed to act as negative electrode (cathode). When electricity is passed between the anode and the cathode, the positively charged water molecules flow towards the negatively charged well point where this water is collected and pumped away to a discharge point. The amount of water collected at the well point is small compared to the conventional methods, but power consumption is substantial. This method of groundwater control is expensive.

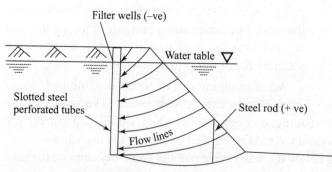

Fig. 4.13 Electro-osmosis

Ground freezing

Ground freezing method is suitable for both supporting earth pressure as well as water exclusion in all kinds of soil having moisture content in excess of 8% of the voids. Groundwater freezing should be considered only when all other types of groundwater control are found ineffective. Ground freezing is suitable for constructing deep shafts, tunnels, or subways. The basic principle involves inserting freeze tubes of about 100 mm diameter into the ground and circulating a refrigerant down the core of the tubes to form ice in the voids around the tubes, thus creating a column of ice. By spacing the tubes at intervals, the columns of ice would merge to act as an impermeable continuous wall. This method would give the soil temporary mechanical strength, but there is slight risk of ground heaving particularly when operating in clays and silts because freezing causes the ground to expand.

There are basically two freezing methods: (i) two phase process, and (ii) direct process. In the two-phase process, a heavy solution of brine containing calcium chloride is circulated through the pipes to a lorry mounted refrigeration unit. The refrigeration unit maintains a temperature of $-15°C$ to $-40°C$ and circulates the brine through the pipes causing the surrounding soil to freeze to create an impermeable barrier. The brine solution is re-chilled in the refrigeration plant for recycling through the freeze pipes as shown in Fig. 4.14. The freeze pipes at one-metre centres would produce a wall of ice in the ground one metre thick in sand and gravel in about ten days of operation, but clay would take 17 days. There is no recycling in the direct process, which is followed in works of short duration where quick freezing is required. In this process, the more expensive liquid nitrogen can be used as the circulating medium. The liquid nitrogen, however, is let out after cooling through the outlet core to the atmosphere. The system of groundwater freezing could be competitive with the other groundwater control systems where excavation work is to be carried out at depths more than nine metres. Ground freezing is to be maintained at all times till the construction work is completed.

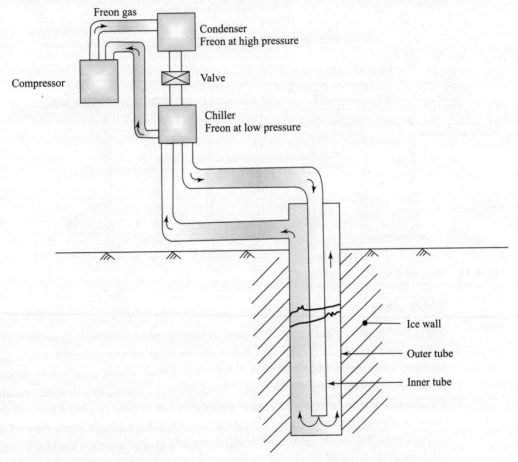

Fig. 4.14 Ground freezing

Permanent Exclusion

Permanent exclusion can be achieved by the following: (a) Drainage/sand drains, (b) Sheet piling, (c) Diaphragm walls, (d) Contiguous piling, (e) Secant piled walls, (f) Slurry trench cut-off, (g) Thin grouted membrane, (h) Pressure grouting, and (i) Caissons

Sand drains

Drainage using sand is easy and inexpensive way of dewatering. This process of water exclusion from soil is used on both sides of road and also under road embankment formation. The horizontal drains are free flowing at shallow depths up to 1.5 m and does not need pumping (vide Fig. 4.15). The vertical drains are used in cohesive soil like clay under embankment formation. This process involves drilling of holes in clay at close intervals and filling the same with sand or permeable materials. Exclusion of water from clay is a very slow process, and that is why embankments are overburdened to hasten the process. Water is squeezed from the clay into the permeable holes for ultimate evacuation, thus consolidating the cohesive soil mass.

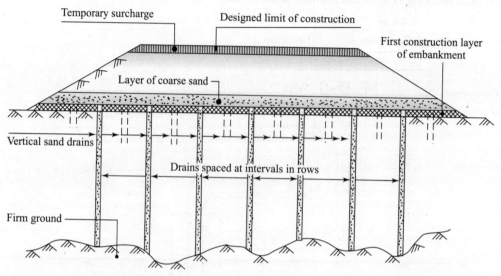

Fig. 4.15 Vertical drains in impervious soil

Sheet piling

Sheet piling using steel sections is used to form permanent retaining walls for riverbank strengthening and in docks and harbours as well as sea defence work. Sheet pile sections have interlocking joint to form a water seal, which may need caulking where high water pressures are encountered. Installation of sheet piles in series would require a guide pile frame or trestle to ensure verticality in pitching piles and prevent them from moving during driving through obstructions. Guide walings form part of trestle or piling frame. For onshore or land cofferdams, the guides are provided by a pair of guide walings at the ground level and another pair of guide walings at the highest level. Over water, guide walings are supported by timber or steel piled staging constructed by floating plant. Sheet piles that are exposed on one side remain in place by cantilever action or by propping across the excavation. Guide walings should comprise heavy timber beams or steel H-sections placed with webs horizontal. The piles are lifted by crane and placed in position for driving in panels of six to twelve pairs of piles. At first, a pair is pitched and driven partly ensuring true verticality in two directions (pitching means lifting and placing of the piles in pile frame with the end of piles touching the ground and the joints interlocked with the adjacent piles). Eccentric driving blows can correct deviation in verticality. Then, more pairs of piles are interlocked and the last pair is driven maintaining verticality. Driving is continued in one or more stages of part driving until the first panel is driven to the required penetration level. Once a panel is driven, another panel of six to twelve pairs is driven in the same way. The clearance in the interlocks of piles would allow the piles to be driven in the form of a circle, as and when required, without bending any of the piles. Interlocks permit a deviation of 10° between adjacent piles. Special pre-bent piles are available for directional angle change up to 90° and also Tee, cross, and Y are fabricated by welding. Hammers, percussion or hydraulic and suspended from the jib of a crane or derrick, are used to drive piles. Percussion hammers are activated by compressed air, steam, or diesel power. Pile driving is noisy except for vibratory pile driver. The vibration system vibrates the piles in high frequency thus liquefying the granular soils as a result of which the piles can be pushed into the ground. The walings should be secured in position, keeping in view that they are to be removed. It is important to protect the heads of sheet piles during driving.

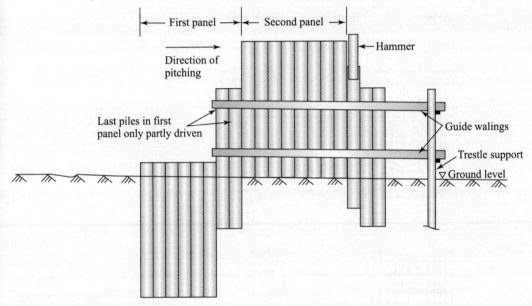

Fig. 4.16 Sheet piling

Diaphragm walls

The alternative solutions to steel sheet piling are to use diaphragm walls, contiguous reinforced concrete piling and secant piled walls. The selection of walling method would depend on whether such wall is required to retain water or not. Groundwater information is particularly important for bored or drilled piles.

A diaphragm wall (Fig. 4.17) is a cast-in-situ reinforced concrete wall. Diaphragm walls can be cast in all kinds of soil, except boulders and rock, as well as close to foundations and substructures because its method of construction is simple. Because of simplicity of construction, diaphragm walls can also be constructed in restricted areas. Apart from water exclusion, diaphragm walls support excavation by using struts and props on the open side. However, construction of diaphragm walls would be a costly proposition, unless such walls form part of the permanent structures. Diaphragm walls are suitable for:

- Construction of foundations or bases
- Construction of basements or underground chambers
- Construction of subways
- Construction of underground tanks
- Coastal defence works and river wall
- Retaining walls

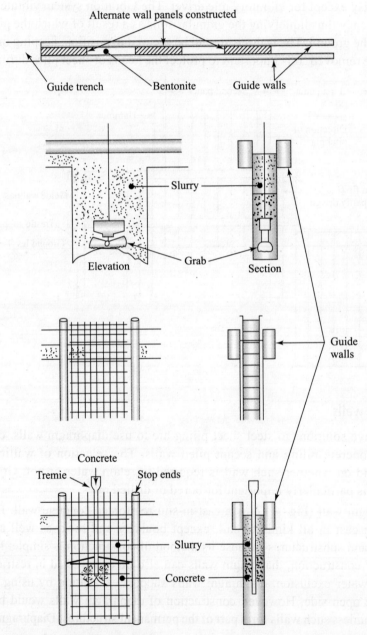

Fig. 4.17 Diaphragm walls

Grab bucket (clamshell) is used to carry out excavation work. Number and size of grab buckets to be deployed would obviously depend on the volume of work involved. About 1 m deep guide walls are cast at the top on either side of the proposed diaphragm wall leaving sufficient clearance for grab to excavate. A panel up to 6 m in length is excavated up to 50 m or more depth to form the first panel of the diaphragm wall. More depth means considerable difficulties of extracting excavated materials from the stop ends. A diaphragm wall is cast in alternate panels. Wall thickness may be 0.5 to 1 m. Accuracy of the verticality of the excavated wall is important. Bentonite slurry, which has to be heavier than water (3–10% bentonite powder in clear water to have specific gravity of 1.2), is continuously pumped into the excavated trench to stop ingress of groundwater and collapse of the excavated sides.

Bentonite is a clay rock largely comprising the mineral sodium montmorillonite. When in contact with water, it has a characteristic ability to swell to many times its dry volume, resulting in jelly-like consistency which, when stirred becomes fluid. But when left undisturbed, revert to jelly form. This property is known as *thixotropy*. The four useful properties of bentonite/water suspension in civil engineering are: (i) lubricity, (ii) plastering or sealing ability, (iii) thickening, and (iv) gelling. Bentonite is a free flowing non-irritant, non-corrosive powder.

Bentonite slurry is filled in the guide trench before commencement of the grab operation. The bottom of the excavated panel is cleaned of slurry/soil deposits before the reinforcement cage is placed. A tremie tube of about 200 mm diameter and sufficient length to reach the bottom of the excavated pit is used to pour high slump concrete to form the diaphragm wall. The tremie tube with a hopper at the top is used so that the bentonite slurry does not contaminate concrete during pouring and concrete is not segregated. As the concrete fills the trench, over-spilled contaminated bentonite slurry is collected in tanks. The top layer of bentonite slurry in the tanks is recycled into next excavation panels.

A diaphragm wall is constructed in panels to allow concrete to set and have thermal movement before the next panel is cast in the same way, thus minimizing the possibility of cracking of concrete wall. However, it would not be possible to have watertight joints between two panels because of the method of concreting. One way of resolving this problem is to place two vertical pipes at the two ends of a panel before concreting work is started. This would create a semicircular joint at each end of a panel. The pipe is to be treated with a debonding coating. Concreting of the adjacent panel would result in a circular void, which could be grouted and filled up.

Contiguous piling

Contiguous piling is reinforced concrete piled walling having similarity in concept to reinforced concrete diaphragm wall, but relatively more simple, quick, and inexpensive. Their main use is in cohesive soil where dewatering is not a problem. The pile is driven down to the required depth (up to 20 m or more) by boring with an auger of diameter equal to the diameter of pile designed on the basis of soil properties and excavation depth. The piles are driven one after another as a result of which the system would not be watertight and dewatering would be necessary to carryout excavation in non-cohesive water-bearing soil. In practice, driving leaves a gap of 50 mm to 100 mm between adjacent piles. These gaps may be attended by grouting. In addition, it is desirable to cast a beam at the top of piles so that a single pile is not overloaded. Diameter of contiguous piles varies between 300 mm to 600 mm. Contiguous cantilevered piles of 500 mm to 600 mm diameter can be used for retained heights of about 5 m. In poor or water bearing soil, bentonite slurry may be required to support the walls of the hole. A tremie tube is used to pour high slump concrete. Reinforcement cage is placed before concreting. Concrete of high slump of about 150 mm is injected down the hollow auger stem to the head of the auger below. The auger is lifted out of the ground as the grout continues to intrude.

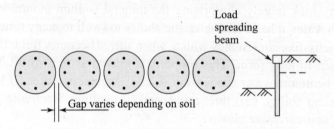

Fig. 4.18 Contiguous piling

Secant piled walls

A secant piled wall is an improved version of contiguous piling towards groundwater control as well as retaining earth pressure. It is more like a continuous watertight diaphragm wall but less expensive. How watertight a secant piled wall would be, however, would depend on constant control of construction tolerances of both initial and intermediate piles. Guide walls at the top, as are constructed in the case of the diaphragm walls, are sometimes constructed to ensure correct plan positions of both the secant and contiguous piling. This reinforced concrete piled wall (sometimes 'I' sections are used for reinforcing instead of round bars) comprises a series of interlocking bored piles driven in a way different from the contiguous piling. In this case, alternate piles are driven down to the required depth (up to 30 m or more) by boring with an auger as is done in contiguous piling. However, special cutting tools would be required to form keys in the alternate piles for interlocking intermediate piles to be driven immediately afterwards to facilitate cutting of weak concrete mixed with a retarding admixture. The initially constructed piles do not necessarily have to be bored to the same depth as the intermediate piles that follow, depending on the way in which the wall has been designed. Once such a piled wall is in place, the excavated face can be treated with another layer of concrete (shotcrete or gunite) to give a fair-faced finish. Another alternative is to cast a reinforced wall in front of the secant piling terminating in a capping beam to the piles.

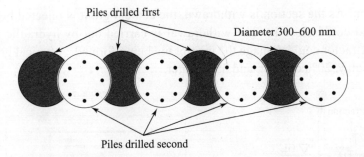

Fig. 4.19 Secant piled walling

Slurry trench cut-off wall

Unlike diaphragm wall, slurry trench cut-off wall is an inexpensive method of excluding groundwater. Execution process is also simple—drilling, augering, or similar simple method of excavation. Slurry trench cut-off wall is designed to act as water barrier without having any capacity to resist earth pressure, as this kind of wall is not designed as structural element. These thin walls without reinforcing steel are suitable for use in such soil as silts, sands, and gravels. Sufficient earth should remain between these walls and the sloped excavation pits so as to support earth pressure from the opposite direction. Slurry trench cut-off wall is constructed around the perimeter of the proposed excavation area.

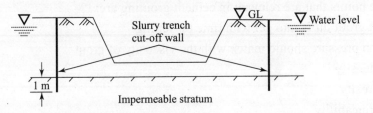

Fig. 4.20 Slurry trench cut-off wall

Thin grouted membrane

A continuous impermeable thin grouted membrane is similar to the slurry trench cut-off wall except for the execution method. This kind of membrane is also formed around the outer excavation perimeter leaving enough soil to counter earth pressure, as this non-structural membrane would only bar water ingress. This groundwater exclusion process is not only suitable for silts, sands, and gravels, but also for more permeable soil where bentonite slurry method would not be effective. The execution method involves driving of structural steel sections, sheet pile sections, or rectangular hollow sections into the ground down to the required depth. A grout injection tube (diameter 20 mm or so) is fixed by welding to the web or face of the selected steel section. This tube is connected to the grout pump at the ground level by means of a flexible pipe. When seven units or so are driven in place, the rearmost section is withdrawn whereby, a void is created below the section. As the section is withdrawn, the cement grout is injected before the sides of the void could close in. The withdrawal is carried out by hydraulic rams as hammer-type extractor could damage the membrane being formed. A typical grout mix comprises 45% clay, 20% cement, and 35% water.

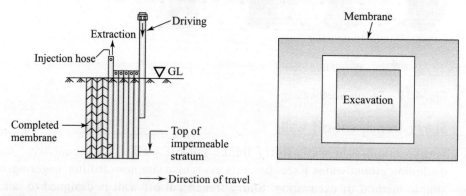

Fig. 4.21 Thin grouted membrane

Pressure grouting

The principle of grouting is to inject a substance to fill the voids in soil or the fissures in rock by pumping the fluid down a small diameter tube placed in a drill hole. Drill hole is made to the full depth as designed with subsequent use of packers to obtain stage grouting. The points that are relevant in cement grouting are:

- Cement grout turns into gel with time on setting
- Injection pressure should match with the viscosity of grout
- Grout density
- Soil porosity
- Soil permeability
- Diameter of grout pipe

Elaborate soil investigation is necessary to locate where pressure grouting (Fig. 4.22) could be effective. Grouting is successful when the materials through which grout is to be injected are relatively large. Fine materials tend to obstruct grout to penetrate. And even a very liquid grout cannot be pumped far into fine material. The intent of pressure grouting in soil, rock, or any other material is for:

- Reduction of the flow of groundwater
- Increasing strength of the material

Grouts may be categorized as follows:

- Suspensions in water of solid particles like clay, cement, bentonite, plaster, lime, and PFA (pulverized fuel ash)
- Emulsions like bitumen in water
- Chemical solutions that react after injection to form insoluble precipitates

The basis of choosing grouts:
- Grout must be able to penetrate the voids of the mass to be injected without being filtered by fine materials
- Grout should be resistant to chemical attack when in place
- Grout should attain sufficient strength to withstand pressure it would be subjected to

Grouting equipment required:
- Pumps
- Grout pipe, packers, casing tubes, flow meters, pressure gauges
- Agitator—for keeping solid particles in suspension
- Measuring tank—to check volumes of grout injected
- Mixer—for mixing the grout ingredients properly

Cement grout acts as barrier for groundwater if the grout is injected extensively in permeable sand or sand and gravel mixture of more than 1 mm size by forming a layer of concrete. In fissured and jointed rock strata, cement grout is injected through a series of holes bored into the ground in lines at close centres. Cement grout is effective when injected extensively. Grout is injected at high pressure, and the pressure builds up as voids, fissures and cavities are filled up. The cement grout can be a mixture of neat cement and water, cement: sand (1:4) in water, and pulverized fuel ash (PFA): cement (1:1) with 2 parts of water by weight. To start with, the grout should be thin in consistency and gradually reduce water content thus increasing viscosity. PFA is the cheapest material available for grouting. Chemical admixtures are available to control water content and setting time (gel formation).

Clay/cement grout is used in fine soil materials. Clay is grouted in medium coarse sand mainly to reduce permeability, as its strength is poor. Cement may be added with clay to increase its shear strength. But cement has relatively large particle size because of which clay-cement grout can be used in coarse-grained soils or fissured rocks. Clay-cement grout has predominant proportion of clay. Where clay is less in proportion, bentonite is added with cement so that cement cannot settle before the total penetration of the grout mix is completed.

Chemical grout is like cement grout except for the use of chemicals instead of cement for grouting. A chemical grout is formed of two different chemicals that react together to produce the gel, which subsequently solidifies into a hard and fairly impermeable barrier one or several hours after injection. This can be done in two ways:
- By injecting the two chemicals one after the other, known as 'two shot' grouting
- By using single injection of a chemical that is diluted with water and mixed with a catalyst to control the gelling time, known as 'single shot' grouting

These methods are not good for fine silt or clay. In *two shot* grouting, an initial injection of sodium silicate is followed by an injection of sodium chloride to form a *silica gel*. Because of high viscosity of the gelled compound, injection is to be done under high pressure and at close intervals. In *single shot grouting*, a liquid base diluted with water is mixed with a catalyst to control setting time (gelling) before being injected into the ground. The liquid base could be 'sodium silicate—sodium bicarbonate', 'chrome-lignin', 'sodium silicate-amides'; or 'resinous polymers'. This kind of grouting is expensive.

Bitumen grouting is suitable for decreasing the permeability of the soil without adding any strength to the soil. Flowing water in permeable soil is likely to wash away clay and cement grout during injection. As hot bitumen at 150°C makes contact with cold water, its outside is covered with cold water but its inside remains hot for injection over extensive areas. Hot bitumen can completely fill the voids in the fractures and fissures in rock strata as well as coarse sand. This way, an effective water seal may be formed. A bitumen emulsion with 50–60% bitumen, 1% emulsifier, and water is good for application in fine to medium sands to seal water permeability.

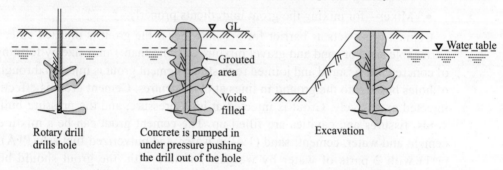

Fig. 4.22 Pressure grouting

Caissons

A caisson is like a reinforced concrete cofferdam within which excavation is carried out by dredging. The caisson is constructed on-shore or offshore to support bridge pier or other structures. It may ultimately form an integral part of the permanent structure like intake water well or a bridge for which the excavation is made. A caisson may have different shapes and sizes:

- Circular
- Twin circular—side by side along an axis
- Double D—with common straight side
- Double octagonal—with a common side
- Rectangular
- Double rectangular—with a common wall
- Multiple dredge hole well—monolithic

The caisson sinks into the ground to the level of the base under its own weight as soil excavation within is continued. If the bottom is located above the water table, or if the water is removed by pumping from an open sump, the excavation can be manually done; otherwise, the soil is removed by dredging using grab (clamshell) and the bottom of the caisson is plugged with mass concrete when the grade is reached. There are four types of caissons:

- Open caisson
- Monolithic caisson
- Box caisson
- Pneumatic caisson

Open caisson

An open caisson comprises a cell, open at both top and bottom ends. The caisson sinks as excavation is carried out within the well. It means sinking would be difficult if large fragments are encountered or if skin friction between soil and caisson is very high. Also, a caisson resting on sloping stratum may slip and get displaced. A circular caisson because of its less surface area encounters less skin friction and is, therefore, preferred. Construction of a circular caisson is also relatively easy. However, size and shape of a caisson is decided for maximum possible open excavation. Care must be taken during construction phase to keep the caisson vertical with smooth surfaces so that skin friction would not hamper sinking. Verticality is controlled by adding kentledge, jetting water or selective excavation carried out on the higher sides. There would be both skin friction and bearing resistance to resist sinking force. The shape of the cutting edge at the bottom should conform to the soil type that a caisson has to sink into. An open caisson may be partially or fully preformed—normal practice is to preform a caisson partially. A level concrete pad is cast at the location where the caisson is to be sunk. Over the pad, the caisson's wall is cast up to 5 m above cutting edge toe. Cutting edges are usually made of steel in combination with reinforced concrete. The triangular-shaped shoe must be strong and rugged to allow the caisson shear through the soil, boulders, buried timber, weathered hard strata and into bedrock. The cutting edge, thus, works as the bearing element of the caisson. Where the soil is stiff, the angle of cutting edge would be steeper.

In water, a temporary sand island is formed for launching a caisson. The difficulty with the sand-island method is the possibility of scour around the island against which constant vigilance is needed. The caisson is then allowed to sink by breaking out the concrete pad. Excavation is then started and continued till it is in a position to allow another lift of concrete to be cast. The process is repeated over and over again till the founding level is reached. The initial few lifts may be preformed but the same process would be followed thereafter. Completely preformed open caisson would involve very heavy handling and dredging using cranes of requisite heights.

Open caissons can be sunk onshore or even inland, if subsoil conditions are favourable for sinking. The cutting edge at the bottom is made wider than the caisson wall by 75 to 100 mm so that bentonite slurry can be pumped into the annular space so as to minimize skin friction against sinking weight. Dredging inside the wells is usually carried out by clamshell buckets for removing the excavated spoils at the same time. Caisson walls are raised above the ground level by slip forming or by conventional methods.

Open caissons are used where construction of cofferdams is not viable or not cost-efficient. Open caissons are used for bridge construction or deep underground silos.

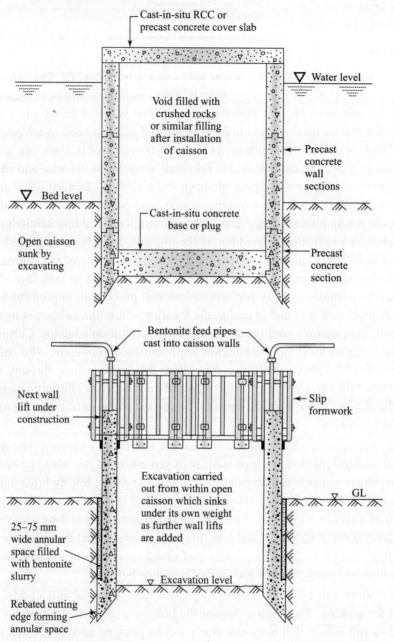

Fig. 4.23 Open caisson

Monolithic caisson

A monolithic caisson is similar to an open caisson, but has a number of wells of different shapes and cross-sections through which excavation is carried out. Verticality

Earthwork 79

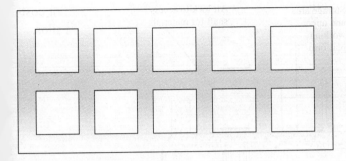

Fig. 4.24 Monolithic caisson

of the caisson is important, and it can be maintained by selectively carrying out excavation in different wells. Because of a number of wells within a well, a monolithic caisson would be heavier than an open caisson. Because of thicker walls and more weight, monolithic caissons would be suitable for structures like quays, which are likely to resist large impact forces.

Box caisson

A box caisson is precast concrete box that is open at the top and closed at the bottom and the sides. The temporary works like islands or falsework, which are necessary for open caissons, are not required in the case of box caissons. A box caisson is constructed on land or dry dock and subsequently floated and towed to the exact location where it is sunk onto its foundation or dredged base. The box caisson is to be held in position using guides or moorings against the current pressure and other lateral forces. A gravel or crushed rock bed is a suitable base if unaffected by such current as may cause erosion. Otherwise, concrete foundation is to be cast by building temporary cofferdam. In fast flowing waters, a box caisson should be founded on piles. Filling the caisson with water or adding kentledge over the walls would be necessary to lower the caisson on the foundation. Thorough structural and stability analysis should be carried out before taking up launching, transporting and sinking of a box caisson. The space inside the caisson is filled with concrete using pumped or tremie concrete. Concrete may be placed using a crane and even a skip. The placing of concrete inside the caisson bottom increases stability. The sides of a box caisson should remain above water level when the caisson is firmly placed on its base or foundation. The outside perimeter of the base must be protected by depositing riprap or rock fill against the danger of scour. Even under or around the pile caps, cement-sand slurry needs to be pumped/placed as a measure against possible scour. Box caissons are used in such construction work as bridge piers, breakwaters, jetties, and other harbour protection work.

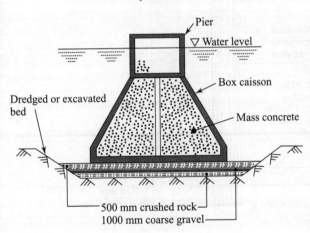

Fig. 4.25 Box caisson

Pneumatic caisson

A pneumatic caisson is similar to open caisson except that there is an airtight working chamber at the cutting edge where positive pressure is maintained. The air pressure is

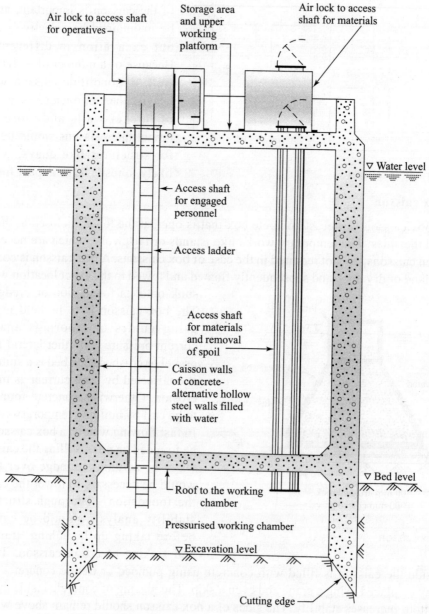

Fig. 4.26 Pneumatic caisson

always maintained at a level higher than the pore water pressure at the bottom of the cutting edge; thus keeping the excavation area free from water, while the sides of the caisson support the sides of excavation. They are used where difficult subsoil exists and where manual excavation in dry conditions is necessary. Where logs and boulders below the cutting edge halt sinking of the caisson, compressed air method would help continue sinking by removing the obstacles manually. The working chamber should have access for entry/exit of workers and lowering of materials, (vide Fig. 4.26). Precaution is to be taken for workers to work in an environment of pressurized air. In Japan, robots are deployed in place of workers. Safe limit of compressed air pressure is 3.1 bar or 310 kN/m^2. Sinking of the caisson is controlled as follows:

- Excavation is to be carried out symmetrically
- Working chamber should be made leaving berm with cutting edge to avoid compressed air leaking
- Air pressure should be maintained depending on the types of soil encountered or to be encountered—sand blowing may cause compressed air leakage resulting in stoppage of work and re-pressurising of the working chamber
- Building up of static friction could be avoided by continuous sinking—this would require mobilisation of standby compressors and decompression facilities

When the required depth is reached and sinking is completed, the floor of the working chamber is plugged with well-vibrated concrete. The sealing is completed by carefully pressure grouting the top gap of the chamber. Reinforced bottom plugging is possible only in pneumatic caissons.

4.6 TRENCHLESS (NO-DIG) TECHNOLOGY

Trenchless technology is adopted for underground utility services in urban or developed areas underneath existing buildings, streets, roads, railways, embankments, installations, or wherever open excavation or trenching is not possible. This technology (no-dig methods) allows underground work to be executed without disrupting the existing utilities/facilities and/or causing inconvenience to the concerned people. This relatively new technology should be exploited to its full potential by more research and development efforts. There are different methods involved in trenchless technology on installation of drains, pipes, or ducts. The method to be selected would depend on:

- Sizes/dimensions of cables/pipes/ducts
- Length involved
- Soil conditions
- Existing underground hindrances

Pipe Jacking

Pipe jacking is based on the principle that involves pushing pipes up to 400 mm in diameter over a distance of up to 500 m by engaging one or several hydraulic jacks from jacking pit (also referred to as launch pit). The first length of pipe is fitted with a sharp closed hood and is set on guide frames. Jack pressure is transmitted to the pipe through a thrust bar with an adjustable rod. The distance between the holes on the thrust bar equals the length of the jack-working stroke. Extensions are attached to the pipes as it is forced in. The wall of the launching pit is reinforced with a shield to provide support for the jack. Pipe jacking is applicable at depths of not less than 3 m in soils free of boulders and other solid inclusions that could cause the pipe to lose direction. But jacking can be done in water bearing granular materials, soil containing cobbles, and even soft rocks by innovating jacking methods, vide Fig. 4.27. Excavation of soil in the pipe, which is pushed in, is carried out manually or by deploying backhoe wherever possible. In case of smaller-diameter pipes, soil inside the pushed pipe is cleaned by rotary slurry cutting heads or by using augers.

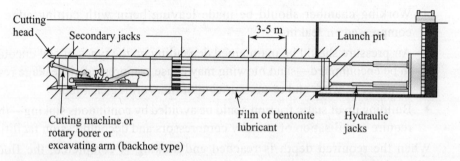

Fig. 4.27 Pipe jacking

Thrust Boring

Thrust boring is another method of jacking pipe (diameter 250 mm up to 1800 mm) as well as sewer line of square cross-section (duct) through ground between the launch pit to the receiving pit (also referred to as exit pit)—a distance that could be up to 200 m. The first piece/section of pipe/duct is equipped with a rotating boring head with tungsten carbide cutters, which cut the soil into small pieces as hydraulic jacks push the pipe forward. The whole system uses laser and computer control located in the launch pit to guide the advance of pipe/duct through the ground. The cut soil is channeled back through the boring head into a crusher wherefrom the resulting mixture is sucked to the surface with a vacuum pump. As the boring makes progress, additional pieces of pipe/duct are added at the launch pit.

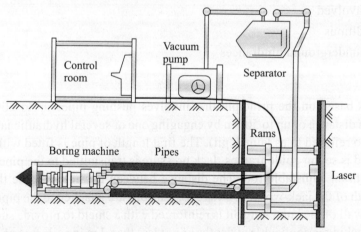

Fig. 4.28 Thrust boring

Horizontal Direction Drilling

Horizontal direction drilling (HDD), unlike pipe jacking, involves both pushing and pulling. The system consists of a self-contained hydraulic bore rig operated by an integrated diesel engine and a bentonite mixing system. A fresh water tank and a high-pressure cleaning system are also an integral part of the bore rig for cleaning the drill pipes during and after the boring operation. Bentonite slurry is forced at high pressure through a small-diameter nozzle for cutting into soil. The cutting head could also be a mechanical device, something like a rotating drill. An electric rotary motor mounted on a drilling rig capable of producing rotation of the drill head is an alternative to the hydraulic system. A transmitter unit (sonde) fitted in the drill head transmits signal that is detected at the surface through a hand-held antenna. The cutting jet/head is thus steered clear of existing obstacles by constant monitoring of direction and depth of cutting head. The hindrances can be circumvented and curves accommodated. HDD is used for installing underground ducts, cables, and service pipes for crossing roads/rivers or meandering through known existing underground services. The drill stem consists of flexible steel tubes of 80 mm diameter. The drill bit is started at the launch pit for termination on completion of boring at the reception/exit pit. The boring is done through clay, but boring through sand and gravels is also possible.

At the reception pit, a back-reaming head replaces the drill head. The back-reamer can work up to bore of 350 mm. With the progress of back reaming, the service pipe is simultaneously pulled ultimately back to the launch pit. Whereas the push force should be limited to so much as not to buckle the drill stem, the pull force can be relatively much more.

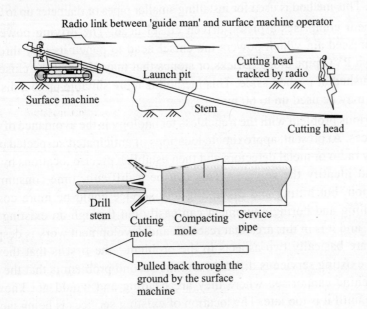

Fig. 4.29 Horizontal direction drilling

Pipe Bursting

Pipe bursting is a trenchless method of replacing any underground pipe that needs to be replaced. To do this, a larger-diameter pipe is pushed along the old pipeline using hydraulic rams. A metal cone is fitted at the head of the new pipe. The hitting of this metal cone into the old pipe results in bursting of the old pipe into pieces. The impact of the ramming consolidates the pieces into the surrounding ground, thus making way for the new pipe to replace the shattered old pipe. Instead of applying hydraulic force, the metal cone can be pulled by a steel wire powered by mechanical energy. The impact and the consequence of the impact would be the same as was in the case of hydraulic rams. The principle of pipe bursting is a simple and quick way of replacing old pipes, but this method cannot be applied if the pipe to be replaced is connected to other pipelines joining it at angles.

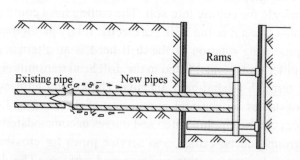

Fig. 4.30 Pipe bursting

Percussive Boring

Percussive boring is based on using a chisel head for punching its way through the ground forming a hole along the way. The diameter of this hole is larger than the diameter of PVC pipe that is to be pulled by the chisel head, which is powered by compressed air. This method is used for installing smaller pipes of diameter up to 200 mm.

The equipment comprises a piston-driven chisel head. The driving power of the piston is compressed air, which powers the chisel head to punch its way through the ground pulling a PVC pipe along. Bricks or stones that may obstruct the chisel head's progress are crushed or pushed aside. This method is more suitable up to a distance of 20 m or so but may be used up to 60 m.

The most serious problem with the trenchless technology is the avoidance of existing pipes and services. At present, approximate locations of anticipated/suspected pipes are to be located by radio or metal detectors and then establish precise locations by manual excavations and identify the services. This can be a difficult, time-consuming and costly proposition, but hitting and disrupting any service could be more costly and hazardous. Drilling and boring equipment can easily cut through an existing service unintentionally, and it is in this area that research and development work is desperately needed. There are basically two aspects to the problem—the first is that the precise location of any existing service is difficult, and the second problem is that the drilling and boring machine cannot see where they are heading and would not know of an existing service until it is too late. The location of existing services is being tackled by an integrated national computerized register of service locations in many countries with the intention of doing the work as accurately as possible, but since many locations are

Earthwork 85

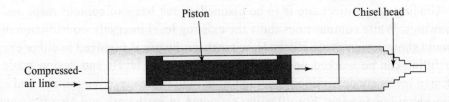

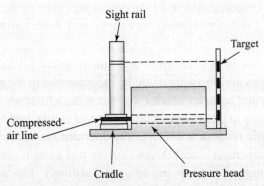

Fig. 4.31 Percussive boring

recorded plus or minus three metres, there is a lot of work to be done in this area before the system can be considered reliable. The detection of underground services from the drill head or surface is an area that requires further research and development. Trenchless technology for pipe and service installation and repair are already very useful, but hold the potential for great benefit to our society in the future if the problem of existing service location can be solved.

4.7 GRADING

Construction project sites need to be properly graded and accurately finished by eliminating natural terrain roughness and providing slopes as shown in design drawings. The site surfaces should be smooth and level without undulations and ridges. In any project, both cutting and filling is usually involved. Grading, therefore, could involve balanced, excessive or deficient amounts of earthwork in excavation and filling. In the first case, the volumes of cutting and filling are balanced, which should be the ideal case. In the second case, the volume of cutting would be in excess of filling, thereby necessitating disposal of the soil mass that is in excess. In the third case, the volume of cutting would be less than the volume of earthwork involved in filling, thereby necessitating hauling of the deficit volume from outside the project site. In site development work, there could be cases where cutting would be nominal or negligible, and there could also be cases where filling would not be required at all. In any case, it is easier to spread several small piles of excavated spoils than spreading a huge pile. Accordingly, excavation should be interrupted from time to time to allow spreading of the excavated spoils.

Grading of a project site is to be planned on the basis of contour maps and design drawings. While contour lines show the existing level intervals, construction drawings would show design grade levels. How much earthwork is involved in either excavation or filling can be worked out on the basis of existing levels and design grade levels. Design levels are determined keeping technology, drainage, environment, and aesthetic considerations in view. For effecting economy in earthwork and also for compacting time schedule for the same, a project site may be divided into different design grade levels on the basis of contour lines. This grading referred to here is related to the site development. This initial problem of site development would be further compounded as and when excavation for foundations, trenches, and reservoirs are taken into consideration.

Grading is important for both excavation and embankment formation for accurate final levelling work. Lasers and software could be incorporated in the grading operation in order to achieve quick and accurate results. To start with, a laser level is erected at the centre or to one side of area to be graded. A laser detector is attached to the blade of the grader and a diagrammatic readout is given to the operator via a liquid crystal display mounted in front of the operator's seat. From this control system, the operator would know if the blade is too high or too low and adjust accordingly. This adjustment can be automatically controlled by software.

About the final grade level, certain issues deserve consideration. There should be different considerations for excavation and filling. Clay and clayey soil is compressible in nature. Underlying compressible layers remain in equilibrium under the weight of the overburden above. As and when the overburden is excavated and removed, there would be heaving due to decompression. This heaving process continues over a period of six months to a year. Final grading should be carried out after total heaving and after having been exposed to weather during the decompression period.

In case of embankment formation, the considerations that deserve attention are: (a) the soil materials with which filling is carried out would settle during and after execution (b) the underlying soil would settle due to the weight of filled-up overburden. Such consolidation could continue for a period of 12 months. The final grading should be carried out after consolidation. More details are given in Chapter 12, Section 12.1; and Chapter 13, Section 13.4.

Compaction is the process whereby particles of soil are packed together using construction equipment to reduce the air content. As for water content, it acts as lubricant to achieve greater compaction. Then as the air content decreases, the water tends to keep soil particles apart, thus hindering compaction. Full compaction can be achieved only when the moisture content is at the optimum level. The optimum moisture content can be found out by laboratory tests.

4.8 DREDGING

Dredging is the process of moving/excavating soil/rock (hard rock by blasting) under water. The object of dredging is generally to remove material from a particular location to place it at another location or let it be moved to a new location by natural forces. It is a highly specialised excavating method carried out for many purposes, as follows:

- Moving/excavating soil or rock under water to deepen lake, reservoir, river, or sea for offshore construction, restoration of water reserve capacity, navigation, and submersible pipe/cable laying
- Filling up to raise level under water or on land with spoil excavated under water (i) to ensure safety of under water foundations/structures and pipeline beds, (ii) to form beaches and dykes, (iii) to raise embankments for roads, (iv) to improve areas around ports
- Replacing poor quality of soil with good quality soil under water for foundation construction or on land for land reclamation
- Mobilizing soil materials for construction purposes like obtaining coarse/fine aggregates for concrete mixing or land reclamation
- Improving environment by dredging and disposing off contaminated materials or sealing up contaminated materials with good materials and filling up wetlands with good dredged materials

A dredger is a vessel or a floating plant fitted with equipment/devices to excavate or move soil or rock under water. The basic design of a dredger comprises equipment mounted on barge/pontoon for floating and moving on the water surface over the soil to be excavated in a bay, harbour, river, lake, or reservoir. On the pontoon or barge, the projecting frame at the forward end supports the excavating part of the dredger. The arrangement or layout may vary, but frame or boom would be an essential item. Also, there would be a number of spuds (pointed steel pipes that are vertically driven into the bottom and used as braces against digging thrust, wind, and waves—they are vertically retractable anchor posts) for maintaining position or pivoting. Power may be diesel or electric (within reach of power lines). Different types of dredgers deployed for dredging operation are:

- Trailing suction hopper dredger
- Bed leveller
- Water injection dredger
- Suction dredger
- Cutting suction dredger
- Dustpan dredger
- Clamshell dredger
- Grab hopper dredger
- Hydraulic backhoe dredger
- Dipper dredger
- Bucket dredger
- Special dredgers

Trailing Suction Hopper Dredger

A trailing suction hopper dredger is a self-propelled (propelled at about 4 km/h) vessel fitted with among others things trailing suction pipe, and hoppers built into the hull. It is normally rated according to its hopper capacity, which is usually in the range of 500 m^3 to 10000 m^3 (modern pumps can fill up 10000 m^3 hopper in 1 hour). Dredged materials are loaded by one or more (up to 4) pumps into the hopper/s. The pumps may be installed onboard or may also be fitted in the trailing suction pipe as submersible pumps. The maximum depth to which dredging is possible depends on the vacuum head generated by the pumps—more depth means less output for a particular pump. Dredging depth could be as much as 80 m or so. Greater suction depth is possible in case of submersible pumps. The output of a dredger depends on:

- Capacity of pump/s
- Depth of digging/dredging
- Height of discharge
- Line friction
- Percentage of solids

Solid dredged materials settle down in the hoppers, top water is let out through the overflow pipes. In case of very fine dredged materials, overflowing water contains fine particles and pumping is discontinued as soon as water starts overflowing. In case of coarse or heavy materials, pumping is continued even as water starts overflowing. The bearing pressure of the draghead at the end of the trailing suction pipe on the bed is usually controlled by an adjustable pressure compensation system that acts between the draghead and the hoisting winch. The type of draghead to be deployed is selected on the basis of type of materials to be dredged. Since the suction draghead does not have a rotating cutter, teeth of various types and shapes are fitted on the bottom of the draghead to help breakup compacted materials in front of the suction inlet. When the hopper is loaded full, the suction pipe is restored onboard and the dredger is moved to the designated offshore/onshore unloading site. Offshore hopper bottom discharge using hydraulically operated dump doors is quick, but pump-discharge onshore/upland for land reclamation takes time. The cycle time of this kind of dredger:

- Loading time
- Sailing time to discharge point
- Time of discharge
- Time for sailing back to next loading area

Approximate limiting factors for this kind of dredger:

- Minimum water depth for operation   4 m
- Maximum water depth for operation   45 m
- Minimum turning circle   75 m
- Maximum wave height   5 m
- Maximum particle size   300 mm

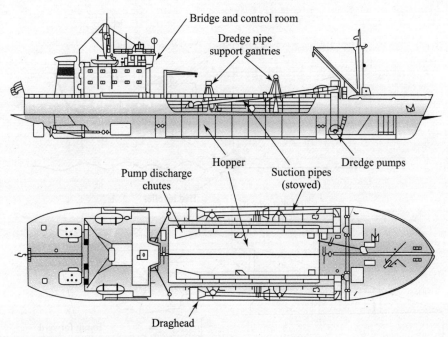

Fig. 4.32 Trailing suction hopper dredger

Bed Leveller

A bed leveller is a very large tray-like cutter with cutting blade on one side. Cutting side is open for loading materials. Self-propelled tugboats tow bed levellers. Bed levellers are used for two applications: (i) as a dredger to move materials over short distances cost-efficiently - materials close to quays, jetties and in entrances or restricted areas can be moved without turning to conventional dredgers to do the job (ii) to improve efficiency of the other types of dredgers. Bed levellers can be deployed in such areas as are inaccessible to the other types of dredgers. The cycle time of a bed leveller:

- Positioning of tug boat
- Lowering of bed leveller with blades
- Towing blades
- Raising blades
- Returning to starting point for repeating the process

Approximate limiting factors for this kind of dredger:

- Minimum water depth for operation   3 m
- Maximum depth of operation   30 m
- Maximum wave height   1 m
- Maximum swell   1 m
- Soil type—very soft and soft clays and silts, loose and medium dense sands

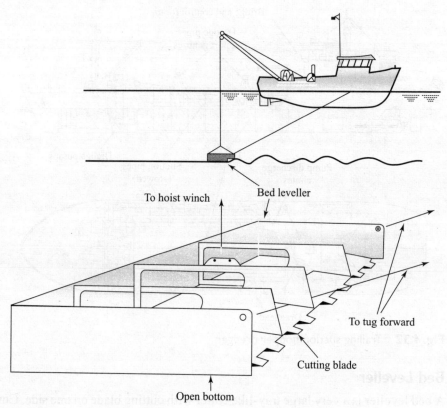

Fig. 4.33 Bed leveller dredger

Water Injection Dredger

A water injection dredger is a self-propelled vessel. It is deployed to inject water on bed soil materials to fluidize the same for disposal purposes. This type of dredger can be successfully deployed in low strength fine-grained materials for the fluidized material to flow to lower levels. A fixed set of water jet nozzles are lowered to initiate water injection at a pre-determined pressure and flow rate to penetrate the bed materials. The vessel moves ahead slowly driving the fluidized materials before it. If the seabed slopes away from the working area, then fluidized materials would move over considerable distances speedily. The fluidized materials would move also at level or undulating ground. The cycle time would be similar to that of a bed leveller. Approximate limiting factors for this kind of dredger:

- Minimum water depth for operation   3 m
- Maximum depth of operation   15 m
- Maximum wave height   0.5 m
- Maximum swell   0.75 m
- Soil type—very soft and soft cohesive soils and very loose and loose fine granular soils

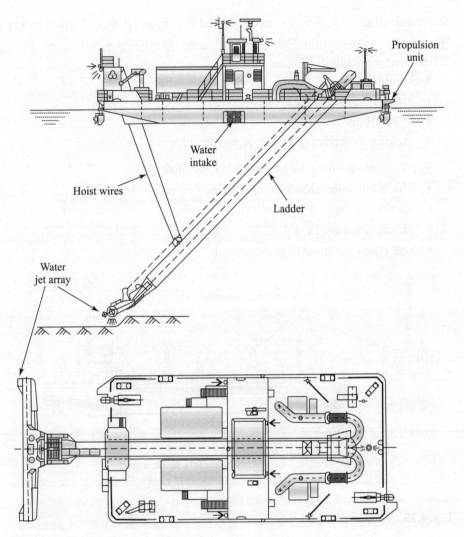

Fig. 4.34 Water injection dredger

Suction Dredger

A stationary suction dredger, unlike a trailing suction hopper dredger, is anchored at the place of dredging before commencement of dredging operation. The result of dredging is generally an inverted shaped cone in the bed. This type of dredger is deployed for obtaining granular materials for land reclamation or aggregates for concrete work. A stationary suction dredger may be fitted with hoppers for discharging dredged materials. Otherwise, dredged materials are loaded on barges or pumped through pipelines to delivery points over distances of 10 km or more. On larger vessels, over 10,000 m^3/h may be pumped for disposal.

At the point of dredging, suction pipe is lowered from the vessel to the bed for starting dredging operation. High-pressure water jets around suction pipes may be used for fluidization of soil to be dredged. The resulting spoil and water mixture is then drawn up the suction pipe for discharging as mentioned above. The cycle time of a stationary suction dredger:

- Loading
- Moving to discharge area
- Discharging
- Moving to dredging area to repeat the process.

Approximate limiting factors for this kind of dredger:

- Minimum water depth for operation   3 m
- Maximum depth of operation   50 m or more
- Maximum wave height   3 m
- Soil type—permeable granular soils

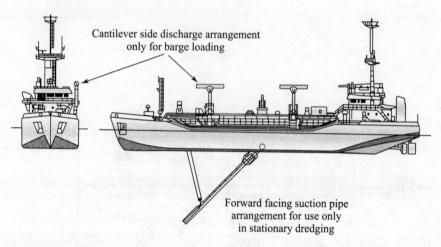

Fig. 4.35 Suction dredger

Cutting Suction Dredger

A cutting suction dredger is also anchored at the place of dredging before commencement of dredging operation. In general, it is a pontoon hull structure with no power unit for propulsion. However, larger version of this type of dredger may be self-propelled. Where the effect of normal suction pressures or jetting is insufficient to loosen the soil, a mechanical cutter is preferred. The equipment comprises a rotating cutter head fixed to the end of a stiff jib on which the drive motor and the suction pipe are also carried. The cutter operates by rotating the arm in an arc. The dredging operation is initiated by powerful cutting followed by suction and pumped discharge to either barges or along a floating pipeline to the disposal points. A wide range of materials including rock can be dredged by this type of cutter for pumping the dredged materials directly to the discharge points. A cutter suction dredger is generally rated by the diameter of discharge pipes (150 mm to 1100 mm) or by the power driving the cutter head (15 kW to 4,500 kW). The soil or rock to be dredged is cut, dislodged or broken by a powerful cutter driven by hydraulic or electric power. The cycle time of a cutter suction dredger:

- Lowering cutter to bed
- Cutting one across face
- Further cutting to finish depth
- Cleaning cut
- Raising cutter
- Advancing back to lowering cutter to bed

Approximate limiting factors for this kind of dredger:

- Minimum water depth for operation — 0.75 m
- Maximum water depth for dredging — 35 m
- Maximum cut width (single pass) — 175 m
- Maximum wave height — 2 m
- Maximum swell — 1 m
- Maximum particle size — 500 mm
- Maximum compressive strength (rocks) — 50 MPa

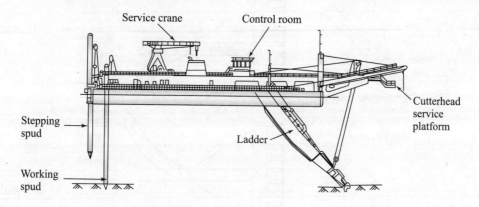

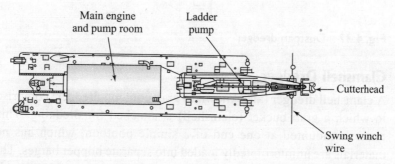

Fig. 4.36 Cutting suction dredger

Dustpan Dredger

A dustpan dredger, which is similar to some extent to the cutter suction dredger, is a suction dredger that has an exceptional form of mouth on its suction head. This suction mouthpiece may be 9 m or more wide. Because of its shape and form, it is called *dustpan*. Its purpose is to agitate and pump for water jetting for loosening and fluidizing materials to be dredged without any mechanical cutter. The dustpan mouthpiece has water jets spaced along its width to assist in dislodging or loosening of the soil being dredged. It is particularly useful for excavating in loose, soft, and free-flowing materials over relatively large areas. Its most effective purpose for maintaining navigation channel is to fluidize and pump shoal materials in large rivers and discharge the same through floating pipeline into currents for deposition elsewhere in the waterbody. Dustpan dredgers are moved and positioned by winch power. Because of the deployment of winches, this type of dredger can move over the ground freely and quickly. There is no definite cycle time of the dustpan dredger. Approximate limiting factors for this kind of dredger:

- Minimum water depth for operation   1.5 m
- Maximum water depth for dredging   20 m
- Maximum cut width (single pass)   10 m
- Maximum discharge distance   500 m

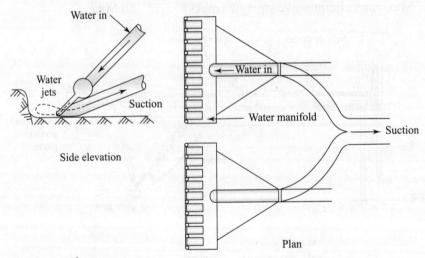

Fig. 4.37 Dustpan dredger

Clamshell Dredger

A clamshell dredger (also known as grab pontoon dredger) comprises a lattice jib crane to which a grab bucket (clamshell) is attached for dredging. This rope-operated jib crane is mounted at one end of a simple pontoon, which has no hopper. Dredged materials are uninterruptedly loaded into separate hopper barges. The pontoon is held in position by anchors and winches when spuds are not used. A clamshell dredger is rated by its clamshell bucket capacity (0.75 m³ to 20 m³). The weight of the bucket itself determines the digging effort. A clamshell dredger is very useful in soft to medium-hard dredging operation. Unlike pumped discharge, clamshell buckets load hoppers with more solids. A clamshell bucket is well suited for deep (beyond the capacity of other dredgers) dredging in confined areas such as wharfs and breakwaters. The cycle time of a clamshell dredger:

- Lowering grab
- Closing grab
- Raising grab
- Swinging for discharging
- Discharging
- Swinging to place of dredging for repeating the process

Approximate limiting factors for this kind of dredger:

- Minimum water depth for operation   1 m
- Maximum water depth for dredging   50 m or more
- Wave height   2 m
- Maximum compressive strength (rocks) 1 MPa

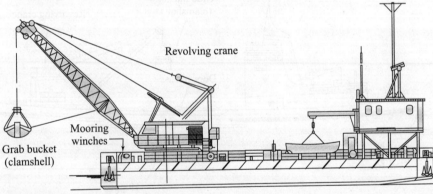

Fig. 4.38 Clamshell dredger

Grab Hopper Dredger

A grab hopper dredger is generally a ship integrally fitted with a hopper. One or more lattice jib cranes are mounted on the ship. The ship is anchored for dredging operation. The hopper capacity rarely exceeds 1500 m³. The hopper is loaded by one or more clamshell buckets. Hydraulic crane is possible alternative to the rope-operated crane. Because this type of dredger is self-propelled and fitted with hopper, its operation covers much larger areas. The cycle time of a grab hopper dredger:

- Mooring
- Swinging to working area

- Lowering grab
- Closing bucket
- Raising grab
- Swinging for discharging
- Discharging

Or

- Releasing anchors
- Sailing to discharge point
- Discharging
- Sailing to working area for repeating the process

Approximate limiting factors for this kind of dredger:

- Minimum water depth for operation   3 m
- Maximum water depth for dredging   45 m or more
- Minimum turning circle   75 m
- Maximum wave height   2 m

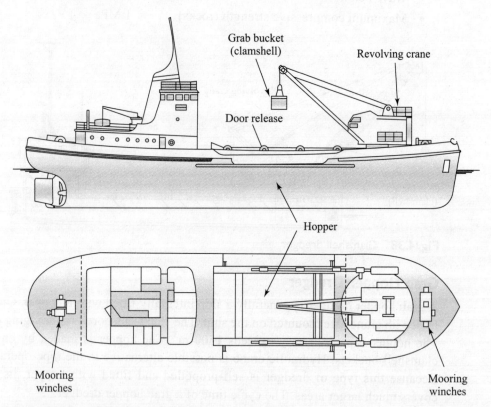

Fig. 4.39 Grab hopper dredger

Hydraulic Backhoe Dredger

A hydraulic backhoe dredger utilizes a hydraulic backhoe excavator for digging towards the dredger. This backhoe dredger is mounted on a fabricated pedestal at one end of a spud-rigged pontoon. Spud location of the pontoon is generally necessary to provide a positive reaction to the hydraulic digging action. This kind of dredger is rated according to the maximum size of the dredging bucket ranging from 1 m^3 to 20 m^3 depending upon the material and depth of dredging. Materials dredged include boulders, debris, stiff clay, or weak rocks. The cycle time of a hydraulic pontoon dredger:

- Lowering bucket
- Shoving backhoe bucket
- Raising bucket
- Swinging for discharging
- Discharging
- Swinging to place of dredging for repeating the process

Approximate limiting factors for this kind of dredger:

Minimum water depth for operation	2 m
Maximum water depth for operation	24 m
Maximum width of cut	25 m
Minimum width of cut	Bucket width
Maximum wave height	1.5 m
Maximum swell	1 m
Maximum compressive strength (rocks)	10 MPa

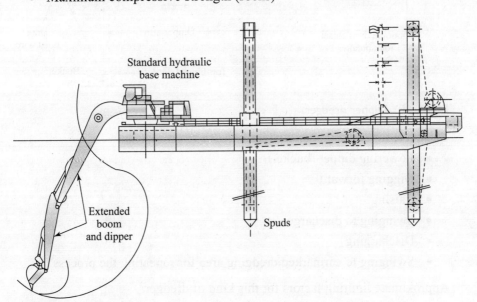

Fig. 4.40 Hydraulic backhoe dredger

Dipper Dredger

A dipper (or dipper-bucket) dredger, cable-operated or hydraulic powered, operates by digging forwards and upwards. Both are mounted on spud-rigged pontoons to balance reaction to the digging action. The dipper (bucket) is mounted at the forward end of the dipper stick. Its operation resembles that of a face shovel. The boom and the dipper stick assembly turns through 180° and the materials are discharged to barges moored on either side of the dredger or on shore if close enough. The main advantage of a dipper dredger is its powerful crowding action and, as such, it can dredge a wide range of compact and oversize materials like rocks, weak rocks, and stiff clay without the need of blasting. Even blasted materials could be handled, if required. The rope-operated type compared to hydraulic variety can dredge to greater depth. Dipper dredgers are suitable for deepening harbours and channels in restricted areas or in the areas where the bottom is too hard. These dredgers are also used for digging canals through swamps.

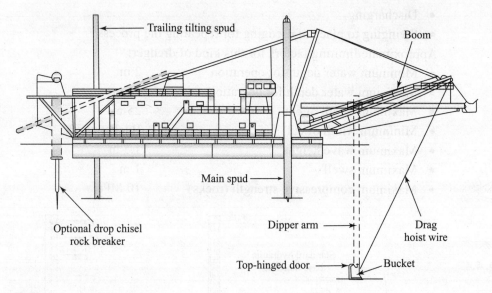

Fig. 4.41 Dipper dredger

The cycle time of a dipper dredger:
- Lowering dipper (bucket)
- Digging forward
- Raising
- Swinging to discharge
- Discharging
- Swinging to earmarked dredging area for repeating the process

Approximate limiting factors for this kind of dredger:
- Minimum water depth for operation   3.5 m
- Maximum water depth for operation   20 m

- Maximum width of cut 30 m
- Minimum width of cut Bucket width
- Maximum wave height 1.5 m
- Maximum swell 1 m
- Maximum compressive strength (rocks) 12 MPa

Bucket Dredger

A bucket dredger operates by using a ladder-mounted endless chain of buckets like a ladder-type trenching machine. The buckets are inverted. They scoop out bed materials (mud or clay, medium sand, gravel, and loose rock fragments) for lifting above water for discharging under gravity onto chutes for conveying to hopper-like containers on the barges moored alongside. The heavy bucket chain is supported by a steel ladder and driven by hydraulic or electric power via a tumbler at the top. The ladder is mounted centrally on a rectangular pontoon. Five or six winches are used for positioning the pontoon. A bucket dredger is generally used only to load barges in quiet waters. The ladder is susceptible to damage by possible shifting of the barge precipitated by currents, passing vessels, or inclement weather. The breakout force, which is dependent on the size and mass of the dredger, is exerted via the bucket cutting edge with/without teeth and may be substantial. The powerful head winch provides reaction to digging force. A bucket dredger has the advantage of a continuous dredging process, but it cannot be deployed in shallow water. It is preferred where working in restricted area is required, for example, along quay walls and in dock systems. The chain and buckets are subjected to considerable wear, and consequently, maintenance costs turn out to be high. The production cycle time of any dredger depends on bed materials, depth of each cut, number of cuts per unit time, time of advancing to new faces for cutting (in case of forward cutting, it is referred to as face cutting), time of replacing barges, and anchoring time of the pontoon at new location.

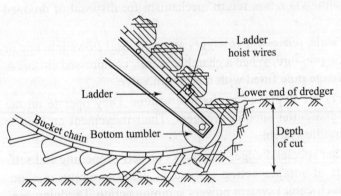

Fig. 4.42 Bucket dredger

Approximate limiting factors for this kind of dredger:

- Minimum water depth for operation 5 m
- Maximum water depth for dredging 35 m
- Maximum cut width (single pass) 150 m
- Maximum wave height 1.5 m
- Maximum swell 1 m

- Maximum particle size   1500 mm
- Maximum compressive strength (rocks)   10 MPa

Special Dredgers

Special dredgers are additional devices/facilities to enhance production of various types of dredgers. They may also be deployed for dredging. Special dredgers include:

A booster pump like a jet pump may be introduced into any dredger that works on the suction process like a stationary suction dredger so as to force water at considerable speed to produce the required suction. A jet pump comprises a high-pressure water system that forces a jet of water through a venturi near the extreme end of the suction pipe. The resulting additional kinetic energy helps in lifting water/solid materials to the required height.

Another booster is the *airlift* that works by forcing compressed air at the submerged extremity of the pipe. Its simplicity lies in not having any submerged moving parts. The jet of compressed air reduces density of water-solid mixture resulting in induced upward flow. This kind of lightweight dredging assembly is suitable for divers.

The exhaust stroke of a positive displacement piston pump forces materials from a hopper into a discharge pipeline via a non-return mechanism for disposal of dredged materials.

For dredging in cohesive soils, *pneumatic dredgers* are used. Soil flows into one or more chambers by hydrostatic pressure. When a chamber is full, compressed air forces soil to the surface via a discharge pipe fitted with non-return valve.

Small *amphibious dredgers* are deployed in shallow waters. They operate on the principle of grab, backhoe, and cutter suction dredgers. Their movement on land is based on track-laying or hydraulic system.

A *scraper dredger* is another special dredger that works on self-propelling and self-loading system using power of scraper action instead of usual pumping method. Hydraulic or mechanical (winch-cable) system powers scraping action. Dredging here is restricted to shallow water.

Selection and performance of a dredger depends, among other things, on:
- Access to the dredging area
- Location of reclamation and discharge areas
- Depth of water
- Length of the dredging area
- Width of the dredging area
- Dredging profiles and accuracy depending on dredging technique and site conditions
- Proximity to structures
- Site conditions: (i) wind (ii) rain (iii) fog (iv) temperature (v) waves and swells—limiting heights for efficient operation (vi) currents (vii) anchorage (viii) disturbing local traffic

To calculate the quantum of dredging done, the area to be dredged is outlined as accurately as possible, and existing bed levels are determined by soundings. Intervals between sounding lines may vary according to the general contours of the area and the nature of the materials to be dredged. In practice, the interval is usually between 7.5 m to 50 m. As precise work is not possible in dredging, it is normal to allow both horizontal and vertical tolerances in over-dredging. The results of dredging operation are checked by a second set of soundings made on completion of dredging work, using the same sounding pattern as for the original survey.

The dredged materials should be dumped in the sea in accordance with the regulations of the appropriate authority having jurisdiction over that part of the sea, or disposed of elsewhere with the written permission of the owners of the disposal area. No dredged materials should be allowed to leak into or be deposited in navigable channels during transportation. Most common methods of disposal of dredged materials are pumping and hauling by trucks. Other modes of disposal of dredged materials are transportation on barges, railways, and belt conveyors. Whereas transportation using barges is economical up to 480 km, truck movement is economical only up to 80 km. Belt conveyor has limited use. Belt specifications vary in width (76 cm to 178 cm), flight length (270 m to 780 m), and speed (11 km/hr to 144 km/hr). Again, while pumping is possible in case of spoil in slurry form, only dry materials can be hauled by the railways.

SUMMARY

Earthwork is so vital in implementation of construction projects that construction technology related to earthmoving is receiving a lot of attention worldwide. Earthmoving equipment and plants are continuously being innovated to execute voluminous earthwork in less and less time. Trenchless (no-dig) technology is now so advanced that it is now possible to dig through road/railway embankments without disrupting existing facilities. This technology can be used in digging under river beds for laying/replacing utility/protection pipes. Groundwater problem does not pose problems anymore, as there are several solutions to this problem. More serious problems exist in construction of marine structures. Dredging is routine work in some countries just for survival.

Dredging should receive adequate attention in India because beds of dam reservoirs, rivers, and canals are being silted up. Regarding earth filling for embankment formation, dredged materials could be conveniently used. Grading during embankment formation especially for highways can be accurately carried out using laser beams. Even setting out has now been made easy by replacing theodolites and dumpy levels with 'total station', which is a fully integrated instrument that captures all the data necessary for a three dimensional position fix and displays it on digital readout systems recorded at the press of a button. All these throw light on how onetime arduous earthwork has now been made easily manageable utilizing innovative technology.

REVIEW QUESTIONS

1. What is meant by cohesion? What are the problems related to cohesive soils?
2. Explain how consolidation is different from compaction.
3. What does site development work comprise of?
4. How grid lines are established at construction sites for setting out?

5. Why mechanized earthwork in both excavation and filling is taken into consideration these days in project execution work?
6. Name the operations involved in earthwork in excavation.
7. There are two ways of groundwater control. Name them.
8. What is the difference between temporary and permanent sheet piling?
9. What is a cofferdam? Name the materials used for building cofferdams.
10. Describe the electro-osmosis process of temporary groundwater exclusion.
11. Define a caisson. Describe briefly the four types of caissons.
12. Name the methods of trenchless technology. Describe how pipe jacking is carried out.
13. Describe how lasers and software could be incorporated in the grading operation in order to achieve quick and accurate results.
14. What is the objective of dredging?
15. Describe the basic design of a dredger.

5
Excavation by Blasting

INTRODUCTION

Rock excavation by blasting is carried out using explosives, both on surface and under water. Blasting is the result of conversion of a chemical substance into a gas that produces instant devastating pressure shattering the rock adjacent to the explosive. The compact explosive, on detonation, is transformed into glowing gas with an enormous pressure. The first cracks are formed in fractions of millisecond. Drilling of blastholes is the prerequisite of blasting.

Drilling is mainly executed by mechanical means and drilling equipment. The *bit* is the most vital part of a drill that must engage and disintegrate the rock by crushing with blows. Drilling method includes (i) surface drilling (ii) down-the-hole drilling.

An explosive is a chemical compound or mixture of compounds that can decompose instantaneously and violently when initiated by energy in the form of heat, impact, friction, or another detonation in difficult conditions like in densely packed holes, under water and so on.

A typical blasting pattern comprises a number of blastholes drilled in one or more rows in such a way as to suit specific project requirement. Conventional blasting work normally results in rough and uneven contours of blasted surfaces. There are techniques for carrying out smooth blasting at extra cost. Noise, dust, ground vibration, air blast and fly-rock resulting from blasting operations create environmental problems that call for remedial action.

5.1 ROCK EXCAVATION

Rock may, depending on its properties, be excavated by mechanical means like ripping. Otherwise, rock is generally excavated by blasting using explosives.

Excavation by blasting is done:

- On the surface
- Under water—for dredging or exploration or construction of bridge/marine sub-structures or sub-marine pipe-laying

Surface excavation is done for the following purposes:
- Stripping—the spoil has no value

- Cutting—removal primarily to lower the surface level
- Quarrying or mining

In stripping, the overburden is excavated by blasting. The spoil of any excavation by blasting has, in most cases, no value. The overburden is stripped so that foundations or substructures may rest on solid bases or at required levels. Drilling and blasting methods would depend on the properties of rock encountered in the overburdens. Loose soils in the overburdens may be removed manually or by deploying bulldozers and/or scrapers.

Cutting is involved in the construction of roads and highways. The cuts may be through types where road construction is carried out through rock masses. This is different from tunneling by controlled blasting or boring. In cutting the hill sides for construction of roads and highways, the spoils may be utilized for filling the opposite sides wherever required. The excavated spoil may be cleaned and crushed to produce aggregates for road surfacing.

Mining is not the subject matter of this book, though mining equipment and machineries are similar to those used in construction work. As for quarrying, rocks, if found suitable, are crushed for production of aggregates as ingredients of concrete (vide Chapter 13, Section 13.6).

5.2 BASIC MECHANICS OF BREAKAGE

Solid rock mass yields and breaks as a result of rock explosion. The energy from an explosion is spent in rock breakage overcoming the resisting forces of compression, shear and tension. Immediate reaction of the forces of explosion in solid rock mass is an enormous compression. The rock is generally unyielding solid mass. It breaks only under the sudden impact of huge compressive forces. Shear is the movement of rock pieces/blocks along the lines of weakness. Tension is developed by reflection of the compressive waves in the first stage against the unconfined free surfaces of rock. Apart from the above, the energy from an explosion is also spent on overcoming gravity and imparting kinetic energy to the lumps of broken rock.

An explosive is a chemical compound that undergoes rapid chemical change and is converted into gas, on being heated or struck, having much greater volume producing high pressure. The compact explosive is transformed into glowing gas with an enormous pressure which, in a densely packed hole, can be equivalent to 100,000 atmospheres on the wall of the drill hole. The amount of energy generated per unit of time in a tiny hole drilled by a hand-held jackhammer exceeds the energy generated by the world's largest power stations. The temperature during detonation process is thousands of degrees centigrade. This is not because of the enormous energy latent in the explosive, but due to the rapidity of the reaction (2000–6000 m/s). The significant characteristic of the explosive is its ability to provide power in a limited part adjacent to it.

The area adjacent to the drill hole subjected to high pressure is shattered and the space beyond is exposed to vast tangential stresses and strains under the influence of the outgoing shock wave that travels in the rock at a velocity of 3000–5000 m/s. The radial cracks originating from the centre of the hole as a result of tangential stresses extend considerably further from a decimetre to about a metre. The first cracks are formed in fractions of a millisecond.

There is hardly any breakage during the first stage of cracking. Only, pressure of the gases widens the radial cracks. The drill hole is slightly widened to less than double the diameter by crushing and plastic deformation.

In rock blasting, a drill hole has a free face in its front and parallel to it. The compressive waves in the first stage reflect against the free surfaces and results in tensile stresses there. This development causes scabbing of a part of the rock near the surface. However, this scabbing is of secondary importance in blasting work. In hard granite, scabbing would require considerably higher charge of explosives. In loose rocks, the free surface would wear out in a millisecond after the detonation.

In the third and final stage of breakage, the primary radial cracks expand under the influence of the pressure of the gases from the explosion. The free rock surface in front of the drill hole yields and is moved forward. As a result, the pressure is released and the tension increases in the primary cracks. Depending on the burden, several of the cracks expand to the exposed surface, and breaking up of the rock is completed.

5.3 BLASTING THEORY

Blasting is the result of conversion of a chemical substance into a gas that produces instant devastating pressure shattering the rock adjacent to the explosive. The equations given below on rock blasting are based on the work of U. Langefors and B. Kihlstrom of Sweden.

A rock mass may be blasted and taken out in one layer, or in a series of benches. Contractors generally prefer benches. The height taken out in each layer is the face height. The explosive in each hole is supposed to break out a section of the rock mass, referred to as the burden, between the line of holes and the face.

On detonation of an explosive charge placed on top of homogeneous rock, a part below the charge and surrounding rock is crushed forming a crater. If the charge is

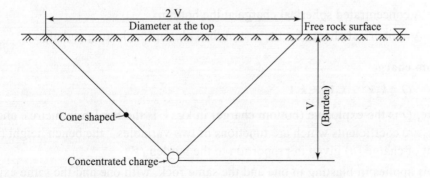

Fig. 5.1 Blasting theory

enlarged, the crater would also be enlarged in all directions in the same proportion as the linear extension of the charge.

Blastability is one way of expressing the measure of the resistance of the blasting with respect to maximum burden, depth of the hole, height of charge, and diameter of the hole. Burden is the horizontal distance from a face back to the first row of drill holes. Face is the approximately vertical surface extending upward from the floor of a pit to the level at which drilling is being done.

A charge placed at a depth V (burden) below a free surface is supposed to throw out crushed materials forming a conical crater with 45° side slopes. For this to happen, the energy input would be proportional to the mass comprising the cone, that is, V^3. To counter the gravitational forces, the energy input required to lift the mass to the surface level is proportional to the mass and the distance moved, that is, V^4. The energy input required to counter frictional resistance between the crater surface and the surrounding rock mass is proportional to the crater surface, that is, V^2.

Thus, the total charge required is represented as follows:

$$Q = k_2 V^2 + k_3 V^3 + k_4 V^4 \tag{a}$$

The coefficients k_2 and k_3 depend on the elasto-plastic properties of the rock, and k_4 on the weight of the rock to be excavated. The above formula is fundamental in rock mechanics and has been found dependable for the burden $V = 0.01$ m to 10 m with charges varying in proportion.

The values of coefficients can be determined by test blasts. The coefficients are functions of two variables, bench height (K) and height of the charge (h) given in proportion to the burden (V).

Benching

The above equation is appropriate for spherical charge and for a section of cylindrical charge of length V, where V is much smaller than the full length of the charge (h). Thus to tear apart the burden in the case of benching, the charge in the drill hole may be deemed to comprise two partial charges:

- A concentrated spherical charge at the bottom
- A cylindrical charge for the rest of the column (neglecting end effects)

Bottom charge

$$Q = k_2 V^2 + k_3 V^3 + k_4 V^4$$

Where, Q is the explosive (bottom charge) in kg, V is the burden in metres, and k_2, k_3 and k_4 are coefficients which are functions of two variables—the bench height (K) and height of charge (h) given in proportion to the burden (V).

This applies in blasting in one and the same rock, with one and the same explosive with the same density and rate of detonation for different cases comparable.

The charge is concentrated at the bottom and stemmed with dry sand to avoid loss of energy from exhaust. With fixed bottom and $K/V = 1$, the above formula may be represented as follows for burdens of different sizes:

$$Q_0 = a_2 V^2 + a_3 V^3 + a_4 V^4 \qquad (b)$$

The coefficients a_i are special cases of k_i (K/V, h/V) for $K/V = 1$ and $h/V \approx 0$, we get

$$a_i = k_i (1, 0)$$

Values of a_2 and a_3 can be determined by test blasts with the burden V within 0.5 m to 1.0 m. Value of a_4 may be ignored as it is less than 1% in the tests.

Column charge

In bench blasting, the bench height may be too much relative to the burden. To obtain a conformal case, a part of the column with a length equal to the burden may be considered. The size of the charge can then be calculated from the relation:

$Q_1 = b_2 V^2 + b_3 V^3 + b_4 V^4$. (Here, b represents another special case of k_i for cylindrical charge.)

The charge per metre (Q_1/bench height) = q

$$q = b_2 V + b_3 V^2 + b_4 V^3 \qquad (c)$$

Total charge

The total charge is, therefore, summation of the concentrated bottom charge plus the column charge. The smallest charge that can cause full breakage is a function of bench height K and burden V is:

$Q_t (K, V) = Q_0 + q (K - V)$ (this holds good for a concentrated bottom charge and a column charge for which no losses have been considered).

$$= a_2 V^2 + a_3 V^3 + a_4 V^4 + (b_2 V + b_3 V^2 + b_4 V^3)(K - V) \qquad (d)$$

The coefficients a_2, a_3, a_4, b_2, b_3, b_4 depend on the rock to be blasted and must be determined by test blasting. Analysis of test results have shown some relation between a_i and b_i like $b_2 = 0.4 a_2$ and $b_3 = 0.4 a_3$. These relations are applicable independent of the rock type. Existence of crevice etc may, however, influence the result.

The total quantity of charge in a hole will be:

$$Q_T = 0.4 a_2 (K/V + 1.5) V^2 + 0.4 a_3 (K/V + 1.5) V^3 + a_4 V^4 + b_4 (K/V - 1) V^4 \qquad (e)$$

In practice, the full effect of concentrated bottom charge cannot be realized as the bottom charge cannot be concentrated and has to be distributed along the drill hole uniformly. The charge effect in this type of loading is only 60% of the theoretical value of a concentrated charge. This 60% effect can be enhanced to 90% if the drill hole is extended to 0.3V below the base, thereby loading the bottom charge to a height of 1.3V. Distribution of the bottom charge in this manner is equivalent to Q_T as worked out above. Thus, the required uniformly distributed bottom charge, Q_b, can be calculated from the equation for Q_T for a bench height of $K = 2V$.

$$Q_b = 1.4a_2 V^2 + 1.4 a_3 V^3 + a_4 V^4 + b_4 V^4 \qquad \text{(f)}$$

As the bottom charge is adequate for breakage of rock up to bench height equivalent to twice the burden, the charge that is required in addition to this when the bench height exceeds $2V$ is defined as column charge. The size of the column charge, $Q_C = Q_T - Q_b$.

Diameter of drill hole

Diameter of the drill hole means the diameter of the drill bit as it is not possible to measure the actual diameter of each drill hole separately. The diameter of drill hole has hardly any effect on the amount of charge that is required for rock breakage. It is the total quantity of charge especially at the bottom, irrespective of larger or smaller diameter of a hole, which determines the breakage. The breaking force is obtained with the required amount of charge if the height of the charge does not exceed $0.3V$ from the bottom level.

If the charge concentration per metre remains unchanged, even with a moderate variation in the diameter of hole, the breaking force would remain unaltered. This would be true as long as the density of charge is maintained at about 1.0 kg/dm^3 (≈ 2.0 kg/m in a 50 mm hole). How large a burden can be loosened is determined by the size of the charge per metre at the bottom of a hole.

Spacing of holes

Generally, the spacing of drill holes is maintained as equal in size as the burden. For a given drill hole, the product of burden (V) and spacing (E) is constant. The spacing has great influence on the fragmentation and the nature of the remaining rock face. When $V = E$ is maintained, the results of individual hole blasting show a degree of overlapping and field tests have indicated that drill hole charge may be reduced by about 20%.

Throw

Limit charge means the charge that does not result in throw. A charge in excess of the limit charge provides the excess energy for throwing. The extent of throwing is directly proportional to the excess energy available for the purpose. The throw that affects the main part of the rock increases with an increased charge. Stones and rock parts are thrown 5–10 times as far or still further in unfavourable conditions. This phenomenon is called *scattering*. Both throw and scattering can now be fully controlled by controlling the variables.

Fragmentation

Fragmentation has bearing upon the sizes of rock pieces obtained on blasting. Fragmentation is often described as the average sizes of the blasted pieces, and sometimes also described as the largest pieces resulting from blasting. The sizes of the broken pieces resulting from blasting are important especially regarding handling and disposal/utilization. The pre-requisite of good fragmentation of rock is the uniform distribution of the explosive in the rock mass to be blasted. The required bottom charge per m^3 in case a single hole being blasted is Q/V^3 where Q is the quantity of explosive in a hole and V is the burden. For a row of holes with spacing of $1.25V$, the required bottom charge would be $Q/1.25V^3$. The volume of the rock blasted by a single hole with the bench height (K) equal to the burden is approximately V^3. The bulk of the rock resulting from blasting a 0.5 m burden with a limit charge may result in a single boulder. With increasing dimensions, the increasing volume to remain together as a single boulder becomes more difficult. If the burden considered is considerably larger than the spacing of holes, then the fragments would be larger.

Both the following factors influence fragmentation:
- Specific charge
- Specific drilling

A blasthole may be divided into three parts:
- Bottom charge
- Column charge
- Uncharged depth of blasthole

The bottom charge portion usually has high specific charge resulting in satisfactory fragmentation. In the column charge, crushing is relatively less because of the charge concentration being lower with constriction. The result of uncharged depth of blasthole is not favourable. If the uncharged depth is considerable, chance of formation of boulders would be more. Apart from the bottom, column or uncharged portions of blastholes, the quality of rocks with faults and cracks would also influence fragmentation.

Precision of blasthole drilling also influences fragmentation. If the blastholes are irregular or faulty, then fragmentation would be considerably poor.

The delay firing has a great effect on the entire blasting procedure and, for this reason, it is very important from the view point of fragmentation. The two essential conditions to provide the best possible fragmentation:
- Accurate planning of blasting operation
- Accuracy in the executed work

5.4 DRILLABILITY OF ROCKS

Igneous rocks, solidified out of molten state, are subdivided into volcanic cooled at the surface, and plutonic hardened deep underground. Sedimentary rocks are formed of soils or plants or animal remains and have been hardened by pressure, time, and depositing of natural cementing materials. Metamorphic rocks were originally igneous or sedimentary rocks, but have been altered by extreme heat and pressure.

The resistance that must be overcome to excavate a formation would be made up of largely hardness, coarseness, friction, adhesion, cohesion, and weight. Hard igneous rocks like granites, basalts, and diorite cannot be ripped because they do not have stratification and cleavage planes that are necessary for ripping. Sedimentary rocks like sandstone, limestone, shale, caliche, and conglomerate rocks are most easily ripped solid materials. Of metamorphic rocks, although slate and thinly bedded schist slate can be ripped, massive rocks like gneiss, quartzite and marble are not rippable. Rocks, which cannot be ripped, need to be blasted.

The strength of rocks vary with their type and location. Rocks that are relatively easy to excavate in the open using rippers are difficult to excavate in confined spaces. Weak rocks like silts and mud stones can be excavated with standard construction equipment. Hard rocks like slates and sandstones can be ripped or smashed with a pneumatic breaker. Excavation of very hard rocks like limestone and granite is to be carried out by blasting.

Drilling of blastholes is the prerequisite of blasting. Drilling rate is the number of metre of hole drilled per hour per drill. The drilling efficiency is specified by the drilling rate that depends on:

- Resistance of rock to breaking by drilling action
- Kind and shape of the drilling tool
- Method of drilling—percussion, rotary, rotary-percussion, etc
- Force and rate with which the drill bit acts upon the blasthole bottom
- Hole diameter and, occasionally, its depth
- How quickly the cuttings are removed
- Organization and mobilization of resources

Selection of drills is important, and the drilling efficiency depends on proper selection. Drillability of rocks depends mainly on the properties of rocks, and the difficulties encountered during the drilling process. Blasting in hard rocks is more critical than with rocks comprising soft materials. More energy is needed to break and displace harder and denser materials. However, if the material is overshot, it can result in fly rock and air blast; and if undershot, it can be very difficult to excavate for disposal. Yielding of rocks to fragmentation at the bottom of a drilled hole is dependent on the forces developed during drilling, and on masses of rock that exist above the bottom of the drilled hole.

On the basis of drillability, rocks are classified as:

- Easy—easily blasted
- Medium—blasted with least difficulty
- Difficult—blasted with average difficulty
- More difficult—blasted with great difficulty
- Extremely difficult—hard to blast rocks

This classification is done on the basis of:
- Compressive force developed during percussion drilling, and shear force developed during rotary drilling
- Jointing in rock mass
- Density of rock

5.5 KINDS OF DRILLING

Holes are drilled for various purposes, such as:
- Subsurface exploration/investigation—core sampling for geological investigation
- Rock blasting using explosives—general construction, tunneling, quarrying, and mining
- Grout injection
- Rock bolting and anchoring

Extensive drilling is required in rock excavation by blasting. Considerable drilling is done for soil investigation and resource (fuels, minerals, etc.) exploration. Grout injection is specialized work, and drilling for such work is not done as extensively as is done in case of investigation/exploration and rock blasting. Drilling for rock bolting and anchoring is done as and when required.

Blasthole or borehole is a hole drilled to receive explosives. Drilling pattern is the spacing of the drill holes. There are endless combination of rock types and structures in different construction projects that would require drilling of holes for excavation by blasting. Drilling is mainly executed by mechanical means. Manual drilling tools may be suitable for limited depths and small volume of work. Of the non-mechanical techniques, only the thermal method involving intense heat concentrated on a confined part of rock has become of any practical importance. The various drilling equipment are grouped into the following categories:
- Percussion
- Rotary
- Abrasion—a drill operation for grinding rock into small particles through the abrasive effect of a rotating bit
- Rotary-percussion

Drilling involves:
- Rock crushing by the bit
- Cuttings' removal

Percussion drilling

The bit is the most vital part of a drill that must engage and disintegrate the rock by crushing with hammer blows. The smashing of the rock is usually done by four, but occasionally two or six, ridges that are radial to a centre blasthole, with their faces at right angles to the centreline of the drill steel. The ridges may be replaced by round buttons of carbide set in almost continuous, slightly convex surface. The success of a drilling operation depends on the ability of the bit to remain sharp under the impact of the drill. Drills of many types of various hardness, shapes, and sizes are available. The bits that are in use now are screwed to the drill steel. Detachable bits have many advantages compared to forged bits that were in use earlier —they may be replaced and re-sharpened. They are relatively inexpensive also. Drill steels (big or deep steels are also called rods or pipes) are connectors between percussion drills and their bits. The number of air holes in percussion bits for removal of cuttings is variable. There may be a single hole for a small bit generally at the centre. In addition, there may be one to five holes in the recesses between the wings (projections).

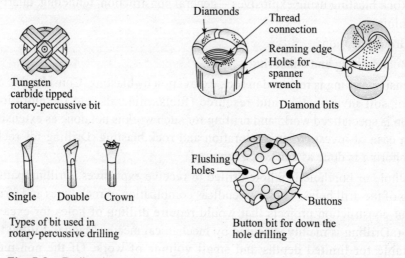

Fig. 5.2 Drilling bits

Regular bits are not suitable for drilling in some types of rocks that are so abrasive that steel bits need to be replaced after drilling only a few centimetre of hole. The increased cost of bits and the time lost in changing bits are so high that it would generally be cost-efficient to use carbide-insert bits consisting of a very hard metal, tungsten carbide or silicon carbide, which is embedded in steel. Despite these bits being more expensive than steel bits, the increased drilling rate and depth of hole obtained per bit would result in overall economy in hard rock. Depth per bit is the depth of hole that can be drilled by a bit before it is replaced.

Percussion drill is a drill that breaks the rock into small particles by the impact from repeated blows and turning of a cutting bit. The bit is usually screwed to the end of a hollow drill steel. The bit contacts the rock and disintegrates it. The bit is periodically pulled out of the hole, and then allowed to fall freely on the bottom. After each impact, the drill bit fall is adjusted to ensure uniform breaking of rock at the bottom. Cuttings are the disintegrated rock particles that are removed from a hole. The cuttings at the bottom are removed from the hole using water or compressed air, which also cools the bit. Dry drill is a drill which uses compressed air to remove the cuttings from a hole. Depending on the type of rock to be drilled, chisel-shaped or cross-shaped or hoof-shaped bits are used. Bits measure 25–250 mm in diameter, 0.8–1.8 m in length and weighs up to about 360 kg. The light ones up to about 15 kg are hand-operated. Jackhammer (Fig. 5.3), which is also known as sinker, is an air-operated percussion type (or rotary percussion type) drill that is small enough to be handled by a worker. Heavier ones are always mounted on tripods or frames or wheeled chassis.

A drifter is a cradle-mounted pneumatic drill, which is used for horizontal drilling. A stoper is an air-operated percussion-type drill, similar to a drifter, which is used for overhead drilling, as in a tunnel.

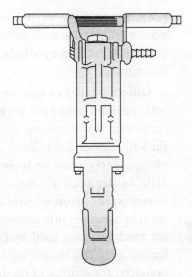

Fig. 5.3 Percussive drill—jackhammer

A churn drill is a percussion-type drill comprising a steel bit, attached to heavy rod that is mechanically lifted and dropped repeatedly to disintegrate the rock. It is used to drill deep holes, usually 150 mm in diameter or larger. Bits are available in diameters varying from approximately 150 to 300 mm. A churn drill may be used to drill rock having any degree of hardness. The bit must be re-sharpened at intervals in order to bring it back to the required gauge.

The bit reduces the rock to chips, sand, and dust. These materials, called cuttings, must be removed quickly to facilitate continuation of drilling. The cuttings may form a layer that would prevent the bit from striking the rock. The cuttings could be removed by forcing compressed air through the hollow drill steel, and the compressed air with cuttings emerging through the holes at the bottom or sides of drill bits. The compressed air must have sufficient momentum in the form of volume and velocity to force the cuttings out of the holes. Water is also used for flushing out the cuttings. Apart from the removal of cuttings, the fluid also cools the bit.

Rotary drilling

Rotary drills provide their drilling action by the application of a heavy down-pressure on a roller-cone bit, which cuts, chips, and/or grinds without hammering. High torque and rotation of the drill shaft cause cracking and chipping, and rocks are fragmented. As the bit rotates, its teeth also knock on the bottom of the hole. The rotary drill is supported on a mast above the drill hole. The derrick frame of the rotary drill provides storage space for additional drill pipes of varying length from 5–15 m for larger drilling operation. These additional pipes are added to extend the depth of borehole. The drill pipes are handled mechanically, and adding or removing a section can be done in a few minutes. Separate variable-speed motors, electric or hydraulic, create the pull-down

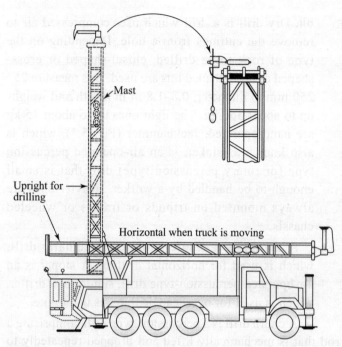

Fig. 5.4 Rotary drilling rig—truck-mounted

pressure and the drill rotation to power the drilling operation. A pneumatic motor may also be used for this purpose.

Different types of bits are used for soft, medium, and hard rocks. Large teeth with wide spacing are suitable for soft rock drilling. Teeth sizes are progressively reduced to be able to drill harder rock formation. Teeth comprise hard-faced steel or full carbide inserts—full carbide inserts are used for most hard and abrasive formations. The button type bits are preferred for drilling in harder rocks. The bits would need high force to be effective against the high compressive strength of rocks. For a 200 mm diameter blasthole, a feed force of 300 kN may be required. And, for a 500 mm blasthole, this force would be as much as 7500 kN. However, smaller bits of size 75 mm are available when feed force requirement would be less. For drills with smaller bit sizes, rigs for drilling would be of truck mounted type for versatility.

Cuttings are removed by flushing the drill hole with water or compressed air. The fluid also cools the bit.

Rotary drilling requires a drilling rig that comprises:

- Rotary drill with variable speed motors, that is, controllable rotation
- Hydraulic or electric power create pull-down thrust to the drill stem and bit plus rotation
- A mast to raise and support the drill stem above the drill hole
- Swivel loading system to position the drill stem insertions/additions
- Winch for raising/lowering the drill stem and bit
- A mechanical device to hold the stem and casing on line when drilling
- A carrier to accommodate the engine, drilling equipment, winches, compressor, and flushing pump

In construction work, medium-to-light rigs, mounted on trucks or wheels or tracks, are generally used. Rigs for drilling boreholes up to 200 mm diameter and 100–200 m depths need to be mounted on larger vehicles/bases. A truck-mounted rig is preferred at construction sites for its mobility and versatility. A rotary drilling rig can be easily converted for drilling in different kinds of soils and rocks of different hardness just by changing the bit or by changing the method of drilling. For example, the method of drilling can be changed from rotary to rotary-percussion drilling by changing the bit accordingly.

Blasthole drill is a rotary drill comprising a drill pipe at the bottom of which is a roller bit that disintegrates the rock as it rotates over it. A blasthole drill is a self-propelled drill that is mounted on a truck or on crawler tracks. Drilling is accomplished with a tri-cone roller-type bit attached to the lower end of a drill pipe. As the bit is rotated in the hole, a continuous blast of compressed air is forced down the pipe and the bit to remove the rock cuttings and cool the bit. Rigs are available to drill holes of different diameters and to depths up to about 100 m. This drill is suitable for drilling soft to medium rock, such as hard dolomite and limestone, but is not suitable for harder igneous rocks.

The rig is equipped with levelling jacks. The standard mast is 10 m to 12 m in length. It is powered by a diesel engine or electric motor. It is also equipped with a dust collector to remove the disintegrated rock as it comes from the hole.

Abrasion drilling

Abrasion drilling is a diamond drill rotary operation whose bit comprises a metal matrix in which a large number of diamonds are embedded. The selection of the size of diamonds depends on the nature of the formation to be drilled. The drilling rig comprises a diamond bit, a core barrel, a jointed driving tube, and a rotary head to supply the driving torque. Very high feed force is required for drill bit rotation. Compressive stresses generated by the abrasive effect of a rotating bit break the rock materials into small fragments that are continuously removed by flushing fluid to prevent obstruction, and also to cool the bit. This drill is extensively used to obtain core samples for exploratory purposes when rock samples need to be collected for physical examination and testing. This kind of drilling is not economical for blasthole drilling because of high cost of the abrasive bit.

Shot drill depends on the abrasive effect of hard chilled steel shot for rock penetration. The equipment comprises a shot bit, core barrel, sludge barrel, drill rod, water pump, and power-driven rotation unit. The bit comprises a steel pipe section with a serrated lower end. As the bit is rotated, shot are discharged to the lower end through the drill rod. These shot, under the pressure of the bit, erode the rock to form a kerf around the core. Water forces the rock cuttings up around the outer drill into a sludge barrel for removal when the whole unit is taken out of the hole. The core is broken off and removed from the hole from time to time so that drilling may be continued. Shot drill is engaged for exploratory purposes.

Rotary-percussion drilling

Rotary percussion drilling is used for blastholes, rock anchors, grouting holes, and wells, as rotary percussion drills are the most versatile of all rock drills. The rig is light and can operate economically over a wide range of hardness or abrasiveness. High feed force is required in rotary drilling. But in this method of rotary percussion drilling, the drill bit is supplied with both percussive and rotary action without resorting to high feed force. A reciprocating hammer (piston) strikes a rotating drill bit or a string of rotating drill rods with a bit at its bottom to form a crater. The rotating motion, however, affects this crater. Rock cuttings are removed by flushing the drill hole with water or compressed air. The fluid also cools the bit. There are two basic methods of rotary-percussion drilling:

- Surface drilling
- Down-the-hole drilling

In surface percussion drilling, percussion hammer is located outside the hole it is drilling and is operated by pneumatic/hydraulic power and connected to the bit by drill steel. Surface drills have the advantage of being light in weight and highly maneuverable. A jackhammer is an example of smaller-type surface drills that can be operated by one person.

Jackhammers are hand-held, air-operated, rotary-percussion-type drills suitable for shallow and small diameter holes of diameter up to 30 mm and depths up to 8 m although mostly used for depths up to 3 m. They are similar in appearance to concrete breakers except for the rotary action. They are also referred to as *sinkers* because they are mainly used for drilling downward. They are classified according to their weight varying from 10–25 kg. A complete drilling unit comprises:

- A hammer
- Drill steel
- Bit—this is the portion of a drill that contacts the rock and disintegrates it

Jackhammers are equipped with springs for eliminating almost 80% of vibration. As compressed air flows through the hammer, it causes the piston to reciprocate at a speed up to 2200 blows per minute causing hammer effect. The energy of this piston is transmitted to a bit through the drill steel. The drill steel is rotated slightly following each blow so that the points of the bit would not keep on striking at the same spots.

Drifter drills are similar to jackhammers in operation, but they are larger requiring mechanical mounting although small in comparison with rotary drills. They are used as mounted tools for drilling up, down, horizontal, and angle holes. In operation, the striking bar transmits the blow from the hammer to the drill rod. As holes are drilled deeper, some of the energy is dissipated in the drill steel and couplings, and the air pressure for removing drill cuttings also drops, thus slowing down the operation. These factors limit the economical depth of a hole. The measure to counter this disadvantage is to have a separate pneumatic motor to provide the turning effect resulting in higher blow rates. About 1500 blows per minute at 100–150 rpm are required to achieve acceptable penetration rates.

Wagon drills comprise drifters mounted on masts that are mounted on three-wheel chassis that can be towed or pushed from place to place. They are exclusively used to drill holes of about 50–125 mm diameter and 3–15 m depth. They give better performance than jackhammers when used on terrain where it is possible for them to operate generally on level ground. The length of drill steel may be 2–5 m or more, depending on the length of the feed of the particular wagon drill.

Piston drill mounted on crawler tracks acts in the same way as a wagon drill except for its drill rod, which is a hollow tube with an outside diameter of about 100 mm. It is attached to a piston and reciprocates with it. The height of the mast is about 15 m. The drill strikes approximately 200 blows per minute. The stroke and rotation of the piston can be adjusted to ensure best performance for the type of rock involved. The diameter of the detachable carbide-insert bit ranges from 125–150 mm. If the bit is re-sharpened at the end of the shift, it is possible to operate with a single bit. The rock cuttings are removed using compressed air, which also cools the bit. One air compressor is mounted on the drill rig, which is self-propelled.

A down-the-hole pneumatic drill operates at the bottom of a rotating drill string, directly above the bit without any steel or rod in between. The outside diameter of the drill is smaller than the hole, and thus enables the drill to follow the bit down the hole. Since the drill string is above the drill, the metre-wise life of the drill rods is very good, and the energy loss between drill and bit is held at a constant minimum.

As the borehole depth increases, the efficiency of drifter starts reducing gradually. Impact energy is dissipated as heat at each drill steel/rod connection. The energy loss is worked out by practical tests as 10% of the energy in the steel/rod just before the joint. There would be loss of impact energy into the drill stem itself. The rate of penetration would decrease with increase of borehole depth.

Down-the-hole (DTH) drilling is evolved to overcome the problem of energy loss. The pneumatic or hydraulic motor remains on the surface powering the rotation of the drill rod. The piston of the pneumatic hammer follows and directly delivers blows on the bit. Compressed air is exhausted through the bit and carries cuttings to the surface. The pneumatic pressure should be such as to be able to remove cuttings from the required depth.

A possible disadvantage of down-the-hole drilling is of losing a whole drill on account of rock fall or formation of mud collars. Because of this, down-the-hole drilling should not be carried out in badly fractured formation or weak strata.

5.6 SELECTION OF THE DRILLING METHOD AND EQUIPMENT

There are many factors which contribute to the selection of the drilling method and the equipment required for drilling holes for blasting. The selection would be based on efficiency and economy within practical limits. The most satisfactory method would produce the blastholes at cost-efficient rates for the particular project. The following factors are relevant and important:

- Project size
- Nature of the terrain
- Rock hardness

- Integrity of the formation - broken or fractured
- Blastholes—required depths
- Extent of breakage—for handling and crushing
- Water availability—if dry drilling is required

For small-diameter holes up to 125 mm maximum for blasting purposes, the selection is to be made between:

- Jackhammers
- Wagon drills

If the nature of the terrain is rough, jackhammers become the obvious choice irrespective of the cost of drilling and longer execution time. If the terrain is such that wagon drills could be used and if the job is substantially large, then wagon drills would be more cost-effective than jackhammers. Using wagon drills, it is possible to drill larger and deeper holes more rapidly than jackhammers. The use of larger holes would allow blastholes to be drilled at greater spacing.

For medium diameter blastholes ranging from 125–175 mm, the choice would be between:

- Down-the-hole drill
- Blasthole drill
- Churn drill

For holes up to 20 m deep in soft to medium rock, any one of the three types of drills could be used. For depths greater than 20 m, down-the-hole drill is not a practical proposition as it involves piston powered rotary percussion down the hole. For rocks harder than limestone and dolomite, the blasthole drill would not perform satisfactorily. Thus for drilling medium-size holes, in excess of 20 m depth, in harder rocks, the churn drill would probably be the most satisfactory equipment. The drilling rate-wise, blasthole would be the fastest, followed by Quarry-master at the intermediate stage, and churn drill would be relatively slower than the two.

If cores up to 75 mm outside diameter are required for investigation, the diamond coring drill would be the equipment suitable for the purpose.

5.7 EXPLOSIVES

An explosive is a chemical compound or mixture of compounds that can decompose instantaneously and violently when initiated by energy in the form of heat, impact, or friction or by another detonation in difficult conditions like in densely packed holes, under water and so on. The effect of an explosion in breaking up rock and hard soils for excavation work, tunneling, quarrying, and demolition work is related to the nature of the explosives used. The explosive detonation produces shock energy and a very large volume of hot gases that generate devastating pressure shattering the rock adjacent to the explosive in the blasthole. Both shock and gases cause the rock to be fragmented and displaced. For this comprehensive process, the explosive must contain all the necessary ingredients, and the process needs to be well-controlled. Oxygen from the air, for example, would not be available for the detonation process.

Oxygen balance is an important factor. In case of oxygen deficiency, carbon monoxide would be formed. On the other hand, oxides of nitrogen would be formed if oxygen is in surplus. To avoid formation of these gases in large quantities, it would be necessary to use well-mixed oxygen-balanced explosives that should be well-made and well-packed. Possible volumes of such gas formation may be checked from the ingredients in the explosives.

The principle ingredients in explosives are fuels and oxidizers. Common fuels are fuel oils, mineral oil, wax, carbon, and aluminium. Common oxidizers are ammonium nitrate, calcium nitrate, and sodium nitrate. Additional ingredients are water, gums, thickeners, and emulsifiers. Nitroglycerin, TNT, and PTEN are molecular explosives and combine the fuel and oxidizer in the same compound, and are used for sensitizing and energy enhancement.

Properties to be considered in selecting an explosive include:

- Sensitivity and sensitiveness
- Density
- Strength
- Velocity of detonation
- Water resistance
- Detonation pressure
- Blasthole pressure
- Fumes
- Store life
- Safety in handling

Sensitivity and sensitiveness

The blasting procedure depends to a great extent on satisfactory initiation, and sensitivity is a measure of the explosives' ease to initiation. Generally, an explosive is initiated by using detonators. However, there are explosives that are so sluggish as to require more powerful initiation. ANFO and slurry explosives are initiated by means of primers. The primer used comprises an explosive with high velocity of detonation. It is also a safety measure—more sensitive means more risk on handling.

Detonation stability means that the detonation of explosive does not stop for any reason until detonation of the entire row of explosive is completed. Sensitiveness is the ability of a product to propagate detonation along its column length once it has been initiated. If it is very sensitive, it may propagate substantially from hole to hole.

Flashover tendency or *sympathetic detonation* must be taken into account in blasting. It means initiation of an explosive charge without a priming device by detonation of another charge in the neighbourhood. In case of high flashover tendency, an explosive could cause flashover between adjacent blastholes if the holes are closely spaced as well as if the rocks are weak with faults and in moist condition. In case of low flashover tendency, there could be possibility of interruptions in detonation if the line-up of explosives in the charged blasthole is not continuous or something has come up between the various units. Flashover tendency decreases considerably at low temperatures.

Density

The charge concentration of a drill hole is determined by the density of an explosive. The density is expressed as the weight per unit volume, normally gm/cm^3. This value helps in determining the weight of explosive in kilogram loaded per metre of the blasthole. The density of most commercial explosives ranges between 0.8–1.7 gm/cm^3. High energy concentration per metre of drill hole is possible with heavy plastic explosives and slurry explosives. Explosive with lower density can be advantageous where it is desirable to spread a column charge. Generally, the higher the density of the explosive, the greater would be the RBS (vide next paragraph) value of that explosive.

Strength

Strength indicates the energy content of an explosive with respect to its weight—that is, its capability of doing useful work. Maximum explosive power can generally be obtained from any particular blasthole by using a high-density, high-strength explosive in it. A rating system has been developed for most commercial explosives, which compares the strength of a mixture of the explosive against ammonium nitrate and fuel oil (ANFO). These are the relative bulk strength (RBS) and the relative weight strength (RWS) values quoted by the manufacturers. The RBS is the measure of energy available per volume of explosive as compared to an equal volume of ANFO. The RWS is the measure of energy available per weight of explosive as compared to an equal weight of ANFO. Since a blasthole represents a volume, the RBS is the more widely used value. If comparison is not made, absolute weight strength (AWS) would indicate the energy content in calories per gram of the explosive. And, absolute bulk strength (ABS) would indicate the energy content in calories per cm^3 of the explosive.

Velocity of Detonation

The velocity of detonation (VOD) is the speed at which the detonation wave moves through a column of explosive. It varies from 300–900 m/s for black powders to 7500 m/s for blasting gelatin. The higher the velocity, the greater would be the shock energy and shattering effect from the explosive. Slower velocity explosives tend to produce their gas energy over a longer period that results in greater heaving action. The composition, density, particle size, degree of mixing, confinement, and the primer, influence the VOD.

The velocity of detonation is important in such cases as secondary blasting or blasting of structural units. High velocity of detonation is essential for the striking effect of the detonation to be powerful so as to result in the necessary tensile stresses.

Water resistance

The ability of an explosive to detonate after exposure to water is of vital importance. Water resistance varies from explosive to explosive depending not only on the characteristics of any particular explosive, but also on how it is packed and wrapped. It is possible for manufacturers to provide water resistance in the explosives instead of providing in the wrapper. Some explosives are very water resistant because of their compositions (emulsions). ANFO is not water resistant by itself. However, ANFO is made water resistant by mixing emulsion with it. All gelled products are having water resistance.

In regular rock blasting work at construction sites, blastholes remain full of water quite often. The plastic explosives usually have high resistance to water especially when packed well. However, explosives in the form of powder in cartridges would remain, under guarantee, in water for an hour or less. The manufacturer should be consulted to determine the water resistance of a product.

Detonation pressure

Detonation pressure, expressed in kilobars, is developed in the reaction zone of an explosive. This pressure is an indicator of the shock energy that is produced by an explosive. The pressure varies from 5 to about 200 kilobars.

Detonation pressure is a function of charge density, VOD and particle velocity of the explosive material. The higher the detonation pressure, the more useful the explosive would be in functioning as a primer for another explosive. Blast design is carried out on the basis of detonation pressure to achieve good fragmentation.

Blasthole pressure

Blasthole pressure is different from the detonation pressure. It is exerted by the expanding gases produced on detonation of the explosive on the blasthole walls. The pressure on the walls would depend on the volume and temperature of the gases and the degree of confinement. Blasthole pressure can vary from 30–70 % of the detonation pressure depending on the type of explosive—in the range of 10–60 kilobars. Blasthole pressure contributes to fragmentation to an extent but largely contributes to the heavy movement of the rock. Low-velocity explosives tend to have higher blasthole pressures than high-velocity explosives.

Fumes

The gases that are produced during blasting explosions vary in toxicity and irritating qualities. Insufficient priming, insufficient charge diameter, water deterioration, wood spacers, paper blasthole plugs, and plastic blasthole lines may add to toxic gases. This could be hazardous in underground work, particularly if ventilation is insufficient. It is important that adequate time for dispersal of the fumes on the surface as well as underground is allowed during the planning of blasting. Explosives are rated by the manufacturers according to fume emission as excellent, good, fair and poor or class I,

II, III, or IV. The release of gases in open does not pose any problem, but gases in underground or tunneling work may cause a lot of problem unless proper ventilation is arranged.

Store life

When any explosive is kept in storage for a long time, it is necessary to ensure that it remains in such condition as to be effective at the time of blasting. Some plastic explosives undergo an ageing process as the air bubbles formed during manufacturing process disappear wholly or partly, thereby affecting flashover characteristics. The plastic explosives should not be stored in a place having high temperatures, which can soften the explosive substance and cause cartridges to be deformed. It is difficult to use deformed cartridges.

The powder-explosive in the form of cartridges are sensitive to moisture. In humid environment, the salts in the explosive would form deposits on the cartridge, which then harden. The mixed explosives in some cases can separate resulting in complete change of their characteristics. This does not happen in shop-made explosives because of precautionary measures taken in the shops.

Safety in handling

The important factor is that the personnel engaged in blasting work should be able to carry out their work without risks. Explosives are subjected to various tests, and the test results are approved by the country's designated authority to permit their handling and use. As for handling, the agencies carrying out the blasting work should follow the safety procedures to avert accidents (vide Chapter 15, Section 15.12).

There are many types of explosives available for blasting operation. They have different characteristics and properties. They may be broadly classified into:

- Low explosives
- High explosives
- Initiating explosives

Low explosives

A low explosive is one that produces pressure by slow burning—the release of energy occurs over a gap of time. Black powder (generally known as gunpowder) is the most commercially important member of this class. Black powder is composed of finely ground sodium nitrate, sulphur, and charcoal in the following proportions:

- Sodium nitrate   75%
- Sulphur   10%
- Charcoal   15%

The ingredients are combined in grains of various sizes. The grains are coated with graphite or similar materials to make the powder free-running. An expensive alternative to this is to use potassium nitrate (saltpetre) instead of the sodium compound.

Fine-grained powders explode faster than the coarse-grained ones. Blasting powder can be ignited by exposure to an open flame and must be closely confined to maximize the effect of the relatively low gas pressure produced. There is very little crushing of the rock near the centre of explosion as the velocity of detonation is slow (300–900 m/s). As the gas is slowly released from the reaction, there is bursting action causing displacement and fragmentation. It is seldom used for rock blasting.

An electric squib comprises a metal tube with a charge of deflagrating mixture like black powder, an electric firing element, and wires sealed into them. A squib is used to ignite blasting powder by passing electric current through a wire bridge within the mixture.

High explosives

A high explosive is one that reacts to detonation as an extremely rapid, almost instantaneous process. Explosion by rapid burning is called *deflagration*, and by almost instantaneous decomposition is called *detonation*. High explosives detonate.

A high explosive is not very sensitive to initiation by spark or flame. It would require initiating explosive for detonation. On detonation, the reaction takes place at up to 9000 m/s with accompanying devastating gas pressure resulting in surrounding rock being crushed, cracked, and shattered. High explosives are, therefore, suitable for use in excavation work, tunneling, quarrying, and demolition at the construction sites. The most common types of high explosives are:

- Nitroglycerin
- Ammonium nitrate plus fuel oil (ANFO)
- Slurries

Nitroglycerin

Nitroglycerin, $[C_3H_5(NO_3)_3]$, is yellow, heavy, oily liquid, which is highly sensitive to impact and temperature. It explodes when subjected to shock or detonated. It is produced by the action of nitric and sulphuric acids on glycerin. Pure nitroglycerin is oily in nature having freezing point of 13°C and specific gravity of 1.6. The sensitivity to shock increases appreciably when nitroglycerin freezes. Nitroglycerin is so sensitive that it must be mixed with other substances before it could be included in commercial explosives. However, all explosives containing nitroglycerin freeze when the temperature falls to 8°C or less. To avoid this hazard, a low freezing agent like di-nitro-glycol, which freezes at −22°C, is used. This agent is as powerful an explosive as nitroglycerin. Dynamite is a high explosive whose primary constituent is nitroglycerin. Nitroglycerin is mixed with *kieselguhr* (a mass of hybrid silica formed from skeletons of minute plants known as diatoms—very porous and absorbent material, used for filtering and absorbing various liquids in the manufacturing of dynamite), wood pulp, starch flour, etc. making nitroglycerin safe to handle until detonated. Dynamite is the most common Indian high explosive. Gelatinized dynamite is water-proof and used in and under water. It is a jelly-like explosive made by dissolving nitro-cotton in nitroglycerin. The approximate strength of dynamite is specified as a percentage, which is an indication of the ratio of the weight of nitroglycerin to the total weight of cartridge. Individual cartridges vary in size from approximately 25–200 mm in diameter and 200–600 mm in length. Dynamite is available in many grades and sizes to meet the requirements of diverse demands in different projects. Types of dynamites available are straight dynamite (straight means no ammonium nitrate), ammonia dynamite, blasting gelatin, straight gelatin, ammonia gelatin, and semi-gelatin. They differ in ingredients, sensitivity, water resistance, performance characteristics, and cost. Blastholes are charged with cartridges of dynamite, especially for the smaller size holes. As a cartridge is placed in a hole, it is tamped sharply with a wooden pole to expand so that it fills the hole. For this purpose, it may be desirable to split the sides of a cartridge, or cartridges with perforated shell be used.

ANFO

Ammonium nitrate (NH_4NO_3) explosives are much safer than dynamite and becoming increasingly popular for blasting purposes, especially for use in medium and large diameter dry blastholes. Ammonium nitrate, called AN in short, is so insensitive that it is not rated as an explosive. It is stable at normal temperatures. AN is hygroscopic and would absorb moisture, which contributes to caking and handling problems. AN decomposes explosively, as follows, when subjected to great heat under confinement or direct detonation of high explosives:

$$2\ NH_4NO_3 = 2N_2 + 4H_2O + O_2$$

The sensitivity and explosive power of AN can be enhanced by using additives. Of the additives tried such as lampblack, powdered coal, saw dust and fuel oil; fuel oil is found to be the ideal material to mix with AN to make a practical explosive called ANFO. AN is mostly prepared in prill form with low clay content, low moisture, free flow and sizing. The sensitivity and performance of ANFO depend on the quality of AN prill. 94.3% of AN and 5.7% of fuel oil provides maximum explosive power. The fuel content is very important. If there is too much of fuel oil, sensitivity and energy output deteriorate; and if there is too little, it results in toxic gases as well as poor energy output. Mixing may be done at the factory, or may be poured directly into the blasthole—pouring bagfuls of AN and then adding correct proportion of fuel oil. Mechanical mixing may be done with engine-powered concrete mixers with fuel-oil added by spraying. Factory mix is, of course, very thorough. ANFO is not suitable for wet blastholes as it is not water resistant. However, this problem can be overcome by using ANFO in gelatinized form. The gelatins are usually packaged in paper or cardboard cartridges for easier handling and loading.

Slurries

Slurry is a mixture of a sensitizer, an oxidizer, water, and thickener. The sensitizer may be any of a number of reducing agents (removes oxygen). This usually is TNT [trinitrotoluene—$C_6H_2(CH_3)(NO_2)_3$—an explosive], aluminium (finely divided—sand size or very fine), smokeless powder and carbonaceous fuels. Accordingly, slurries are identified as TNT slurry or aluminized slurry or smokeless powder slurry. Aluminized slurry is in common use. The oxidizer is an agent such as ammonium nitrate or sodium nitrate that brings about an oxidation reaction. The mixture is thickened with guar gum or jelling agent (syrup or jelly). The consistency of the mixture is regulated by the amount of thickener that is used.

Water resistance would depend on the flow of water. Flow should not be such as to wash away the slurry. Slurry should be sealed in cylindrical plastic packages where water flow condition is severe.

Slurries are less expensive than dynamite, but cost more than ANFO. The higher density slurry compared to ANFO could allow smaller diameter blastholes or wider spacing of holes to obtain same fragmentation. This would reduce cost of drilling, whereby the cost of using slurry would be comparable to ANFO. In wet holes, ANFO is not suitable. So, either dynamite or slurry is to be used.

Initiating explosives

Initiation can be categorized into four main groups:
- Fuse ignition
- Electric ignition
- Detonating fuse
- Cord detonators

A rational and reliable method of causing explosive to detonate is an important aspect of blasting technology and the basic requirement for safe and efficient blasting is to ensure that explosives are initiated effectively so as to instantaneously convert chemical energy into heat and mechanical energy resulting in propagation of violent shock waves. The chemical reaction in the explosive would ensure the release of the maximum amount of effective energy to achieve the violent blasting result. Despite design, drilling, and explosive selection being alright in the cases of poor blasts, the explosives do not release their full potential energy because of inadequate initiation.

What is required for initiation of explosives is powerful detonation, which is carried out using initiating devices. The initiators are either electric or non-electric depending on the source of energy required for initiation.

Safety fuse

Safety fuse comprises black blasting powder sheathed in narrow cord made of several plies of yarn, and insulated from water and other damages by double coatings of insulating materials. The cord is so designed as to burn at the rate of a metre in about 100 seconds. This rate of burning should be considered approximate as it is affected by weather, altitude, storage conditions, and damage to the fuse, if any. The fuse may be damaged by oil, kerosene, or petrol. A detonator is not required as black powder itself is an explosive compound.

The fuse is water-resistant except for the cut ends, which get spoiled in contact with moisture. The cut end should not be used if the same is not shot on the same day of cutting. Safety fuse can be lit by using a match or special fuse igniters having pyrotechnic heads. It is possible to achieve the effect of delay detonation by using different lengths of safety fuse.

Fuse ignition is not suitable for large scale rock blasting. It is used to fire a single charge, but has earlier been used largely for drifting work underground.

Blasting caps

It is a capsule of sensitive explosive to initiate a charge of high explosive for both fuse and electric firing.

Fuse/Non-electric—it comprises a small metal tube loaded with a charge of powerful initiating explosive of pentaerythritol tetranitrate $C(CH_2ONO_2)_4$ (PETN in abbreviation) at the bottom of the tube and a very sensitive primary charge of lead azide (heavy metal azides are explosives – azides are salts of hydrazoic acid) or mercuric fulminate (a white crystalline substance which explodes on being struck). Fuses are inserted into the hollow tube and fastened in by crimping the metals with special tools very carefully so that neither water nor moisture can penetrate into the primary charge/detonator. Crimping can be carried out with special pliers. The burning of the safety fuse converts the primary charge into detonation, and the high explosive base charge is initiated. The aluminium tube is 6 mm in diameter and 38–50 mm in length. Caps are manufactured in different strengths.

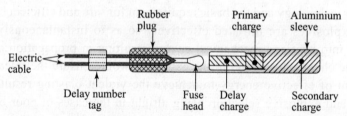

Fig. 5.5 Detonating (blasting) caps

With this kind of blasting caps, time delays are incorporated by varying the length of the safety fuse cord. The lighting of a large number of fuses from one single point can be carried out by using an igniter cord which is lit by using a match. Several fuses can be lit at the same time by using a circuit igniter which consists of a cardboard sleeve with the igniter charge in the bottom.

Electric detonators—An electrical initiation is provided by electrical power source with an associated circuit to transmit the impulse to the electric detonators. The primary charge and base charge are similar to the charges used in the non-electric system. Electrical current is passed through a bridge in the cap as a result of which the bridge is heated. The heat energy in the bridge eventually ignites the primary charge, which then detonates the secondary charge of high explosives. For analysis of an electric circuit used to fire blasting caps, it is necessary to know the resistance of the caps and the leading wires, which conduct the current to the caps.

Delay blasting

The reaction due to explosion by blasting is rapid. This reaction sets up stress wave that could be as high as 2000–6000 m/s. The rapidity is more in harder rock. The rock fracture due to stress radiates at 0.15–0.4 times the stress wave in the rock. This amounts to 300–2400 m/s or 0.3–2.4 m/millisecond.

When the explosive charges in two or more rows of holes parallel to a face are fired at the same time, the rock broken by the blasting of first row of holes that are nearest the face should move and clear out of the way before the second row of holes are detonated. This procedure would reduce the burden on the holes in the second row, and thereby permit the explosive in the second row to break the rock more effectively. Each row of holes should have free face for better throw and desired fragmentation. The delay firing sequence may be achieved by using delay electric blasting caps. The delay electric blasting cap is a cap that is designed to delay the detonation for a predetermined time after the electric current is passed through it. The period of delay may be several seconds. Delay intervals depend on the type of rock and burdens. MS connectors with 5–25 millisecond delays can effectively:

- Reduce cost of blasting
- Increase blasted rock, tonnage or volume wise, per metre of blasthole
- Make possible increase of blast size

In electrical firing, there should be parallel circuits so that circuit would be maintained after each blasting till the blasting operation is over. Electrical wires should be long enough to reach the wires from the charges with the lead wire to the electrical power source. The ends of the wires are connected into a bridge or shunt to prevent premature firing through contact with live electrical connection. The bridge or shunt diverts the power to the alternative path of least resistance.

It is possible to use instantaneous and delay caps in the same round. However, caps made by different manufacturers should not be fired together as their current requirement would be different from each other. If instantaneous and delay caps are fired in the same series, it would be better to increase current by one third.

In order to reduce overbreak, the charges should be loaded up to the top of the hole, or as far as can be permitted while having regard to the scattering and throw of small pieces of rock from the free face. Excellent results especially for the reduction of backbreak are reported with charges reaching up to the top and with covering material to prevent the scattering of stones. This is of importance when blasting close to buildings. Further, if the empty space in the holes around the charge is filled with dry sand, the flow of gas into cracks in the surrounding rock is prevented and the tearing in the cracks is reduced.

5.8 BLASTING PATTERNS AND FIRING SEQUENCE

Drilling pattern is the spacing of the drill holes. A typical blasting pattern comprises a number of blastholes drilled in one or more rows in such a way as to suit specific project requirement. The simplest pattern is a straight line of vertical blastholes parallel to the vertical face. Surface blasting may be of three types:

- Sinking cut—for new area or level without any face existing
- Box type cut—for extending an excavation lengthwise
- Corner cut

When a level site or quarry is to be opened up, there would be no vertical face to facilitate blasting. As a result, the first charges of explosives would, when detonated, throw the loosened rock materials upward. Obviously, this initial sinking cut would not be very effective. However, this would result in some kind of face for subsequent blasting. A proper face should be formed as fast as possible.

There are two standard types of blasthole patterns:

- Square—the blastholes are positioned directly behind each other
- Staggered—the blastholes are staggered by half the spacing from the row in front/ behind

The blastholes are drilled according to the drilling pattern (Fig. 5.6). But the true burden would depend not only on the drilling pattern dimensions, but also on the sequence of delay blasting. In the sketch below, blastholes are drilled in the square pattern with 3 m spacing and 3 m burden. However, the firing is carried out on the basis of staggered echelon sequence. The firing sequence changes the blasting direction. The true spacing considering the blasting direction becomes 4.24 m with a burden of 2.12 m. The same principle holds good in case of rectangular or staggered drilling setups.

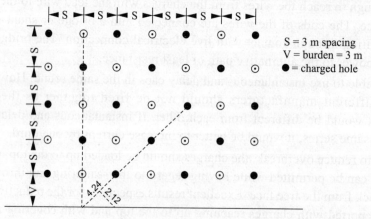

Fig. 5.6 Drilling pattern

Selection of blasthole drilling pattern as well as selection of sequence of delay blasting should be done thoughtfully keeping in view the result of blasting – good fragmentation, desired displacement, and easy disposal of the spoils (mucking).

Generally, square drilling pattern with delay echelon sequence results in spoils thrown near the bench for removal by a shovel. In case of staggered pattern setups, blasted materials are moved forward resulting in low, well-displaced spoils that can be picked up by loaders for disposal.

There are innumerable drilling patterns and delay firing sequences. The sketch below illustrates a square and a tunnel pattern delay timing sequences.

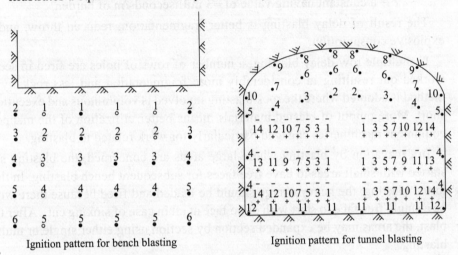

Ignition pattern for bench blasting   Ignition pattern for tunnel blasting

Fig. 5.7 Delay timing sequences

The following rules are pertinent in choosing drilling pattern and firing sequence:

- Drilling pattern should be such as to take advantage of the geological structure of rocks to ensure good results of blasting including clean break up lines at the edges
- Each blasthole should have the right burden at the time of firing
- Drilling pattern should be such as to allow adequate space for each blasthole to account for swell

After calculating the charge requirement, the blasthole pattern is designed. The method of firing is carried out in one of the following two ways:

- Instantaneous blasting
- Delay blasting

In instantaneous blasting, a number of blastholes are fired simultaneously. This method is suitable for blasting out a long bench of rock. Accuracy of drilling is not important in this method compared to the delay blasting. Minor variation along the direction of drilling and spacing of blastholes may be allowed.

In delay blasting, there would be:

- Single row delay blasting
- Multiple row delay blasting

In single row delay blasting, time interval may be calculated using the following equation:

$$t = cV$$

where t = time in milliseconds

V = burden in m

c = a constant having value of ≈ 3 milliseconds/m of burden

The result of delay blasting is better fragmentation, reduced throw, and reduced explosive consumption.

In multiple row delay blasting, a number of rows of holes are fired in sequence as worked out resulting in considerably more fragmentation and less rock flying. This method is adopted where the rock blasting involved is voluminous and execution time is short. More output of blasted materials means better utilization of the manpower and equipment/machineries engaged particularly for work related to blasting.

In excavation by blasting where large areas are concerned, the blasting should be started with small areas to have free faces for subsequent bench blasting. In the first or initial blasting, the heavy charge should be loaded and fired because there would only be a single free face to start with or no face at all in case of sinking cut. After the initial blast, the areas may be expanded section by section using either single or multiple row blasting.

5.9 SMOOTH BLASTING

Conventional blasting work normally results in rough and uneven contours of blasted surfaces. Such surfaces may require removal of blocks of stone or dressing up of roof surfaces. Explosives may penetrate far into the rock resulting in great overbreak. All these would require additional expenditure including cost of filling up cavities. Conventional blasting also destroys the quality of remaining and surrounding rock. Accurate controlled blasting techniques are important methods of achieving rock walls that are smoother and structurally stronger than those that are left by redesigned production blasts. These techniques are most effective in tough massive rocks and in tight, horizontally bedded formations which are relatively undisturbed by faults, joints, etc. Smooth blasting is very important to obtain a rock construction with good strength characteristics. In particular cases like underground blasting, it is of utmost importance for the surrounding rock to be free from cracks as cracking would result in rock losing either all or part of its self-supporting properties. The design of blasthole patterns and charge loads is dependent on the properties of rock encountered in smooth blasting area.

Due to the increased amount of drilling required, all the controlled blasting techniques are more costly than regular or redesigned production blasting. It is essential that drilling be carefully supervised so that blastholes are at the designed spacing, in the intended direction, and parallel to each other. Drilling accuracy becomes more important as the bench height increases.

Differentiation is made between two different types of blasting:
- Conventional smooth blasting
- Pre-splitting

Conventional Smooth Blasting

Conventional smooth blasting is carried out in practice with explosive of low charge concentration per metre and other characteristics that result in mild effect. A drill hole filled with explosive may result in heavy cracking of surrounding rock when detonated. This type of heavy cracking can be significantly suppressed when the same charge is detonated by increasing the diameter of drill hole. Because of larger diameter, the blasthole pressure on the walls would be reduced and only a few cracks would extend. To keep pressure on the remaining and surrounding rock to the minimum, it is necessary to use the correct charge of low concentration thus avoiding overcharging of the blasthole. At closely spaced holes along a line parallel to the free surface, detonation of low charge in a hole would result in a line of cracks along the close by empty holes without cracks in the other direction. The pressure at the walls of the blasthole is below the compressive strength of the rock and insufficient for developing tension cracks around the blasthole. The remaining and surrounding rock, thus, remains unaffected.

The quality of the remaining rock face is largely dependent on the relation between the spacing (E) of the drill holes and the burden (V). The relationship between the spacing and burden (E/V) should be ≤ 0.8 implying that the burden should not be too small compared to spacing.

Time-spread in detonating a row of blastholes for smooth blasting should be as little as possible between the rows of holes. Ignition can be done with instantaneous caps or detonating fuse for best result. The contours created by instantaneous and short delay blasting are better. Delay time of more than 0.1 second means that every charge has to work separately as in blasting shot by shot. This effect is to be counted on with most half-second delay detonators. Keeping the ignition time lag below 0.1 second is important for another point of view. Charges may be ejected by the detonation gases in adjacent holes in fissured rock.

Smooth blasting is required for the following blasting operations:
- Tunnel construction
- Construction of underground chambers
- Road cuttings
- Slopes for building foundations
- Pipeline ditches
- Blasting of stone blocks in quarries
- Cautious blasting

For smooth blasting to be effective, the blastholes should have practically free burden as otherwise a good part of charge would be used up to strike back into the rock. Consequently, smooth blasting result may not be good.

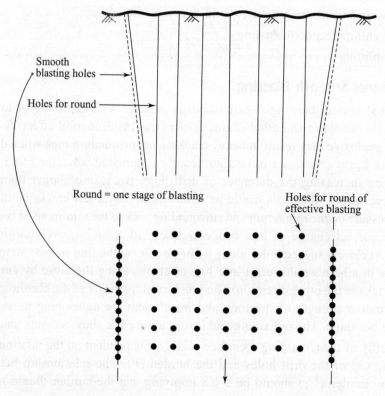

Fig. 5.8 Smooth blasting

Smooth blasting in cuttings is carried out by drilling holes at close spacing as shown in Fig. 5.8. In deep cuttings in road construction, the initial bottom charges could be heavy. However, excessive heavy charges should be avoided. Also, the holes close to the smooth blasting row should be charged carefully so as not to ruin smooth blasting operation. For road construction, pre-splitting would be the better alternative to achieve satisfactory results.

It is to be ensured that charges are loaded properly with sufficient stemming or plugged in the holes after loading so that charge of any blasthole is not blown out by the detonation of charge in the adjacent holes.

In excavation for foundation construction of buildings, conventional blasting can be carried out up to certain distance from the proposed excavation edges followed by smooth blasting towards the edges when there would be free burden in front of the holes.

In excavation by blasting in case of tunnel construction, drilling precision is very important for achieving good results. However, holes drilled in continuing tunnel excavation by blasting should be angled so that there would be space for drilling and manoeuvring in the next round of drilling. The extent of angling would depend on the types of rock and sizes of drilling equipment. Apart from drilling and loading of holes, initiation is also an important factor in tunneling. After blasting the production round of holes, smooth blasting holes should be detonated using milli-second detonators. Instantaneous detonation of large charges influences the surroundings and open up existing faults and cracks—which could increase the extent of work involved.

In the construction of underground chambers, large roof areas compared to tunnels would be involved. This would require smooth blasting. Large roof areas would also need some kind of reinforcement. Underground chamber, however, can be located in selected area after testing core samples obtained by borehole drilling. Excavation for underground chamber can be carried out by adopting a combination of bench and tunnel blasting.

It could be concluded in the final analysis that smooth blasting results show improvement in the rock contour as well as the remaining and surrounding rocks. Reinforcement may still be necessary to take care of faults and weak zones in spite of the fact that smooth blasting, by retaining curvature, can hold the rock just for a while. Advantages of smooth blasting are as follows:

- Less overbreak
- Less concrete needed for rendering the surfaces
- Smoother rock faces
- Stronger rock surfaces requiring less reinforcement
- Less cracking – less water penetration
- Close conformity with complicated profile in foundation work

Pre-splitting

It is another technique of smooth blasting for reducing excessive overbreak and ground vibrations especially in soft rock. Pre-splitting involves drilling a row of closely spaced, about 40 to 150 mm in diameter, full-depth blastholes along the design limit/periphery. These blastholes are very lightly charged and then detonated simultaneously sufficiently in advance of the rest of the holes for the main production blasting. Instantaneous electric ignition or ignition with detonating fuse with no or minimum of time lag can then be initiated thus creating an internal surface to which the production blast can then break. Once the crack is made, it insulates the surrounding rock to some extent from ground vibration during main production blasting. As the blastholes are closely spaced, crack formation follows the rows of blastholes. If the pre-split blastholes are spaced quite closely, uncharged holes between charged holes would improve the result if so designed considering rock characteristics and spacing of blastholes for such rock. However, if the blastholes are spaced too closely or charged too heavily, the pre-split charges would themselves create overbreak.

The spacing between blastholes in pre-splitting normally increases with the blasthole diameters. But because rock characteristics are very pertinent on blasthole spacing and explosive load, both blasthole spacing and explosive load should be worked out on the basis of rock characteristics. Drilling precision is very important in pre-splitting. Initiation has also bearing on the results of pre-splitting.

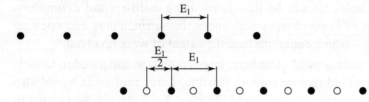

Fig.5.9 Pre-splitting

The blastholes are charged up to about 75% of the depth of the holes. This depth of charge should be reduced to about 55% should the rocks be very much cracked. If there is any evidence of fault found during drilling, it would be advisable to load the charge past such fault.

Pre-split holes are stemmed by pushing down a wad of paper or plugged to the top of the charge and then backfilling above with drill cuttings. The blastholes are connected together in groups using detonating cord, and the groups are initiated by detonators with corresponding delay numbers. The charge per group is designed keeping in view ground vibration as the charges explode simultaneously. The best pre-splitting result is achieved when pre-split charges are fired simultaneously. Where, however, ground vibrations are likely to cause environmental problem, millisecond delays should be utilized so as to obtain the consecutive firing of instantaneous groups of holes to minimize noise. Pre-split holes should be initiated sufficiently in advance of the first hole in the production blast to enable the pre-split fracture to develop to its fullest extent. The charges are initiated from the above.

The face obtained by pre-splitting would be affected or even ruined if production blastholes are drilled too close to the face. If the distance between the pre-split face and the last row of production blastholes is too much, rock is left in front of the pre-split face that could make disposal of spoils hazardous. It may even make it necessary to blast in the area again. The distance between the pre-split face and last row of production blastholes should be maintained at about 30–50 % of the normal burden for production blastholes for achieving satisfactory results.

Line drilling

Line drilling involves drilling closely spaced drill holes of diameter between 40–75 mm along the design limit/periphery (contour holes) to protect existing buildings located nearby. The spacing between these uncharged holes is maintained at 2 to 4 times the diameter of holes. These uncharged holes form a plane of weakness along which the production blasting round would break. The spacing between these uncharged holes and the nearest charged blastholes would be 50–75% of the spacing of rows of production blastholes. The charges in the blastholes adjacent to the uncharged holes should be limited so as not to damage the final contour. Line drilling is practically outdated as drilling empty holes is costly exercise.

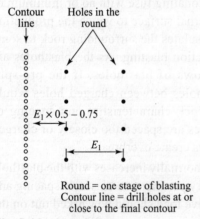

Round = one stage of blasting
Contour line = drill holes at or close to the final contour

Fig. 5.10 Line drilling

Cushion blasting

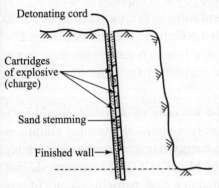

Fig. 5.11 Cushion blasting

Cushion blasting is similar to line drilling except that blastholes along the design limit are lightly charged with spaced cartridges or undersize cartridges. The spaced cartridges are attached to detonating cord. The space in a drill hole not occupied by cartridge/s is filled with *stemming* – usually sand. Time lag between ignition of cartridges in the blastholes should be minimum possible. There should be some charge at the bottom of the hole as is done in case of the conventional smooth blasting. The production blasting is carried out close to the cushion blasting line before or after the cushion blasting. The blasting in the area close to the cushion blasting line should be carried out with utmost care. Compared to line drilling, cushion blasting is more economical and accurate because of the smaller number of larger diameter holes are required.

5.10 ENVIRONMENTAL EFFECT OF BLASTING

Noise pollution is a serious problem at construction sites where drilling and blasting work is involved. Even compressors create noise problem when in operation. Percussion drilling is very noisy. Mufflers can be fitted on the direct exhausts of compressed air from the drill piston. By using water mixed with foaming agents for removal of cuttings, noise level of the exhaust air from the drill holes may be reduced. Nevertheless, noise pollution would remain. And, the people living in the vicinity of the construction sites would have reasons to complain.

Shallow blasts, overloaded holes, fractured rock, and other such deficiencies and shortcomings allow blasting to take place before expending total inherent energies causing noise pollution. Boulders and oversize fragments should be drilled before blasting. This action would reduce noise drastically. Brittle rocks may be broken by hammering, whereby blasting and associated noise may be avoided.

Blasting should be carried out in favourable weather. Air is more or less still from the ground up to high altitude in unfavourable weather conditions. This kind of situation may arise when the days are foggy, gloomy, hazy, or smoky showing no signs of air movement. Sound travels rather slowly and dissipates in thin air, but its travel is affected by unfavourable weather conditions.

Apart from noise, the other environmental issues related to drilling and blasting are:

- Dust
- Ground vibration
- Air blast
- Fly-rock

Dust in the cuttings becomes a major nuisance as the cuttings emerge from the drill holes. This dust would pollute the environment. This polluted air would be a health hazard for the personnel involved in drilling. The dust would also be harmful for engines and other machineries. This dust may be trapped in *mechanical dust collectors* by cyclonic separators. Alternatively, the cuttings may be dampened using water mixed with detergent. This process is widely used, but more power is required in removing wet cuttings. In place of detergent, special foaming agent along with diesel oil may be added with water for mixing with the cuttings against dust pollution. The bubbles formed by the foaming agent hold and buoy up rock particles of all sizes so that the particles do not tend to fall down. Consequently, compressed air consumed for removal of cuttings would be less.

Blasting would invariably cause vibration—the intensity of vibration would vary from annoying to damaging. Vibration to an extent is controllable. Ground vibrations are caused by poor blast design such as too much energy released at inadequate delay time between detonations, etc. The US Bureau of Mines (USBM) recommends that vibration levels due to blasting operations be limited to a peak particle velocity of less than 51 mm/s to avoid damage to nearby structures. The USBM's formula for working out the likely particle velocity resulting from an explosive charge:

$$V = k \frac{D}{\sqrt{E}} - B$$

where E = explosive charge (kg)
V = particle velocity (mm/s)
D = distance from shot to recording stations (m)
k and B are empirical coefficient, which can be determined from field trials with instrumentation investigation

If part of the explosion escapes to the air, the expanding gases push back the air causing a local increase in air pressure resulting in shock waves called air-blast. Concussion is air-blast that has a frequency of less than 20 Hz that cannot be heard. Frequencies above 20 Hz called sound waves that are audible. Concussion comprises one or more waves of highly compressed air moving outward from the explosion and is caused by:

- Unconfined explosives such as uncovered surface detonating cord
- Venting of explosive charges by inadequate stemming and/or inadequate burden
- Gas escaping through cracks or soft seams
- Movement of the bench face during a blast can act like a large piston and create concussion in front of the shot

Generally, if the intensity level is kept below 128 dB at any structure, the air-blast is unlikely to cause as much damage as ground vibrations even though people around the place would still complain. Temperature and wind can enhance the effect of air blast and can focus it. Windows can be damaged when the air-blast lies in the range of 0.05–0.14 kg/cm^2.

The leading cause of injury and property damage are mainly fragments of rocks flown through the air by blasting. It is generally attributable to shallow blasts, overloaded holes, shots in rock with irregular resistance, etc. Care in the selection of the burden, stemming height, and correct loading of explosives can reduce the chances of fly-rock causing damage. To compensate for unknown geologic factors, an adequate blast area safety zone should be cleared of personnel and equipment prior to blasting (vide Chapter 15, Section 15.12). Danger of damage by fly-rock may be reduced by increasing the number of boreholes so that smaller explosive charges may be made. Also, throw of fly-rocks can be controlled by covering blastholes with heavy materials. Thus, the areas to be covered should be much more than the borehole-areas.

SUMMARY

Unlike earthmoving, excavation by blasting is not a common feature at construction project sites in India. Nevertheless, almost all the states of India are covered with rocky strata here and there, and blasting cannot be avoided in such areas. Blasting is a costly and time-consuming process of construction work which is not an environmental friendly operation either. It is rather hazardous for people and adjoining localities. All these call for extra precautionary measures so as to ensure continuation of construction activities without causing interruption. It has been possible to photograph what happens in milliseconds on initiation of the detonation process. Such extraordinary vivid pictures help in planning precautionary measures. Sweden is a country where a number of people are engaged in research work on blasting. They are contributing significantly on improving blasting techniques. Blasting is the result of conversion of a compact explosive containing fuels and oxidizers into a glowing gas that produces instant devastating pressure shattering the rock adjacent to the explosive, as its significant characteristic is its ability to provide power in a limited part adjacent to it. The first cracks are formed in fractions of a millisecond. Drilling of blastholes is the prerequisite of blasting. Drillability of rocks depends mainly on the properties of rocks, and the difficulties encountered during the drilling process. Selection of drills is important, and the drilling efficiency depends on proper selection. Drilling is mainly executed by mechanical means by deploying drilling equipment starting from jackhammers to down-the-hole drills. The bit is the most vital part of a drill that must engage and disintegrate rock by percussion, rotary, abrasion, or rotary-percussion action. Different kinds of drilling equipment are used for vertical, horizontal, and overhead drilling. A typical blasting pattern comprises a number of blastholes drilled in one or more rows so as to produce the desired results. Conventional blasting work normally results in rough and uneven contours of blasted surfaces. The trend now is to adopt smooth blasting techniques to achieve desired results without disturbing the environment too much. Smooth blasting is essential in tunneling work.

REVIEW QUESTIONS

1. Drilling of blastholes is the prerequisite of blasting. Why?
2. What does happen on detonation of explosives in a tiny drilled blasthole?
3. Holes are drilled in rock for various purposes. What are the purposes?
4. What is meant by diameter of drill holes? What is burden? What is spacing of drill holes?
5. Explain why the prerequisite of good fragmentation of rock is the uniform distribution of the explosive in the rock mass to be blasted.

6. What is involved in drilling? Explain how the cuttings are removed.
7. In excavation by blasting, is it possible to control throw and scattering?
8. Explain how desired fragmentation could be achieved in blasting operation.
9. What is drilling rate? What does efficient drilling rate depend on?
10. There are four categories of drilling equipment. Describe their salient features.
11. Name the factors that are important for selection of the equipment and the drilling method.
12. What are the principle ingredients of any explosive? Give examples of each ingredient.
13. What are the properties that need to be considered before selecting an explosive for use?
14. What is deflagration? What is detonation?
15. Draw a sketch of blasting caps.
16. Explain how delay firing sequence may be achieved by using delay electric blasting caps.
17. Elaborate how pre-splitting is different from conventional smooth blasting.
18. How does excavation by blasting pollute the environment?

6
Piling

INTRODUCTION

Piling is intended to transfer both vertical load to the underlying hard soil strata as well as to resist heavy uplift and horizontal forces. This allows off-shore heavy structures to be founded conveniently on piles. It is necessary to carry out detailed soil investigation to better understand the existing soil quality and decide whether piling is necessary or not. If soil investigation results require piling, then it would be up to the design engineers to decide what kind of piling would be appropriate.

There are different kinds of piles made of different materials having different shapes and cross sections. Considering driving methods, piles have been broadly classified into two categories: (i) displacement, and (ii) replacement types. Quality assurance is essential at every stage; from soil investigation right up to the installation of piles. Ultimately, representative piles are selected and load tested to make sure that the piles driven conform to the specifications.

6.1 BASIC CONCEPT

Piles are generally used if settlement is a problem in transferring load to the underlying firmer soil bed and also to enhance load-bearing capacity of weak soil or to limit settlement to inconsequential dimension. They are also used in normal ground conditions to counter heavy uplift forces or in poor soil conditions to resist horizontal loads. Piles are conveniently used for offshore foundation construction over water, such as piers, breakwaters, and bridges. Sheet piles are used as supporting members to earth or water as in cofferdams, foundation excavations, or retaining walls. Piles are usually used for:

- Transferring both vertical and lateral superstructure loads through soil strata of low bearing capacity to deeper soil or hard/rock strata of high bearing capacity
- Resisting uplift forces such as: basement slab below the water table or overturning forces caused by lateral forces
- Enhancing load-bearing capacity of soil by compacting and consolidating surrounding soils
- Controlling settlement of spread footings or mat foundations on weak or compressible soils by transferring load, for example, from a column to hard soil strata at lower depths

- Stiffening the soil under machine foundations against dynamic forces
- Providing additional safety below the bridge abutments and/or piers where scour poses problem
- Transmitting loads through water into the underlying soil in offshore structures

A pile is a structural member that is installed to transfer load by end bearing, frictional resistance, or a combination of both. The total load to be supported must be carried in such a way that the distribution of the load over the area of the subsoil is regulated to suit the properties of supporting soil strata. On the basis of soil investigation reports, a designer has to determine the safe load that a pile or a group of piles would be able to support.

Detailed soil investigation is carried out for mapping of different soil strata below the ground level, up to such depths as required for proper design of piles. Contour lines drawn on the basis of site survey would provide information on the topography of the site. Samples collected during soil investigation are tested in laboratories to determine various types of soil lying below the ground level and their properties.

On the basis of the site survey, soil investigation (Fig. 6.1), test results, and estimated loads to be transmitted to the soil mass, an engineer decides whether piling would be necessary or not. Great care should be taken so that neither excessive number of piles nor extra lengths are specified. A pile foundation is much more expensive compared to spread footings and mat foundations.

6.2 CLASSIFICATION OF PILES

Piles are structural members made of timber, concrete, reinforced concrete, and steel. Piles may have different cross-sections like circular, square, square with round cavity, rectangular, tubular, or polygonal. They may be load bearing piles, raker piles or batter piles, fender piles or sheet piles. They may have uniform cross-section or may be tapering. Use of timber piles is mostly confined to temporary structures to facilitate construction work.

A pile is to be predominantly one of the two generic types: end-bearing or friction. End-bearing piles are driven to refusal and derive their strength by simply sitting on hard soil mass/strata. They must be strong enough to transmit load through weak or soft soil to a stronger stratum or rock.

Friction piles derive support from frictional resistance generated by the soil in direct contact with the peripheral skin of the piles. Usually a pile transfers its load, partly through end-bearing and partly through skin friction as in clay with negligible resistance from the tip.

Soils are of two types – cohesive and non-cohesive. Pile foundations may, therefore, be classified into four groups as:
- End-bearing pile in or over cohesive soil
- End-bearing pile in non-cohesive soil
- Floating foundations on piles in cohesive soils
- Floating foundations on piles in non-cohesive soils

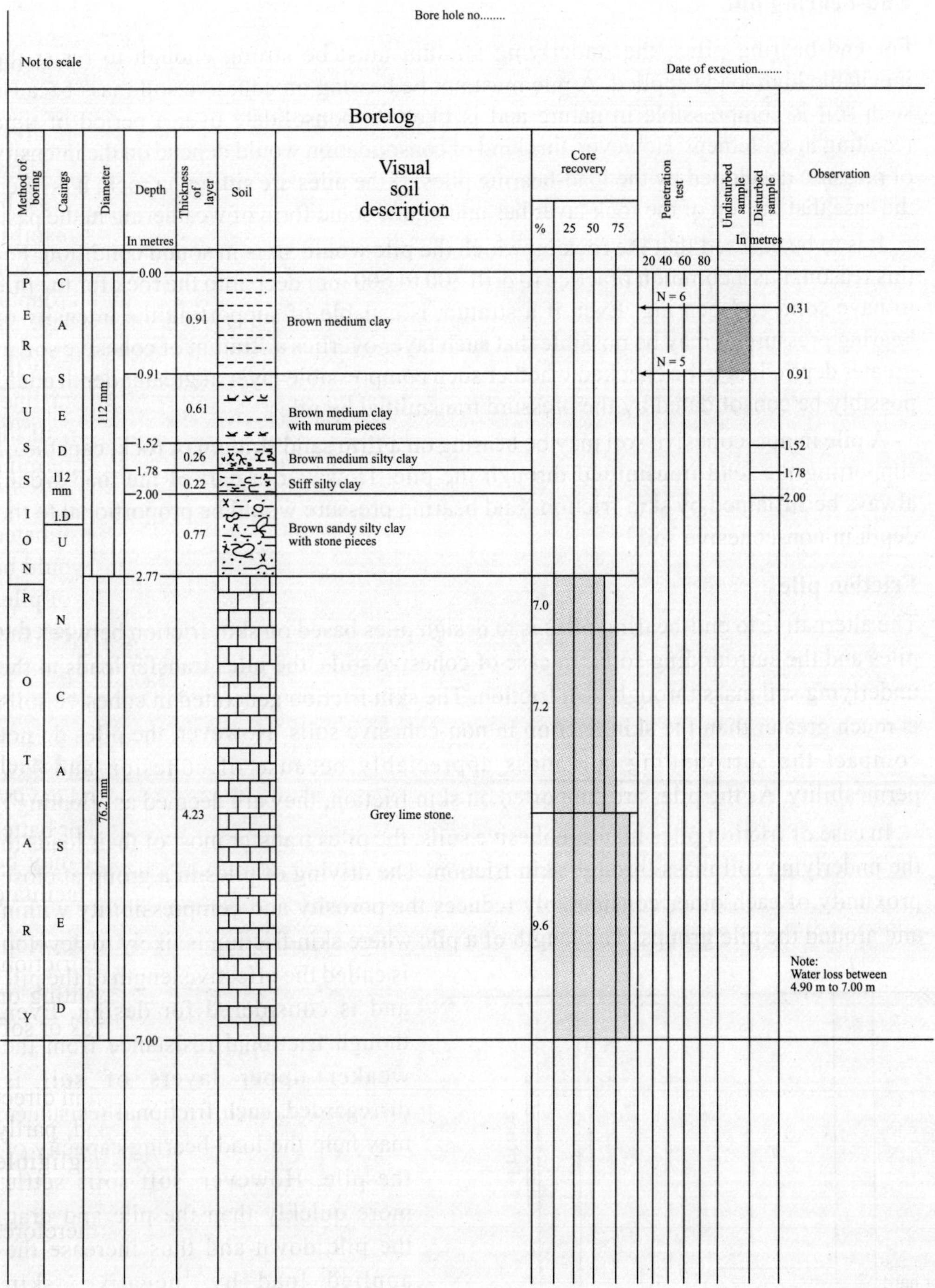

Piling 141

Method of boring	Casings	Diameter	Depth In metres	Thickness of layer	Soil	Visual soil description	Core recovery % / 25 50 75	Penetration test 20 40 60 80	Undisturbed sample	Disturbed sample	Observation In metres
P E R C U S S I O N	C A S E D 112 mm I.D	112 mm	0.00 — 0.91 — 1.52 — 1.78 — 2.00 — 2.77 —	0.91 0.61 0.26 0.22 0.77		Brown medium clay Brown medium clay with murum pieces Brown medium silty clay Stiff silty clay Brown sandy silty clay with stone pieces		N = 6 N = 5			0.31 0.91 1.52 1.78 2.00
R O T A R Y	U N C A S E D	76.2 mm	7.00 —	4.23		Grey lime stone.	7.0 7.2 9.6				Note: Water loss between 4.90 m to 7.00 m

Bore hole no........
Date of execution........
Borelog
Not to scale

Fig. 6.1 Soil investigation results

End-bearing pile

For end-bearing piles, the underlying stratum must be strong enough to resist the inevitable high loads applied. A pile must not be bearing on cohesive soil mass because such soil is compressible in nature and is likely to consolidate over a period of time resulting in settlement. However, this kind of consolidation would depend on the intensity of pressure developed by the load-bearing piles. If the piles are sitting on rock, it is often the case that the top of the rock layer has undergone some form of weathering in the past.

It is to be ensured that the rock on which the pile would sit is in sound condition. For this reason, it is a common practice to drill 300 to 600 mm deep into the rock for the pile to have solid end-bearing. Even if a stratum is capable of supporting the intensity of bearing pressure, it may be possible that such layer overlies a stratum of cohesive soil at greater depth. It is to be checked whether such compressible layer at greater depth could possibly be consolidated by the pressure transmitted by piles.

A pile in non-cohesive soil may be bearing on a firm sand stratum or rock, capable of supporting the load transmitted through the pile. However, a part of the load would always be sustained by skin friction. End bearing pressure would be proportional to the depth in non-cohesive soil.

Friction pile

The alternative to end-bearing piles is to design piles based on skin friction between the piles and the surrounding soils. In case of cohesive soils, the piles transfer loads to the underlying soil mass through skin friction. The skin friction generated in cohesive soils is much greater than the skin friction in non-cohesive soils. However, the piles do not compact the surrounding soil mass appreciably because of cohesion and low permeability. As the piles are supported on skin friction, they are deemed as 'floating'.

In case of friction piles in non-cohesive soils, the piles transfer most of their loads to the underlying soil mass through skin friction. The driving of piles in a group at close proximity of each other considerably reduces the porosity and compressibility within and around the pile groups. The length of a pile where skin friction is likely to develop is called the effective length of the pile and is considered for design. Even though frictional resistance from the weaker upper layers of soil is disregarded, such frictional resistance may help the load bearing capacity of the pile. However, soft soils settle more quickly than the pile and drag the pile down and thus increase the applied load by 'negative' skin friction. Plastic sleeves are used to protect piles from developing negative skin friction.

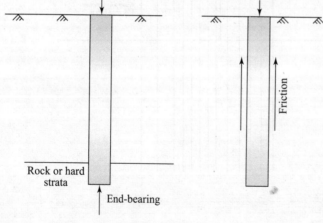

Fig. 6.2 End-bearing and friction piles

Piles cannot be simply classified into 'bored' or 'driven' categories. There are many 'hybrid' types of piles that cannot be categorized into these two categories. To overcome this problem, piles are broadly classified into the following two categories:
- Displacement piles
- Non-displacement or replacement piles

Both these categories are shown in Fig. 6.3.

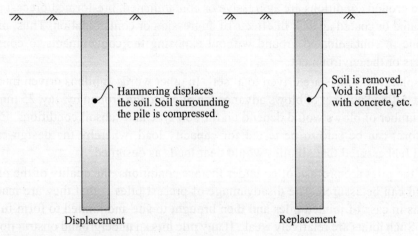

Fig. 6.3 Displacement/replacement piles

Displacement Piles

The types of various displacement piles may be classified as follows:

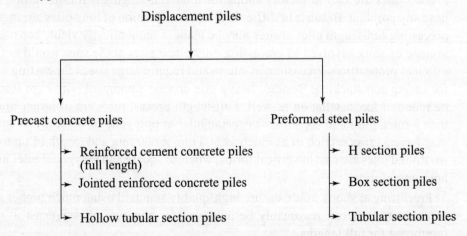

Displacement piles are driven into the ground by hammering, thus displacing soil sideways to allow room for the piles to penetrate. This simply compacts the soil around, allowing them to be more load-bearing in non-cohesive soils. However, this may cause heaving in cohesive soils and so cannot be used adjacent to any existing foundations.

Displacement piles are not recommended in fill areas or gravel where large obstructions may be encountered. Displacement piles are prone to deflection underground or even cracking in case of precast concrete piles if any obstruction is confronted during installation, and there is no way of checking for such damage from the surface. Also during piling, underground obstructions can cause the piles to move off their correct positions or move out of plumb, or both. However, displacement piles can be installed in poor ground conditions and in aggressive ground water conditions. Where ground conditions are aggressive or contaminated, pre-formed/precast piles can be treated or coated against ill effects of aggression or contamination. Thus, piling can be done in contaminated ground without exposing the contaminants to construction workers or the environment.

Displacement piles are driven to a 'set'. In other words, a pile is driven into the soil until a defined number of blows advance the pile into the ground by, say, 25 mm or less. The number of blows would depend on the soil type and subsoil conditions. Each pile, therefore, can be said to be tested for capacity load, whereby the design engineers would feel assured that all piles would bear loads as designed.

As the piles are precast, often under factory conditions, the quality of the pile in the ground can be assured. The disadvantage of precast piles is that they are made to set lengths in case of jointed piles and then brought to site and jointed to form full-length piles. Such joints are relatively weak. If any pile hits an underground obstruction during installation, then this can damage the joints and dislocate the pile without any sign on the surface as mentioned earlier.

Precast concrete piles

Precast piles are cast in factory shops for their full lengths if transportation does not pose any problem. Because of difficulties in transportation of long piles, arrangement of precasting full-length piles at sites may be made if economically viable considering the volume of work involved as precasting only a few piles at the sites would not be cost-efficient proposition. Precasting at site would require large space for casting as well as for curing and stacking. Besides, heavy pile driving equipment with high leads would be required for installation as well. Full-length precast piles are of larger dimensions than jointed piles and are square or rectangular or polygonal in sections. A square pile may have a cross section of as much as 600 mm × 600 mm and length of up to 30 m or so. Round piles are cast in vertical forms, whereas square and polygonal piles are cast in horizontal forms.

Precasting at shops would ensure high-quality standard using much higher strengths of concrete than can reasonably be achieved for concrete piles precast at sites and reinforced for full lengths.

Prestressing of reinforcement is done for many reasons. Transverse strengthening of reinforcing bars is not required. Piles withstand stresses developed during handling, pitching, and driving. Prestressed concrete piles require high-quality concrete. As such, bending stresses developed during driving are unlikely to cause any cracking in the piles. However, piles are pre-compressed because of prestressing and, as such, are

Piling 145

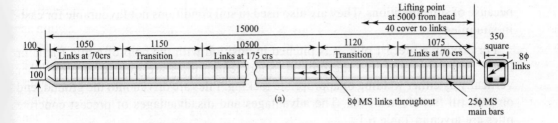

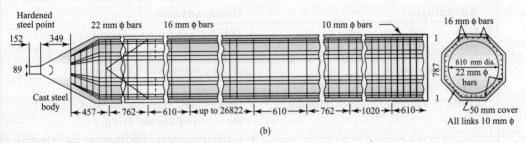

Fig. 6.4 Precast concrete piles

vulnerable to damage if obstructions are encountered underground during installation by hammering. Also, shortening of prestressed pile length is quite difficult.

As an alternative, piles may be precast at shops in small lengths to be jointed during installation. Jointed piles are not prestressed. Short length piles are used to avoid damage during transportation and handling. Special fittings are embedded at the ends for jointing pile lengths. The fitting should be perfect so that more energy input is not required for driving due to problems related to bad joints. Imperfect joints may affect verticality of piles. Tolerance in verticality should remain within 1 in 300. Driving stress would be more even in piles of polygonal section, in case verticality is not maintained.

The piles are generally driven using hammers weighing 3 to 4 tonnes or more. Otherwise hydraulic hammers, vibrators, or diesel hammers are sometimes used. Flat or pointed pile shoes are fitted for driving piles through granular soils or soft rock. Shoes are not required in uniform clays or sands. Pile heads are cushioned as precautionary measure against adverse effects of driving.

Hollow precast tubular piles may be of large diameters for special purpose driving. Tubular piles are made with plain or prestressed reinforcement. For marine structures such as jetties or wharfs/quays, large-diameter tubular piles (900 mm to 1,400 mm diameter) of around 200 tonnes are designed to withstand combined axial and bending stresses due to lateral forces like structural columns. They are prestressed by post-tensioning. Open-end tubular piles are used in cohesive soils and closed-end piles are used in weak soils. Precast piles are usually of square section for short and moderate lengths, but polygonal and circular sections are preferred for longer pile lengths. Tubular precast piles are used where soil conditions require large cross-sectional area because of soil conditions. They are also used in soil conditions not favourable for cast-in-situ piles.

For driving a precast pile, it is pitched in position with the help of a pile frame or leader and hammered into the ground. Short-length piles are jointed to each other vertically by either welding or appropriate locking. Piles are driven into the ground, end on end, till 'set' is achieved. The advantages and disadvantages of precast concrete piles are given in Table 6.1.

Table 6.1 Advantages and disadvantages of precast concrete piles

Advantages	Disadvantages
• High strength with assured quality • High resistance to chemical and biological attacks • A pipe may be embedded along the centre of a pile to facilitate jetting	• Difficulty in reducing/increasing lengths • Deployment of expensive handling and driving construction equipment is required for piles of larger sizes • Delay due to possible breakage of piles during handling and driving • Commencement of piling may be delayed if specified piles are not available at workshops

Preformed steel piles

The advantages and disadvantages of preformed steel piles are given in Table 6.2.

Table 6.2 Advantages and disadvangages of preformed steel piles

Advantages	Disadvantages
• Handling is easier • Overstressing during lifting or pitching is unlikely • Can withstand hard driving without shattering • Extension or reduction of length is possible and relatively easy • Have high-carrying capacity if driven to hard stratum • Resistance to impact forces by their resilience and column strength against impact from berthing of ships in marine structures	• Corrosion • Cost

Totally preformed steel piles are manufactured of various steel sections and pipes. Steel sections comprise both 'H' and 'I' sections. Web and flange of H-piles are made of the same thickness and about the same dimensions. Wide flange beams or I-section beams may also be used in place of H-piles. Continuous butt-welding of two identical rolled steel sections, edge to edge, forms *box piles*. *Tubular section piles* are manufactured in seamless spirally welded or lap welded forms. Tubes up to 4 m diameter are fabricated for marine structures. Tubular piles may or may not be filled up

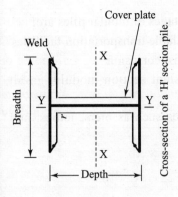

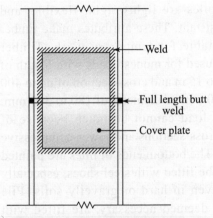

Fig. 6.5 H-piles showing splicing

with concrete after driving. A tubular pile may be driven open-end or closed-end. Closed-end piles are usually filled up with concrete on completion of driving. H-piles (Fig. 6.5) are used as bearing piles. A disadvantage of H-piles is the tendency to bend on weak axis during driving. However, the advantages of H-piles are more as indicated below:

- Can be easily handled
- Can be easily extended by welded splices and shortened by flame cutting
- Can withstand prolonged hard driving without damage
- Accidental eccentric driving does not affect these piles
- Can be designed for high-bearing loads when driven to rock strata
- As H-piles of small cross-sectional area cause less volume displacement, they can be driven close to existing structures
- Can either break smaller boulders or displace them to one side because of their rigidity

Special rolled sections are also available to form box piles (Fig. 6.6) of octagonal section. Advantages of box piles are:

- Do not need any special care in handling
- Can be driven open ended but can be provided with flat or conical shoe at the bottom either by welding or by bolting
- May be inspected after driving by lowering down electric lamp inside
- Can be designed for high bearing loads when driven to rock strata
- Can be used as struts as they have high column strength
- Suitable for resisting the bending stresses caused by impact and horizontal forces as in marine structures such as piers, jetties, and mooring dolphins
- Heaving in cohesive soil may be reduced considerably by grabbing out soil from inside
- Over-compaction may be of advantage in loose granular soils

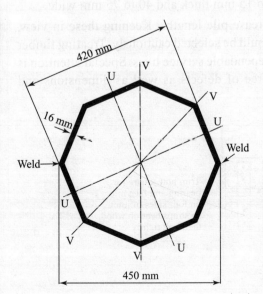

Fig. 6.6 Box piles

Both tubular and box piles are hollow steel pipes. Advantages of tubular piles are:
- Pipes can be supplied in any length, subject to available transportation facilities
- Pipes can be coated externally and internally to resist corrosion
- Pipes have a high radius of gyration and a constant section modulus in all directions
- Large diameter pipes are used in marine structures such as piers, jetties, and dolphins to resist berthing impact and wave action
- Other advantages are similar to those of box piles

Timber piles

Timber piles are lightweight, flexible, and shock-resistant. These attributes make timber piles suitable for temporary work. Timber piles are used for modest loads with length of piles up to 15 m and cross-section of up to 400 mm × 400 mm or diameter of 180 to 300 mm. Working load cannot be much because of smaller cross-sections and lower compressive strength. The bottom ends of piles are pointed and may be fitted with steel shoes, especially when driven in hard or gravelly soils. Pile heads, if deemed necessary, are fitted with metal bands to prevent damage due to driving because of fibre crushing (brooming) in dense

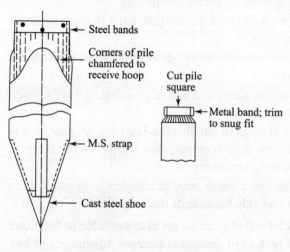

Fig. 6.7 Timber piles head and toe

sand and hard materials. The bands are 10 to 15 mm thick and 40 to 75 mm wide.

Steel collars are used for splicing to increase pile lengths. Keeping these in view, driving equipment and method of driving should be selected cautiously. Treating timber piles with preservatives would add to their dependable service lives. Special attention is necessary to ensure that timber piles are free of defects as well as dimensions and straightness is within approved tolerances.

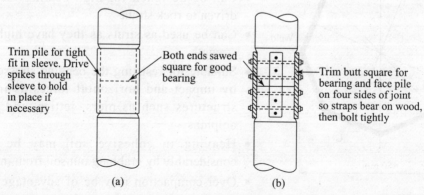

Fig. 6.8 Timber pile joint

Timber piles would have a very long life as long as they are driven totally above (permanently dry) or totally below (permanently wet) the water table. They are likely to decay in case of fluctuating water table. The immersed portions of the piles are liable to severe attack by marine organisms. Care in selection and treatment of timber can prevent deterioration from such attacks. Timber piles lose strength in an environment of high temperature and should not be used where heat is generated from furnaces.

Table 6.3 Advantages and disadvantages of timber piles

Advantages	Disadvantages
• Cheaper cost-wise	• Difficulty in obtaining sufficiently long and straight piles
• Can be handled easily with hardly any danger of breaking	• Difficulty in driving in hard formations
• Can be cut off to the required length after they are driven	• Difficulty in splicing them to increase their lengths
• Can be easily removed by pulling up if such removal is required	• Although satisfactory as friction piles, they are not suitable for use as end-bearing piles
	• They are short-lived unless treated with appropriate preservatives

Driven cast-in-situ displacement piles

These piles may be of two forms:
- A tube is driven to form void and filling the void with concrete as the tube is withdrawn—this form is for lower load range (vide Fig. 6.9)
- A tube is driven to form permanent casing—this form is for supporting considerable loads in suitable ground conditions (vide Fig. 6.10)

In the first form, concrete fills up the void as the steel tube is retrieved by withdrawing. The steel tube may either be bottom driven or top driven.

In the bottom driven form, the tube casing is not stressed much. The bottom of the tube is closed with a steel plate. A plug of dry mix concrete is compacted by an internal drop hammer. The hammer may then be used to drive into the ground via the compacted plug. When the required depth is reached, the tube is restrained at the ground level and the plug is hammered to form a larger base.

In the top-driven form, the tube is placed on expendable shoe or blank plate and hammered to the required depth. Concrete is placed to fill the voids in both the cases as the tube is retrieved. A timber dolly on a helmet is placed at the top of the tube to cushion the blow of driving hammer in case of top driving. Reinforcing steel cage is lowered in the tube before filling the void with concrete. High slump concrete is compacted as the tube is withdrawn. It is to be ensured that already driven piles are not damaged during driving of adjacent piles.

The second form uses an outer concrete or steel shell threaded on to steel inner core at the end of which is a steel or concrete shoe. The shell and the shoe form the outer

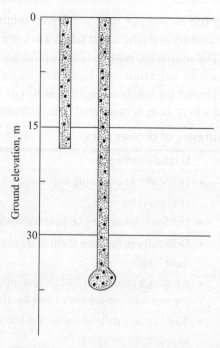

Fig. 6.9 Cast-in-situ pile with tube withdrawn

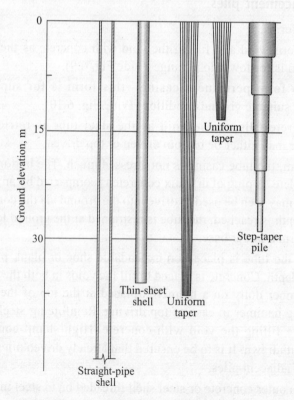

Fig. 6.10 Cast-in-situ pile with tube

shape of the pile and will remain in the ground, forming the outside of the pile. The piling force is applied by a conventional hammer and is delivered to the shoe via the steel core. The hammer drives the shell and steel core into the ground and is driven to a set. More shell sections and extensions to the steel core can be added to extend the pile until it achieves a set. Removing the steel core leaves a stable, watertight hollow concrete column. Reinforcement is lowered into the hole. The shell is filled with concrete and compacted forming a solid concrete pile.

This form of pile combines the advantages of the precast system and the in-situ system; ease of installation and guaranteed quality of concrete outer skin to resist aggressive ground or ground contamination. The steel core solves the disadvantages of precast system due to deflection and cracking underground. However, the disadvantages of ground heave and noise remain, and the system cannot be used where the groundwater pressure is high.

The advantages of cast-in-situ piles are:

- Easier handling and driving of the lightweight tubes
- The length of a tube may be increased or decreased easily
- The tube lengths may be transported in short lengths and assembled at the site
- Extra reinforcing steel to resist handling stresses is not required
- Possibility of pile breaking during driving is eliminated
- Additional piles, if required, may be driven easily

Helical cast-in-situ displacement piles

This is an auger type operation that does not involve removal of soil. Instead, the soil is compacted. A heavy-duty single-start auger head with a short flight is screwed into the ground. The auger head is carried on a hollow stem, which transmits the necessary torque and compressive forces to screw the auger head into the ground to the required depth. The end of the hollow stem is sealed with a disposable tip. When the screwing process is completed and the required depth is reached, the reinforcement cage is inserted through the hollow stem. Concrete pouring through this stem follows. As concreting operation progresses, the auger is unscrewed and withdrawn. A screw-threaded cast-in-situ pile is thus formed. A rig provides hydraulic power for auger rotation and downward push. The threads formed by augering action are of robust dimensions. In this method of displacement piling, noise is low and vibration is less. This kind of piling is good for most cohesive and granular strata up to a depth of 22 m for piles ranging in diameter from 360 to 560 mm.

Replacement Piles

The replacement methods of piling are:

- Bored cast-in-situ piles
- Partially preformed piles
- Grout or concrete intruded piles

Bored cast-in-situ piles

By excavating a borehole in the ground, a pile can be produced by filling the void with concrete. In case of stiff clay, borehole walls do not need any support except for the portion near the ground level. The excavation process usually produces a borehole of circular cross-section. This kind of piles varies considerably in sizes. The piles of diameter up to 600 mm are referred to as small-diameter piles. Piles of sizes above 600 mm up to 2100 mm are large diameter piles.

Small-diameter percussion bored cast-in-situ piles

Diameter of small cast-in-situ piles range between 300 to 600 mm. The rigs or tripods used for small-diameter piles are similar to those used in soil investigation.

In cohesive soils, 'clay cutter' is used, often adding weight to facilitate penetration. A cutter comprises an open cylinder with hardened cutting edge. Energy for raising the cutter is obtained from diesel engine or pneumatic motor (vide Chapter 13, Section 13.2). The cutter is dropped repeatedly to form the hole. A winch is engaged to raise the cutter with its burden of soil.

In granular soils, 'shell' is used for boring. A shell comprises a heavy top open tube with a flap valve at its base. It is important to have water in the bore to prevent granular soil to loosen and collapse. As the shell is dropped, the flap valve at the base opens. The flap closes as the shell is raised. The spoil from the shell is simply tipped out on the ground surface. In case of very loose granular soil, temporary casing may be used and advanced by driving.

Large-diameter percussion bored piles

Diameter of large piles ranges between 600 to 2100 mm. Semi-rotary down-the-hole percussive hammer is used for boring of wells. Similar hammer is now available for percussive boring of large-diameter piles in hard strata including weak sedimentary rock. The rig is mounted on chassis supported on wheels or skids and employs an oscillatory drive to affect penetration of casing. The boring tube has a hardened cutting edge and is hydraulically clamped in a collar that transmits the hydraulic ram's semi-rotary action and vertical motion. The spoil inside the casing is removed by grabbing (vide Chapter 4, Section 4.5).

In hard strata, rock jaws are fitted to a heavy hammer grab (weighing as much as 1400 kg). The grab chisels through obstructions below the temporary casing. Other impact devices can be used to penetrate the hard soil strata. Reinforcement is lowered when boring is completed. Pouring of concrete is easy in the dry, but properly designed skips are used to place concrete in water. Piles up to a length of about 40 m can be formed at a rake of up to 1 in 5.

Rotary bored cast-in-situ piles

Large-diameter bored piles are formed by augers. The plant for augering is mounted on either a truck or a crane. The truck-mounted rig is advantageous for movement between sites. But crane mounted rig has more manoeuvrability in confined sites, although transporting such crane-mounted augering plant is expensive. A second crane is sometimes required for handling temporary casing.

In case of smaller-diameter piles, a winch may be used for handling temporary casing. A crane-mounted auger can bore to a diameter of at least 3000 mm. Depths of boring vary from 25 m to as much as 60 m. Rigs for auger piles can form piles at a rake up to 1 in 3.5. A special feature of auger-bored piles is of providing enlarged base of the pile by under-reaming. A worker in a cage is lowered inside the pile to inspect the shaft and the result of under-reaming before concrete pouring is done. Under-reaming is not possible under water.

Continuous flight auger piles

Continuous flight auger (CFA) piles are clearly replacement piles. Installation of CFA piles have many advantages, such as noise is low and vibration is minimal. In permeable soil with high water table, no special arrangement is necessary for concreting operation, and casing or bentonite slurry is not required to support borehole walls. CFA piles can be installed close to existing foundations and in all kinds of soils like sands, gravels, and clays. As they can be drilled past weathered layers to hard rock, they may be used as bearing piles. They are available in various diameters starting from 300 to 1500 mm.

The piles are constructed by screwing the continuous flight auger into the ground down to the required depth up to 35 m. The auger has a central stem through which concrete is pumped at high pressure. A plasticizing admixture is added in the concrete to improve the pumpability and an expanding admixture is added to resist shrinkage during the setting and hardening phases. The pressure of concrete helps withdrawal of auger and lifting of the spoil to the surface. Utmost attention is necessary at this stage not to withdraw the auger too fast, otherwise a void would form at the end of the auger and the soil would collapse into this void. Reinforcement cage is pushed in after concreting operation is completed. Reinforcement cage is pushed down into the concrete shaft up to 15 m. A small vibrator attached at the top of the reinforcement cage helps insertion in the concrete. Full-length reinforcement is required if any pile is likely to be subjected to lateral forces and tension.

Strict quality control of workmanship is required in CFA pile driving, especially when substantial load is to be carried in end bearing. When regular augers are used, it is possible to examine drill cuttings to check soil conditions before any concrete is poured. But in case of CFA piling, the drill cuttings do not reach the surface until the shaft concreting operation is complete. It would be possible to ascertain the soil strength by measuring and recording the torque on the drill stem over the full depth of drilling.

It is possible to enhance the bearing capacity of appropriate soil strata by providing an enlarged base in auger-bored piles. This can be done in non-caving stiff to hard clays where soils would remain open unsupported as and when under-reaming operation is completed. The under-reaming tool would fit inside the straight section of a pile shaft in its closed position. The tool can be expanded at the base of the pile wherever under-reaming is to be done.

Excavation for under-reaming (Fig. 6.11) is achieved by a belling bucket rotated by drill rods. Two types of belling buckets are available. The one generally favoured has arms hinged at the top of the bucket and are actuated by the drill rods. The arms are provided with cutting teeth and the bucket removes the excavated spoil. This type cuts to a conical shape. The other type has arms hinged at the bottom of the bucket. This type has the advantage of being capable of cutting a larger bell than the top-hinged type. It produces a cleaner base with less loose and softened materials. However, the hemispherical upper surface of the bell is less stable than the conical surface. Besides, the bottom-hinged arms have a tendency to jam in the hole when raising the bucket.

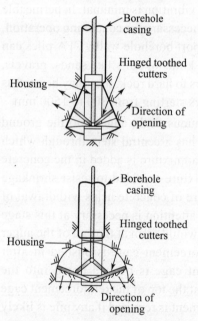

Fig. 6.11 Under-reaming of pile

The under-reamed bore is to be inspected for checking if there are any loose materials. Concreting is to be done after removing loose soil materials, if any. For inspection, a person is to be lowered down to the bore level in a cage if the diameter of pile shaft is sufficient for the purpose. All safety precautions should be taken for the person's safe return to the surface. Otherwise, remote controlled cameras/sensors may be used for the inspection. All loose materials must be removed before concreting work is taken up. The minimum length of pile from ground level to pile toe should not be less than 3.5 m so that seasonal ground movement does not affect the piles. The skin friction of the top 1.5 m of an under-reamed pile should be neglected for design purposes.

The diameter of the under-reamed bulb is normally two-and-half times the diameter of a pile's stem. There may be one or more bulbs in under-reamed piles. In case of more than one bulb, the spacing of bulbs should not exceed 1.5 times the diameter of the bulb. Also, the topmost bulb should not be reamed close to the ground level, especially for piles likely to be subjected to uplift pressure. The minimum depth at which under-reaming should be done under all loading and ground conditions should be about 2 times the bulb diameter. The minimum spacing in case of a group of piles should be normally 2 times the bulb diameter. There is no upper limit. If the piles are formed at a spacing closer than two times the bulb diameter, loading capacity of each pile should be appropriately reduced.

Replacement piles are columns of concrete cast in the ground in a pre-drilled hole. The drilling action removes the soil where the pile is supposed to be and replaces it with concrete. The advantages of this system are that it is relatively quiet and underground obstructions can be dealt with during the drilling. Disadvantages are that all piles cannot be load-tested and a separate testing procedure must be instigated. The walls of the drilled hole must also remain stable when unsupported to allow concrete to be placed.

The walls of the hole may bulge in during the concrete pour or when the concrete is wet, causing a restriction in the diameter of pile; this is called *necking*. If severe, necking can weaken the pile to an extent that it may even fail under load. The solution to this problem is to use a temporary steel casing inserted in the hole during drilling to support the ground whilst the concrete is placed, and remove it before the concrete gains strength.

Micro piles and mini piles

Micro piles' diameters range up to 150 mm and mini piles' diameters cover the range from 150 to 250 mm. They do not have sufficient bearing area and, therefore, work primarily by shaft friction.

Micro piles are used as:

- Bearing piles
- Soil-retaining and reinforcing piles
- Tension piles

Bearing piles are used:

- In light industrial or domestic units in (i) weak soils, (ii) swelling soils, (iii) shrinkable soils, and (iv) for minimizing differential settlement between old and new construction
- For underpinning as well as strengthening of structure where necessary
- For support of isolated machine foundations
- For support of floor slabs as a rehabilitation measure where structures are supported on piles
- As settlement reducers

Piles for soil-retaining and reinforcing are used for:

- Forming reinforced soil walls
- Stabilization of slopes
- Bored micro pile wall

Tension piles are used:

- Below buoyant structures
- Below towers and masts

Micro piles are installed by:

- Drilling/boring cast-in-situ methods
- Displacement methods

For installation of micro piles, conventional piling methods are modified to suit micro piling operation.

Corrosion

Corrosion of steel piles could be a serious problem in the following two situations:
- Piles driven into disturbed soils or fill soils due to higher oxygen concentration
- Piles exposed to seawater or effluents with a pH value above 9.5 or below 4.0

The measures to combat corrosion are:
- Use of copper bearing steel against atmospheric corrosion
- Use of high-strength steel where mild steel is considered in design, thus keeping margin for corrosion loss
- Increasing steel thickness by using next heavier sections over sections as per design
- Use of protective coating, particularly for the exposed sections of piles—the coating may be metallic type (galvanizing or metal spraying) or just painting
- Cathodic protection of piles below water table or in marine conditions—costly proposition
- Concrete encasement above the water level

Attack on concrete piles

Attack on both precast and cast-in-situ concrete piles may occur due to:
- Sulphate attack
- Chloride attack
- Acidic groundwater

The sulphates commonly encountered are those of sodium, potassium, magnesium, and calcium. Against sulphate concentration of 200 to 400 ppm, cement content in concrete should be increased with low water:cement ratio in case of ordinary Portland cement. Otherwise, sulphate-resistant cement or 25% pozzolonic material cement replacement may be used. At concentrations of 1200 ppm or more, sulphate resistant cement is considered necessary.

Chlorides can cause accelerated corrosion of reinforcing steel. This may result in bursting of concrete due to scale formation. Stainless steel, galvanized or epoxy-coated reinforcing steel bars are used against corrosion. Another protective measure is to wrap a protective membrane around the pile shaft.

Acidic groundwater exists in peaty soils or in areas contaminated by industrial wastes. Possible measure against acidic groundwater is to take care of permeability of concrete by using dense concrete with as low water:cement ratio as possible maintaining workability by adding plasticizer. In extreme cases, external protection of concrete surfaces may be necessary.

6.3 PILE DRIVING METHODS

Pile driving methods may broadly be categorized as:

- Driving by the impact of a steady succession of blows on the pile top using a hammer. This method is not environment friendly as hammering generates both noise and vibration.
- Forming a void in the ground and either inserting a pile into it or filling the void with concrete with/without reinforcement. A number of methods are there for this kind of piling.
- Driving a vibratory device attached to the pile top, generally in soils with less cohesion. This method is not noisy and vibration is also not excessive.
- Jacking the pile against reaction. This method is applicable for short stiff piles.

Pile driving equipment

The piling rig as normally used on ground comprises a leader mounted on a standard crane base. The leader comprises a sturdy lattice mast or tubular member, which carries and guides the hammer and the pile as it is driven into the ground. Two adjustable struts, which allow forward and backward raking, support the pile leader. This backward and forward raking is made possible by screw or hydraulic adjustment of the backstay and lower attachment to the crane base. A group of piles can be driven without moving the rig by swiveling the base and adjusting the position of the leader. The base frame rests on four swivel steel castors so that a pile rig may be moved around on hard ground or pads.

Impact pile driving

Impact pile driving is done with hammer, which imparts dynamic energy by the impact of a falling ram. This energy is mostly generated by gravity due to free fall of the ram. In some types of hammers, mechanical energy is added on top of the ram to increase the rate of fall. During pile driving, both pile and hammer must be temporarily held in position from a pile rig or a crane jib or a leader. Also, pile extractors are required for extracting piles installed for temporary works. The piling method includes:

- Installing a pile driver
- Transporting a pile to the pile driver
- Lifting, pitching, and placing the pile under the hammer
- Hammering the pile to the required depth

Types of hammer:

1. Impact hammers
   - Drop
   - Single-acting steam, pneumatic or diesel-operated
   - Double-acting steam, pneumatic or diesel-operated
   - Hydraulic

2. Vibratory pile-drivers
3. Hydraulic sheet pile-driver

Impact hammers

The selection of suitable pile hammer depends on a number of factors including the size and weight of the pile, the driving resistance that has to be overcome for the required penetration, the available space and headroom at the site, the availability of cranes, and local noise pollution control restrictions. Hand-driven simple drop hammers are generally used for driving timber piles to shallow depths in soft soil.

Drop hammer

The simplest form of hammer is the drop hammer. A drop hammer comprising metal weight is suspended with wire rope over a pulley from a strong frame or rig. The impact energy is generated by free fall of the hammer along the guides, which ensures that the pile is struck axially and concentrically. The power for raising the hammer is transmitted through a winch. The energy input is proportional to the product of the weight of hammer, height of fall, and efficiency factor; and energy input can be enhanced/reduced by changing either the weight or the height of fall.

Hammer weight varies from 250 to 5000 kg—a weight approximately equal to that of the pile. Drop height can be as much as 3 m. Number of blows varies from 5 to 15 or more for checking 'set'. The advantages and disadvantages of drop hammers are given in Table 6.4.

Table 6.4 Advantages and disadvantages of drop hammers

Advantages	Disadvantages
• Small investment	• Slow striking rate during driving
• Simplicity of driving operation	• Possibility of damaging piles by lifting a hammer too high
• Easy variation of energy input by adjusting the height of fall	• Heavy vibrations caused by a hammer may damage adjacent buildings/structures

Generally, a power operated hammer forms part of a plant deployed for driving of piles instead of drop hammer.

Single-acting hammers

Single-acting steam or pneumatic hammers drive piles like drop hammers do. For hammering, steam or compressed air is used. Wire rope is not required. However, a rope from a winch is attached to the piston rod for raising or lowering the hammer when changing piles. A single-acting steam/pneumatic hammer comprises a massive weight in the form of a cylinder with part hollow piston rod and a sliding ram. The ram is raised up to the piston shaft as steam or compressed air is forced in the chamber. As the steam/compressed air is released by operating a valve or opening up of a port, the ram falls onto the anvil at the base of the piston rod, thereby producing the hammer blow.

Table 6.5   Advantages and disadvantages of single-acting power operated hammers

Advantages	Disadvantages
• Faster driving due to greater number of blows per minute	• More investment is necessary to generate power
• Faster driving reduces the increase of skin friction between blows	• More complicated with higher maintenance cost
• Heavier ram falling at lower speed transmits a greater portion of the energy to driving piles	• Require more time to install as well as to dismantle
• Lower speed of ram decreases the danger of damage to piles during driving	• More skilled manpower is needed
• The enclosed types may be used for underwater pile driving	

The hammer length must be such as to obtain a reasonable impact velocity. This driving unit is attached to a leader or pile frame to accomplish directional control during driving. Compared to drop hammer, higher strike rate of up to 60 strokes/minute may be achieved by single-acting hammer. However, this type of hammer is characterized by relatively short stroke and low impact velocity. Heavier version of single-acting hammers is used in driving steel tubular piles in marine structures.

Single-acting diesel hammer is a better alternative to a drop hammer or a single-acting steam/pneumatic hammer. For a comparable weight of ram, input energy in case of single-acting diesel hammer is roughly double. The process is similar to that of a single-acting steam/pneumatic hammer. Raising the ram is the starting point of piling operation. The ram falls automatically at the top of the upward stroke. During downward stroke, the ram closes the exhaust port. Diesel fuel is injected and trapped in the cylinder. The falling ram compresses the air-fuel mixture and the mixture becomes hot due to compression. The impact of the ram, apart from driving the pile, ignites the compressed air-diesel mixture, thus imparting further energy for driving. The

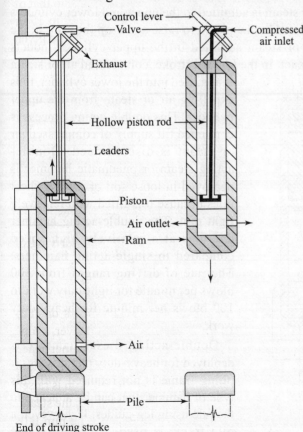

Fig. 6.12   Single-acting hammer

explosive combustion pushes the ram upwards. The burnt up gases are expelled through the exhaust port. This type of impact driving would continue till the diesel injection is cut off.

Boiler for steam generation or compressor for compressed air is not required for single-acting diesel hammer, which is relatively light compared to other hammers. However, impact may not be sufficient in soft soil to ignite air-fuel mixture for imparting additional driving energy. In contrast, impact would be too much if there is high penetration resistance. The rated energy of the single-acting diesel hammer is the product of the ram weight and the stroke. The stroke varies with pile resistance.

Usually, the weight of a drop or single-acting hammer should be the same as that of the pile. In case of heavier reinforced concrete piles, the hammer weight should be half of the pile weight. However, the hammer weight should never be less than one-third of that of the pile. In order to avoid damage to the pile, the fall should be limited to 1.2 m. A blow delivered by a heavy hammer with short drop is more effective and less damaging than a blow from a light hammer with high drop. The weight of the ram should be at least equal to the weight of the pile for the pile driving operation to be efficient.

Double-acting hammer

Double-acting steam or pneumatic hammer has typically short stroke, even less than 300 mm, because compressed air or steam is admitted to the upper and lower cylinders alternately by means of a valve actuated by the piston to raise and drop the ram. In the downward stroke, compressed air or steam admitted in the upper cylinder inducts additional energy to the free fall impact. In the upward stroke, compressed air or steam is admitted into the lower cylinder, thus expelling air or steam from the upper cylinder. This double-acting process is continued till supply of compressed air or steam is discontinued. Double-acting steam or pneumatic hammer is deployed in loose soil or driving sheet piles because it is fitted with relatively light ram. The double-acting hammer operates at a relatively high speed compared to single-acting hammers. The rate of driving ranges from 300 blows per minute for light-duty work to 100 blows per minute for heavy-duty work.

Double-acting diesel hammer is deployed for heavy-duty pile driving. A piling frame is not required with this type of hammer that can be attached to the pile using leg-guides. If, however, a pile frame is used, back-guides are

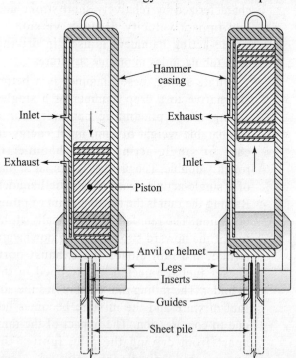

Fig. 6.13 Double-acting hammer

bolted to the hammer to engage with the leaders and short-leg guides are used to prevent the hammer moving relative to the pile top. Driving method of a double-acting diesel hammer is similar to that of single-acting diesel hammer. The hammer compresses air in the top of the cylinder on the upward stroke, which causes acceleration to the ram on its downward stroke.

As already mentioned, double-acting hammers work at relatively very high speed compared to single-acting or hydraulic hammers. This ensures continuous penetration of pile, thereby possibility of developing static friction in granular soil is obviated.

Table 6.6   Advantages and disadvantages of double-acting hammer

Advantages	Disadvantages
• Greater number of blows per minute reduces the time required to drive piles and ensures continuous motion in driving • Greater number of blows per minute reduces the increase of skin friction between blows, especially in granular soils • Piles can be driven easily without leads • Can be deployed for underwater piling after fitting with an exhaust extension	• The relatively light weight and high velocity of the ram make this type of hammer less effective in cohesive soils such as clay • Rapid blow delivery rate may damage the heads of concrete piles • The hammer design is more complicated

Hydraulic impact hammer

Hydraulic impact hammer tends to be more efficient and less noisy in operation, producing less vibration than other types of impact hammers. Hydraulic power can be used for both raising the ram as well as enhancing the free-fall impact. The weight of ram varies from 3 to 10 tonnes. Blow rate may be increased to about 50 blows per minute with stroke lengths similar to single-acting hammers. Hydraulic hammers can be operated in the water up to depths of 1000 m. Hydraulic hammers are manufactured in very large sizes, obviously for marine construction work.

Except for the simple drop hammer, the efficiency of all types of hammers depends on mechanical condition (wear and tear) and maintenance.

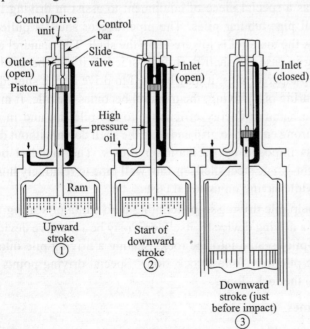

Fig. 6.14 Hydraulic impact hammer

Pile helmets and dollies

A pile helmet is required for:
- Distributing the blow from the hammer evenly to the head of the pile
- Cushioning the blow
- Protecting the pile head

The cast steel helmet should fit loosely around the pile so as to allow the pile to rotate without binding on the helmet.

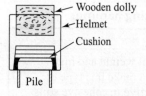

Fig. 6.15 Helmet, dolly, cushion

A dolly is placed in the recess in the helmet to cushion the blow. For light to medium driving, a simple wooden dolly is used. In heavy-duty driving, a wooden dolly with a steel plate may be used. A plastic or resin-bonded fabric dolly with a steel plate is also used for heavier driving. Longer dollies are used in case the piles are driven below the bottom level of the leaders.

Concrete piles and other materials are likely to sustain damage from impact and require a cushion placed between the pile head and the underside of helmet. Packing plates of minimum 25 to 30 mm thickness made of soft plywood are suitable for soft ground. For hard-driving conditions, asbestos fibre has been found effective. Other types of materials used for packing include hessian and paper sacking, coconut matting, and sawdust in bags.

Selection of materials and thickness of dollies and packing demands utmost attention as lack of resilience could lead to excessive damage to the pile head or, in extreme cases, to breakage of the hammer.

A mandrel is used as a special piece of equipment to assist in driving light-gauge shell piles or thin wall pipe/tubular piles. The pipe/tube or shell is pulled over the mandrel or, conversely, the mandrel is lowered into the shell. The mandrel either grips the shell by expanding in case the shell is straight sided or it may be used to bottom drive or, in case of step-tapered piles, it may be used to drive on the shoulder of short segments of casing. During pile driving, the mandrel becomes the pile. It must be stiff enough to transmit the hammer energy efficiently to the pile tip, and in case of an expandable mandrel, strong enough to grip and drag the shell down without damaging it while under the stress imposed by the hammer blows. The mandrel permits the contractor to install light-gauge steel shells or thin wall pipe to relatively high capacity loads (> 100 tonnes) while saving on material costs.

Spuds are used to help pile driving in penetrating hard strata or seating the pile on rock. The spud may be a driving device by itself, or it may be a massive device attached to a pile (generally H-pile) seated into the rock. Seating a driving pile into a sloping rock is difficult as the pile may follow rock slope. Special driving points help piles seating adequately into the rock slope.

Leaders and pile frames

For accurate driving, it would be necessary:

- To pitch and hold the pile in position
- To support the hammer
- To guide both the pile and the hammer vertical or along the required angle

The driving method would depend on:
- Pile and hammer weight
- Ground conditions
- Pile length
- Manoeuvrability

Based on the above, pile frames or hanging leaders or rope-suspended leaders may be deployed for piling.

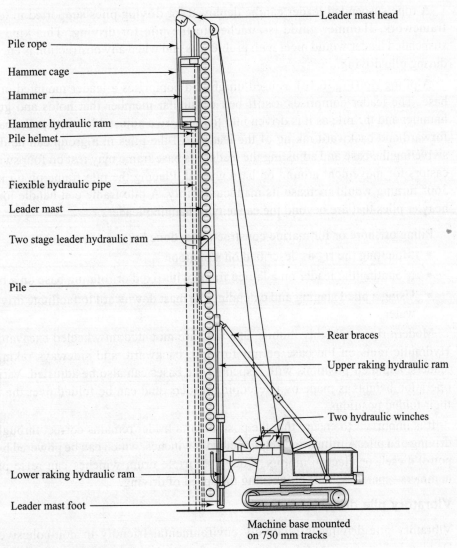

Fig. 6.16 Piling rig

A hanging leader comprises a lattice mast attached to a crane jib. They are designed for suspension from the jib of a crane or excavator. A steel strut, whose length can be adjusted, forms a rigid attachment from the foot of a leader to the base of the pile rig. A hydraulic ram can also be used in place of a steel strut to provide fixity to the leader and move the leader into the required position. The arrangement allows adjustment of the mast angles in forward, backward, and lateral directions; thus allowing piling to be carried out in uneven or sloping ground. Two winches are required on the base of the crane or excavator—one for lifting and pitching the pile and another for holding the hammer, which is also attached to guides on the leader to control the direction of pile during driving. A crawler-mounted hanging leader is suitable for driving on uneven or poor ground. The leader must be held firmly and rigidly to prevent the pile from drifting out of plumb during driving.

A rope-suspended leader can be deployed for driving piles supported in temporary framework. Hammer guide is attached to the pile for driving. This kind of rope-suspended leader would meet with problem as and when any obstruction is encountered during pile driving.

A piling rig (Fig. 6.16) as used on ground comprises a leader mounted on a crane base. The leader comprises a stiff box or tubular member that holds and guides the hammer and the pile as it is driven into the soil. Two adjustable struts at the base allow forward and backward raking of the leader. All the piles in a group can be driven by swiveling the base and adjusting the leader. Its base frame may rest on four swivel steel castors for movement around on hard ground. Placing the pile frame on turntable for 360° turning would increase its manoeuvrability. A pile frame can handle longer and heavier piles that are beyond the capacity of hanging leaders.

Piling offshore or for marine construction is done by:
- Mounting the rig as described on a pontoon
- Mounting the leader on a lattice frame with fixed or rotating base on a barge
- Using a piled staging and extending the mast downward to facilitate driving over water

Modern rigs are highly mobile, and they are mounted on wheeled excavator bases. Hydraulic rams on the base permit forward, backward, and sideways raking of the leaders, and their positions with respect to the bases can also be adjusted. Variation in operation heights is made by telescoping leaders, and can be folded over the bases in their folded positions.

It is important to ensure that the position of a leader remains correct throughout the driving of a pile. A piling rig is equipped with winches, which can be powered by steam, petrol/diesel, or electric motors. Double or triple drum winches can raise piles and hammers separately, thus quickening the speed of driving.

Vibratory pile drivers

Vibratory pile driving is relatively environmental friendly in non-cohesive sands. Driving is less noisy and free of exhaust fumes. As no hammering is involved, pile

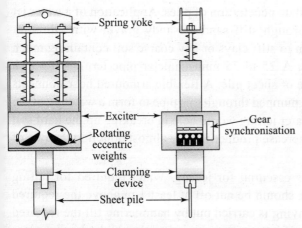

Fig. 6.17 Vibratory pile driver

heads are not likely to be damaged. This vibratory method may also be used for extracting piles. The vibratory drivers operate at either low (20 Hz) or high (40 Hz) frequency.

A vibratory pile driver's operation is based on two eccentrically and separately mounted cams (counter-rotating eccentric weights) rotating in unison—the two cams powered by a motor rotate in synchronized opposition to each other. During each cycle, the centrifugal forces act downwards at 0° and 360°, and upwards at 180° from the vertical; while at all other positions, the opposing horizontal components of their respective forces are cancelled. The whole unit is housed in a steel casing, which is suspended from the lifting wire rope of a crane. A suspension bracket is spring-mounted to the steel casing, thereby eliminating its upward vibration. The bracket is hung with a wire rope from a crane. The vibrator is attached to a sheet pile with remote controlled grips. The two eccentric cams produce the rotation in unison resulting in the oscillatory force shaking the pile up and down. This motion generated at low frequencies breaks down the static friction between the pile and the soil adjacent to it because of the soil being fluidized. The pile penetration is accomplished by:

- The push-pull of the counter-rotating cams—push (+ pile weight) > upward pull
- Fluidization of the soil in the immediate vicinity of the pile

The end bearing area of the pile is small. The pile sinks by its own weight and the weight of the pile driver. It is possible to control the effectiveness of the vibrator by regulating the frequency of vibration to suit particular soil conditions and to avoid resonance in nearby structures. It is also possible to adjust the amplitude of vibration by altering the mass of the eccentric weights.

Vibratory pile driving is most effective in saturated sands, and is also effective in loose to medium sands and gravels. Some difficulty may be encountered in very dense granular materials, where the downward dynamic force may not be sufficient to produce adequate penetration of the pile. This method of piling is less effective in cohesive soil as fluidization would not occur in cohesive soils. Penetration of piles in cohesive soils may be speeded up by electro-osmosis, which would result in precipitation of water in the proximity of the cathode pile. There are pile-drivers that operate at such high frequencies (up to 150 hertz) as would put the pile at its resonant frequency. Such pile-drivers could be used in stiff clay.

The principal advantages of vibratory pile drivers are:

- Reduced driving vibrations
- Reduced noise
- High speed of penetration

Water jetting may be used to facilitate penetration of piles. Application of a water jet assists the driving of sheet piles through stiff sand or sandy gravel with vibratory hammer. Jetting is ineffective in firm to stiff clays or any coarse soil containing much coarse gravels, cobbles or boulders. A 25 to 75 mm diameter pipe terminating in a tapered nozzle is attached to the side of sheet pile. A flexible armoured hose connects the pipe to the jetting pump. Water is pumped through this pipe to form a water jet at the base spreading at an angle of 30°. Water jetting would help in breaking up the hard soil at the tip of sheet pile. This would increase productivity by almost doubling the rate of penetration.

An adequate quantity of water is essential for jetting. Water required for jetting would be about 5 to 20 m^3/h. Jetting should be cut off at least 1 m above the required founding level and the remaining driving is carried out by hammering till the required resistance to penetration is achieved. If jetting is envisaged to start with, then jet pipes should form part of the piles. For jetting of particular piles in special cases, separate jet pipes should be used. In difficult conditions, two or more jet pipes may be used for a single pile.

Hydraulic pile driver

Hydraulic pile driver is suitable for relatively noiseless driving in fine granular sands or cohesive clay. This method tends to be restricted to micro piling (having diameter less than 250 mm, but more commonly 150 mm) as the reaction loads then are provided by the structure being underpinned. However, 'Taywood Pilemaster' system is used to drive sheet piles. In this system, adjacent piles provide reaction. The Pilemaster (pile driver) comprises electric motor, hydraulic pumps, fuel tank, etc. Eight hydraulic jacks

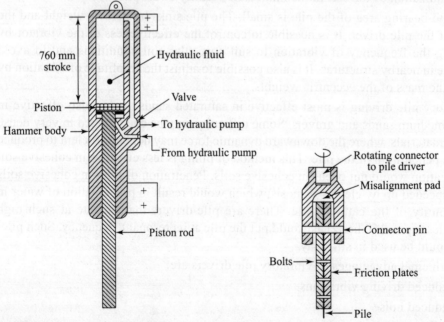

Fig. 6.18 Hydraulic pile driver

are attached to the base flange of crossheads. Oil at high pressure is pumped into the cylinder through a valve for raising or forcing the piston downwards for pile driving. Each jack has a fork-shaped steel connector connecting it to a friction plate, which is bolted to the sheet pile. The connector head may be rotated to fit any shapes of piles.

Piles are normally pitched in a temporary frame to provide initial support. The pile driver, which weighs about 10 tonnes and is hung from a crane, drives piles in panels of seven or eight piles. Pile driving commences with the central piles in a panel and working out to the end piles. When all the piles are driven to the jacks' capacity level, the pile driver is lowered for repeating the process at the lower level. The initial reaction for driving the central piles is derived from the pile driver's weight and the weight of the adjacent piles in the panel. The friction and restraint of the already driven first pair then provide additional restraint to drive the next pair of piles in the panel. A progressive accumulation of frictional restraint is, therefore, generated as the piles penetrate deeper into the soil, until the restraint is almost entirely frictional. When the panel of piles becomes self-standing, the crane connection is freed for pitching the next panel of piles.

Pile extractors

Pile extractors generally work using a hammering system in reverse and are specifically fabricated for pulling piles. The extractors are powered by compressed air or steam. Air or steam is admitted from the top to the upper cylinder until the base of the cylinder meets the underside of the ram. This would result in upward thrust to the pile. The whole unit is hung from a crane or frame via a spring mechanism to prevent vibrations being transferred to the lifting rope. This method can be used for extraction of light timber or sheet piles to heavy H-piles or even sheet piles. Vibration or hydraulic jacks can also be used for pile extraction.

6.4 LOAD TESTS AND QUALITY CONTROL

Quality of piles driven determines whether any pile foundation would measure up to the design requirements or not. Based on soil investigation results, pile foundations are designed. Pile driving results are recorded in standard pile driving log sheets.

Quality of materials and products is checked to assure conformance to the specifications and design requirement. Quality assurance is essential at every stage from the soil investigation right up to the driving of piles. A pile cannot be inspected once its driving/construction is completed. Variations in the bearing stratum are difficult to detect.

Therefore, the pile test method should be considered as part of the design and execution process.

Installation of piles, especially of those driven preformed types, disturbs, and re-moulds soils as the driving progresses. Driving may produce even high water pressure adjacent to the piles. Thus, the bearing pressure of piles may vary with time. A pile, therefore, should be tested after 1 to 3 weeks or more of driving. This time-lag between driving/forming and testing should conform to the specifications.

The two main aspects of pile quality to be considered are:

- The integrity of the pile and its capability as a structural unit to carry the applied load
- The load bearing and deformation characteristics of both the soil and pile

Pile testing may involve integrity testing to test quality of construction including workmanship and load testing to check its behaviour when subjected to loads as per specifications. All displacement piles are tested as they are installed by virtue of being driven to a set. This is a kind of load testing that is not applicable in case of replacement piles. However, load testing is done for both displacement and replacement piles. Load testing is a costly and time-consuming event, but it is the most reliable method of determining the load bearing capacity of a pile.

Load test

It is generally accepted that a load test is the most reliable means of checking the designer's estimate of safe working load and determining the actual load bearing capacity of a pile. A pile is driven by dynamic means; its driving history is recorded; and then the load testing is done. While destructive load tests are carried out on special test piles, non-destructive tests are carried out on working piles.

The objectives of preliminary destructive pile load tests are:

- To determine bearing capacity of piles
- To assess the pile resistance contributed by friction and end bearing capacity for design purposes
- To determine the stiffness of soil/pile combination at design load.

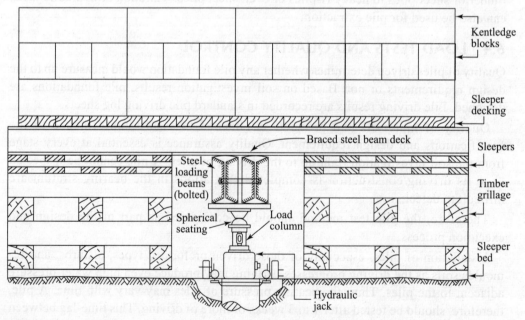

Fig. 6.19 Load test

Analysis of data collected during piling would help in evaluating the soil properties and ascertaining pile behaviour. This would further help in predicting possible deformation of pile groups

Destructive load test, which is subjected to 2 times the working load, is expensive because a special pile is to be installed only for this purpose. Load test is generally carried out on a pile that is installed for the structure by loading it with 1.5 times working load. At least 1% of the piles driven should be load tested. The preparation of a test pile involves cutting it down to the required level, thus exposing reinforcement and then casting a small pile cap over the pile for testing purposes.

Load is applied to the test pile by jacking for which provision of reaction is to be made. The ultimate load of a pile may range from a few tonnes to as much as 2500 tonnes in special construction work. Kentledge is most commonly used to provide the necessary reaction for jacking. Concrete blocks or steel plates/sections are used to form kentledge. It is to be ensured that load materials do not topple/slip during testing.

An alternative means of loading the pile is sometimes used by jacking against a rigid beam across the test pile and securely held down by two tension piles. Similarly, rock/ground anchors may also be used for providing reaction. A high capacity jack is placed between the kentledge/reaction beam and the test pile to generate the test load increments. Load measurement is carried out using at least two dial gauges of at least 0.02 mm sensitivity. Load is applied in steady increments at one-fifth to one-fourth of the working load and released gradually to establish dynamic characteristics, and then the pile is loaded to the test load limit while settlement is measured. Alternatively, load should be applied up to test load in 6 or 7 increments and unloaded in four decrements. Load should be maintained for 30 minutes after each increment/decrement except for the test load when the load should be maintained for 24 hours. Settlement should not exceed 12 mm, and settlement greater than 25 mm under 1.5 times working load is considered a failure.

The load test on its own is also an unreliable test because only one test is carried out in one location. This may not reflect the performance of other piles due to varying ground conditions. It is essential that all the relevant data is recorded and it is useful to use a standard *pro forma* for this purpose.

Integrity testing

Load test on all piles is not possible, but it is necessary to ensure that all the piles are in sound conditions. The non-destructive integrity test gives an indication of the soundness of the shafts of either bored or driven and cast-in-situ piles and is cheap enough to be carried out on all piles. The basic important methods followed for integrity tests are:

- Acoustic
- Radiometric
- Seismic
- Dynamic response

For the tests to be carried out, sufficient time should be allowed for concrete to mature for testing. All these methods are quite reliable to indicate a picture of the length of a pile or to readily identify gross imperfections, such as cracks, necking, honeycombing, segregation or large soil inclusions. However, such tests do not give conclusive evidence on cracks.

A probe is lowered down a small diameter hole drilled or formed in the pile in case of acoustic and radiometric methods. The acoustic probe transmits ultrasonic pulses that pass through the mass of concrete down the pile and are picked up by a receiver mounted at the bottom of the probe or in an adjacent test hole in the same pile. Radiometric probes employ gamma ray backscatter equipment.

In seismic method, sound waves are sent down the pile by hitting the top of the pile with a hammer or mechanical impulse device. Reflected waves from the pile toe are picked up by accelerometer mounted on the pile head and recorded by an oscillograph. The dynamic response method employs a vibrator attached to the pile head and a velocity transducer monitors the response.

SUMMARY

Apart from transferring load to the underlying firmer soil strata, piling enhances load-bearing capacity of surrounding soil. This has eliminated the necessity of designing buildings and structures based on favourable soil investigation reports. There are different types of piles from simple timber piles to giant piles required for marine construction or building bridges across rivers and bays. Piling is not environment friendly. Hydraulic system is sure way of overcoming this problem. CFA piles also can be augured into the soil without disturbing the environment. Pneumatic equipments disturb environment in general. As more and more high-rise buildings or bridges spanning rivers and bays continue to be built, piling cannot be avoided. If there still remain problems associated with installation of piles, those problems are to be solved. Displacement piles are driven into the ground by hammering thus displacing soil sideways to allow room for them to penetrate. This simply compacts the surrounding soil allowing them to be more load-bearing in non-cohesive soils. However, this may cause heaving in cohesive soils and so cannot be used adjacent to any existing foundations. Replacement piles are produced by excavating boreholes in the ground and filling the void with concrete. Replacement piles are thus columns of concrete cast in the ground in a pre-drilled hole. Large diameter bored piles are formed by augers as well. Continuous flight auger (CFA) piles are clearly replacement piles. Vibratory pile driving compared to impact driving is environmental friendly. Hydraulic pile driver is operated without making noise. Quality assurance is essential at every stage of piling from soil investigation right up to actual execution of pile driving. A pile cannot be inspected once it's driving/construction is completed. Variations in the bearing stratum are difficult to detect. The pile test method, therefore, should be considered as part of the design and execution process.

REVIEW QUESTIONS

1. Apart from transferring loads to firmer strata, what are the other purposes for which piles are used?
2. What are the parameters to decide whether piling would be necessary or not?
3. Is sheet pile a pile in conventional sense? Justify your answer.
4. Piles are of two generic types based on soil mechanics. Define these two types.
5. Piling is classified into two categories. Explain these two categories in details.
6. Displacement piles can be driven in aggressive or contaminated ground conditions. How?
7. Displacement piles are driven to a 'set'. What is this set and what does it signify?
8. Prestressing of reinforcement of precast piles is done for many reasons. What are the reasons?
9. What are the disadvantages of precast piles?
10. What are the advantages of preformed steel H-piles?
11. Driven cast-in-situ displacement piles are of two types. Briefly describe them.
12. Describe the methods of producing replacement piles.
13. What are the advantage of using CFA piles?
14. Describe a piling rig and how it works.
15. Single-acting diesel hammer is a better alternative to drop hammer or pneumatic hammer. Why?
16. Why are hydraulic impact hammers manufactured in large sizes?
17. Describe how a pile is load tested in-situ.

7
Concrete and Concreting

INTRODUCTION

Concrete is a widely used construction material. It is produced by mixing properly proportioned quantities of aggregates, cement, and water by weight. Admixtures may be added in the mixture to produce concrete of specified quality. Concrete may be moulded into any desired shape or size. Concrete is reinforced with steel to make a composite material that combines the ceramic properties of concrete with the tensile strength of steel. The integrity of reinforced cement concrete (RCC) depends to a large extent on the reinforcing steel, plain or deformed.

Concrete in its plastic state must remain workable until it is placed in position and compacted. Hardened concrete must be durable against the process of deterioration. There are different types of concrete, such as shotcrete, light-/heavyweight concrete, ready-mixed concrete, high performance concrete, self-compacting concrete, polymer-modified concrete, fibre reinforced concrete, etc. The influence of the elastic incompatibility of steel and concrete is avoidable in prestressed concrete.

There are specifications and guidelines for production, transportation, and placing of concrete in extreme hot/cold weather. Underwater concreting is a special operation that may have to be carried out in remote and difficult areas in unusual environment. Timber or metal forms are used to mould the concrete as per the requirement. Stripping of forms from the set concrete demands the same care that goes into the fabrication and erection of forms. Curing process is a very important factor in determining the strength of concrete.

Quality of concrete is to be assured right from the stage of procurement of the ingredients of concrete including testing of water available at the site. Quality of concrete is to be checked by carrying out both destructive and non-destructive tests.

7.1 DEFINITION OF CONCRETE

Concrete is a versatile construction material. It is defined as a properly proportioned, homogeneous, and dense mixture of fine and coarse aggregates, cement, and water with or without admixtures. The aggregates are considered as economic filler materials, generally inert in nature. Cement and water comprise a continuous binder phase that, after hardening, holds the aggregates together into a compact mass having load-bearing capacity. There are many other binders, such as asphalt, sulphur, epoxy resin, and so on—but the unqualified term 'concrete' means that the binder is principally hardened cement paste, which is the product of the chemical reaction of the ordinary Portland cement and water. In common parlance, the term concrete implies ordinary Portland cement concrete despite wide acceptance of the blended cements.

An admixture is a material other than the essential ingredients such as water, aggregates, and Portland cement (ordinary/blended) used as an ingredient of concrete and added to it immediately before or during its production.

Concrete is the most widely used material, second only to water, which is used in larger quantities. Concrete is a material that has been used in some form since the ancient times because:

(a) It possesses excellent resistance to water unlike wood and steel.
(b) It can be easily cast or formed into any predetermined shape or size.

But the era of modern concrete dates from the middle of the nineteenth century with the advent of the first truly 'Portland' cement.

Although aggregates constitute about 70% of the produced concrete, it is the cement paste that is responsible for most of the good and bad qualities of concrete. For a given aggregate, physical and chemical characteristics of the cement paste determine the workability of plastic concrete as well as the most significant engineering characteristics of hardened concrete such as strength, durability and dimensional stability. However, it has certain weaknesses. It is brittle and very poor in tension. Ductility and toughness are also poor.

Despite its deceptive *uniformity*, cement is not a unified chemical entity. It comprises at least four major phases, which retain their different chemical identities during the hydration process (vide Section 7.3). Concrete strength and durability are affected significantly by the relative proportions of these four phases of cement with some minor constituents of cement influencing concrete durability.

It is well known that reinforced concrete is a composite material, combining the ceramic properties of concrete with the tensile strength of steel. The quality of both these components is essential to increase the maintenance–free life of reinforced concrete structures in marine conditions, severe atmospheric pollution, and other extreme conditions.

7.2 IMPORTANT PROPERTIES OF CONCRETE

Both plastic concrete and hardened concrete have distinct desirable properties. The essential properties of plastic concrete are:

- Workability
- Non-segregating
- Setting in specified time

The essential properties of hardened concrete are:

- Strength

- Water-tightness
- Durability
- Volume stability
- Abrasion resistance
- Economy

Plastic Concrete

Workability

Workability is that property of plastic concrete mixture which determines the ease with which it can be placed and the degree to which it resists segregation to produce full compaction. It embodies the combined effect of mobility and cohesiveness. It is important that workability of fresh concrete be such that the concrete can be properly compacted and easily transported and placed without segregation. The production of workable concrete depends on the working out of the appropriate water:cement ratio based on the grading and shapes of the aggregates. Water, apart from reacting with the cement and forming a gel that binds the aggregates together, provides mobility to the concrete mix thereby facilitating placement and compaction of concrete.

The ability of concrete to flow would depend on overcoming the internal friction between the individual particles in concrete as well as the surface friction between the concrete and the formwork. However, overcoming internal friction is the characteristic of the concrete alone. The importance of adequate compaction cannot be over emphasized, for tests have shown that for every 1 percent of entrapped air, the compressive strength of concrete tends to fall by 5 to 6 percent. For any given method of compaction, the optimum water content of the concrete mix would be such as to have the sum of volumes of air bubbles and of water space to be minimum and the density to be maximum. Workability is also relevant in pumping concrete (vide Chapter 13, Section 13.6).

Aggregate particles having sharp edges and rough surfaces need more water compared to smooth and rounded particles. The cement content of a mix made with crushed aggregates or irregularly shaped gravels should be increased to allow water to be added to make the concrete sufficiently workable without reducing the strength below the required level. However, a crushed aggregate concrete may have a higher strength than a smooth or rounded aggregate concrete of the same water:cement ratio, and this extra strength may be sufficient to offset the effect of extra water.

The fine and coarse aggregates should be proportioned to obtain the required degree of workability with the minimum amount of water. If the fine and coarse aggregates are badly proportioned, an excessive amount of water is required to maintain adequate workability. The excess water would result in low strength and poor durability.

Workability depends on a number of interacting factors as follows:

- Water content
- Fineness of cement
- Types and grading of aggregates

- Aggregate:cement ratio
- Presence of admixtures

Over the years, many devices have been developed for measuring workability as follows:
- Slump test
- Veebee consistometer test
- Ball penetration test
- Compacting factor test
- Flow test

Slump test

The usual test for measuring workability at a construction site globally is the slump test because of its simplicity and usefulness, but this test has limitations. A conical mould, frustum of a cone, of 300 mm height (Fig. 7.1) is used in this test. The mould is placed on smooth and level surface with the smaller opening (100 mm diameter) at the top with the bottom opening of 200 mm diameter. The mould is filled with concrete (produced with coarse aggregates of up to 38 mm size) in four equal layers of approximately one-fourth height each and tamped 25 times per layer with a 600 mm long steel rod (rounded at the end) of 16 mm diameter. The top is struck level, the mould is removed and the unsupported concrete is allowed to slump. The subsidence of the specimen, slump, is measured as compared to the mould height. Slump of 25 mm or less implies low workability. Slump of 25 to 75 mm implies medium workability and over 75 mm implies high workability or collapse. Mixes of stiff consistency show zero slump—that is, in rather dry range, no variation would be observed between mixes of different workability. The details of carrying out slump test along with a sketch of a mould are given in IS: 1199–1959 (reaffirmed 1999).

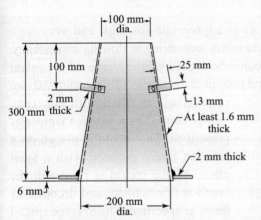

Fig. 7.1 Slump cone

Compacting factor test

The shortcomings of the slump test, particularly with regard to drier concrete, may be overcome by using compacting factor test in which concrete is allowed to fall under gravity through hoppers into a cylinder from standard height. The hoppers and cylinder shall be of rigid construction, true to shape and smooth inside. The hoppers should be closed at the bottom with hinged trap doors. The ratio of the weight of concrete compacted in the lowest cylinder divided by the same volume of fully compacted concrete is the *compacting factor* and this value, naturally, is less than one.

The apparatus (Fig. 7.2) comprises two hoppers each in the shape of a frustum of a cone mounted above a cylinder at the bottom—the three vertically in a row. Concrete is dropped from the top hopper to the one below and finally into the cylinder. The compacting factor of the value of 0.85 is low, 0.92 is medium, and 0.95 is high. The procedure of this test, normally carried out in a laboratory, along with the sketch of a compacting factor apparatus, are given in IS: 1199.

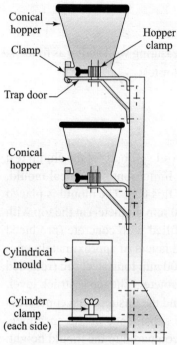

Fig. 7.2 Compacting factor test arrangement

Veebee consistometer test

The Veebee consistometer (Fig. 7.3) test measures the time required to change the shape of the concrete from a standard compacted slump cone to a flat cylinder under vibration. The name Veebee is derived from the initials of V. Bahrner of Sweden, who developed the test. Veebee tests determine the time required to achieve compaction, which is related to the total work done. Compaction is achieved using a vibrating table. The workability is expressed in seconds of time required to level the cone—10 mm slump is approximately equivalent to 12 seconds, 30 mm equivalent to 6 seconds, 60 mm to 3 seconds, and 180 mm to less than 1 second.

Veebee consistometer test is suitable especially for dry mixes. The procedure of this test as carried out along with proper sketch is given in IS: 1199.

Flow test

Flow test (Fig. 7.4) is appropriate for high and very high workability concrete mixes amounting to flowing consistency. A 120 mm high concrete mould of the shape of a conical frustum with top and bottom diameters as 170 mm and 250 mm respectively is filled with concrete, lightly tamped by a wooden tamper as prescribed. The mould is placed on a wooden board covered with a steel plate. As the mould is removed, the board is lifted, tilted, and dropped 15 times as specified. The average spread of concrete denotes the workability measure. The details of the procedure for flow table test are given in IS: 1199.

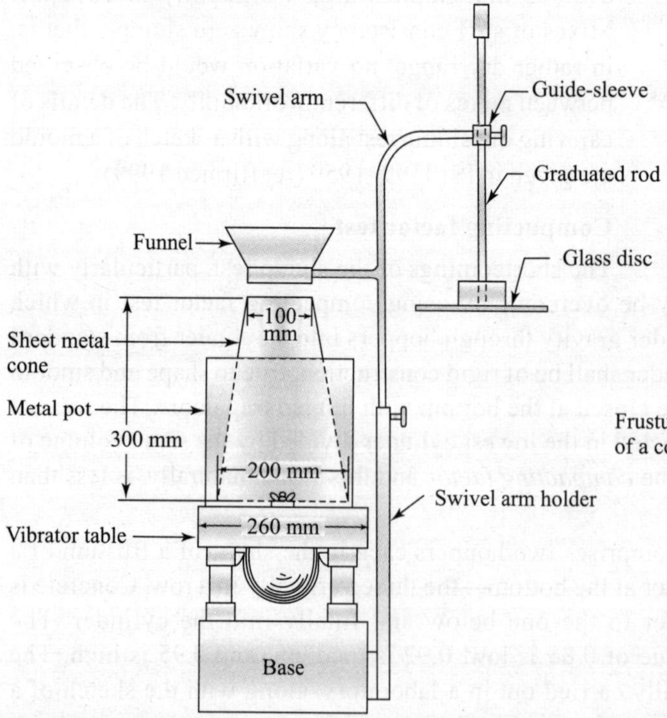

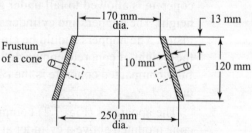

Fig. 7.4 Mould for flow test

Fig. 7.3 Veebee Consistometer

Ball penetration test

Ball penetration test is a simple test carried out at the construction sites. A

152 mm (6") diameter metal hemispherical ball weighing 13.6 kg (30 lb) is allowed to sink in fresh concrete under its own weight. This test is conducted in the USA as routine checking, just as slump test is conducted all over the world. For this test, depth of concrete should not be less than 200 mm (8") and the least lateral dimension should be 457 mm (18").

Segregation

As concrete, by definition, is a properly proportioned, homogeneous, and dense mixture of heterogeneous ingredients; segregation of ingredients is one of the most detrimental characteristics of plastic concrete. It is much easier and safer to take precautionary measures against segregation than to repair its damaging effects. It is the difference in the sizes of particles of solid ingredients and specific gravity of the constituents of the mixture that cause segregation. Higher viscosity of the cement paste in plastic form with low water:cement ratio would hinder downward movement of heavier solid particles, thereby hindering segregation.

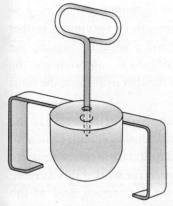

Fig. 7.5 Ball penetration test

In case of dry mix with less water, coarser ingredients would settle down faster; and in case of too wet mix, cement paste would separate out. Segregation could be avoided by producing cohesive mix using properly graded aggregates using optimum amount of water maintaining appropriate water:cement ratio.

The last opportunity for controlling segregation of plastic concrete is provided during its consolidation in the forms. Total consolidation is essential to achieve maximum durability, strength, economy, and uniformity. Availability of better vibration tools and devices have made it possible to use stiffer mixes containing less cement and water, resulting in production of higher quality of concrete avoiding segregation.

Setting time

Setting, which is different from hardening, has been discussed in detail in Section 7.3. Hardening refers to the gain of strength of set cement paste. Setting, on the other hand, refers to the change from a fluid state to rigid state. Of the four compounds that are formed on hydration of cement with water, the reaction of pure C_3A starts first violently leading to immediate stiffening of the cement paste. This is termed as 'flash set'. The addition of gypsum in the cement production allows C_3S to set first, and let it initiate the setting process with 'initial set'. C_2S sets in a more gradual manner. 'False set', another type of abnormal premature setting within a few minutes of mixing with water, occurs under certain conditions. False set affects both the hydration process and gain of strength. It is now accepted that false set is caused by dehydration of gypsum, and this problem may be overcome during cement manufacture.

Hardened Concrete
Strength

The strength of concrete is usually measured by the resistance it offers to axial compression as all concrete structures in general are subjected to compressive forces.

The most important overall measure of quality of concrete is strength. There may be other critical characteristics as well. The compressive strength of specimens (either cubes or cylinders) made with concrete has traditionally and practically become the criterion of the acceptance of the concrete so as to be considered as conforming to the design requirement. The strength of concrete depends on a number of factors, the most important of which are discussed below.

The most critical factor affecting the compressive strength of concrete is the water:cement ratio. The function of water in concrete: (i) it forms a gel with cement to bind the aggregates together, and (ii) it provides mobility to the mixture for placement and compaction of concrete. The actual quantity of water required for workability is more than the quantity required for hydration and depends on the surface area of the aggregates to be wetted and the type of equipment available for compaction.

The extra water increases porosity in the cement paste and reduces concrete strength. The higher the water:cement ratio, the lower is the compressive strength. Even with the same materials and fixed water:cement ratio, increase in the aggregate:cement ratio would result in the increase of compressive strength subject to the concrete being fully compacted. Thus, for any water:cement ratio, there is also an optimum aggregate content that would give the maximum compressive strength.

The cement and aggregates used have significant influence on the strength properties. The use of unsound aggregates would cause large variations in the strength of concrete. However, quality aggregates would develop the full strength of the cementing matrix, thus causing less variation.

On the basis of their shape and surface texture, the aggregates influence the water demand for workability and thus influence the strength of concrete indirectly through change in water:cement ratio. The usual alternative is to fix the water:cement ratio for the required strength, and then adjust the aggregate: cement ratio for the desired workability.

The surface texture and mechanical strength of the aggregates directly influence strength. As already mentioned, the bond between the aggregate particles and the cement paste depends largely on the surface texture of the aggregates. The smoother is the surface, the poorer is the bond as would be evident from the use of gravels in concrete. Regarding mechanical strength of the aggregates, it is significant at higher strength concrete. The failure at higher strength level may occur due to crushing of aggregates rather than through the failure of the cement paste or bond. Aggregates having flaws cause micro-cracking at the interface with cement paste. It is not possible to quantify porosity in cement paste and micro-cracking in any meaningful manner. For engineering purposes, empirical studies of the effects of various factors need to be resorted to. For any particular type of aggregates, there should be a ceiling strength of concrete.

Water-tightness

Water seeps through concrete:
- When under pressure
- By capillary forces

Well produced, placed, and compacted concrete could be impervious, had it been composed of completely solid matter. But concrete is produced with more water than is required for hydration of cement for workability. The excess water eventually evaporates creating voids and cavities, which may be interconnected and form continuous passages. Entrained and entrapped air also produces voids in the concrete. However, permeability of concrete may be controlled to make construction of water-tight concrete structures possible.

Water-tight concrete is produced using non-porous aggregates surrounded by an impervious cement paste. The mix should have a low water:cement ratio and should be appropriately cured for at least two weeks or up to specified curing period. Newly placed concrete should be protected from the adverse effects of rain, sun, and wind by approved means. Leaks are confined to small areas and around construction joints. Construction joints are particular sections which are often porous, unless special preparations for these areas are made. As such, placing of concrete should be continuous whenever and wherever possible. Where expansion joints are necessary, the concrete should be continuous between joints. However, where construction joints are made, it is necessary to obtain a good bond between adjacent sections.

Capillary forces will cause a slight passage of water whenever such water comes in contact with concrete for considerably long time. Water seeps through concrete when under pressure and by capillary forces.

Where even a slight amount of moisture is objectionable, damp-proofing and water-proofing should be assured by using admixtures, coatings, or felts.

Durability

Concrete, after the first 28 days, continues to mature and age depending on the original material composition and properties as well as environmental exposure and aggression during its service life. Nevertheless, concrete is highly durable material, requiring very little maintenance and performing very well under a wide range of service conditions. Some of the more common adverse service conditions which good concrete needs to endure are weathering which includes erosion, freezing and thawing, and chemical conditions which exist in surface water and sea water. Durability under these adverse conditions depends on many different factors and is highly variable.

Durability is not a property of concrete but its behaviour in certain exposure condition. It is its capability to withstand the test of time against the processes of deterioration by the adverse nature, environmental conditions, and aggressive forces including man-made forces. The relevance of the term 'durability' assumes far greater significance for concrete because concrete is a homogeneous mixture of heterogeneous materials produced at site under difficult and varying working conditions.

The most important factor that is relevant for durability of any concrete structure is the transportation mechanism of fluids within microstructure, pores, and cracks. The ease with which a fluid can flow through a solid is called permeability. Pores are classified as capillary pores, gel pores, and pores due to compaction and entrained air. Gel pores are micropores. Aggregate may also have pores, but they are discontinuous and enveloped by the cement paste.

Of transportation mechanism, permeability refers to pressure differential, diffusion refers to differential in concentration, and sorption is the result of capillary movement. The resistance to permeability is caused by the viscosity of the fluid, the friction at the pore and crack walls, and the narrowness and tortuous paths of the pores and cracks. Permeability of cement paste varies with the progress of hydration, and it is lower for pastes having higher cement content, that is, low water:cement ratio. The ease with which air, some gases, and water vapour can penetrate into concrete is relevant for durability of concrete under different exposure conditions. Permeability is, therefore, a primary factor which influences both concrete and steel behaviour. The extent of permeability varies considerably and depends on factors such as cement, water:cement ratio, and age. Electromigration is a transport mechanism that affects all ion transport in concrete.

Concrete cracking occurs when the concrete is subjected to tensile strain that exceeds its resisting capacity. Types of cracks:

After hardening:
- Physical — (i) shrinkable aggregates (ii) drying shrinkage (iii) crazing — minute hair cracks that sometimes appear on the surface of precast concrete work
- Chemical — (i) corrosion of reinforcement (ii) alkali aggregate reaction (iii) carbonation
- Thermal — (i) freeze/thaw cycle (ii) external seasonal temperature variation (iii) early thermal contraction
- Structural — (i) accidental overload (ii) creep (iii) design loads

Before hardening:
- Plastic — (i) plastic shrinkage (ii) plastic settlement
- Constructional movement — (i) formwork movement (ii) subgrade movement

Discussed below are the factors which affect durability.

1. Carbonation

CO_2 present in the atmosphere reacts with calcium hydroxide solution in gel pores to form neutral calcium carbonate. However, the actual agent is carbonic acid (H_2CO_3) because gaseous CO_2 is not reactive. Of the hydrates in the cement paste, the one that reacts readily is $Ca(OH)_2$.

- $Ca(OH)_2 + CO_2 \rightarrow CaCO_3 + H_2O$
- $Ca(OH)_2 + H_2CO_3 \rightarrow CaCO_3 + 2H_2O$
- $CaCO_3 + CO_2 + H_2O \rightarrow Ca(HCO_3)_2$

Due to the effect of the above process, the pH around steel which is generally 12 to 13 is gradually reduced to a level of 9 to 9.5. On carbonation of all $Ca(OH)_2$, the pH value is reduced to 8.3. This alkaline environment is no more sufficient to ensure passivity of the oxide film and the presence of moisture and oxygen causes the reinforcement to corrode.

Concrete in which the pH value of the capillary water is below 9.5 is known as carbonated concrete.

In highly alkaline concrete (pH value > 12.4), passivity film of ferric oxide (Fe_2O_3) is formed over reinforcing steel providing complete protection from reaction with oxygen and water, that is, corrosion.

The processes of carbonation proceed from the surface to the inside of concrete. The rate of carbonation broadly depends on:

- Relative humidity of the atmosphere—the rate of carbonation is maximum at a relative humidity of 50 to 70 percent
- CO_2 content in the air—varies from 0.04 to 0.4 percent
- Moisture content of concrete
- Cement factor/grade of concrete
- Quality of concrete—rate of carbonation depends mainly on permeability of concrete—lower the pore size, lower will be the rate of carbonation
- Time

The depth of carbonation can be calculated in mm from the following formula:

$$C = k \sqrt{T}$$

where

C = Depth of carbonation

T = Time in years

k = Coefficient depending on environmental and physical condition of concrete (value varies from 3 to 4 mm/year)

As environmental and site conditions are not changing much, the only way to protect the concrete from carbonation and ultimately corrosion is to take precautionary measures. The following points must be taken care of, both by the design engineers during detailing and by the construction engineers during execution:

- Adequate concrete cover to reinforcement—The thickness of concrete cover to steel will also influence the steel corrosion. As with the permeability, it affects the time taken for salts to penetrate the steel.
- Dense and well compacted concrete free from honeycomb
- Low water:cement ratio in saline environment, the water:cement ratio must not exceed 0.4—lower the water:cement ratio better is the corrosion protection.
- High cement content—High cement content in concrete is one of the means of high durability and corrosion protection. Such concrete having high cement content will contain few porous capillary voids caused by the mix water.
- Good grouting of pre-stressing steel
- Use of aggregates that do not react with cement
- Coating to reinforcement if economically viable—Central Electro chemical and Research Institute, Karaikudi, Tamilnadu have developed painting for steel

reinforcement, which is reported to give an excellent protection. This was used in some bridges in Maharashtra.
- Cathodic protection—Cathodic protection prevents corrosion of steel by supplying current flow that suppresses the galvanic corrosion cell. This can be achieved by applying direct electrical current or by using sacrificial anodes (+ anode, –cathode).
- Surface coatings, membranes, and sealants—on concrete surfaces

Use of galvanized steel with high quality concrete with proper water:cement ratio, good compaction and workmanship will not allow the rebar to rust even in a generally aggressive environment. The use of galvanized re-bars, therefore, helps assuring the quality and long-maintenance free life of reinforced cement concrete constructions.

Of the positive aspects of carbonation, $CaCO_3$ occupies greater volume than $Ca(OH)_2$ that it replaces. Thus, pores in carbonated concrete are reduced. And, water released on carbonation may help hydration of hitherto unhydrated cement.

2. Acid attack

Acid attack occurs by decomposing the products of hydration and forming new compounds. Such new compounds, if soluble, may be leached out. If not, the result may be disruptive. The attacking compounds need to be in the form of solution. The calcium compounds such as calcium hydroxide, calcium silicate hydrate, and calcium aluminate hydrate (Section 7.3) are converted to calcium salts. The most vulnerable cement hydrate is calcium hydroxide.

Liquids with a pH value below 6.5 can attack concrete, but such attack could be severe if the pH value is below 5.5 and very severe if the value is lower than 4.5.

Acid rain comprising mainly sulphuric acid and nitric acid having a pH value between 4.0 and 4.5 may cause surface weathering of exposed or unprotected concrete.

Sulphuric acid is particularly aggressive as, in addition to the sulphate attack of the aluminate phase, acid attack, on calcium hydroxide and calcium silicate hydrate takes place.

In rare tropical conditions, some algae, fungi and bacteria can convert nitrogen present in the air into nitric acid that may cause acid attack on concrete.

Use of blended cements is beneficial in reducing the ingress of aggressive substances.

3. Sulphate attack

When sulphate solutions gain entry into concrete mass, crystalline salt formation takes place. These voluminous salt formations give rise to rupture in concrete. Solid salts as such do not affect concrete. But when in solution, they can react with hydrated cement paste. Disruption may occur due to sulphate expansion because of such reaction.

Sulphates of sodium, potassium, magnesium, and calcium occur generally in adjacent soil or groundwater. Sulphates in groundwater may be of natural origin or external origin like fertilizers or industrial effluents. Soil of abandoned industrial units may contain sulphates or other aggressive constituents.

The patterns of reactions of the sulphates with hardened cement paste are indicated below. The reactions involved are complex.

- $Ca(OH)_2 + Na_2SO_4 \cdot 10H_2O \rightarrow CaSO_4 \cdot 2H_2O + 2NaOH + 8H_2O$
- $2(3CaO \cdot Al_2O_3 \cdot 12H_2O) + 3(Na_2SO_4 \cdot 10H_2O) \rightarrow (3CaO \cdot Al_2O_3 \cdot CaSO_4 \cdot 31H_2O) + 2Al(OH)_3 + 6NaOH + 178H_2O$
- $3CaO \cdot 2SiO_2 \cdot (aq) + MgSO_4 \cdot 7H_2O \rightarrow CaSO_4 \cdot 2H_2O + Mg(OH)_2 + SiO_2(aq)$
- $MgSO_4 + Ca(OH)_2 \rightarrow CaSO_4 + Mg(OH)_2$

In the first equation, sodium sulphate attacks calcium hydroxide totally leaching out calcium hydroxide in flowing water unless accumulation of sodium hydroxide brings about equilibrium. In the second equation, calcium sulphate attacks only calcium aluminate hydrate readily forming calcium sulphoaluminate ($3CaO \cdot Al_2O_3 \cdot 3CaSO_4 \cdot 31H_2O$) known as *ettringite*. In the third equation, magnesium sulphate attacks calcium silicate hydrate. Magnesium sulphate also attacks calcium hydroxide and calcium aluminate hydrate. Under certain conditions, the attack by magnesium sulphate is more severe compared to other sulphates.

This damage on concrete itself reduces the cover to reinforcing steel and thus reinforcement becomes susceptible to corrosion. Concrete contains free lime which is soluble. As a result of penetration of calcium sulphates, calcium sulphoaluminates (ettringites) are formed, most readily in the presence of concentrated $Ca(OH)_2$ solution, which are of expanding type. Effects on the parts which are above the ground are also prominent. By wicking action, sulphate solution rises in concrete.

In arid climate or in areas subjected to alternate drying and wetting, as the solution evaporates from concrete just above ground, the crystals of sulphoaluminates or ettringites develop in pores under the surface with sufficient force to flake off. Concrete attacked by SO_3 has characteristic whitish appearance. The damage usually starts at edges and corners and is followed by progressive cracking and spalling. The sulphate solution has deep penetrating capacity due to high wetting power and combines with aluminate of set cement to form ettringites with increased volume. This is also now being eliminated by using low alumina or sulphate resistant cement.

Apart from disruptive expansion and cracking, concrete also loses strength on account of loss of cohesion in the hydrated cement paste and of adhesion between it and the aggregates.

The precautions to be taken against sulphate action:
- Use of sulphate resistant cement with controlled C_3A content
- Use of blended cements containing blast furnace slag or pozzolona
- Use of rich and dense concrete having low permeability to prevent ingress of sulphates
- Increased cover for reinforcements
- Use of truly potable water for concrete production

4. Efflorescence

Efflorescence is the deposit of leached lime compounds on the concrete surface. Concrete contains a small amount of $Ca(OH)_2$, which when brought to the surface, combine with CO_2 in the air to form $CaCO_3$ to remain as white deposit on the surface. Water percolates through poorly compacted concrete or through cracks or via badly formed joints, and when surface water can evaporate from concrete. Use of unwashed seashore aggregates having salt coating on the surface can also cause efflorescence. Efflorescence can be controlled by producing dense concrete and forming watertight joints.

5. Effects of sea water in concrete

The causes detrimental to the question of durability of concrete exposed to sea water and marine environment could be physical, mechanical and chemical action of aggressive media. The causes:

- Chemical attack
- Salt weathering
- Chloride-induced corrosion
- Freeze-thaw attack

Chemical attack

Chemical action of sea water is related to its containing a number of dissolved salts. The total salinity on the average is 3.5 percent. Sea water also contains some dissolved CO_2. The pH value of sea water varying between 7.5 and 8.4 is almost in equilibrium with atmospheric CO_2. Ingress of sea water as such does not affect the pH value of pore water of hardened cement paste. Because of large quantities of sulphates in sea water, there exists a possibility of sulphate attack on concrete. Sulphate ions react with both C_3A and C-S-H to form ettringite, though without causing damage as both ettringite and gypsum are soluble in water in the presence of chlorides facilitating easy leaching out by the sea water.

As regards the use of sulphate resistant cement in concrete exposed to the sea, it would not be necessary if C_3A content remains within 10 % with SO_3 content not exceeding 2.5 %. Apparently, excess SO_3 causes delayed concrete expansion. C_4AF also leads to formation of ettringite which would require the content of $2C_3A + C_4AF$ to be less than 25 % of the clinker for sulphate resistant cement.

What is discussed on the effects of sea water is applicable to structures permanently immersed in water. This condition is better than alternating wetting and drying because salt deposits can build-up within concrete as a result of the ingress of sea water followed by evaporation of pure water leaving salts behind in the form of crystals, mainly sulphates. The chemical action of sea water on concrete is as follows. The magnesium ion present in the sea water substitute for calcium ion:

- $MgSO_4 + Ca(OH)_2 \rightarrow CaSO_4 + Mg(OH)_2$

$Mg(OH)_2$ formed is known as *brucite*. It precipitates in the pores at the surface of the concrete and forms a protective surface layer that impedes further reaction. Some precipitated $CaCO_3$, in the form of aragonite, arising from the reaction of $Ca(OH)_2$ with CO_2, may also be present. It has been observed in a number of fully submerged sea structures that the precipitated deposits form rapidly. The blocking nature of brucite makes its nature self-limiting. However, if abrasion can remove the surface deposit, then the reaction by the magnesium ion freely available in the sea water continues.

Salt weathering

When concrete is wetted repeatedly by sea water, some of the salts, mainly sulphates, dissolved in sea water are left behind in the form of crystals. These crystals re-hydrate and grow upon subsequent wetting and exert an expansive force on the surrounding hardened cement paste. This kind of progressive surface weathering is known as *salt weathering*. Such weathering takes place when rapid drying occurs in the pores over some depth from the surface due to temperature rise because of strong insolation. Therefore, intermittently wetted surfaces are vulnerable; these are surfaces of concrete in tidal zone and splash zone. Horizontal or inclined surfaces are particularly prone to salt weathering, and so are surfaces wetted repeatedly but at long intervals so that thorough drying can take place. Salt water can also rise by capillary action causing similar disruption.

There are indirect methods of salt weathering also. This happens when deposits of airborne salt on the concrete surfaces get dissolved by dew and subsequently get evaporated. This kind of weathering happens particularly in arid desert regions where condensation in the form of dew occurs in the early hours of the morning.

Salt weathering can ravage concrete surfaces. It can extend to a depth of several millimetres and extend further leaving protruding coarse aggregate particles in course of time. Eventually, protruding particles get loosened, thereby exposing more hardened cement paste, which would in turn be liable to salt weathering. Unless coarse aggregates are dense having low absorption, there would always be the possibility of aggregates being liable to deterioration.

It is important for the concrete to be well cured prior to exposure to sea water; unless concrete, once immersed in sea water, remains permanently submerged. It is wrong to believe that sea water also provides curing. Tests on mortar have led to a recommendation of a minimum period of seven days of curing in fresh water, regardless of the type of cement used. Concrete exposed to deterioration by sea water should have low water:cement ratio, appropriate cementitious materials, adequate compaction, absence of cracking due to shrinkage, thermal effects, or stresses in service.

Chloride induced corrosion

Chloride primarily attacks reinforcing steel and causes corrosion. Concrete surrounding the reinforcing steel is affected as a result of corrosion.

Immediately after the starting of hydration of cement, a passivity film of ferric oxide (Fe_2O_3) is formed over the reinforcing steel. This film tightly adheres to the reinforcing steel providing complete protection from corrosion. However, chloride ions destroy this film and corrosion occurs in the presence of water and oxygen.

The mechanism of corrosion resembles that of a galvanic cell. To get activated, it requires the presence of three elements, namely (i) oxygen, which exists in abundance in air; (ii) an electrolyte—most gases, salts, acids, and alkalis, when dissolved in water, form an electrolyte; and (iii) electrical potential differences may result from the variations in oxygen availability, concentration of electrolyte, heterogeneity in steel composition, and surface conditions. Two points of a reinforcing steel bar in concrete along which a difference in electrical potential has developed—anodic and cathodic regions are connected by the electrolyte in the form of pore water in the hardened cement paste. At anode, positively charged ferrous ions are dissolved in the electrolyte and electrons are liberated.

- $Fe \rightarrow Fe^{++} + 2e^-$
- $Fe^{++} + 2(OH)^- \rightarrow Fe(OH)_2$
- $4Fe(OH)_2 + 2H_2O + O_2 = 4Fe(OH)_3$

At cathode, hydroxyl ions $(OH)^-$ are formed due to the reaction of moisture, oxygen, and liberated electrons.

- $4e^- + O_2 + 2H_2O \rightarrow 4(OH)^-$

Hydroxyl ions travel through electrolyte to form ferric hydroxide $[4Fe(OH)_3]$, which is converted to rust by further oxidation. There is no corrosion in dry concrete. The optimum level of relative humidity for corrosion is 70 to 80 percent.

The reaction product $Fe(OH)_3$, which is brown in colour, occupies four to six times the volume of iron because of which splitting, spalling, and cracking etc of iron occurs. The ionization of iron metal would not continue much, unless the electron flow to the cathode is maintained by consumption of the electrons at the cathode for which, as mentioned above, the availability of both oxygen and water at the surface of the cathode is essential. Chloride corrosion is not possible in dry condition. Chloride ions activate the surface of steel to form anode, the surface of the passivity film being cathode.

- $Fe^{++} + 2Cl^- \rightarrow FeCl_2$
- $FeCl_2 + 2H_2O \rightarrow Fe(OH)_2 + 2HCl$

On diffusion into concrete, chloride ions compete with hydroxyl ions for ferrous ions and form $FeCl_2$. The dissolved iron diffuses in moist concrete away from the anode. At some distance away from the anode, complex chlorides break and chloride ion is free to transport more ferrous ions. In the presence of oxygen, more Ferrous ions continue to migrate and react with oxygen to form higher oxides of iron and corrosion continues.

Chlorides can either be in the concrete mix, or there may be ingress of chloride ions from outside. The reasons for chlorides being in the concrete mix are quite a few—use of contaminated aggregates or sea water or brackish water or admixtures containing chlorides. For reinforced concrete, chloride content should be limited to 0.05% (0.03% if sulphate resistant cement is used); and for prestressed concrete, this percentage should be limited to 0.01. However, the real problem of chloride attack occurs on ingress of chloride ions from outside on account of the use of de-icing salts in cold areas or sea water in contact with concrete (very fine droplets of sea water raised by turbulence may also get deposited on concrete surfaces) or airborne chlorides getting deposited and getting wet by dew on concrete surfaces. Chloride ingress into concrete can occur by a number of mechanisms—diffusion due to concentration gradient, migration in an electric field and water flow. Concentration gradient is related to the quantity passing through unit area per unit time. Migration is determined by ionic mobility. It is indicative of the average velocity per unit of the electricity field. Water flow may result from pressure gradient, absorption into partially dry concrete, wick action (absorption at one location with drying at another location) or electro-osmosis.

The total chloride content is not relevant to corrosion. A part of the chlorides are chemically bound as being products of the process of hydration of cement. Another part of the chlorides are physically bound as being adsorbed on the surface of gel pores. The rest, free chlorides, are available for aggressive reaction with steel. The distribution of chloride ions among the three forms is not permanent as there is an equilibrium situation with some chloride ions always present in the pore water.

It has been reported that the corrosion of reinforcement steel is reduced due to high content of C_3A in cement, since chloride ions react with hydrated tricalcium sulphoaluminate hydrates in the hydrated cement paste to produce tricalcium chloro aluminates. Thus, the ordinary Portland cement is three to four times more effective in removing chloride ions than sulphate resistant cement. In ordinary Portland cement, the C_3A content is between 8.5 to 12 %. However, where chloride attack is a possibility, the cement containing higher C_3A content should be used.

In case of corrosion, two types of reactions are recognized:
- Anodic process – characterized by pitting
- Cathodic process – characterized by rust formation

In case of pitting, iron becomes detached from the steel and gets converted into ferric oxide (Fe_2O_3)

Freeze-thaw action

The climate would be deemed as cold if the average of maximum and minimum temperatures of three consecutive days is less than 5°C and also the air temperature during at least 12 hours in any 24-hour period is 10°C or lower. In such climate, concrete should not be placed, unless its temperature is at least 13°C for thin sections (300 mm) or at least 5°C for larger sections starting from 1800 mm. For this, ingredients should be heated and concrete placed should be insulated from the cold. Anti-freeze admixtures should be added to stop water from freezing. While there are measures for concreting in cold weather, exposure of hardened concrete from alternating freezing and thawing cannot be avoided.

Freezing is a gradual process because of:
- Rate of heat transfer through concrete
- Progressive increase in the concentration of dissolved salts in the still unfrozen pore water (which depresses the freezing point)
- Freezing point varies with the size of the pore

Freezing starts in the largest pores gradually extending to smaller ones. Gel pores do not freeze at temperatures higher than -78°C. In other words, no ice is formed in gel pores even in freezing conditions. However, water in large cavities turns into ice. There is difference in entropy of gel water and ice. As the temperature falls, the gel water acquires an energy potential enabling it to move to the capillary pores containing ice. As a result, ice body grows and expands till there is no room in the cavities to form more ice. There would thus be two possible sources of dilating pressure.

Freezing of water in the cavities causes 9% increase in volume. As a result of such volume expansion, excess water in the cavities is expelled. The expelled water would fill up air voids that can accommodate such water. The rate of freezing would determine the rate of flow of displaced water.

The second dilating force in concrete results in growth of a relatively small number of bodies of ice. This is caused by diffusion of water caused by osmotic pressure brought about by local increases in solute concentration due to separation of frozen water (pure) from the pore water, which contains soluble substances such as alkalis, chlorides, and calcium hydroxide.

When salts are used for de-icing concrete surfaces along roads and bridges, some of the salts are absorbed by concrete. This generates high osmotic pressure, whereby cold water is drawn towards the coldest zone where freezing takes place.

Concrete is damaged when the dilating pressure exceeds tensile strength of concrete. Scaling occurs on the surface. Ice is formed at the exposed surface of the concrete, and then progress through the depth—which may ultimately result in total disintegration.

Regarding progressive damage caused by alternate freezing and thawing, each cycle of freezing causes migration of water to locations where it can freeze. Such locations include fine cracks that are enlarged by the pressure of the ice and remain enlarged during thawing, which results in filling up of the cracks with water. Subsequent freezing results in only repetition of what happened before.

Selection of proper aggregates, which constitute important ingredients of concrete, is of vital importance. Aggregates should not hamper resistance to frost attack. Such aggregates (certain crushed cherts, sandstone, limestone, and shale, etc.) having high porosity and low average pore size cannot resist adverse action of frost. The aggregates become saturated because of fine pore structure, and the pressures due to the movement of water when ice forms are higher than the tensile strength of aggregates. This is particularly the case with large aggregate particles, since the distance that water must travel during freezing is high. Other types of aggregates, even if frost resistant, can have a negative influence by expelling water in the surrounding paste upon freezing. These considerations have given rise to the concept of critical aggregate size related to frost damage.

With a given pore size distribution, permeability, degree of saturation, and freezing rate, a large aggregate may cause damage but the smaller particles of the same aggregate would not. There is no critical size for an aggregate type as this would depend on, as already mentioned, freezing rate, degree of saturation, and permeability of aggregate. The permeability has dual role: (i) the rate at which water would be absorbed in a given period of time, and (ii) the rate at which water would be expelled on freezing. Access to water is again a very important criterion, and low paste porosity, therefore, helps to reduce the degree of saturation of the aggregates at the time of freezing. Air-entrainment is also important, since the air voids close to the paste-aggregate interface can help to reduce the pressures that are due to the expulsion of water by the aggregate into the surrounding paste. Entrained air is defined as air that is intentionally incorporated by means of a suitable agent. It is, however, necessary to check whether air-entrainment alleviates the effects of freezing of coarse aggregates or not.

These damages are due to frost and use of de-icing salts. This is not of significance in India.

6. Deterioration due to surface wear

Concrete loses mass progressively under many circumstances when subjected to wear due to abrasion, erosion, and cavitation. *Abrasion* refers to attrition – wearing away of concrete surface by sliding, scraping, or percussion. *Erosion* refers to wearing away by the abrasive action of fluids containing solid particles in suspension. And, the term *cavitation* refers to the formation of vapour bubbles and their subsequent collapse due to sudden change of direction in rapidly flowing water.

Abrasion involves application of local high intensity stress. Resistance to abrasion would, therefore, depend on strength and hardness of concrete. In other words, compressive strength of concrete is the principal factor that would exhibit high resistance to abrasion. Well-produced high-strength (150 MPa) concrete can be harder than high quality granite. High-strength concrete may be produced by using properly graded fine and coarse (hard) aggregates, low water:cement ratio, lowest consistency practicable for proper placing and compaction, and minimum air content consistent with exposure conditions. Concrete should be adequately cured before exposure to the aggressive environment.

Concrete properties in the surface zone are affected by the finishing operations because of which proper attention should be paid to ensure that at least the concrete at the surface is of highest quality. Water:cement ratio may be reduced and compaction improved by well-planned finishing operation. Vacuum de-watering could be beneficial, or otherwise finishing with float and trowel should be delayed till the concrete has lost its surface bleed water. The presence of laitance must be avoided.

Concrete toppings containing latex or super plasticizing admixtures are in increasing use for abrasion or erosion resistance because of low water:cement ratio. In this regard, the use of mineral admixtures/additives, such as condensed silica fume, shows striking possibilities. Apart from considerable reduction in the porosity of concrete after moist curing, the fresh concrete containing mineral admixtures is less susceptible to bleeding. Surface hardening solutions can be applied to well-cured new floors and abraded old floors to resist deterioration by permeating fluids and to reduce dusting on account of attrition. Hardening solutions most commonly used are magnesium or zinc fluorosilicate or sodium silicate that reacts with calcium hydroxide to form insoluble reaction products, thereby sealing the capillary pores at or near the surface.

Erosion is surface wear which may occur in concrete in contact with flowing water due to solid particles carried by the water. The rate at which erosion takes place depends on the quantity, shape, size, and hardness of the particles being transported, on the velocity of their movement, on the presence of eddies, and also on the concrete quality. Like in the case of abrasion, erosion can also be resisted by the compressive strength of concrete. But the mix is also relevant in case of erosion. Concrete produced with large coarse aggregates erodes less than mortar of equal strength. Moreover, hard aggregates improve the erosion resistance. However, smaller-size aggregates lead to a more uniform erosion of the surface. Slump remaining constant, the erosion resistance would increase when the cement content is decreased. This would be advantageous as formation of weak surface called *laitance* would be reduced. When cement content is maintained at a constant level, the erosion resistance would improve with a decrease in slump in line with the general influence of compressive strength.

Irrespective of the quality of concrete, it is difficult to prevent damage due to cavitation, which results from the formation of vapour bubbles. Nonlinear flow at velocities in excess of 12 m/sec may cause severe damage by the process of cavitation. On surfaces of concrete structures subjected to high velocity flow, an obstruction or abrupt change in surface alignment causes a zone of sub-atmospheric pressure to be developed against the surface immediately downstream from the obstruction or abrupt change. This affected zone is rapidly filled with turbulent water interspersed with tiny fast-moving bubble-like cavities of water vapour. The cavities of water vapour formed at the upstream edge of the zone pass through the zone and collapse by condensation at a point just downstream with great impact. The collapse means entry of high velocity water into previously vapour occupied space. Extremely high pressure (~ 7000 kg/cm^2) on a small area is generated in a short span of time interval, and it is the repeated collapse over a particular part of concrete that causes pitting. The greatest damage is caused by clouds of minute cavities found in eddies, which momentarily coalesce into large amorphous cavity that collapses extremely rapidly. Many of the cavities pulsate at a high frequency, and consequently the damage is aggravated over an extended area. Cavitation may occur on horizontal water flows. Cavitation may also occur on vertical surfaces past which water flows.

In all cases of surface wear, it is only the quality of the concrete in the surface zone that is relevant. Even the best concrete would rarely withstand severe erosion over prolonged periods. Vacuum de-watering and use of permeable formwork are possible resisting measures.

7. Fire exposure

Concrete performs well when exposed to fire because of its good properties with respect to fire resistance, that is, incombustible, poor conductor of heat (the period of time under fire during which concrete continues to perform satisfactorily is relatively high), and it does not emit toxic fumes on exposure to high temperature. The relevant criteria of performance are:
- Strength loss—load-carrying capacity

- Barrier—resistance to flame penetration
- Heat conduction—resistance to heat transfer when concrete is used as a protective material for steel

In practice, what is required of structural concrete is that it preserves structural integrity over a desired period of time (known as fire rating). This is distinctly different from being heat resisting.

Considering the behaviour of concrete as a material, the following should be noted:
- Strength loss by degradation of hydrate structure which occurs at various stages from 300°C upwards.
- Spalling of outermost concrete layers—fire originates high-temperature gradients as a result of which hot surface layers tend to separate and spall from the cooler interior. Spalling can be either localized or widespread, depending upon the condition of fire and/or concrete, especially moisture content and susceptibility of breaking up of aggregates. Spalling is very common consequence of fire.
- Formation of cracks is promoted at joints, in poorly compacted parts of the concrete or in the planes of reinforcing bars—on exposure of the reinforcing steel, aggression of high temperature is accelerated by conduction of heat.

Fire-resistance ratings of concrete depend on the type of aggregates used, the thickness of material, and the particular application. Siliceous gravels containing a large percentage of chert or flint are badly disrupted by exposure to fire because of markedly different coefficient of thermal expansion between the aggregates and cement paste. Limestone aggregates have typically low co-efficient of thermal expansion compared to siliceous aggregates and they are closer to cement paste, giving lower internal stresses on heating.

Both calcium carbonates and magnesium carbonates (dolomite is double carbonate of magnesium and calcium) begin to break down on heating at temperatures in excess of 660°C and 740°C respectively. On breaking down, the minerals release CO_2, thereby providing blanketing protection against heat transfer. The residual aggregate particles also have lower thermal conductivity – heat transfer into the concrete is thus further reduced. Concrete with a low thermal conductivity has a better fire resistance so that, for instance, lightweight concrete stands up better to fire than ordinary concrete. The loss of strength is considerably lower when the aggregates do not contain silica.

8. Thermal influences

Jet deflector ducts at rocket launching pads are made of concrete and constructed underground. In one such rocket launching case, jet of hot gases impinge on the top of the wedge portion of the deflector duct on taking off of the jet with a thrust of 800 tonnes and temperature of about 1400°C. The concrete surfaces in such ducts are protected by providing 100 mm thick heat-resistant lining applied over the surfaces by guniting. Hot flue gas concrete ducts and chimney stacks are also protected by lining with insulating materials. Determination of thickness and type/quality of insulating materials depends on the flue gas and temperature involved as well as on the type and quality of concrete.

9. Acoustic pressure

Ordinary concrete is not a particularly good sound-absorbing material since the properties desired in structural concrete work to a disadvantage as a soundproofing material. At rocket launching pads, the outside noise level at take off is about 170 decibels (dB) which need to be weakened to 120 dB inside control room. This is done by converting acoustic pressure to equivalent static loading/pressure. All sidewalls are designed to take care of this additional load. In an acoustic testing centre, 500 mm thick wall and roof slabs are constructed for expected attenuation of 55 dB while a specimen is subjected to acoustic pressure of 155 dB.

10. Blast pressure

Concrete structures constructed at locations close to a rocket launching pad need to be designed for impact loads and blast pressure in the possible event of a rocket exploding on the launch pad. A building housing people monitoring different parameters related to launching could be located as close as 150 m of the launch pad. The safety of the building and the personnel engaged therein is to be assured. In case of an explosion on the launch pad, the building structure would be hit with rapidly moving shock waves which may exert pressures many times more than those experienced under hurricanes. Such a structure was conceived and designed in the form of toroidal shell to withstand the blast due to rocket explosion at launch pad as well as impact by rocket fragments due to explosion on take off—partly buried with the top cushioned with around 150 mm deep earth to absorb the shock of flying splinters, if any. A rocket fragment could be as heavy as 700 kg and may be moving at 100 m/sec at the time of impact.

11. Alkali Silica Reaction (ASR)

The chemical reaction involving alkali ions from Portland cement (or from other sources), hydroxyl ions, and certain siliceous constituents that may be present in the aggregates is disruptive in nature. To start with, relatively high content of alkali hydroxides mainly from cement react with aggregates containing certain potentially reactive forms of silica in the presence of moisture to form 'swelling type' alkali-silicate gel. Subsequently, this gel can swell very much on the absorption of further moisture giving rise to expansive stresses on the enveloping concrete.

As the tensile strength of the concrete matrix is relatively low, the expanding reaction sites generate radiating micro-cracks. Solubility of the gels in water/moisture accounts for their mobility from the interior of aggregates to the micro-cracked regions both within the aggregate and the concrete. Continued availability of water/moisture to the concrete causes enlargement and extension of the micro-cracks, which eventually reach the outer surfaces of concrete. The irregular pattern of cracks is referred as *map cracking*. In view of the reactive aggregates distributed throughout the concrete, to some extent – greater or lesser, micro-cracking from individual reaction sites can become linked into a network of cracking that can affect strength, elasticity and durability of concrete. The cracks in turn open path for ingress of oxygen, moisture, chlorides, CO_2 etc. Hence, it is necessary to be careful while using alkali reactive aggregates.

This expansive phenomenon is not common as we use cement with low alkali content. When it becomes necessary to use the reactive aggregates, it is necessary to use cement with the alkali content not exceeding 0.6 % (designated as low-alkali cement) so as to prevent damaging effects of alkali aggregate reaction irrespective of the type of reactive aggregates. Raw materials used in the ordinary Portland cement production account for the typical presence of alkalis in cement in the range of 0.2 to 1.5% equivalent Na_2O. It is a convention in cement and concrete chemistry to assess alkalis on the basis of acid-soluble 'equivalent soda' (Na_2O eq.), which is expressed as $Na_2O + 0.658 K_2O$. The pH value of the pore fluid in concrete is normally 12.5 to 13.5 depending on the alkali content of the cement. Such pH values characterize strongly alkaline liquid in which some rocks composed of silica and siliceous minerals do not remain stable for long. On economical considerations, if such reactive aggregates are required to be used, the alkali silicate reaction can be reduced by mineral admixtures like ground granulated blast furnace slag or pulverized fuel ash. The beneficial effects of blended cements in controlling ASR expansion are attributable to:

- Reduction of alkali content in the mix
- Lowering of hydroxyl ion concentration
- Densification of the cement matrix

Water is essential for the alkali-silica reaction to continue. Drying out the concrete and stopping of further contact with water is the only preventive measure against this reaction to take place.

The alkali aggregate reaction creates cracks in concrete as alkali-silicate gel, which is of unlimited swelling type, is formed.

Stress corrosion

Usually high strength steel is more susceptible to stress corrosion where the damage caused by stress and corrosion acting together exceeds the damage produced when they act separately. This does not happen in low strength steel. Stress corrosion is the result of: (a) small imperfections in steel, and (b) inter granular steel corrosion caused by absorption of hydrogen, hydrogen sulphide, high concentration of ammonia and nitrite salts.

At the time of initial prestressing, IS: 1343 permits maximum tensile stress not exceeding 89% of ultimate tensile strength (UTS) of wire or bar or strand.

Resistivity

When corrosion is in progress, the corrosion current which controls the rate of corrosion would depend on the potential developed and the resistivity of the electrolyte, that is, the concrete itself.

Resistivity is primarily dependent upon the moisture content of the concrete, but at higher level of moisture, the chloride content has a marked influence. Concrete itself has a surface skin of higher resistivity. This is caused by deposits of insoluble salts such as magnesium hydroxide and calcium carbonate formed by reaction with sea salt at the surface of the concrete.

Concrete cover to reinforcements plays one of the most important roles in preventing corrosion of reinforcement since the degree of carbonation and penetration of chlorides depends mostly on thickness of concrete cover to reinforcement. The thickness of cover depends on the grade of concrete and the denseness of concrete. Not much is known on the thickness of concrete cover versus durability of RCC structures.

Corrosion can be resisted by providing coating on concrete with asphalt and coal tar, chlorinated rubber, epoxy resin, sand filled epoxy or polyester or urethane resin, procured neoprene sheet, plasticized PVC sheet, polyester resins, etc. Success is to be checked with respect to the following—brittleness of paint with time, elasticity of paint with extensibility matching that of concrete under loading, chemical reaction with atmosphere, washing away with rainwater.

7.3 COMPOSITION AND FINENESS OF CEMENT

The function of cement, the most chemically active component of concrete, is:
- To fill up the voids between fine (sand) and coarse (stone) aggregates and develop specified strength
- To bind the fine and coarse aggregates on addition of the requisite amount of water and exhibit the appropriate rheological behaviour

There are two types of cements:
- Hydraulic—not only harden by reacting with water but also form a water resistant product
- Non-hydraulic—their products of hydration are not resistant to water

Natural hydraulic cement is produced by calcination of limestone, clay, and other materials, such as aluminium and iron ores in rotary kiln. The raw materials required for the manufacture of cement are mainly lime (CaO), silica (SiO_2), alumina (Al_2O_3), and iron oxide (Fe_2O_3). The argillaceous (SiO_2 and Al_2O_3) materials contain silica and alumina, and calcareous (CaO) materials contain lime. The argillaceous materials are derived from clay, shale, and sand. The clinkers resulting as a result of calcination are ground with gypsum to fine powder. Gypsum (2.5–5 %) is added to retard the setting time of cement.

Chemically, four major complex compounds are formed in the kiln. Cement chemists have adopted abbreviated notation in which molecular entities are given a single-letter symbol like: $C = CaO$; $S = SiO_2$; $A = Al_2O_3$; $F = Fe_2O_3$, and $H = H_2O$. The major compounds in cement are usually referred to by the abbreviated symbols. The tricalcium silicate phase is called C_3S; the dicalcium silicate phase is called C_2S; the tricalcium aluminate phase is called C_3A; and the tetracalcium aluminoferrite phase is called C_4AF.

Compound	Composition	Abbreviated symbol	Content (%)
Tricalcium silicate	$3CaO.SiO_2$	C_3S	49
Dicalcium silicate	$2CaO.SiO_2$	C_2S	25
Tricalcium aluminate	$3CaO.Al_2O_3$	C_3A	12
Tetracalcium alumina-ferrite	$4CaO.Al_2O_3.Fe_2O_3$	C_4AF	8

In addition to the above compounds, a few more compounds, such as MgO, TiO_2, Mn_2O_3, K_2O and Na_2O are formed which would amount to only a few percent of the weight of cement. Of these, Na_2O and K_2O, known as alkalis react with some aggregates adversely (vide Sections 7.2 and 7.4). In most clinkers, K_2O is usually more abundant than Na_2O. The products of reaction cause disintegration of concrete and affect the rate of gain of strength of cement.

The sum of the percentages of C_3S and C_2S in various types of cements ranges from 70 to 80 and are responsible for the strength of hydrated cement paste and also of the concrete. C_3A and C_4AF contribute very little to the development of strength. The four complex compounds contribute variously on strength and other characteristics of hydrated cement paste. By adjusting the proportions of the four complex compounds, different types of Portland cement have been developed to be used under diverse conditions. Also, blast furnace slag, fly ash, or silica fumes that may contribute to the cementing action are blended with Portland cement to produce blended cement for high-strength or high performance concrete. Concrete produced with blended cement is usually designated as special concrete.

A particular physical property of cement that influences the rate of hydration and gain of strength of the cement paste is the fineness of the cement particles. Fineness is the measure of average particle size of the cement achieved by grinding; it has units of surface area per unit weight, and a larger value means smaller particles. Modern cements are much more finely ground than the ones in earlier days, because the larger area exposed to the mixing water results in a more rapid hydration of the cement and strength gain of the concrete. Also, fine sand bleeds less than coarser cement. However, finer cement would rapidly deteriorate on exposure to the atmosphere. Finer cement leads to higher shrinkage and greater proneness to cracking. Fineness of cement should be maintained at the optimum level. Fineness of cement should conform to the relevant specification issued by the Bureau of Indian Standards.

Hydration is the process by which the silicates and aluminates in the cement form hydrated compounds that, in due time, harden into solid mass. Both C_3S and C_2S produce calcium silicate hydrates and calcium hydroxide in the course of reaction as follows:

$$2C_3S + 6H \rightarrow C_3S_2H_3 + 3Ca(OH)_2$$

$$2C_2S + 4H \rightarrow C_3S_2H_3 + Ca(OH)_2$$

The compound $Ca(OH)_2$ acts as the binding material between the hydrated products, which eventually form a hard mass, commonly called hydrated cement paste (hcp). But being soluble in water, this may be leached out due to ingress of water into concrete. This leaching of $Ca(OH)_2$ affects concrete resulting in loss of strength.

The hydration process continues for a long time and then slows down progressively with time.

Hardening is related to the gaining of strength of set cement paste. Setting, although part of a continuous process with hardening, is different from hardening and refers to the change from fluid state to rigid state. Setting is a gradual process and the two terms are in use to signify this are *initial set* and *final set*. C_3A first reacts with water, rather violently, resulting in immediate stiffening leading to *flash set*. Because of the addition of gypsum, C_3S sets first and is responsible for the initial set. C_2S sets in a more gradual manner. The setting time is determined by penetration tests of cement mortars made with standard sand.

Hydration of cement is an exothermic process, and the heat liberated during hydration is called *heat of hydration*. Heat liberated may be as much as 120 calories per gm of cement. Excessive heat of hydration may lead to thermal cracking, especially in mass concrete.

For determining the strength potential of cement, mortar is made with standard sand maintaining given water:cement ratio, since that is the primary factor that determines the strength of a cement-based mixture. Cubes are cured for a specified time, and then tested in compression in a laboratory.

The water:cement ratio by weight is important for the strength of concrete, which also depends on the type of cement. A very low water:cement ratio causes incomplete hydration whereas a very high water:cement ratio would result in excess amount of free water in the capillary pores. Both these conditions are detrimental for development of strength of concrete. The optimum value of water:cement ratio should be determined to achieve best results. The total requirement of water would depend on:

- Hydration @ 0.253 gm of water per gm of cement—water:cement ratio need to be 0.253 for only hydration of cement
- Surface area of the aggregates to be wetted, including possible absorption
- Construction equipment and plants available for mixing, transporting, placing, and compaction of concrete

Excess water increases the porosity of the cement paste, and thereby reduces strength. For good strength, the basic tenet of construction practice is the use of a workable mix with water:cement ratio as low as possible. It is, however, equally important to ensure good compaction by having the right degree of workability of concrete.

The different compounds present in cement have varying influence on strength, rate of hardening, heat evolution, and other characteristics of hydrated cement. By suitably varying the percentage contents of the compounds, various types of Portland cements with special properties have been developed.

- Ordinary Portland cement is the type manufactured for use in general concrete construction when the special properties of other types are not required.
- Low-heat Portland cement is obtained by lowering the content of C_3S and C_3A. Because of low C_3A content, concrete would have moderate resistance to sulphate solutions; and because of lower C_3S content, strength development in concrete would be slower and heat evolution would be less.
- High-early-strength Portland cement is obtained by increasing the percentage of C_3S in the cement, whereby heat evolution would also be high. An increase in early strength of any cement may be obtained by finer grinding of the clinker. This type should not be used in mass concrete.

- Sulphate resistant Portland cement is obtained by lower content of C_3A and C_4AF. The combination of lower C_3A and C_4AF compounds imparts much greater resistance to sulphate attack compared to using any other types of cement.

7.4 QUALITY OF FINE AND COARSE AGGREGATES

Aggregates form the bulk of the volume of concrete – about 60 to 75% of the total volume. And by far the largest amount of aggregates used in concrete is mineral aggregates, such as gravels, crushed stones, and sands. Aggregates are obtained by crushing granite, basalt, harder types of limestone, and sandstone. Aggregates differ in quality, and in some locality, good quality aggregates are in short supply. Sometimes non-mineral aggregates are used. Aggregates are supposed to be inert materials. However, their physical, thermal, and even chemical properties may influence the performance of concrete. Aggregates should be free from deleterious substances such as iron pyrites, coal, mica, and organic impurities which affect hydration of cement and durability of concrete.

Aggregates are classified into two groups:
- Fine aggregates passing through the 4.75 mm sieve and within the grading limits
- Coarse aggregates retained by the 4.75 mm sieve ranging up to 150 mm

Fine aggregates again may be divided into two groups:
- Those consisting principally of rock fragments resulting from natural processing of weathering—by disintegration or by glacial action
- Those consisting of fine aggregates obtained by crushing of natural rocks and the subsequent screening of crushed materials into several specified sizes (vide Chapter 13, Section 13.6)

Sand deposits are the result of weathering of natural rock minerals which have been transported, collected, and sorted by the long action of streams. Natural sands, compared to other types of fine aggregates, are extensively used. The quality of materials found in sand pits may vary considerably as the sand partakes the characteristics of the rock from which the same is derived. Natural sands are quite often the result of weathering of rock masses by breaking down as a result of the action of alternate freezing and thawing, by erosion resulting from the continuous action of wind or water, or by disintegration because of other natural processes.

The mineral composition largely determines the suitability of the quality of aggregates for producing durable concrete. As such, mineral composition of the particles is a matter of considerable importance. For structures which are to remain exposed to aggression affecting durability, only such aggregates as of proven resistance to the particular condition of exposure should be used.

Sands may be divided into two general classifications:
- Calcareous—when calcium carbonate is present in large quantities
- Siliceous—constituted largely of quartz or silicates—considered best for concrete production

In many cases, however, sand quite often comprises of a mixture of both calcareous and siliceous materials.

Natural siliceous sands from river beds or pits are the major source of fine aggregates. In the absence of natural sands, crushed rocks are used as fine aggregates. To assure quality of such crushed rocks, the structure and strength of the original rocks need to be examined. If the results are satisfactory, only then such fine aggregates should be used in concrete production. The practice of blending crushed fine aggregates with finer sands to produce a coarser grading is growing in many countries.

Gravels and crushed stones are the principal sources of coarse aggregates. Blast furnace slag, air-cooled and properly crushed, is also used as coarse aggregate where normal aggregates are not easily available. The basic requirement is that the coarse aggregates should be clean, strong, and durable, devoid of any friable and laminated particles. The common natural coarse aggregates are obtained from two principal sources—sedimentary rocks and igneous rocks. Sandstone, limestone, dolomite, marble, and shale are some of the common types of sedimentary rocks. However, calcareous rocks are liable to destruction from attack by acidic liquids and gases. High porosity of limestone makes it vulnerable to chemical attacks.

All aggregates should be graded for use in concrete. Graded aggregates contain appropriate amounts of the finer size particles to fill in the gaps between the larger sizes and thus reduce the void content. Grading and shape of aggregates and maximum size of coarse aggregates influence workability. Natural sands with rounded grains produce more workable concrete than crushed sands with angular, flat, or elongated pieces. Crushed coarse aggregates having cubical shapes, on the other hand, are good for concrete's workability. Smaller size coarse aggregates would result in higher sand requirement with resulting higher water demand with all its accompanying undesirable effects. It has been found that from the point of strength, there is no significant improvement from the use of aggregates larger than 20 mm size. Coarse aggregates up to the size of 40 mm may be used to develop the optimum properties of strength, durability, and shrinkage because larger aggregates enable the use of minimum water content.

Next to water: cement ratio, it is the aggregates that have the most significant effect on the workability of concrete. This is primarily a function of the surface area of the aggregates in a given volume of concrete and consequently of the demand of mixing water that aggregates create. The lower the surface area of the aggregates, the lower is the demand for mixing water for a given workability. Conversely, if the water content of a mix is fixed, as it normally is from the consideration of strength, workability will increase when:

- Maximum size of aggregates is increased
- A coarser overall grading of the aggregates is available as the ratio of coarse to fine aggregates increases
- The particle shape approaches a sphere—in other words, rounded particles increasing workability most, followed by irregular ones, then by angular ones, and finally by flaky and elongated ones

- The surface texture of the aggregates becomes smoother

None of the above factors would affect workability to the same extent as a change in the water: cement ratio that directly controls the fluidity and thickness of the cement paste.

The problems of concrete durability arise from the reaction between cement and certain alkali reactive aggregates. This reaction is apparent in the expansion, extensive map cracking, loss of strength, pop outs and oozing out of gel. All these due to the chemical reaction between cement alkalis and the reactive form of silica in some aggregates. Opal, obsidian, cristobalite, tridymite, chalcedony, cherts, and certain cryptocrystalline volcanic rocks are the usual reactive siliceous sediments that make the aggregates reactive with the cement alkalis. The difference between reactive and unreactive forms of silica lies in their crystal structures.

The cement compositional components responsible for generating alkali-silica reaction (ASR) are the alkalis present in ordinary Portland cement. The ASR reaction finally leads to formation of an alkali silicate gel, which sets up expansive forces by imbibing water, causing cracking and disintegration of concrete. The severity of alkali silica reactions depends on:
- Alkali content of the concrete from all sources
- Cement content of the mix
- Amount, size, and type of the reactive siliceous sediment in the aggregates
- Availability of moisture
- Ambient temperature

When reactive aggregates are unavailable, cement composition with alkali content lower than 0.6% (equivalent Na_2O) usually insures control against expansion. If significant alkalis are introduced into the concrete from other sources as well, the composition of cement in terms of alkalis should be such that the total alkali content of the concrete does not exceed $3kg/m^3$. If reactive aggregates need to be used in conjunction with locally made high-alkali ordinary Portland cement, the most effective protective measure against ASR deterioration would be the use of modified cements formulated by partial replacement of high-alkali ordinary Portland cement with such materials as fly ash, blast furnace slag, or silica fume in highly powdered form.

7.5 QUALITY OF WATER

The function of water, the active component of concrete, is two fold:
- To react with cement chemically (hydration) to form a cement gel wherein the aggregates remain in suspension till hardening of cement paste
- To serve as lubricant between the fine and coarse aggregates so that the concrete may be easily placed and compacted—to make concrete workable for specific use

Water that is to be used in concrete shall be clean and free from such impurities as suspended solids, organic matter and dissolved salts which are frequently contained in natural water and which may adversely affect the properties of concrete, especially setting and hardening. Water should, for the same reason, be free from injurious oils, acids, alkalis, organic matters, salts, silts or other deleterious impurities. Normally mixing water for concrete is required to be fit for drinking or to be taken from an approved source. The relevant codes issued by the Bureau of Indian Standards are followed for routine tests. In case of doubt, it can be assessed by comparing the setting time of cement paste and compressive strength of the concrete made with it and with distilled water under similar conditions.

Water is essential for hydration of cement. Generally, potable water is considered suitable for concreting. Water reacts chemically with the cement and forms a gel that binds the aggregate particles together. In addition, water provides mobility to the concrete mix, thus facilitating placement and compaction of the same.

Though the use of saline sea water may be regarded as adding sodium chloride as an admixture to the concrete mix, it does not adversely affect the strength or durability of Portland cement concrete. Sea water containing up to 35,000 ppm of dissolved salts is generally suitable for un-reinforced concrete. In reinforced concrete, possible corrosive effect of the sea water on the reinforcement must be considered. Also to be considered is where efflorescence could mar the appearance of the work. In USA, sea water is not allowed for prestressed concrete. Again, water containing mineral oil may affect the strength of concrete. Water containing algae has the effect of entraining considerable amounts of air in concrete with an accompanying decrease in strength.

The sources of satisfactory water can be lakes, streams, or wells. Groundwater can be another source of water. Water from streams with excessive suspended solids is clarified in settling tanks with the aid of coagulants. Limits on the amounts of suspended matters are usually set between 1000 to 2000 ppm.

Water is an essential ingredient of concrete, but excessive water is the primary cause of drying-shrinkage cracks. The additional water beyond what is needed for hydration of the cement creates an excessive number of bleed channels to the exposed surfaces. When the cement paste undergoes normal drying shrinkage, these channels cannot provide any resistance to the ingress of aggressive chemicals. Drying shrinkage causes the most undesirable volume changes, because it produces cracks on the surfaces of concrete.

7.6 USE OF ADMIXTURES

Admixtures are defined as materials other than aggregates, cement and water, which are added to the concrete batch immediately before or during mixing to achieve special properties and requirements of concrete in the fresh and/or hardened state. An admixture is, however, not an essential ingredient of the concrete mix. The use of admixtures is now widespread because of many physical as well as economic benefits that may be derived from the use of such admixtures. Because of multifarious benefits, admixtures have become regular and important ingredient of concrete mix. It is possible, for example, to modify the setting and hardening characteristics/properties of the cement paste by using chemical admixtures by influencing the rate of cement hydration. Water-reducing admixtures can increase plasticity of fresh concrete mixtures by reducing the surface tension of water, air-entraining admixtures can improve the durability of concrete exposed to cold weather, and mineral admixtures such as pozzolona can reduce thermal cracking in mass concrete. However, before using any admixture in the concrete mix, its performance should be evaluated.

Admixtures are used for many purposes, for example:
- To increase plasticity of concrete without increasing the water content
- To reduce bleeding and segregation
- To retard or accelerate the setting time
- To accelerate the development of strength at early ages
- To reduce the rate of heat evolution
- To increase the durability of concrete to specific exposure conditions

The hydrated cement paste (hcp) is first formed as gels comprising thin fibrous hydrated crystals bound by $Ca(OH)_2$. The gels form hard mass in due course of time. As the crystals are formed, some water is locked in the inter-crystal small spaces which are called *gel pores*, while the locked water is called *gel water*. Apart from gel pores, there would be larger voids or capillary pores – empty or filled with water. Water required for hydration is small because water:cement ratio for the total hydration is only 0.253. More water, however, is required for workability. The extra water fills the voids. More porosity means less strength, and less water:cement ratio means more strength. The strength of concrete depends on:
- Strength of hcp depending on the quality of cement, mix proportion, water:cement ratio, degree of hydration and porosity
- Strength of hcp-aggregate interface which is a zone of weakness
- Strength of aggregates

The causes of deterioration of concrete may be of three different categories:
- Mechanical—high stress due to load or temperature, wearing due to abrasion, erosion, cavitation etc
- Physical—freezing and thawing, water evaporation, shrinkage and swelling
- Chemical—leaching of $Ca(OH)_2$, replacement of $Ca(OH)_2$ by the formation of non-cementitious compounds due to reaction with aggressive substances, corrosion of steel

Water-reducing admixtures

Surface-active agents, known also as *surfactants*, cover admixtures that are generally used for reduction of water in concrete as well as air-entrainment. The *surfactants* involve a physico-chemical process occurring at the surface of constituent materials in a system. The surfactants are better known through their application by the detergent industry.

Water-reducing admixtures are used to reduce the quantity of mixing water to produce concrete of a given consistency, that is, concrete with desired slump with low water:cement ratio with associated high early mechanical strength and better durability. Plasticizers and superplasticizers are common water-reducing admixtures. The difference between plasticized and superplasticized concrete is that of water reduction – 5–15% to 20–30 % respectively. However, plasticizers are cheaper. The main ingredients of plasticizers are lignosulfonic acids and their salts, hydroxylated carboxylic acids and their salts, or polymers of derivatives of melamine or naphthalene or sulphonated hydrocarbons. The combination of admixtures used in a concrete mix should be carefully evaluated and tested to ensure that the desired properties are achieved. The main ingredients of superplasticizers are sulphonated melamine formaldehyde condensates, sulphonated naphthalene formaldehyde condensates, modified lignosulphonates, or synthetic polymers.

The superplasticizers have fluidizing effect which is basically related to three physico-chemical phenomena—adsorption, electrostatic repulsion and dispersion. Recent developments, however, indicate that polymer adsorption rather than electrostatic repulsion is responsible for the dispersion of large agglomerates of cement particles into smaller ones resulting in remarkable increase in the fluidity of the mix. Apart from improving strength and durability, superplasticizers can also be used for reducing the cement content which is especially important in case of prestressed concrete structures.

The increase of slump of concrete due to the use of superplasticizers is transient for about 30 – 60 minutes. Because of this, transport of ready mixed concrete (RMC) could become a serious problem.

Extended set control admixtures

Ready mixed concrete (RMC) is beset with the possible problem of disposal of set concrete. There is now a cost efficient technique to keep concrete fresh for longer periods. In other words, it is now possible to store concrete for later use. The system uses two non-chloride admixtures – a stabilizer and another activator which alternately suspends and reactivates hydration of cement.

The stabilizer comprises carboxylic acids and phosphorous containing organic acid salts. The action mechanism of the stabilizer admixture is thought to be related to the inhibition of C-S-H and C-H nucleation. Cement hydration is arrested by the admixture acting on all phases of cement hydration including C_3A fraction.

Set accelerating admixtures

These admixtures are used to: (i) reduce the setting time from the start of addition of water, (ii) accelerate hardening of hcp (iii) increase the rate of gain of strength. The most commonly used set accelerating admixture was once calcium chloride, which can still be used in plain concrete. As calcium chloride in concrete, even in small percentage, causes corrosion of reinforcing or prestressing steel, calcium formate, $Ca(HCOO)_2$, is used as a chloride-free accelerator.

Strong alkalis are powerful accelerators. Organic compounds such as diethanolamine $NH(CH_2CH_2OH)_2$, triethanolamine $NH(CH_2CH_2OH)_3$, urea $CO(NH_2)_2$ etc., are also used as accelerators.

Retarding admixtures

All water-reducing admixtures retard initial set of concrete to some extent. Admixtures that lengthen setting time and workability time are known as retarders. Extensive use is made of retarding admixtures in hot climate to facilitate proper placing and finishing and to overcome damaging and accelerating effects of high temperature.

In case of placing of concrete for large structural units, it is necessary to eliminate possibilities of formation of cold joints and discontinuities. With retarding admixtures, it is possible to control setting of large pours, thereby eliminating cold joints and discontinuities.

The organic compounds that are used as retarders include unrefined Na, Ca, or NH_4 salts of lignosulfonic acids, hydroxylated carboxylic acids, and carbohydrates. Inorganic salts like phosphates and the oxides of Pb and Zn also act as retarders. Initial set can be retarded for several hours to several days depending on the dosage and base chemicals used in the admixture.

Air-entraining admixtures

Air-entraining admixtures are surfactants used to create a controlled number of microscopic air bubbles in the concrete to protect the same from deterioration due to repeated freezing and thawing (weathering) or exposure to aggressive chemicals. Air gaps provide room for expansion of external or internal water of concrete exposed to repeated freezing and thawing, which otherwise would be damaging to the concrete.

There would be fewer avenues available for ingress of aggressive chemicals into concrete as air-entrained concrete bleeds to a lesser extent compared to non air-entrained concrete.

Air-entrainment improves workability of concrete because of which water content may be reduced. For lean and low-strength mixes, the improved workability would permit a large reduction in water content, sand content, and water:cementitious materials ratio, which tends to increase concrete strength. The strength gained as a result offsets the stress-reducing effect of the air itself, and net increase in concrete strength is achieved.

Corrosion inhibiting admixtures

Corrosion-inhibiting admixtures are added in small amounts in concrete to control or prevent the corrosion of reinforcing steel. Corrosion inhibitors are broadly classified as: (i) anodic, (ii) cathodic, and (iii) mixed depending on whether they interfere with the corrosion at the anodic or cathodic or both sites. Calcium nitrite, sodium nitrite, potassium chromate, sodium benzoate are anodic inhibitors, while NaOH, Na_2CO_3 or NH_4OH are a few of the cathodic inhibitors. In cracked concrete, corrosion increases with calcium nitrite. Sodium nitrite is considered to be deleterious to strength development for the efflorescence and expansion due to alkali-aggregate reaction. NaOH may cause flash setting. Calcium nitrite used as an accelerator may also be used to control corrosion by acting as anodic inhibitor. A mixture of calcium nitrite and sodium molybdate (1: 1) is more effective than calcium nitrite.

Anti-washout admixtures

These are cohesion-inducing anti-washout admixtures, especially made for placing concrete under water frequently in combination with a superplasticizer. The advantages of these admixtures are:

- Elimination of costly dewatering systems
- Increased washout resistance of cement and fine aggregates
- Inducing and improving cohesiveness
- Improved pumpability
- Prevention of segregation

Most anti-washout admixtures are composed of high-molecular weight, water-soluble polymers like natural gums, cellulose ethers or water-soluble acrylic type polymers as the main components. The action of the admixture is to increase viscosity of water in the mix. This results in increased thixotropy of concrete and improved resistance to segregation. When this admixture is added in concrete, it resists attrition or washout from flowing water; or when placed under water, remains integrated and compact due to its cohesive and high-viscous character. The problem of enhancement of viscosity affecting workability can be eliminated by using this admixture in combination with superplasticizer.

Self-curing concrete admixtures

Curing is the most common procedure of assuring that there is enough water present in the concrete for continuous hydration of the cement so that development of strength is not hampered. Efforts are on to produce concrete without the need for such curing, that is, self-curing concrete. This has been made possible by incorporating chemical admixture in the concrete. Poly (ethylene oxide) polymer is an admixture that retains water within concrete, thus reducing the chance of drying out of water within the capillary.

Alkali aggregate expansion reducing admixtures (AAERA)

High-alkali hydroxides content mainly from cement react with aggregates containing certain potentially reactive forms of silica in the presence of moisture form 'swelling type' alkali-silicate gel causing cracking of the concrete. Another type of reaction known as alkali-carbonate reaction takes place between alkali hydroxides and aggregates largely composed of carbonates (dolomite carbonates). Admixtures like soluble salts of lithium, barium, sodium, and aluminium powder reduce the expansion of the mortar specimens due to alkali-silica reaction. Lithium salts and ferric chloride are known to have reduced expansion due to alkali-carbonate reduction. The salts generally used as alkali aggregate expansion reducing admixtures – lithium carbonate, lithium hydroxide, barium acetate, barium hydroxide, sodium, and potassium nitrate. The individual admixtures may be having drawbacks that can be eliminated by combining with retarders or superplasticizers appropriately.

Pumping aid

For enhancing concrete's pumpability, admixtures are used so as to (i) avoid segregation (ii) control movement of water within the cement matrix so that more water is available for hydration, less loss of moisture, and improved curing; and (iii) prevent local turbulence and pressure loss. The admixtures which are used as viscosity modifiers and thickening agents like cellulose ethers, alginates, polyacrylomides, and polyvinyl alcohol can be used as pumping aid for concrete. Besides, air-entraining agents, water reducers, and superplasticizers are sometimes used to impart good pumpability.

Pumping aids based on cellulose ethers and alginates, however, tend to reduce the concrete strength to an extent. Most of the pumping aids in powder form are hygroscopic and tend to agglomerate in the presence of moisture. It is, therefore, necessary that such powder be dissolved in water before adding to the mix. Because of the possible retarding and air-entraining potential, trial mixes should be produced and tests conducted to fix the upper limit of the dosage.

Gas-forming admixtures

These gas-producing admixtures with some stabilizer can be used in two ways: (i) in small dosage to offset volume reduction due to plastic, drying shrinkage, and settlement, and (ii) in high dosage to make self-stressed concrete for special application. The basic chemical composition of gas-forming admixtures is a variety of metals such as Al, Mg, or Zn, and other materials, such as H_2O_2 or NH_3-based compounds. Hydrogen is produced by the reaction that takes place between Al powder and the alkaline constituents of the hydrating cement paste. Minute bubbles produced throughout the mass cause it to expand. The gas forming reaction becomes complete before setting of cement. Oxygen is released by the reaction between calcium hypochlorite and H_2O_2 with $CaCl_2$ as byproduct. As minute bubbles of Oxygen gas causes expansion by dispersing throughout the mass, $CaCl_2$ accelerates the setting. Nitrogen is liberated when Cu salts activate NH_3 compound. Nitrogen also causes expansion.

Of the three gases, hydrogen may cause corrosion of steel in the presence of oxygen and moisture. The process of oxygen production with $CaCl_2$ as byproduct may also cause corrosion of steel.

Blended admixtures

Blended admixtures, combination of two or more common or uncommon types of formulations, are turning out to be very popular because of their wide range of properties. The advantage of a combination of the blended admixtures is that it reduces the number of site trial since most of the trials are already conducted at the research laboratory level.

Blending of two admixtures

- Accelerating—water reducing admixture
- Retarding—water reducing admixture
- Retarding—superplasticizer

- Air-entraining – superplasticizer: to compensate strength reduction
- Air-entraining – water reducing admixture: to compensate strength reduction
- Anti-washout – superplasticizer: to offset possibility of strength loss
- Pumping aids – water reducing admixture: to eliminate strength reduction possibility
- Pumping aids – compatible superplasticizer: to eliminate strength reduction possibility

Blending of three admixtures

- Air-entraining – water reducing – retarding: to utilize their benefits in combination

Application of any blended admixture is limited because of its highly specific properties. Admixtures, if used, should conform to IS: 9103 – Indian Standard Specification for Chemical Admixtures issued by the Bureau of Indian Standards.

7.7 FORMWORK INCLUDING ENABLING WORK

For in-situ concrete construction work, all concrete because of the plastic state of its first stage requires some kind of form to mould it to the required shape. Wet concrete is placed into the mould and compacted and it is set to the inner profile of the mould. A form is a mould which ensures that the dimensions of the structure or element to be constructed conform to the drawings within specified tolerances. Shuttering comprises form proper, supporting scaffolding and fasteners. Shuttering accounts for bulk of the cost of the finished concrete up to as much as 60–75 % depending on what is under construction. Where the cost of formwork is likely to be high, form design merits careful study. Cost of formwork goes up due to the following reasons:

- Design work is carried out without taking the sizes of available forms into consideration
- Possible reuse of forms is overlooked by not making concrete members of similar sizes

Proper design of forms provides:

- Strength – strong enough to support all vertical and lateral loads including adequate margin for unexpected loads till such loads can be supported by the ground, the concrete structure, or other construction with adequate strength and stability
- Rigidity – must maintain shapes shown in the drawing/s and not deflect under the load of wet concrete, self-weight and possible superimposed loads
- Tightness – formwork must have cement paste tight joints – leakage of cement paste causes honeycombing or produces ugly fins that need to be removed – leakage could be stopped by sealing all joints or by using sheet materials or adhesive tapes
- Good alignment – because of the fluidity of wet concrete, it takes up the shape of formwork that needs to be of correct shape and size and also in the correct position – setting out, therefore, should be accurate
- Reasonable economy – minor changes in design may reduce cost of formwork without jeopardizing the adequacy of design – such changes would require the approval of the concerned design engineer

- Desired texture on exposed concrete surfaces – form or its lining, if any, must be designed to produce the desired concrete surface texture for maintaining desired architectural finish
- Ease of handling – design should be such that even the maximum size forms can be easily handled manually or by mechanical means
- Ease of stripping – the design of the formwork should be such that they may be assembled and stripped from the hardened concrete easily so as to protect the concrete as well

In arriving at a particular way of executing a job, keeping in view that no two jobs are alike, an executing agency must weigh the following points involved in every job:

- Quantity of forms necessary for the jobs – this would require thorough knowledge of the jobs involved in totality
- Number of reuses that can be made of the forms – this is pertinent as shuttering accounts for bulk of the cost of the finished concrete work
- Process of erection – some forms may be handled manually but some would require mechanical handling and lifting
- Order of erection – this is related to the speed of erection as well as the ease of erection
- Speed of erection – to make shuttering ready for placing concrete within the time schedule
- Manpower to be deployed – to maintain the time schedule
- Location of the construction joints – as per drawings or as per the available resources
- Speed of placing concrete – this depends on the mobilization of resources, infrastructures, quantum of job involved in cubic metre and height/location of placing
- Grade of concrete and its consistency
- Method of concrete placing – manually, mechanically or by a combination of mechanical and manual means
- Method of compaction – vibration, tamping, etc.
- Method of curing – how soon the forms would be available for reuse
- Stripping of forms – method, speed and order of stripping
- Finishing after stripping of forms – specified/architectural finish, repair or rework
- Maintenance – care of forms and hardware

Materials

The basic methods of using timber and plywood have been evolved over the years. Many of the methods and materials have been propagated by the movement of labour and supervisory personnel from site to site under multifarious contracts. Each of the principal materials has not only been applied to a wide variety of uses, but also ways and means have been devised to overcome the problems such as those of scale and geometry.

Timber and timber derived materials

Individual boards of timber are hardly used as formwork except for those cases that require some particular finish. Timber is often used in the production of feature and fillet forms, box outs, through holes, and so on. Where the material is to be used structurally rather than in sheathing, the material should be sufficiently strong for the particular application.

The soft timbers are usually preferred as they are cheaper, easier to work, generally lightweight and do not warp so much when wet. Because timber forms are bound to absorb some moisture in use, kiln-dried timbers should not be given preference as such timbers swell with resultant loss of profile and even loss of strength. The moisture content of the timber should remain within 15 to 20 percent so that moisture movement is reduced to a minimum. If timber with a high-moisture content is used, it would shrink resulting in open joints and leakage of cement paste.

Common sections of timber used in formwork construction are 100×75, 100×50, and 150×50. Timber is best used uncut in available lengths, ends being lapped or allowed to project from the structure. As timber is progressively reduced in length, the shorter pieces should be set aside for various uses. Hardwood is good for use as bearing pads and wedges.

Higher grades of timber are required for use on contact surfaces rather than for studding, waling, and heavy structural members. However, the effects of lower-grade timber on the concrete surface finish may call for the use of higher-grade timbers.

Plywood

Because plywood is strong, light, and available in standard lengths, it is widely used as forms. Plywood is supplied in sheets of 1.2 m width with standard lengths of 2.4 m, 2.7 m and 3.0 m. The thickness of plywood should be such as to be able to withstand the pressure of vibrated wet concrete with the minimum strengthening required on the back. The advantages of using plywood panels are:

- Economical in large panels
- Available in various thicknesses
- Large plywood sheets would reduce the number of joints
- Predictable strength
- Creates smooth finished surfaces on concrete

Besides thickness and strength, the uniformity of plywood forms tends to reduce the amount of face finishing of concrete, especially because plastic facing material is generally incorporated in proprietary formwork panels.

Chipboard or particle board

Chipboard can also be used as formwork material; but because of its lower strength, it will require more supports and stiffeners. The number of uses which can be obtained from chipboard forms is generally less than plywood, softwood boards or steel.

However, development of laminated plastics for facings has resulted in the material becoming capable of offering 12 or more uses when casting concrete.

Metal

Steel forms are proprietary forms based upon a manufacturer's patent system and within the constraints of that system are excellent materials. Steel is not as adaptable as timber but if treated with care, it would give 30 to 40 uses, which is approximately double that of similar timber forms. Steel forms are found only in special uses, such as round column forms or concrete conduits or walls made up of small panels, especially where many reuses may be expected.

Job-built forms

Built-up forms are very common in India because labour is cheap and carpenters are used to build up forms from the basic materials at site. If built-up forms are used at the site, same forms may be used and reused for different construction work. In case of small jobs, use of built-up forms at site would be most cost-effective. The built-up forms would be particularly cost-effective if materials are available locally for the purpose.

Prefabricated forms

Prefabricated or patented forms are preferable if sufficient reuses of the forms justify their use. Prefabricated forms do have a much longer life than job-built forms, which may be built for any particular job. Prefabricated forms, on the other hand, are built especially to be used over a longer period. Additional benefits are derived in erecting and stripping of proprietary patented forms. These forms are manufactured in small units so that a single worker would be able to handle a unit easily. Prefabricated forms are also available in metal.

Footing forms

Column loads are transferred to the soil through foundation footings, which are the only structural members that can occasionally be constructed directly against the excavated earth walls without using any forms, especially in case of excavation in rock by blasting. Otherwise, earth should stand up the forces involved in construction work as forms would be strutted against the earth faces (Fig. 7.6).

As the footings would remain buried, most ordinary rough timber would be all right as forms. Width of the timber should be the depth of the footings. For deeper footings, planks should be spliced together using battens at about 600 mm centres.

For footings of depth up to 1 m, the forms may be supported

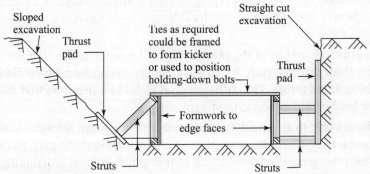

Fig. 7.6 Formwork for footings

by steel ties in tension instead of bracings. Column footings are generally rectangular, and the form corners are generally designed against possible opening out. Deep footings are stepped in many cases to effect economy. For construction purposes, forms for the upper portions are made and supported on the lower formwork.

Column forms

The shapes of columns that are generally in use are square, rectangular, circular, octagonal, etc. Because of the cross-section of a column being relatively small, the rise of level of concrete in a column during placing is rapid. As a result, column forms are subjected to more hydrostatic pressure than most. Concrete is placed in columns through windows at one or more points to avoid segregation or honeycombs. Concrete may either be placed through elephant trunk chute or by pumping. The recent development of self-compacting concrete does not require all these extra efforts. It is a good practice to locate column formwork against a kicker (~75 mm high) that is usually cast monolithic with the foundation/base/floor. The forms can be made of full height leaving cut-outs to accommodate beam forms. Additional pieces should be fixed around the cut outs to add extra bearing for beam forms.

Round columns are made of curved pieces of wood, steel or reinforced fibre tubes. Round wooden forms are expensive and may be used only when a few such columns are required. For a large number of round column forms, heavy-gauge pressed steel sheets should be the best choice.

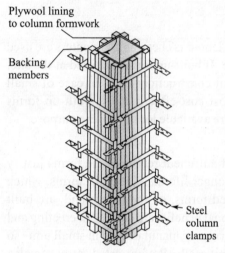

Fig. 7.7 Formwork for columns

Timber yokes or metal clamps (Fig. 7.7) are used to hold the column forms together. The spacing of the yokes and clamps should vary with the estimated pressures of wet concrete, the highest pressure occurring at the base.

Slab and beam forms

Formwork for the construction of slab and beam (Fig. 7.8) is rather complicated. The scheme of the formwork should be developed keeping in view both economy and ease of stripping. Only those materials should be used which can assure good finish and multiple uses.

Planks for a beam or girder should be of the exact width as the concrete member with the beam-soffit propped from the underside by cross-members. Beam forms may also receive support from the column forms. The soffit material should be strong so that the same may withstand the load of wet concrete until hardening.

Generally, the slab formwork, to an extent, is supported from the beam forms so that beams and slab could be cast together. This way, placing of concrete is made easy. Slab soffits are supported from the ground or floor using timber poles or steel scaffolding pipes. The slab supports are left in place for three weeks or more as specified. Steel forms are also manufactured for the floor slab and are economical where several reuses are expected.

Permanent formwork for slab construction, unlike traditional formwork, is left in place for the life of the element it is supporting. There are, however, two types of permanent formwork used in composite slabs:

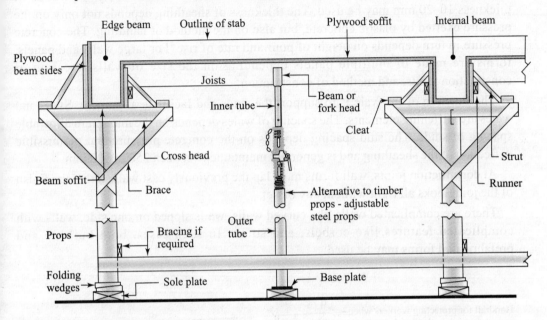

Fig. 7.8 Formwork for slabs and beams

- Structurally participating, which is designed to provide temporary support for plastic concrete as well as loads caused due to construction activities, and then become part of the permanent works contributing to the strength of the completed slab/deck
- Structurally non-participating, which is designed solely to support plastic concrete as well as loads caused due to construction activities—hardened roof slab needs to have adequate strength to bear all loads

The use of permanent formwork has the following benefits:

- Eliminates the necessity for falsework installation
- Reduces the need of skilled manpower to be deployed for erecting/stripping of forms
- Increases the potential for standardization related to roof construction among other possibilities
- Speeds up fixing time and eliminates stripping of forms on hardening of concrete

Specific benefits for the construction of bridge decks include:

- Provision of a safe working platform early in the deck construction process, which is a major advantage when the bridge is spanning a rail track, river or live carriageway.
- Elimination of the need to strip formwork in difficult and confined spaces when working at heights

Wall forms

Forms for concrete walls are made of timber, metal, or even hardened concrete. In case of timber forms, sheathing is stiffened by studs and wales. Sheathing is assembled by using 25 mm × 100 mm or 25 mm × 150 mm boards. Alternatively, plywoods of thickness 10–20 mm may be used. The thickness of sheathing depends not only on the pressure exerted by plastic concrete, but also on the method of handling. The concrete pressure in turn depends on height of pour and rate of rise. For large walls and panels, forms are made of multiple panels depending on the size of walls, location of construction joints, and method of construction.

The function of the wale is to support the studs and facilitate alignment. Studs and wales are stiffening elements. The spacing of wales depends on the maximum allowable spacing of studs. The stud spacing depends on the concrete pressure and permissible deflection of the sheathing and is generally maintained between 250 to 600 mm.

At construction joints, wall forms must lap the previously cast wall so that the finish of the joint looks all right without any offset.

There are complicated walls like curved walls, walls sloped on one side, walls with complicated features like corbels, and so on. In such cases, both built-up and prefabricated forms may be used.

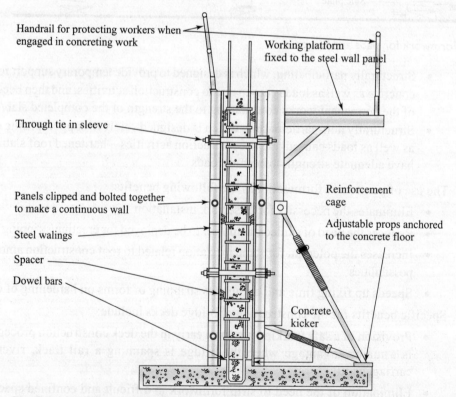

Fig. 7.9 Formwork for walls

Special forms

Special forms are required for:
- Arches
- Shells
- Folded plates
- Post-tensioned cast-in-situ structures
- Special method construction like slip-forming
- Well sinking—finished concrete sinks as construction progresses
- Permanent forms of any type
- Underwater concreting
- Combination of precast and cast-in-situ concreting

An arrangement of slip-formwork is shown in Fig. 7.10.

There are no limitations on the plan shape or extent of the structure to be constructed using slipform. The shape can be circular, rectangular, cruciform, curved, irregular, solid, hollow, or cellular. The requirement is that this area must be projected straight upward to a height that would be cost-effective. As slip forms slide past the face of concrete, no projections beyond the face would be possible during slip-forming. If floors are required to be cast, then keys, beam pockets, weld plates, reinforcement dowels or other devices should be kept flush with concrete surfaces for casting of floors later.

Slip-formed concrete surfaces are vertical in most cases; as otherwise expenses could be prohibitive. Even though slip-forming of stepped, tapered, or even vertically curved surfaces is possible, but such construction work would be taken up only if found cost-effective and timely executionable.

Slip-forming, apart from its use in bins and silos, is also used in high-rise buildings including their cores, such as lift shafts, stairwells, toilets etc. Slip-forms are also used in water towers, dams, missile silos, cooling towers, air-traffic control towers, chimneys, tanks, bins, bridge piers etc. Slipforming is now used in road pavement construction also.

The slipform as shown in Fig.7.10 comprises forms, wales/walings, and yokes/trestle legs. These forms are subjected to both vertical and lateral loading. The sketch shows two platforms—upper one for general purposes and lower one for curing and finishing work. The entire load of forms, platforms (with men and materials) and lifting the forms is carried by the climbing pole. The only function of concrete is to support self-weight and prevent the climbing pole/jack rod from buckling. The frictional drag force between the forms and hardening/hardened concrete is to be carried ultimately by the climbing pole. Apart from the vertical drag, the forms must carry the hydrostatic lateral pressure of the plastic concrete. The form material used for slip-forming is timber.

A very important aspect of slip-forming is propulsion and support of the forms and is effected by means of jacks and the climbing pole on which these jacks are supported. These jacks are operated by electric, hydraulic, or pneumatic jacking. The jacks are

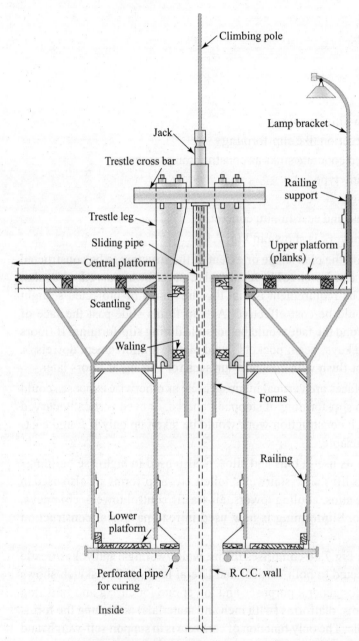

Fig. 7.10 Formwork for slip-forming

generally cylindrical in shape with a hole at the centre for climbing pole/jack rod to pass through. The extent of each climb is governed by the stroke of the jack, which is generally 25 mm. The spacing of climbing poles depend on span of wales, curvature of wall, capacity of jacks, capacity of climbing poles/jack rods, distribution of loads and placing of jacks. The rate of rise of form depends on the rate of strength gain of the concrete. The rate of pouring versus the rate of rise of forms should be set so as to keep the forms filled at all times.

Vibration of concrete in slip-forming work was not allowed at one time, but not any more. Vibrator is used to penetrate as deeply as it would under its own weight, but it should not be forced deeper into the concrete and should not be placed too long at one place.

The two main methods of curing slip-formed concrete are water curing and membrane curing.

7.8 REINFORCING STEEL

Development of RCC as a structural medium is based on the fundamental fact that the coefficient of expansion of both concrete and reinforcing steel is approximately the same ($\sim 11.7 \times 10^{-6}/°C$). Concrete may be reinforced with either plain or deformed steel bars to resist tensile stresses to which a structural member is subjected while the concrete resists compressive stresses. Reinforcing steel, if required, may also assist concrete in resisting compressive stresses. Moreover, concrete and steel in combination resist shear stresses. Bond between the concrete and the steel must be adequate so as to prevent slipping when subjected to load. Deformed steel bars develop more bond stress resistance compared to plain steel bars.

The integrity of RCC largely depends to a large extent on the reinforcing steel. It is, therefore, essential that the reinforcing steel bars are placed accurately in accordance with the approved drawings. The design should strictly be followed. This would require:

- Delivery of reinforcing steel bars on the basis of the test certificates confirming that the delivered materials conform to the relevant specification
- Test samples from the delivered materials should be sent to a recognized laboratory of repute to further confirm that the quality of materials conform to the specification—achieving tensile strength without undue strain, can be easily bent to any required shape, developing sufficient bond strength between concrete and reinforcing steel, coefficient of thermal expansion as required, and so on
- Reinforcing steel bars as fixed and ready for placing concrete should be supervised to make sure that correct sizes and shapes of steel bars have been used and they are placed in the correct positions as shown in the approved detailed drawings maintaining adequate and clear cover on all sides
- Provision of adequate lap lengths at the points of splicing
- Site supervision should be carried out to ensure that the steel bars have been supported so rigidly as to prevent displacement or distortion from the approved positions under the loads of construction activities
- Substitution of steel bars from the details shown in the drawings should be done with the approval of design engineers

Materials

The materials to meet the above requirements are as per IS: 456 (latest revision) issued by the Bureau of Indian Standards

- Mild steel and medium tensile steel bars conforming to IS: 432 (Part 1)
- High strength deformed steel bars conforming to IS: 1786
- Hard drawn steel wire fabric conforming to IS: 1566
- Structural steel conforming to Grade A of IS: 2062

Site storage

Approved reinforcing steel should be stored at the site much above the ground level over pedestals or wooden/precast concrete sleepers so as to avoid any possibility of contamination due to water logging or other harmful materials. Reinforcing steel must be free of mud, grease, oil, coats of paint, loose mill scales, rust, and any other deleterious materials. Reinforcing steel, if contaminated, should be thoroughly cleaned using wire brushes before being put to use. Storage should be so planned as to ease selection and avoid deformation.

Fabrication

Fabrication of reinforcing steel bars involves cutting to lengths and bending as required. The preparation of bar marking/placing drawings including bar bending schedules is termed 'detailing'. The concerned contractor prepares detailed drawings and bar bending schedules for approval by the design engineers. Bars are cut, bent, and placed on the basis of approved drawings and schedules. For detailing of reinforcing steel bars, IS: 2502 is followed.

The storage area should be so arranged that the steel bars can be easily drawn for cutting to lengths as per the approved bar bending schedules. Steel bars can be handled or transported manually or mechanically. For fabrication, bars can be sheared by machines or chiselled manually. It is a normal practice to cut the longest lengths from each size bar first, leaving the smaller lengths for stirrups and links until last, thus reducing off-cuts to minimum. Cut lengths are labelled for identification on the basis of markings on the approved detailed drawings.

Cutting is a simple linear operation. Bending operation in comparison is not simple and requires greater area for manoeuvring. To overcome the requirement of space, some of the automatic link benders have inclined working tables, some of which are almost perpendicular. The use of modern equipment makes it possible to prejig or so programme the machine as to produce the bends in pre-determined sequence. With such machines, it is essential that due allowances be made in cutting for losses in bending and drawing of bars from the lots. The whole process needs to be checked by trial bends. The mandrels on the machine are interchangeable to form and maintain correct bending radius as per the approved bending schedules. Where mechanized construction methods are adopted or bending of heavy rods is involved, machines are used for bending reinforcing steel bars in India. Shop fabrication of steel bars is not in practice here. What is in common practice is bending steel bars around simple jigs such as a board with dowels fixed to give the required profile.

When full length bars are not available, splicing of bars becomes a necessity. The three general types of splices are: (i) lapped, (ii) welded, (iii) mechanically coupled. In general, lapped splices are provided at construction projects as they are more economical than the other types. Where spacing of bars is very close, it may not be possible to provide lap splices and, therefore either welded or mechanically coupled splices need to be provided. Proprietary mechanical coupling systems are readily available. For welded splicing, reinforcing steel bars are either butt or lap welded (vide Chapter 8).

Placing

Reinforcing steel bars should be brushed and cleaned of all loose rust, mill scale, grease, and oil so that strength and bond of steel bars remain unaffected.

The placing or setting of reinforcing steel bars is to be carried out on the basis of the approved detailed drawings. Every bar should be placed at the exact location. The bar chairs and spacers are set and spaced first, followed by placing of each bar, stirrup and tie positioning them accurately by tying with wire or occasionally by tack welding. Reinforcement may either be located and tied in-situ or pre-assembled into cages ready for installation in the formwork. The hourly output of placing bars in terms of tonnage would depend, apart from the efficiency of the personnel deployed, entirely on the type of reinforced concrete structure involved.

For maintaining the necessary and adequate concrete cover during construction, small concrete blocks may be placed between the reinforcement and the formwork. Occasionally, plastic clips or spacer rings can be used for this purpose. In case of slab, mat or footings, the top layer of reinforcement should be retained in position using chairs or cradles made of reinforcing steel bar cut pieces. All such spacers should be placed in such a way as not to affect the durability of concrete.

7.9 SHOTCRETE

Shotcrete is also known as sprayed concrete or *gunite*. Sprayed concrete is a general term. Shotcrete is pneumatically and continuously spraying concrete at high speed onto a backup surface or substrate using purpose-made process equipment so powerfully that a fully compacted self-supporting concrete structure is formed instantaneously without sagging or sloughing. Depending on the setting acceleration, it can be applied to any elevation including on vertical or inclined face or even overhead.

There are two basic processes for shotcrete application:
- Dry process
- Wet process

The mix requirement on the workability and durability are:
- Pumpable characteristic
- Cohesion and minimum rebound to avoid downtime and delays
- Conformance to specified strength including high early strength
- Environmentally acceptable in case of confined areas, e.g., in tunnel construction
- Good sprayability (pliability)

There is no difference between the properties of shotcrete and normal concrete of similar proportions. The difference lies in the method of placing. Shotcrete has advantages in many applications because of its method of placing. Such advantages, however, depend totally on the experience, skill, and expertise of the operators deployed for actual placing by the nozzle. Sprayed concrete has many advantages:
- As shotcrete adheres immediately and bears its own weight, it can be applied at any elevation
- Can be applied on uneven surfaces
- Satisfactory adhesion to the substrate
- Totally flexible configuration of the layer thickness on site
- Shotcrete may be reinforced with mesh or fibres
- Rapid load-bearing skin can be achieved without formwork or long waiting

Compared to normal concrete, cement content of shotcrete is high apart from the high cost of deploying the necessary equipment and mode of placing. Because of these reasons, use of shotcrete is limited in specific types of construction work:

- Excavation stabilization in tunneling and underground construction
- Tunnel and underground chamber lining
- Prestressed tanks
- Lightweight reinforced thin sections like shell or folded-plate roofs
- Repair of deteriorated concrete – concrete replacement and strengthening
- Stabilization of rock slopes
- Encasing steel for fire proofing
- Protective lining – thin overlay on concrete, masonry, or steel
- Restoration of historic buildings

Shotcrete by dry process has been in use successfully, and is being continuously developed and improved. The dry mix comprising cement and damp aggregates are intimately mixed and fed into a mechanical feeder or *gun*. The mixture is then transferred by a feed wheel or distributor into a stream of compressed air in a hose, and carried up to the delivery nozzle that is fitted inside with perforated manifold through which water is introduced under pressure and intimately mixed with the other ingredients. Water added should just be sufficient for hydration and fluidity for spraying. Water:cement ratio maintained is 0.3 to 0.5. The mixture is then sprayed at high speed onto a backup surface or substrate. Admixtures, if required, may be added either dry with the mix or wet at the nozzle. In tunnels, it is possible to use remotely controlled robotic spraying, whereby hazards to the operators is reduced and requirement for platforms for access to the higher levels is also reduced.

Shotcrete by wet process demands more work at the beginning (start-up) and at the end of spraying (cleaning) than the dry process. In this process, all ingredients including water are mixed together prior to being fed into the chamber of delivery equipment and from there conveyed pneumatically and continuously at high speed onto the place to be filled with shotcrete. Water:cement ratio maintained is 0.40 to 0.55.

Of the two processes, dry process is better suited for: (i) use with lightweight aggregates, (ii) use of flash set accelerators (iii) greater delivery lengths, and (iv) intermittent operation. The consistency of mix can be directly controlled at the nozzle and higher strengths up to 50 MPa can be achieved. As regards wet process, better control can be exercised on the quantity of mixing water and admixture, if used. Less dust is produced by the wet process as well as lower rebound. Wet process is suitable for large volume operation. For both the processes, curing is important as the large surface-volume ratio of concrete can be the cause of rapid drying.

Some materials rebound because all of the shotcrete do not remain in position because of the high speed of the impacting jet. The rebound materials comprise the coarsest particles in the mix. Thus, the mix remaining in place without rebounding is richer than the mix proportions as batched. This may cause shrinkage. Rebound is greatest in the initial layers but becomes less and less because of the cushioning effect of the plastic concrete already in place. The danger of rebound is not so much in waste of materials as in the possibility of incorporation in the subsequent layers of shotcrete. However, more

rebound means more rework. Factors influencing the extent of rebound:
- Spraying process—dry or wet
- Spraying unit—air pressure, nozzle, spray output
- Spraying direction—horizontally, up, down
- Proportioning of the ingredients, additives etc
- Sprayed concrete—design, strength requirement/ development
- Substrate condition—evenness, adhesion

7.10 LIGHTWEIGHT AND HEAVYWEIGHT CONCRETE

Lightweight concrete

Lightweight concrete can be produced with a dry density range of approximately 300 to a maximum of 1600 kg/m^3, though products up to 2000 kg/m^3 are used for structural concrete, compared to 2100 to 2500 kg/m^3 for normal weight concrete. Lightweight concrete is produced principally for:
- Reducing the dead load of a structure
- Lowering the cost of foundations
- Adding fire-resisting capability, insulation, etc.

The principle techniques of producing lightweight concretes are:
- Excluding fine constituents of the normal aggregates, thus creating air-filled voids
- Including bubbles of gas in a cement paste to form a cellular structure containing 30–50 voids (aerated or foamed concrete) where strength is not the basic requirement but resistance, insulation, and light weight are major considerations
- Replacing natural aggregates as a whole or partly with available lightweight aggregates of low density containing a large proportion of voids

Structural members made with lightweight concrete have reduced self-weight as a result of which structural members would have thinner sections costing less. With thinner structural members, foundation cost would also be reduced, thereby reducing the cost of handling and centering. It can, therefore, be concluded that lightweight concrete would be advantageous for precast construction work because of the following:
- Larger units can be produced compared to normal weight concrete without causing any problems in transporting, handling, and lifting
- Because of larger units, number of joints would be fewer
- Larger units and fewer joints can expedite the construction work.

Specific applications include precast masonry units and slabs, concrete fill and insulation, nonbearing walls, and floors/slabs. Favourable factors for lightweight concretes are weight reduction, better insulating and fire-resisting properties, and savings in the cost of material handling and forms. Types of structures where the use of lightweight concrete can be economical include tall buildings and long-span bridges.

Among other things, the properties of lightweight concrete are greatly influenced by the types of aggregates, their strength and, of course, water:cement ratio.

Aggregates

Lightweight aggregates can be divided into the following categories:

- Structural—(i) natural (n): scoria and pumice (ii) manufactured: expanded clay, shale or slate (n + p → processed natural) or blast furnace slag (bp → byproduct from industries)
- Non-structural—vermiculite and perlite although scoria and pumice may also be used—these materials are used in insulating concretes for soundproofing and non-structural floor toppings.

Properties of a few lightweight aggregates are given below.

Aggregate	Type	Bulk density (Kg/m^3)	Water absorption % by volume in 24 hours	Compressive strength (N/mm^2)
Low strength (0.5 - 3.5 N/mm^2)				
Perlite	n + p	40–200	–	1.2
Vermiculite	n + p	60–200	–	0.2
Medium strength (3.5 - 15 N/mm^2)				
Pumice	n	350–650	50	5–15
High strength (above 15 N/mm^2)				
Volcanic tuff	n	700–1100	7–30	5–20
Expanded shale	n + p	400–1200	5–15	20–60
Sintered fly ash	bp	600–1100	20	30–50
Foamed blast furnace slag	bp	400–1100	10–35	10–45
Sintered colliery waste	bp	500–1000	15	10–40

The lightweight aggregates are not heavy due to their cellular or highly porous microstructure. Porosity has an important role on the strength of the aggregates, and therefore, on the strength of lightweight concrete. The particles with high porosity are weaker than those with lower porosity for a particular material. The size and distribution of the pores and the form of the pore walls also play an important role. The strength of aggregates generally increases with increasing particle density and bulk density.

Unlike normal weight aggregates, lightweight aggregates have an important characteristic of water absorption due to porous structure. Water absorption is of the principal difficulties of mix design as it affects the workability of the fresh concrete. The effective water:cement ratio as well as some important properties of the hardened concrete like shrinkage and creep are also affected.

The lightweight aggregates, most suitable for structural lightweight concrete, are manufactured products because they are more uniform than natural products. Expanded clay, shale, slate, pulverized fuel ash and sintered slate and colliery waste are used for structural lightweight concrete work. Adequate strength of structural concrete can be obtained with foamed blast furnace slag and foamed lava.

Pumice, the light-coloured and froth-like volcanic glass having bulk density of $500 - 900$ kg/m^3, produces low strength concrete. It is used primarily for the manufacture of lightweight concrete where thermal insulation is required. It is generally not meant for load-bearing functions. Volcanic tuff is used to give structural and heat-insulating lightweight concretes.

Lightweight concrete is classified by the following two methods:
- Method of manufacture
- Purpose for which it is to be used

It is produced by three methods:
- Lightweight aggregate concrete is produced by using various types of porous lightweight aggregates:(i) made with both coarse and fine lightweight aggregates (ii) made with lightweight coarse and normal fine aggregates. Interpolation between these two classes is permitted.
- Lightweight concrete is produced by entraining voids in the concrete mass. This type of concrete is variously known as aerated, cellular, gas, foamed or foam concrete. There are two basic approaches for producing aeration:(i) gas (hydrogen) bubbles are introduced into the cement paste matrix generally by aluminium powder that reacts with $Ca(OH)_2$ and alkalis released into the solution (ii) foamed concrete is produced by adding preformed foam from a foaming generator or by adding to the mix a synthetic/protein-based foam producing admixture which introduces and stabilizes air bubbles during mixing at high speed.
- Lightweight concrete is produced by the mix comprising cement, water, and coarse aggregates with fines (sand) omitted. A large number of interstitial voids are created by this method. Coarse aggregates of normal weight are used. This type is known as 'no-fines concrete' (NFC).

Lightweight concrete can also be classified on the basis of the purpose for which it is to be used. Accordingly, we distinguish between structural lightweight concrete, non-load bearing lightweight concrete, and thermal insulation lightweight concrete.

Mix design

The processes used for designing ordinary concrete also apply to structural lightweight concrete, but three additional factors deserve consideration:
- The need to design for a particular density
- The influence of lightweight aggregates on the properties of concrete
- The water absorption of porous aggregates

Based on the above considerations, (i) type and grading of aggregate, (ii) cement content, (iii) water:cement ratio need to be determined.

Type and grading of aggregate

The requirement is to design a lightweight concrete for the specified strength and maximum density. According to the experimental studies on achieving higher concrete strength with lower cement consumption and particular type of aggregate, the heavier aggregates are generally preferable. Should a lighter aggregate be desired for the purpose of reducing the density of the concrete, the strength of the mortar mix must be increased. Determination of the type of aggregate is made on the consideration of bulk density-strength relation depending on availability and economy.

The aggregate content can be determined by any of the following methods:
- Absolute volume method
- Volumetric method
- Specific gravity factor method—trial mix basis
- Weight method
- Effective water:cement ratio

Cement content

For a particular workability, strength increases with cement content, the increase depending on the type of aggregates used. Regarding different types of aggregates, the cement content requirement corresponding to any particular concrete strength can vary considerably. It depends, in essence, on the strength and modulus of deformation of the aggregate particles apart from the free water content required for workability. The cement content, as required, can be determined with any accuracy by conducting tests on trial mixes. On an average in case of lightweight aggregates, 10 percent higher cement content would raise the strength by approximately 5 percent.

Water content

Free water content is the same as for normal weight concrete, but water absorption by aggregates requires higher total water content due to the pores of the aggregate particles. It is difficult to determine the required amount of free water as it depends primarily on the size, grading, shape and surface characteristics of the aggregate particles and to a lesser extent on the cement content. The free water—by far the greater part generally—is in the cement paste. It determines the workability and the strength of concrete.

The properties of structural lightweight concrete is discussed below.

Compressive strength and density

Higher compressive strengths require very high contents of cement and cementitious materials. Silica fume improves the strength development of lightweight concrete. However, compressive strength of lightweight concrete is generally controlled by the strength of porous aggregates which is lower than mortar strength. Compressive strength and density are two of the most important parameters of lightweight concrete as strength increases with an increase in density. With suitable lightweight aggregates, structural lightweight concrete can be made with densities that are 25 – 40% lower but with strengths equal to the maximum generally achieved by normal concrete. The density of concrete is mainly controlled by the particle density of the aggregate mixture. For a continuously graded mix, a grading richer in fines or a smaller maximum diameter leads to a concrete of higher density and strength.

Tensile strength

The factors influencing compressive strength also influence tensile strength. The relation between the tensile strength and compressive strength is of the same order as for ordinary concrete in case of moist cured lightweight concrete whereas the tensile strength is lower in air cured lightweight concrete. The values are found to be within the following limits in most cases:

Cube compressive strength N/mm^2	Splitting tensile strength N/mm^2
10	0.9 – 1.3
20	1.4 – 2.0
30	1.8 – 2.7
40	2.2 – 3.3

The durability of lightweight concrete is discussed below.

Permeability

Despite the lightweight concrete being more porous and absorbent, the water permeability of lightweight and normal concretes is of the same order. However, the resistance against diffusion of water vapour is lower in lightweight concrete compared to that of normal concrete.

Fire resistance

Lightweight concrete provides better protection against excessive heat generated by fire as its thermal conductivity is lower compared to that of normal concrete. Volume changes occur in lightweight concrete as in normal weight concrete, but lightweight concrete is more stable when exposed to heat.

Chemical stability

The nature and quality of cement governs the chemical stability and durability of concrete of all types including lightweight concrete. Despite the aggregate particles being porous in lightweight concrete, the hardened cement paste is usually stronger and denser. Although these two opposing factors vary, lightweight concrete is generally as resistant to chemical attack as normal concrete.

Abrasion resistance

In general, the abrasion resistance of lightweight concrete is low and may be less than that of the hardened cement paste even though lightweight aggregates may comprise hard materials. The abrasion resistance of lightweight concrete, therefore, is poorer than that of normal concrete. The abrasion resistance of lightweight concrete produced with expanded clay and shale and having compressive strength of 25 – 55 N/mm², however, is very good. Resistance can be improved by mixing normal weight fine aggregates with low-density coarse aggregates as well as improving the quality of the matrix and the use of surface treatment.

Resistance against freezing and thawing

The frost resistance of some lightweight aggregates is poor despite their being embedded in the hardened cement paste. Air entrainment improves resistance of lightweight concrete against freezing and thawing.

Heavyweight concrete

Heavyweight concrete produced with both heavyweight coarse and fine aggregates are used in the construction of nuclear reactors and other structures exposed to nuclear particles (neutron, proton, alpha, and beta) and high intensity radiation (gamma rays and X-rays) having high-power of penetration but can be absorbed adequately by an appropriate mass of any material. Heavyweight aggregates increase the weight as well as density of concrete. Also, absorption of radiation from nuclear reactors is proportional to the density because of which heavy aggregates have greater capacity for absorption compared to aggregates generally used in producing normal concrete. Apart from mass and density, neutrons require appreciable amounts of iron, hydrogen, boron or cadmium. Hydrogen contained in the moisture present in concrete is sometimes sufficient for neutron attenuation. Boron or cadmium, however, is mixed purposefully as aggregate or admixture. Heavyweight or high-density concrete accounts for a rather small section of concrete production within the construction industry mainly within the nuclear industries or utilities.

Production of heavyweight concrete for shielding is not as simple as for normal weight concrete because those involved need to be knowledgeable on (i) the source, nature and intensity of the nuclear particles and rays that are to be stopped or attenuated to acceptable limit (ii) choice of aggregates and admixtures required to produce the concrete to do the job. Quality and gradation of aggregates as well as water:cement ratio are important for achieving appropriate strength. However, heavyweight concrete for shielding purposes has been produced with strength good enough only for holding their shape. Strengths of 7 MPa to 14 MPa are easily achieved for concrete to be used in massive slabs on grade, walls, or deep walls, where strength is secondary to shielding capacity. Heavyweight aggregates are used selectively, where strength of 28 MPa or more is required. Strength apart, volume change and freedom from cracking are also issues of prime concern where structural members form part or the entire shield. To avoid possible alkali-silica reaction, both cement and aggregate should be selected only after appropriate laboratory tests.

Heavyweight concrete is useful to be used as counterweight or simply to enhance deadweight to reduce bulk volume of normal concrete. Heavyweight concrete is required in ship's ballast and encasement of underwater/submarine pipelines for air, gas or even liquid. Some heavy industrial wear-resistant floor-surfacing is done with heavyweight mortar or concrete where heavy iron aggregates resist abrasion. Density of heavyweight aggregate concrete varies between 3200 to 5600 kg/m^3.

Materials

Heavyweight aggregates include magnetite with specific gravity δ of (4.2 to 4.8), barite (δ = 4.0 to 4.5), limonite (δ = 3.4 to 3.8), ferro phosphorus (δ = 5.8 to 6.5) and steel shot or punching (δ = 7.5 to 7.8). The ranges in specific gravity indicated for each are much dependent upon the purity of the ore or that of the processed or manufactured material. Ferro phosphorous, for example, is a slag and, therefore, is subject to variation in density as are the ferrous aggregates or iron. Such heavyweight aggregates may be used instead of gravel or crushed stone to produce dense concrete for shielding of nuclear reactors.

Mix

Because of the high density of heavyweight aggregates, segregation of plastic concrete is an important aspect to be given due consideration in mix proportioning. Accordingly, both fine and coarse aggregates should be produced by crushing heavyweight rocks and minerals. Because of rough shape and texture of crushed aggregate particles, heavyweight concrete mixtures tend to be harsh. To overcome this problem, it is common practice to use finer sand, a greater proportion of sand in aggregate compared to normal concrete and higher cement contents (360 kg/m^3). As an alternative to placing plastic heavyweight concrete, formwork mould can be filled by aggregates coarser than 6 mm and consolidated so as to fill the mould completely. The grout is then pumped in starting at the bottom, and in amount, manner and sequence sufficient to fill all the voids in the mass of aggregates. To make this intrusion of grout as complete and thorough as possible, admixtures/additives are invariably used, and these may include plasticizers, pozzolona, air-entraining agents, water-reducers, retarders, and pumping aids (vide Section 7.6).

Important properties

As already made clear, heavyweight concrete is beset with the problem of workability because of the tendency of coarse aggregate to segregate. Heavyweight concrete, therefore, can be placed by chutes or pumped over short distances only. Strength of the order of 22 to 41 MPa is achievable with high cement contents. Strength is, however, of principal concern in the design of heavyweight concrete suitable for use in prestressed concrete reactor vessels (PCRV). Compared to conventional structures, these pressure vessels operate at higher stress levels and temperatures because of which concrete is subjected to appreciable thermal and moisture gradients. In such cases, care is needed to minimize inelastic deformations such as creep and thermal shrinkage because they can cause micro-cracking and loss of prestress. To minimize micro-cracking, the elastic modulus of aggregate and compatibility of coefficients of thermal expansion between aggregate and cement paste should be taken into consideration.

Strength loss can occur considerably when concrete is subjected to wide and frequent fluctuation in temperature. Concrete for PCRV is, therefore, designed not only for high density but also for high strength. The reactor vessels are usually designed to operate with concrete temperatures of up to 71°C, but higher accidental temperatures and some thermal cycling is expected during the service life.

7.11 READY-MIXED CONCRETE

The most significant advantage of ready-mixed concrete (RMC) is that it is produced under controlled conditions unlike at most construction sites except for large sites where developing the required controlled conditions is possible. And the main difference between RMC and site-produced concrete is the time and method of transportation from a central plant to the site of placing. Proper care during transportation is ensured by the agitator trucks (vide Chapter 13, Section 13.6), but the responsibility of proper placing and compaction rests with the site management. Central batching provides flexibility of operation, freedom from waste and contamination, and accuracy of proportioning that cannot be matched at the construction sites with piles of materials dumped on the ground.

There are mainly two kinds of RMC:
- Concrete mixed in a central plant and then transported to the site for placing
- Concrete ingredients are batched in a central plant but are mixed in a truck mixer either in transit or immediately before placing – permits longer haul or less vulnerable to delay

During transportation, the results of evaporation and hydration are the most significant while absorption is pertinent only with dry or highly absorptive aggregates. Abrasion is relevant only in case of aggregates susceptible to abrasion resulting in increased fineness. The effects of all these are:
- Hydration affects workability without affecting the effective water:cement ratio
- Evaporation and absorption reduces water:cement ratio in the paste thereby affecting workability, with possibility of increased strength
- Abrasion increases fines thereby affecting workability—over-mixing results in increase of fines because of grinding action and more fines would need more water to maintain consistency of concrete

The main problem related to the production of ready-mixed concrete right up to placing is workability, which is affected by a number of factors during transportation:
- Mixes of lower cement content lose workability at a lower rate as less water would be consumed in hydration
- Mixes of higher water content lose workability at a lower rate
- Mixes with admixtures incorporated may lose workability faster if such admixture as made for water-reducing purposes is used in the mix because the initial water content would obviously be low
- Mixes may lose workability if ambient weather condition is very hot and dry
- Mixes, if delivered in large volume, rarely lose workability as the surface area would be less compared to the volume of concrete in a truck mixer

The concrete is required to be placed within two hours of coming into contact with water. Within these two hours, the total operation from batching to placing is to be completed. If about 1 hour is required for complete placing and about 15 minutes for batching, then concrete is to be transported in approximately 45 minutes. The delivery distance within this time would vary between rural and urban areas, infrastructures and traffic. The time limit can be extended up to 3 to 4 hours using retarding admixtures. Ready-mixed concrete is placed as close as possible to its final position. Generally, concrete is placed in horizontal layers of more or less uniform thickness, and each layer is thoroughly compacted before the next layer is placed to avoid cold joints. The rate of placement should be so rapid as to ensure that the layer immediate below is still in plastic state.

Retrospective quality control, accurate batching records, improved security and improved consistency can now be made possible as most of the central plants are equipped with automatic or semi-automatic batching systems and controls made possible by the use of microprocessors and computers. For the purpose of achieving better quality control, central plants are preferred over truck-mixing.

7.12 HIGH PERFORMANCE CONCRETE

High performance concrete (HPC), as the name suggests, is a product developed to satisfy effectively the performance requirements for the desired application and is essentially a modification of ordinary concrete. HPC is still in developing stage. A lot of research is in progress globally to develop HPC to meet specific performance requirements of structures in terms of high modulus of elasticity, high density, low permeability, tensile strength, ductility, fatigue strength, thermal resistance, durability, etc. HPC has a wide range of prospective applications. In many cases, it is the high durability that is the required property; in others it is high strength, either very early or at 28 days or even later; in still others, it is the high modulus of elasticity. Regarding the use of HPC, it has been considered suitable for use in high-rise buildings, high dams, prestressed bridges, airport pavements, off-shore structures, structures exposed to severe exposure conditions, etc.

Materials

HPC contains, besides water, the same ingredients as used in normal high strength concrete – good quality aggregates, cement, silica fume, other cementitious materials like fly ash or blast furnace slag and superplasticizer. Their proportions, however, are different in HPC. This is particularly true of the water content of the mix along with a large dosage of superplasticizer in HPC. Ordinary Portland cement is used, though rapid hardening cement may also be used if high early strength is required. Silica fume content ranges from 5 to 15 percent of the total cementitious materials by mass when silica fume is not an essential ingredient of HPC. However, silica fume is essential for achieving higher strength. Superplasticizer dosage is as high as 5 to 15 litres per m^3 on the basis of solid contents in the superplasticizer. It is essential that HPC be placed by conventional methods and also cured normally keeping in view that moist curing is required. What is particularly important in case of HPC is its very low water:cement ratio ranging from 0.25 to 0.35, occasionally the lower limit is required to be as low as 0.2 (extra cement after hydration would constitute fines in the cementitious materials), still having slump of 180 to 200 mm. For HPC, aggregate strength and cement-aggregate bond are the two important controlling factors, and the role of water:cement ratio is less clear. To be sure, however, very low water:cement ratio is maintained to produce HPC. It is better to determine strength versus water:cement ratio for any given set of raw materials to start with. As regards cement content, it is very high ranging from 450 to 550 kg/m^3 when the content of the cementitious materials range from 450 to 650 kg/m^3. It is apparent from all these that HPC may be deemed to be logical development of concrete containing silica fume and superplasticizer.

High strength and low permeability are the two distinguishing features of HPC compared to normal concrete. The strength development of concrete primarily depends on the characteristics of hardened cement paste, aggregates and the transition zone between the hardened cement paste and the aggregates. The strength of hardened cement paste depends on its volume of pores. High strength requires a low volume of pores, especially of the larger capillary pores for achieving high performance. Use of the properly graded particles down to the finest size is the only way to reduce pores to low volume. This could be made possible by the use of silica fume, which fills the spaces between the cement particles, and also between the cement particles and the aggregate. The mix must, however, be sufficiently workable for the solids to be dispersed to facilitate dense packing. This is achievable with the use of large dosage of such superplasticizer as is compatible with the cement being used – compatibility based on rheology. Superplasticizer works against the tendency of cement grains to flocculate and thus to hold water, offering resistance during compaction. In other words, superplasticizer deflocculates the cement particles and thus assuring fluidity of the mix so that very low water:cement ratio is good enough for adequate workability. Typically, 5 to 15 litres per m^3 of superplasticizer can effectively replace 45 to 75 litres per m^3 of water resulting in reduced distance between the cement particles and denser cement matrix.

When the transition zone between the paste and aggregate is improved, the transfer of stresses from the paste to the aggregate particles becomes more effective. The strength of coarse aggregates could then be the limiting factor. Good quality graded aggregates having high intrinsic strength, therefore, should be used though what controls high strength is the strength of hydrated cement paste. Crushed rock aggregates are generally preferred to smooth gravels as there is evidence that the strength of transition zone is weakened by smooth aggregates. Aggregate particles may be severely micro-cracked during crushing. The number of such micro-cracks would be more in larger particles. It is, therefore, a common practice to use smaller particles (10–14 mm nominal size) for achieving high strength. Fine aggregates should be selected to reduce water demand. Rounded, uniformly graded and rather course particles are preferred to crushed rock particles wherever possible because, the rich mixes used in HPC have a high content of fine particles. Ranges and values mentioned relating to HPC are only indicative as these are not yet standardized. HPC production is based on experience and published data.

High strength of HPC is conditional on the full compaction of concrete, and full compaction requires appropriate rheological properties (mainly workability) at the time of compaction. There would be difficulty in maintaining adequate workability if the superplasticizer becomes fixed by C_3A in the cement. This reaction between the superplasticizer and C_3A, which can lead to rapid slump loss, must be avoided by ensuring compatibility between the superplasticizer and the Portland cement to be used. For this, laboratory testing on a trial and error basis of a number of cement pastes containing combinations of different cements and superplasticizers is done for the purpose of establishing the best combination from the rheological point of view.

High strength of HPC is partly due to the presence of a dense matrix. Accordingly, a portion of the Portland cement may be replaced by one or more of supplementary cementitious materials. This would not unduly depress the early strength of the concrete. Besides, the lower reactivity of supplementary cementitious materials amounts to a partial replacement of cement which would be in favour of rheological compatibility.

Durability

Dense structure of hydrated cement paste with discontinuous pore system of HPC and its low permeability imparts resistance to external attack. This is very true regarding the ingress of chlorides into the concrete. Low permeability and very low water content limits the mobility of ions. Water is essential for alkali-silica reaction to take place. About resistance to freezing and thawing, nothing can be said definitely. The resistance of HPC against abrasion is very good because of the strong bond between the coarse aggregate and the matrix. HPC has poor resistance to fire as low permeability does not allow the escape of steam formed with water from the hydrated cement paste. Cementitious material like fly ash can be incorporated in HPC in order to reduce the early development of the heat of hydration of cement. The influence of silica fume is important as it reduces drying creep.

7.13 SELF-COMPACTING CONCRETE

Since its introduction in mid 1980s based on new technology, the concept of self-compacting concrete (SCC) has attracted the attention of technologists and engineers all over the world. The concrete mix cast, except for the exceptional cases, need to be compacted to ensure that sufficient strength and durability is achieved. Inadequate compaction would lead to the inclusion of voids, which could affect both compressive strength of concrete and durability related to the corrosion of reinforcing steel. Concrete in general is compacted by manually operated vibrators in most cases by unskilled workers. The question as to how good is such compaction concerning durability would remain unanswered as only compressive strength of concrete is regularly checked at construction sites. The concept of self-compacting concrete has, therefore, captured the imagination of technologists and constructors. Nevertheless, the overall production of SCC is still relatively small compared to conventional concreting.

Self-compacting concrete is a high performance composite that flows under its own weight over a long distance without segregation and without the use of vibrators. The focus on SCC since its practical use in projects in Japan has been on its fresh properties especially those of hardened concrete particularly in underwater concrete work, in-situ piling work and the filling of other inaccessible areas. Earlier, the cost of concrete production for such work, before the advent of superplasticizers and admixtures, was expensive because more cement was needed for maintaining water: cement ratio to ensure workability. But excessive segregation and bleeding restricted the use of admixtures to flowing concrete having high slumps of up to 150 mm and still requiring vibration for compaction.

Since the introduction of self-compacting concrete in practical use in difficult sites relating to access and congested reinforcements, SCC has been developed further utilizing various materials such as pulverized fly ash, ground granulated blast furnace slag, and condensed silica fume. The supply cost of SCC could be from 10–50% higher than that of conventional concrete of the similar grade. Nevertheless, total elimination of the consolidation process in SCC can lead to many benefits:

- Improved concrete quality in difficult sites
- Increased productivity, shortening concrete construction time—faster placing
- Improved working environment—eliminating unskilled manpower, vibration, and noise
- Reduction in overall construction costs – approximately 2 to 5 %
- Improved in-situ concrete quality in difficult casting conditions even around congested reinforcing steel bars and edges
- Improved surface quality—smooth and uniform

Materials

The two main requirements of self-compacting concrete are a highly fluid material having significant resistance to segregation. Fluidity and deformability of concrete means (i) ability of the flowing concrete to fill every corner of the mould (ii) ability to pass through small openings or gaps between reinforcing steel bars.

Portland cement is used for the production of self-compacting concrete. The total fines content of the mix is balanced against aggregate size and grading. In general, for stability, the fines content is much higher than normal concrete. The requirements of high fines content lead to high cement contents, often in the range of 450–500 kg/m^3. The inert cementitious fillers incorporated usually in the range of 150–250 kg/m^3 are limestone powder, pulverized fly ash, and ground granulated blast furnace slag.

The grading of fine aggregates is important as the fine aggregates play a major role in the workability and stability in the mix. The total fines content in the mix is a function of both the fine aggregates and the cementitious filler materials. The grading of fine aggregates in the mortar should be such that both workability and stability are simultaneously maintained. Standard sands used in normal concrete are also used in self-compacting concrete.

As regards coarse aggregates, both gravels and crushed rocks are used, with the maximum size of aggregates being 20–25 mm, though smaller size aggregates up to 10 mm have also been used. Grading of coarse aggregates should be similar to the grading used in normal concrete.

Admixtures

Superplasticizer is required to lower the water demand while achieving high fluidity. The common superplasticizer used is a new generation type, based on polycarboxylated polyether, which is considerably more expensive than the traditional type used in normal concrete.

Viscosity agent is sometimes incorporated to minimize the addition of fillers. This admixture is similar to that used in underwater concreting. It increases the viscosity of water, thereby increasing resistance to segregation, while maintaining high fluidity allowing concrete to flow through narrow spaces. Too much water can reduce the viscosity to such an extent as to facilitate segregation. The incorporation of superplasticizer causes limited reduction in viscosity. Viscosity agents are typically high molecular weight soluble polymers, which increase viscosity in aqueous medium because of their interaction with water. These admixtures are effective in stabilizing the rheology of the fresh concrete and preventing segregation of the coarse aggregates from other mix constituents.

Mix

Self-compacting concrete mix is based on (i) continuous phase comprising water, admixture, cement and fillers with a particle size less than 0.1 mm, and (ii) particle phase comprising fine and coarse aggregates. The three basic properties of self-compacting concrete in plastic state are: (i) filling ability, (ii) resistance to segregation, and (iii) passing ability.

Filling ability is totally related to the mobility of concrete to change shape under its own weight to mould itself to formwork. This can be made possible in two ways (a) surface tension can be reduced by adding superplasticizers, and (b) optimizing the packing of fine particles by using fillers or segregation control admixtures.

Resistance to segregation is related to stability of mobile concrete which can be made possible in two ways (a) minimizing water demand by using superplasticizers to avoid bleeding, and (b) maintaining coarse particles in suspension by using a high volume of fines in the mix and/or adding viscosity-modifying admixture.

Passing ability depends on the reinforcement arrangement for concrete to pass around immovable objects like reinforcing steel bars. Size and shape of coarse aggregate and the volumes of mortar paste should be selected on the basis of steel bar spacing. Higher volume of paste would be required for more congested reinforcement arrangement.

Quality control

For site quality control, tests requiring simple equipment are often performed to check the three basic properties of self-compacting concrete as mentioned above—filling ability, resistance to segregation, and passing ability. Slump flow test is the most popular test because of its simplicity. A representative sample of concrete is placed continuously into an ordinary slump cone without tampering. The mean spread value (in mm) is recorded. The time to reach a flow diameter of 500 mm and final flow diameter are also noted. The degree of segregation can be judged to a certain extent by visual inspection. L-box, U-box, and V-funnel are other common tests available to asses one or more of the basic properties. These are common tests—not very definitive and not yet recognized by any standard.

7.14 EXTREME WEATHER CONCRETING

Good quality concrete is produced around the world including some regions where the climates are typified by prolonged periods of extreme weather, hot or cold. In many regions where extreme weather prevails, there are specifications and guideline documents for the production of concrete. The specifications and guidelines furnish the details and methods of combating the extreme weather conditions for building concrete elements and structures.

Hot Weather Concreting

Both production and placing of concrete can be affected by high temperatures. Long-term strength and durability will be affected as well. Some of the problems related to high temperature are discussed below.

Concrete production
- Higher water demand for workability – if only water is used for providing the necessary workability at high temperature, there would be loss of strength and durability as a result. Increased water content also causes drying shrinkage.
- Increased difficulty in controlling entrained air content

Transit
- Loss of water by evaporation
- Increased rate of loss of workability – rapid loss of workability due to combined effects of loss of water through evaporation and the more rapid rate of hydration. Hydration of cement in concrete is an exothermic reaction. The rate at which an exothermic reaction takes place doubles for each 10°C rise in temperature. The cement paste would tend to stiffen earlier, should the hydration reaction proceed more rapidly. Concrete would thus lose workability.

Placing, finishing, and curing
- Loss of water by evaporation
- Increased rate of loss of workability
- Increased rate of setting—acceleration of setting resulting in decrease in setting time. This also results in shortening of the time available for transport, placing, and finishing of concrete.
- Increased tendency to plastic shrinkage cracking - more solid ingredients tend to settle down, while water being light tends to move upwards, which in case of

freshly placed concrete is called *bleeding*. Water from bleeding on the surface of slabs may be lost by evaporation. As the temperature and wind speed rise with corresponding decrease in humidity, the rate of evaporation increases. If surface water is lost at a greater rate than the rate at which it is replaced by bleeding from below, there would be reduction in volume of the surface layer. This change in volume is resisted by the mass of concrete below which does not experience any volume change. The restraint from the underlying concrete can cause enough tensile stresses in the surface layer resulting in cracking of the surface layer of the immature concrete. This is plastic shrinkage cracking.

- Higher peak temperature during hydration leading to increased tendency to cracking and lower long term strength – the hydration reaction generates heat and the temperature of concrete rises. In case of the initial higher temperature, hydration proceeds more rapidly and the rate of heat evolution is increased, that is, the peak temperature reached is also increased. Therefore, there would be a tendency of cracking as the concrete shrinks on cooling from the peak temperature.

Long-term problems
- Lower strength—despite the higher early strength of concrete in hot weather conditions, this is not reflected in long term strength. In hot weather conditions where temperatures rise above 30°C, there can be considerable reduction in long term strength. Because of difficulties in achieving proper compaction of concrete due to workability problem on account of heat, concrete strength is impaired.
- Decreased durability—more durable concrete needs to be less permeable. This could be done by maintaining low water:cement ratio. In hot weather conditions, workability is a problem due to evaporation of water. The pores in hcp affects durability and consequent problems.
- Variable appearance

The above problems are the consequences of high temperature which increases the rate of hydration, and the movement of moisture within and from the surface of concrete.

Cooling measures

Production
- Shading the stockpiles of aggregates
- Sprinkling water on the aggregates
- Increasing the cement silo capacity
- Painting the batching plant white
- Shading the water storage tank
- Insulating water pipelines
- Using chilled water as well as ice as part of the mixing water

- Using admixtures to counter slump loss—for retarding hydration or increasing initial workability or increasing admixture dosage to compensate for slump loss (slump tests to be carried out on each delivery of concrete)
- Using cement or blended cement with low heat evolution
- Minimizing mixing time – specifying maximum fresh concrete temperature

The temperature T of the freshly mixed concrete can easily be worked out approximately using the following expression:

$$T = \frac{0.22(T_a W_a + T_c E_c) + T_w W_w}{0.22(W_a + W_c) + W_w}$$

where

$T \rightarrow$ temperature in °C of freshly mixed concrete

$W \rightarrow$ mass of ingredient per unit volume of concrete

a, c and $w \rightarrow$ aggregates, cement and water (water - both added and in aggregates)

The amount of heat contained in a body or the mass of material is the product of its mass, specific heat, and temperature. The value of 0.22 is the approximate ratio of the specific heat of the dry ingredients to that of water.

Transit
- Painting the truck mixers white
- Minimizing the travel time
- Batching dry and adding water at site

Placing and curing
- Careful planning of operations
- Matching production to placing rates—by maintaining constant communication
- Reducing layer thickness
- Providing adequate number of standby vibrators
- Placing concrete at night
- Minimizing the placing time
- Shading the work places
- Using windbreaks
- Starting early curing—appropriate curing procedure should be put in place as soon as possible

Cold Weather Concreting

Both production and placing of concrete can be affected by cold weather at all stages. However, there can be some benefits from low initial temperature. Concrete that is placed at low temperatures but not allowed to freeze and cured properly develops higher ultimate strength, greater durability, and is less subject to thermal cracking than similar concrete placed at higher temperature. Some of the problems related to cold weather are discussed below.

Production
- Incorporation of frost-bound materials

Transit
- Cooling of mix

Placing, finishing and curing
- Formation of ice crystals in concrete
- Increased thermal gradients/increased tendency to thermal cracking
- Delayed removal of forms
- Slower gain in strength
- Greater chance of formwork stripping damage
- Bleed water may remain on surface

Long-term problems
- Slower setting
- Slower gain in strength
- Freeze-thaw damage—free water in the mix during hydration is frozen to ice and hardening of concrete ceases. If it so happens that hardening is not started before freezing, then it would not start after freezing. But if freezing has already started, then it remains practically suspended so long as free water remains frozen. As the frozen water starts thawing, hardening of concrete begins afresh. On freezing, the volume of water increases and icy films are formed on the surfaces of aggregates and reinforcing steel; and on thawing, ice gets reduced to the original volume and films disappear thereby affecting strength, bond and density of concrete, that is, durability. Such harmful effects are more the earlier the age at which concrete freezes. Concrete can be permanently damaged by the pressures exerted by ice crystal growth if this occurs after the concrete has hardened but before it has gained adequate maturity.
- Variable appearance

The main problems associated with cold weather are frost damage to immature concrete and slow gain in strength leading to delayed stripping of forms and the possibility of increased damage when the forms are removed. The consequences of cold weather mean increasing the rate of hydration reaction, and the movement of moisture within and from the surface of concrete. Concrete is vulnerable to freezing temperatures both before and after it has hardened. There are two stages: (i) Expansion of water as it freezes in the plastic concrete causing such severe damage that the concrete becomes unusable (ii) Concrete can be permanently damaged by the pressure exerted by the icy crystal growth if this occurs after the concrete has hardened but before it has matured adequately. This weakens the paste-aggregate bond and may reduce the strength of concrete by up to 50%. The porosity in concrete may ultimately affect durability.

Heating measures

The control measures are taken depending on the project size and the mode of concrete production (ready-mixed or otherwise). The measures are aimed at:
- Producing such concrete as may be delivered for placing at right temperature
- Protecting and maintaining the already placed concrete at appropriate temperature until it gets matured to withstand exposure to freezing conditions

Production
- Building up stockpiles—the required materials and plants and machineries need to be procured and stockpiled before the onset of cold weather.
- Using warm water—the temperature of concrete at the time of delivery should not be less than 5°C. The simplest and most effective method of doing this is to use hot water heated to about 40°C to 60°C. Steam can be generated in boilers and used either by injection or by passing through coils. Water retains more heat than any other materials like aggregates and cement.
- Using faster reacting cement—so that the possibility of damage before the concrete has gained sufficient strength is reduced and stripping of forms is not delayed.
- Using higher grade concrete.
- Reducing slump – bleeding is minimized and earlier setting is achieved by reducing slump. Reduction of slump is pertinent in case of slab construction involving large areas.
- Sheltering the batching plant—mixing plant, ingredients, and the mixing water should be free of snow, ice, and frost. The batching plant should be sheltered as much as possible for protection from the wind, driving rain, sleet, and snow. Open chutes could result in significant heat losses.
- Insulating aggregates—to be protected from the action of frost by covering with tarpaulins or by an insulating layer covered with tarpaulins, gunny bags or by placing other waterproof sheets and to be stored on wooden platforms.
- Heating aggregates—when cold weather persists for a prolonged period, it may also be required to heat the aggregates using steam or hot water (60° or more) for thawing out the surface of aggregate stock piles. Aggregates, however, cannot be heated uniformly resulting in variation of slump of concrete.
- Using accelerating admixtures—to increase the rate of strength gain. Air-entrained admixtures are used to combat the detrimental effects of freeze-thaw cycles on hardened concrete.
- Hot water should never be mixed with cement alone, as in that case 'flash set' may take place. About half of the required quantity of hot water should first be mixed with coarse aggregates in the mixer drum for a few rotations, and then fine aggregates and cement should be added followed by the addition of the balance quantity of hot water. Mixing time of concrete should be more than normal and not less than 3–4 minutes in any case.

Transit
- Minimizing transit time – heat losses could occur during transit – to prevent damage from freezing at early stages
- Reducing the truck mixer revolutions
- Matching the rates of production and delivery for placing

Placing and curing
- Freeing forms and reinforcement of frost—this can be accomplished by hot air blowers
- Thawing of subgrade—concrete should never be placed on frozen subgrade or old concrete as subsequent thawing of frozen masses may result in the settlement of concrete structure or element. All frozen materials should be washed out with hot water before any fresh concrete is placed.
- Insulating formwork—formwork and falsework should be kept in position for longer period due to the reduced rate of gain of concrete strength in cold weather. Insulation should be removed only when the concrete should be in a position to withstand temperature stresses. Timber is a reasonably good insulator, and timber in combination with insulating materials would be able to protect concrete in severe cold weather. If any chute is used for unloading or placing concrete, the chute should also be insulated properly and adequately.
- Heating of enclosed area—an alternative to insulated formwork is to use heated enclosure; in which case, metal formwork would be advantageous, as heat transfer to concrete would be easier.
- Unless tight construction schedule demands otherwise, concrete should be placed during day time when the outside temperature is more than 5°C and should be discontinued as soon as the temperature starts falling below 5°C.
- Providing protection to completed work – protection should remain in place until concrete has achieved a strength of 5 N/mm^2.

7.15 FIBRE-REINFORCED CONCRETE

Fibre-reinforced concrete is comprised of cement, fine or coarse aggregates and discontinuous fibres with or without admixtures as used in normal concrete. Fibres used are of various sizes and shapes made of steel, plastic, glass, and other materials. Fibres made of steel, however, are the most commonly used fibres in most structural and non-structural purposes.

Normal concrete comprising inert aggregates and active cement paste possesses high compressive strength, but weak in tension and shear. Further, normal concrete is not impermeable and resistant to deleterious fluids due to the presence of voids, capillary cavities, and microcracks, etc. The rapid propagation of microcracks under imposed stress is responsible for the low tensile strength of normal concrete. However, fibre-reinforced products do not offer any substantial improvement in strength over equivalent mix without fibres. Although the ultimate tensile strength of fibre-reinforced concrete do not increase appreciably, the tensile strains do. In comparison to normal concrete, fibre reinforced concrete scores higher in toughness, and resistance to impact. Fibre reinforcing has added versatility into concrete so as to overcome its brittleness.

Normal concrete fails abruptly as soon as the deflection corresponding to its ultimate flexural strength is exceeded. Comparatively, fibre-reinforced concrete keeps on sustaining considerable loads even at deflections considerably in excess of the fracture deflections of normal concrete and fails only on debonding or pull-out of fibres. In other words, a fibre-reinforced concrete specimen does not break immediately after initiation of the first crack, which is referred to as *toughness*. The magnitude of improved toughness is strongly influenced by fibre concentration and resistance of fibres to pull-out. This is governed by the length:diameter ratio (aspect ratio) and other factors like shape or surface texture. Fibres bridging the cracks contribute to the increase in strength, failure strain, and toughness of the composite.

Materials

Steel fibres of different shapes and sizes are used as reinforcing materials. Round steel fibres have diameters in the range 0.25 to 0.75 mm. Flat steel fibres have cross-sections ranging from 0.15 – 0.4 mm thickness by 0.25 – 0.9 mm width. Crimped and deformed steel fibres are available, both in full length or crimped at the ends only. To facilitate handling and mixing, fibres collated into bundles of 10 to 30 with water soluble glue are also available. Typical aspect ratios range from about 30 to 150. Typical glass fibres (chopped strain) have diameters of 0.005 – 0.015 mm, but these fibres may be bonded together to produce glass fibre elements with diameters of 0.013 – 1.3 mm. Since ordinary glass is not durable to chemical attack by Portland cement paste, alkali-resistant glass fibres with better durability have been developed. Fibrillated and woven polypropylene fibres are also used as reinforcement.

Properties

The addition of any type of fibres in normal concrete reduces its workability. The loss of workability is proportional to the volume concentration of the fibres in concrete. The slump cone test is not a good index of workability because fibres impart considerable stability to a fresh concrete mass. The Veebee test is considered more appropriate for evaluating the workability of fibre-reinforced concrete.

The contribution of fibre-reinforcement in concrete is the flexural toughness of the product more than its strength—the greatest advantage of fibre-reinforcement is the improvement in flexural toughness. The impact and fatigue resistance are related to the flexural toughness. Even low-modulus fibres like nylon and polypropylene can be used for producing precast concrete elements liable to be under severe impact. Incorporation of steel fibres in concrete has hardly any effect on the modulus of elasticity, drying shrinkage, and compressive creep. The flexural strength being the prime consideration, fibre reinforcement cannot be a substitute for conventional reinforcement.

Well-compacted and cured concrete reinforced with steel fibres possess excellent durability when produced with high cement content and low water:cement ratio.

Applications

Steel fibre-reinforced concrete should only be used for supplementary roles such as:
- Inhibition of cracking
- Improving resistance to impact or dynamic loading
- Resisting material disintegration
- Fibre-reinforced concrete can be used where continuous reinforcement is not essential to assure safety and integrity of the structures, e.g., pavements, overlays, and shotcrete linings. Fibre-reinforced concrete can also be used to improve the flexural strength associated with fibres for reducing section thickness or improving performance or both.

7.16 PRESTRESSED CONCRETE

Conventional reinforced concrete cannot take full advantage of the high strength possessed by both reinforcing steel and concrete. Excessive deformation of steel and absence of tensile strength and low modulus of elasticity of concrete produce a great number of large cracks and excessive deflections. High strength of both reinforcing steel and concrete cannot be availed in conventional concrete because of the incompatibility of the elastic behaviour of steel and concrete. The influence of the elastic incompatibility of steel and concrete is avoidable in prestressed concrete. Prestressing provides an alternative means of increasing the load-bearing capacity of the concrete and the spans of the elements. High tensile steel is used in conjunction with high quality concrete and, by using some means of jacking, a considerable amount of compressive force is introduced into the element to combat the tensile and shear forces. The resulting prestressed elements are, for load-bearing capacity, lighter and more slender than the equivalent reinforced concrete elements.

The advantages of prestressed concrete are:
- High degree of reduction in cracking
- Durability relating to freezing and thawing is higher due to reduced cracking and pre-compression which keeps the cracks closed
- Precasting of prestressed concrete under highly controlled conditions in permanent shops with proper curing arrangement results in superior products, which can be assembled into structures at any construction site very easily if handling and lifting equipment are available
- Prestressing results in maximum economy in size and weight of structural members, thereby economizing transportation and storage
- Time schedule can be squeezed as precasting, site development and preparatory work can continue simultaneously

Prestressing can be introduced into most type of concrete elements, including columns, beams, walls, tanks, and cantilever members, such as retaining walls and floors. Prestressing can provide economic solutions where the design calls for large spans, slender sections, substantial load-bearing capacity or where complicated structures, such as extremely long cantilever members are used.

There are two modern methods for carrying out prestressing of concrete:
- Pre-tensioning
- Post-tensioning

No matter which of the two systems is used, there would be some losses in the prestressing force that is initially applied or transferred to the unit. The losses are due to the properties of concrete, such as elasticity, creep, shrinkage characteristics to friction in the anchorage and the ducting system used. Losses which occur after the transfer could be of the order of 20 % of the force achieved at transfer. Losses that occur before or during transfer are of the order of 5–10% depending on the system employed in pre-tensioning and post-tensioning systems.

Pre-tensioning

In pre-tensioning of prestressed concrete, prestress is applied before the concrete has hardened. Pre-tensioning techniques generally use cables that are fixed or locked at the two ends. To start with, the cables are placed properly threading through end plates and distribution plates. The cables are anchored at one end of the prestressing bed or mould using grips of some form. Similar grips are installed at the other end, which is the stressing end. A small uniform amount of force is applied using a load limiting device so as to make sure that all cables are similarly stretched to the proper degree and eventually force in each cable would be equal. The main prestressing force is then applied by jacks or jacking beam, the wires being stressed individually or in groups. The operation is carried out in such a way as to ensure the building up of such force as specified by design engineer. Concrete is then placed carefully, compacted, and cured to assure adequate bond. The tension in the prestressing cables is released only after the concrete has adequately hardened.

For a bridge at Bhagalpur in Bihar, wires were high tensile bars of specific diameter (4.25 mm ± tolerance) conforming to IS: 6006. A strand was made up of a number of wires (7 for this bridge having diameter of 12.7 mm ± tolerance). A cable or tendon consisted of a number of strands (19 for this bridge) with duct and anchorage. Duct/sheathing is the enclosure around the prestressing cable to avoid temporary or permanent bond between the prestressing steel and surrounding concrete.

Anchorage is the device by which prestressing force is permanently transmitted from the prestressing steel to the concrete. Cables that are bonded to the concrete via grouting and, therefore, not free to move are called *bonded cables*. Grout is a mixture of cement and water with or without admixtures. It provides permanent protection to the post-tensioned steel and develops a bond between the prestressed steel and the surrounding concrete. Grout is pumped into the duct so as to ensure complete filling of the void between the cable and the duct.

The pre-tensioned method of prestressing is most suitable for precast production of prestressed elements in which a length of prestressing cable is pre-tensioned with a series of moulds placed along its length to contain the concrete. A number of proprietary systems are there for pre-tensioning of concrete elements. Pre-tensioning is especially suitable and economical for the mass production of similar units, such as railway sleepers, floor joists, beams, poles, piles, etc.

Post-tensioning

In post-tensioning of prestressed concrete, concrete is cast in the same way as is done in case of cast-in-situ or precast concrete. The alignment of the prestressing cables is fixed by the provision of ducts or tubes or sheathing placed in the concrete. Post-tensioned cables can take non-linear profiles. It is essential that ducts or tubes or sheathing are secured against vertical or horizontal displacement due to concrete placement and the application of vibratory force or uplift due to floatation. In post-tensioning, it is necessary to provide anchorages or bearing plates which transfer the force from the cable into the concrete.

More efficient and versatile use of prestressing forces can be made by the post-tensioning process. The losses are smaller and by curving the cables upwards at supports, the prestressing system enhances the shear resistance of the elements. But more cost would be involved in providing ducts and anchorages in post-tensioning that may make smaller units uneconomical. Post-tensioning is especially suitable for buildings, bridges, off-shore oil platforms, and other engineering structures.

The prestressing steel should conform to either of the following:

- Plain and hard drawn steel wire conforming to IS: 1785 (Part I and II)
- Cold drawn indented wire conforming to IS: 6003
- High-tensile steel bar conforming to IS: 2090
- Uncoated stress relieved strands conforming to IS: 6006

7.17 UNDERWATER CONCRETING

Underwater concreting is a special operation that may have to be carried out in remote and difficult areas in unusual environment. If incorrectly carried out, the results could be serious and remain undetected. Underwater concreting, therefore, should be well-planned and only experienced personnel should be deployed for such operation.

Although it is always preferable to place concrete in air, but sometimes in special situations, it is expedient to place concrete under water. However, underwater concreting presents various problems. Transporting, placing, compacting, quality control, finishing, and accuracy are the factors which must be taken into consideration before launching this operation. Concrete sets well under water as air is not required for setting and hardening. But concrete must be designed to be plastic and dense and still flow readily into position and be self-compacting as conventional vibration cannot be effected under water. It is better not to puddle concrete under water as concrete would mature better if left undisturbed.

Of the various problems, the most obvious one is the washout of concrete by the water. This can be avoided by placing the concrete by tremie (vide Chapter 13, Section 13.6) pipe buried within the already placed concrete. The flow of concrete takes place under its own weight due to the gravitational force. It is essential that the concrete mix has appropriate flow characteristics. A slump of 150–350 mm is necessary depending on the presence of embedded items. Before pouring of concrete is started, a plug is inserted in the pipe to stop the intermixing of water and concrete. The plug can be purpose-made, sponge rubber ball, or exfoliated vermiculite. The pipe has to remain full of concrete throughout the continuous pouring operation with the bottom of the tube remaining buried in the already placed concrete. For large area concreting, a number of tremies are used at 4–6 m centres depending on the flatness required at the top level.

Another method of placing concrete under water is to use specially made watertight box or bucket or *skip* instead of the tremie. The watertight box is filled with concrete and lowered to the required level and discharged from the bottom. For underwater concreting, operation is to be continued without interruption.

For underwater concrete, the quality of concrete should be rich. The content of cementitious material should be at least 360 kg/m^3 with at least 15% pozzolona included to improve the flow of concrete. In large underwater pours, the heat generated by hydration can raise the temperature near the centre of concrete mass to as much as 70°C to 95°C. Concrete is likely to crack on subsequent cooling. A possible remedial measure against such temperature rise and cracking is to use blended cement and silica fume. Concrete can also be pre-cooled to as low as 4°C prior to its placing in the tremie.

The grouted aggregate method of concreting is particularly useful where water is flowing. This method is useful when undercuts are to be filled which are inaccessible to tremies or skips. This method must be executed only by highly skilled agency specialized in such work.

The operation is carried out in two stages. In the first stage, the space to be filled with concrete is filled with single-sized or uniformly graded aggregates, either rounded or crushed, in the formwork. The aggregate represents about 65 – 70% of the volume to be concreted. The remaining voids need to be filled in the second stage by pressure pumping of colloidal grout through pipes which are slotted where grouting is required. The grout pipes are generally of 35 mm diameter and spaced at 2 m apart from each other with their ends close to the bottom of the pour.

Aggregates placed in the formwork would reduce the water flow through them. This would allow the grout to penetrate rather than be washed out. But prior to grouting, sediment could deposit on the aggregates if water is laden with silt and impair bond. The entire operation should be carried out in the same tide so as to avoid settlement of silt.

A typical grout comprises a blend of Portland cement and pozzolona (at a ratio of between 2.5:1 and 3.5:1 by mass) mixed with sand (at a ratio between 1:1 and 1:1.5 by mass) and with water:cement ratio of 0.42 to 0.5. An intrusion aid is added to improve the fluidity, suspending and cohesive qualities of the grout apart from delaying stiffening of the grout. The intrusion aid contains a small amount of aluminium powder that causes a slight expansion before setting takes place. Strengths of 40 MPa are usual, but attaining higher strengths is possible.

The drying shrinkage of grouted aggregate concrete is lower than that of normal concrete due to the contact between large aggregate particles. This contact restrains the amount of shrinkage that can take place, but occasionally shrinkage cracking can develop.

7.18 POLYMERS IN CONCRETE

Polymers or epoxy resins are used in several ways to enhance the properties of concrete:

- Impregnating hardened concrete – Polymer-impregnated concrete
- Addition as an admixture – Polymer-modified cement concrete
- Replacement of Portland cement – Polymer concrete
- Protective coating
- Bonding agent
- Other possible applications

While natural polymers such as rubber, wool, silk, cellulose fibre are of little or no use in concrete production, synthetic polymers like urethanes, acrylics, styrene butadiene resins, vinyl, epoxies can be used for various purposes in concrete.

Because of rapid coagulation of natural rubber at the high pH (alkaline) of Portland cement pastes, attempts in the past to modify the properties of cementitious compositions by the addition of natural rubber were hardly successful. With the advent of synthetic rubbers and other polymers, the desired benefits as follows can now be achieved:

- Improved tensile and flexural strength
- Improved bonding to steel and mature concrete surfaces
- Lower modulus of elasticity resulting in reduced cracking tendency
- Lower water permeability
- Reduced chloride diffusivity
- Improved durability

Considerable research is in progress on improvement of concrete properties by the addition of polymers. Tensile strength of concrete can be significantly improved by the use of polymers and the use of concrete in flexure without the use of reinforcing steel has become a distinct possibility in not-so-distant future.

Polymer-impregnated concrete

The method of impregnating monomers, which have low viscosity, involves creation of vacuum. The concrete is thoroughly dried generally by heating. Monomer impregnation is forced on the concrete by pressure. Concrete is immersed in the desired monomer and then pressure is applied. The impregnated concrete is sealed so as not to lose monomer. The monomer is converted to polymer, either by ionizing radiation like gamma radiation or thermal-catalytic method. The full operation can be carried out in concrete precasting shop. The impregnation can be full or partial. Full impregnation implies 85% of the available void is filled up with monomer after drying. Less than 85% filling of voids would be deemed as partially impregnated.

There is a wide range of monomers that are used for impregnation—methyl methacrylate (MMA), acryl nitrite, ethyl acrylate, furans, polyester styrene, styrene-butadiene rubber (SBR), styrene and vinyl chloride. After impregnation, monomers undergo polymerization by thermal catalytic or promoted catalytic or by ionizing radiation as already mentioned.

Polymer impregnation improves compressive strength and tensile strength. However, the extent of impregnation would depend on the porosity of concrete—impregnation would be more in weak porous concrete. Compared to ordinary Portland cement concrete, polymer-impregnated concrete achieves more strength. However, the strength of polymer-impregnated concrete remains stable whereas ordinary concrete keeps on gaining strength with time. Considering the cost of polymers plus that of the process of impregnation, polymer-impregnated concrete is expensive.

Polymer-impregnated concrete achieves crushing strength in excess of 100 N/mm^2 and such strength is not dependent on the original untreated ordinary concrete. The flexural strengths are usually about 15 N/mm^2, which is a bit higher than that of ordinary concrete of highest strength could achieve.

The failure and cracking strains do not differ significantly in ordinary and polymer-impregnated concrete as the strength and elastic modulus of either are not very different. High strength concrete of both types tend to be brittle. Once cracks develop, the same propagate rapidly.

Polymer-impregnated concrete has still not been used in construction projects except for special repairing activities. The objectives of polymer addition and limitations thereof are to be understood before one opts for its use. Impregnation of concrete with polymers is sometimes used to harden surfaces exposed to heavy traffic.

Polymer-modified cement concrete

Polymer-modified cement concrete is ordinary Portland cement concrete in which soluble or emulsified polymer is added. In other words, the bonding material in this type of concrete is both cement as well as polymer. The hardening of concrete is affected by both hydration of cement and polymerization of polymer/epoxies. The polymerization commences as and when the water from the polymer emulsion is consumed by cement. The polymerization would not start until the polymer coalesces. In certain instances, the polymerization is accomplished by external source such as heat.

The polymer used in polymer-modified cement concrete is mainly latexes. The latex is normally formed directly by emulsion polymerization of the monomer and typically contains 40–50% solids. One of the earliest polymers used is polyvinyl acetate (PVA), but the range of polymers that have been tried is extremely wide and includes PVA copolymers, acrylics, vinyl, natural rubber, and SBR. The proportions of polymer incorporated vary widely, ranging from 1% to over 30% of the solid volume of cement.

The incorporation of the polymers result in certain common effects such as concrete mixes becoming more workable, thus allowing reduction of water content. However, more air is entrained in polymer cement concrete in spite of the higher workability.

Reduction in water content increases the crushing strength, but this increase is rather marginal because of the extra voids. However, there normally is a significant increase in flexural strength attributable to an improved bond between the aggregate and the matrix.

Due to the elastic moduli of polymers being generally lower compared to those of cement pastes and concrete, the elastic moduli of polymer-modified cement concrete are lower than those of the equivalent ordinary cement concrete. The additional entrained air affects the moduli more.

Hydrated Portland cement is alkaline (pH > 12.5) and some of the polymers hydrolyse in damp cement environment and rapidly deteriorate. PVA is such a polymer, which is still used widely where the mortar dries out permanently. New PVA copolymers, such as vinyl acetate-ethylene, styrene-butyl acrylate, butyl acrylate-methylacrylate, and styrene-butadiene have been developed for more demanding situations, but these copolymers cost more.

For achieving substantial changes in the properties of the hardened concrete, at least 5% polymer by weight of cement needs to be added as special admixture.

The coefficient of thermal expansion of polymer-modified concrete is the same as that of the ordinary cement concrete.

The polymer-modified cement concrete has much higher wear resistance compared to that of the ordinary cement concrete. Because of the improved durability on account of the reasons mentioned above, polymer-modified concrete has applications, as enumerated below.

- To use on industrial floors especially where chemicals and oils are likely to smear such floors
- To repair of old and damaged concrete
- As overlay on bridge decks, ship decks and parking garages
- As protective floors on frozen food factories/stores
- On loading ramps where wear due to abrasion is high
- For cementing ceramic tiles to concrete
- For protection of concrete subjected to large doses of de-icing salts due to hostile weather

Polymer concrete

Polymer concrete comprises aggregate or mineral filler bound together in a dense matrix with a polymer binder, which replaces cement. Polymer concrete is produced in the same way as ordinary cement concrete is produced. The binders generally used are methyl methacrylate, polyesters, and epoxy resins. The mix is vibrated for compaction. The other method of producing polymer concrete is to pack graded dry aggregates in moulds and polymer is poured into the voids and, if so required, impregnated by vacuum process. Grading, shape and size of aggregate play an important role on concrete production. The grading and shape of aggregate should be so chosen as to achieve minimum void so that maximum beneficial properties could be achieved using minimum quantity of polymer. The aggregate used in polymer concrete should be dry with less than 1% moisture.

Polymer concrete has found wide range of applications in construction industry and as coating system for protective layer and as bonding layer for old to new concrete. The following physical properties help on coating and bonding:

- High compressive strength that is difficult to achieve with ordinary cement concrete
- High strength adhesion in almost all building materials
- Very low shrinkage during curing period
- Exceptional dimensional stability on curing
- Natural gap filling qualities
- Resistance to chemical attack
- Resistance to freeze-thaw cycles is higher
- Creep is less

After curing for a short time like overnight curing in room temperature, polymer concrete would be ready for use compared to a week's curing required in case of cement concrete before exposure to services. In most polymer concretes, 70 – 75% strength is achieved in a day or two.

Polymer composites are produced using the same ingredients as ordinary concrete along with polymers. Due to addition of polymers, the product gains in: (i) compressive strength; (ii) resistance to wear and tear, fatigue, impact, and chemicals; (iii) impermeability; and (iv) durability. The polymer composites, due to their improved properties, have found applications in the following areas.

- Precast products such as pavement kerb stone
- Bridge ducts
- Manhole and drainage covers
- Sewer pipes, drainage channel
- Tunnel linings
- Pipes carrying chemicals
- Chequered plates for industrial structures

Polymer concrete has never been used as a replacement of conventional concrete at any construction site except for a few special precast units. Attention should be focused on the use of polymers as:

- Repairing material—Polymer concrete/mortar is used for patch repairing or cavity filling. For any polymer concrete to be used successfully, the material has to be made compatible to the physical parameters of the concrete. The three basic physical parameters are: (i) bond—bonding of the repair material to the substrate on which the same is to be applied (ii) modulus of elasticity—the moduli of elasticity of polymer concrete and cement concrete are different. The moduli can be made the same by adding mineral fillers (iii) shrinkage cracks—polymer emulsion has appreciable shrinkage although most of the epoxies have negligible shrinkage cracks. For recommending any protective and patch repairing polymer material, the depth and thickness must be checked; as otherwise, the purpose of using such expensive material would turn out to be meaningless for failing in bond, shrinkage cracks or cohesion.

- Bonding agent—The polymer must be capable of polymerization in wet condition. As a result of extensive research, the epoxies are now available for application on wet surface and polymerize under water. Consequently, bonding agents are now available that can bond fresh concrete to old concrete substrate.
- Grouting—Polymer grout materials in most cases have very low viscosity, low shrinkage, high bond and compressive strength. Accordingly, polymers are extensively used for sealing fine cracks, filling of base/sole plates of heavy machines, grouting of bolts.
- Corrosion inhibitor—Reinforcing steel may be coated with polymer with corrosion inhibitor. Both old reinforcements for rehabilitation work and the new reinforcement for new construction are now coated with polymer-based anti-corrosion paint (fusion bonded).

7.19 STRIPPING OF FORMS

Stripping of forms from the finished concrete demands as much care as goes into the fabrication and erection of forms. Such care is essential for the protection of:

- Concrete cast
- Forms—for reuse without extensive repair

All forms should be designed keeping in view the ease of stripping so as to save the forms and stripping cost. To start with, all props, braces, and ties are removed. If the forms seem to adhere to the concrete, then stripping should be carried out by inserting wooden wedges and never by prying with bars directly against the concrete. In removing the first piece of a form, the new concrete should be checked to ensure that hardened concrete is hard enough so that pieces would not spall out.

The specification should clearly spell out the number of days a form should remain in place particularly those long span members in flexure. The time of stripping in terms of days varies considerably depending on the structural function of the member and the rate of strength gain of the concrete, which depends on the type of cement used, water:cement ratio, and the ambient temperature during curing. Due to these reasons, it is not possible to specify how long each type of forms should remain in place. Also it is difficult to assign fixed time intervals each type of forms must remain in place. The most rational approach would be to make use of test specimens curing them in conditions similar to the ambient conditions of the concrete in question. When the strength indicated by the test specimen is double the stress the member must sustain, the forms may be removed. This logical method requires time and effort and is followed for large and important structures where quick reuse of forms is of the utmost necessity. However, IS: 456, Clause 11.3 shows time intervals of retaining forms in place if the ambient temperature is 15°C or above.

7.20 CURING OF CONCRETE

Curing process, natural or artificial, is a very important factor in determining the strength of concrete, which depends on the continuous hydration of the cement. It is to be assured that there is enough water present in the concrete for the hydration to continue. There is ample quantity of water in the finished concrete product as mixing water. If this water disappears due to evaporation, hydration of the cement would stop; and there would be no further gain in strength of concrete, thereby affecting its durability.

Curing concerns three basic elements:
- Moisture
- Time
- Temperature

Different methods are used for curing concrete by controlling its moisture content or its temperature. In actual construction, curing is done by conserving the moisture within newly placed concrete by providing additional moisture so as to replenish water lost by evaporation. Usually, not much attention is paid to temperature except in extreme weather conditions

Proper curing to a large extent assure the required development of the surface zone of fresh, newly cast concrete into strong, impermeable, crack-free and durable hardened concrete. The objective is to keep the concrete as much saturated as practicable for sufficient time for the original water-filled space to become filled to the desired extent by the products of hydration of cement.

As long as moisture is present in the concrete, the strength of concrete increases with time. The rate of gain of strength, however, decreases as the hydration process approaches completion stage. As concrete dries, it shrinks and if drying occurs when concrete has little strength, the result would be cracking of the concrete. Moreover, as drying occurs first on the surface due to evaporation on account of ambient temperature and humidity, hydration of cement there would practically cease. The surface zone is often relied upon to provide many of the essential requisites of a concrete member/structure like abrasion and chemical resistance and protection of reinforcing steel. Most effective curing is beneficial in that it makes the concrete more watertight and increases the strength.

Temperature of the concrete is very important for strength gain during the hydration process. The problems of curing concrete are much greater during hot weather than during cold weather. Cracks in the surface zone of concrete are likely to develop on hot summer days when temperature hovers around 38°C due to faster setting reactions taking place. Also, the internal heat due to hydration causes expansion of the interior greater than at the surface. On the other hand, temperatures near the freezing points of water retard the setting or hardening of cement to almost zero. Below freezing point, free water in the concrete would turn into ice crystals. Since volume of water expands on freezing to ice, the concrete is disrupted and loses strength on thawing.

Methods of curing may be classified as:
- Those that supply water throughout the early hydration process and tend to maintain a uniform temperature

- Those tend to prevent loss of water but having hardly any influence on maintaining a uniform temperature

The methods generally followed for curing concrete at construction sites:
- Spraying/sprinkling of water
- Ponding with water
- Retention of formwork even after hardening of concrete
- Suspension of covering above horizontal concrete surface until setting of concrete
- Covering with wet sand, sawdust, earth, straw, hay, hessian or cotton or burlap mats
- Absorbent covering with access to water
- Application of impermeable membrane – sprayed clear or bituminous compound
- Waterproof reinforced paper or plastic sheeting
- Tenting or other shelter against drying winds
- Covering with insulating layer or heated enclosure
- Wrapping of protruding steel bars or inserts

The period of curing depends on:
- The requirements of the specification
- The reason for curing
- The size of the concrete member
- The type of concrete
- The ambient conditions prevailing during curing
- The likely exposure conditions after curing
- The thumb rule is to cure concrete for 15 days or more

7.21 INSPECTION AND ACCEPTANCE OF FINISHED CONCRETE

Reinforcing steel bars

This involves approval of reinforcing steel bar materials for conformance to the physical properties required as per the ASTM specifications:
- Strength grade specified
- Approval of the bar details and placing details – gross errors should be corrected within the specified tolerances
- Approval of fabrication to meet the approved details within the permissible tolerances
- Approval of placing of reinforcing steel bars

Bar supports are available in three general types of material:
- Wire – plain, stainless steel or plastic-coated
- Precast concrete
- All-plastic

Bars should be free of: (i) Oils, (ii) Paints, (iii) Form coatings, and (iv) Mud.

Formwork

Formwork should be supervised prior to erection/placing of reinforcing steel or placing of concrete to ensure:

- Positional correctness – true to the centrelines
- Dimensional correctness
- Horizontal, vertical and angular correctness

Concrete

- Segregation must be avoided during all operations between the mixer and the point of placement, including final consolidation and finishing
- Concrete must be thoroughly consolidated—around all embedded items and fill all angles and corners of the forms
- Where fresh concrete is placed on or against hardened concrete, a good bond must be developed
- Unconfined concrete must not be placed under water
- The temperature of fresh concrete must be controlled from the time of mixing through final placement, and protected after placement
- The result of excess water like vertical segregation (water rise), water flow (horizontal), leakage through form openings, planes between vertical layers, laitance on horizontal layers – leakage causes honeycomb, sand-streaks, colour variations or soft spots on the surface – laitance suffers from low strength, low abrasion resistance, high shrinkage, and generally poor quality

7.22 MECHANIZATION OF CONCRETING

Mechanization of concrete work is covered in Chapter 13, Section 13.6. Erection of precast reinforced concrete structures as well as bridges is covered in Chapter 8, Sections 8.7 and 8.8.

7.23 LABORATORY TESTING FACILITIES AT A SITE

For concrete construction work to be of such good and consistent quality as to conform to the specifications, testing facilities and services should be available at construction sites. The bidding documents should have clauses asking the bidders to quote prices keeping in view that they would have to set up site laboratories. Bidding documents should clearly spell out the tests that need to be carried out regularly for materials to be used and products in both plastic and hardened stage. The list of tests should cover both destructive and non-destructive tests. Tests on fresh and hardened concrete must be conducted accurately and the interpretation of test results should be done carefully.

All materials used for testing should be as representative as possible of those to be used for the actual construction work. Small-scale batches would lose consistency due to evaporation (and absorption, if the aggregates are dried) at a much faster rate than large batches used in construction. Results from an initial test are only representative of the qualities of the materials used in the test. It is necessary to consider making allowances for normal variations in larger-scale concrete production.

Cement

Cement tests are carried out for construction work of importance and also where satisfactory performance of the concrete used is to be ensured. To start with, cement should be stored in weatherproof shed on raised platform (vide IS: 4082). If cement is stored for more than 60 days, compressive strength and setting time should be re-tested. Cement samples must be representative of the cement to be used in construction work. Equipment/apparatus required are:

- Vicat apparatus—to check consistency using a 10 mm diameter plunger fitted into the needle holder as per IS: 269.
- Vicat apparatus is also used for determining initial and final setting time as per IS: 4031
- Le Chatelier flask and mould for soundness test as per IS: 4031
- Compression testing machine—standard sand compression testing to check suitability for the particular construction work as per IS: 4031
- Blaine air permeability apparatus for determination of fineness of cements
- Autoclave for determination of soundness of cement by measuring expansion of neat cement paste prisms
- Air content—air content of cement-sand mortar of a standard consistency is determined
- Heat of hydration—amount and rate of evolution of the heat of hydration

Aggregates

- Hot plate for determining surface moisture content of both fine and coarse aggregates—other requirements include a balance, a small shallow pan and stirring rod or spoon
- A balance of 2 kg capacity with sensitivity of 0.1 gm required for the tests of surface moisture content, specific gravity, bulk density and water absorption
- A volumetric flask of 500 millilitre capacity and calibrated to 0.15 mm along with a 6 mm diameter pipette of sufficient length and a balance for quick determination of surface moisture content of fine aggregates
- Electric oven 150°C, 500 mm × 500 mm × 500 mm
- Stop watch or clock with second hand
- Trowel
- Sieve analysis – all specified sieves placed one above the other in order of size with the largest size at the top, and the material retained on each sieve after shaking represents the fraction of aggregates coarser than the sieve in question, but finer than the sieve above – I. S. Sieve Sts 80, 63, 50, 40, 25, 20, 16, 12.5, 10, 6.3, 4.75,

3.35, 2.36, 1.70, and 1.18 mm; 850, 600, 300, 150, 90 and 75 microns. Sieves are mounted one above the other on sieve shaker.

- Sieve mechanical shaker—may be hand shaken if power for mechanical shaker is not available. IS: 2386 (Part I) should be followed for sieve analysis.
- The flakiness index of an aggregate is the percentage by weight of particles in it whose least dimension (thickness) is less than 0.6 times of their mean dimension. The flakiness index can be checked following the method described in IS: 2386 (Part I).
- Los Angeles abrasion testing machine is used for testing abrasion resistance value of coarse aggregates as per IS: 2386 (Part IV), which contains a sketch of this testing machine.
- Cement-aggregate reactions—there is no simple way of knowing whether there would be excessive expansion due to alkali silica reaction. IS: 2386 (Part VII) recommends the mortar bar method of testing for the determination of the potential alkali reactivity (ASR) of the cement–aggregate combination.

Reinforcing steel

- Yield strength of reinforcing steel—determine the ultimate load
- Weighing scales, wire-brush and calipers—for checking suitability for use of rusted bars

Concrete

- Soundness—must not cause swelling, disintegration or deterioration
- Tensile strength—to check if tensile strength is good enough to resist tension due to its own weight
- Slump cone apparatus for checking workability as per IS: 1199
- Flexural test beams—For structural concrete subjected to flexural stresses as in pavements, it is not uncommon to make beam tests rather than compression tests. Beams have two advantages: They are made under stress conditions more like those in the member they represent, and the test is easily made with a simple inexpensive apparatus at the site.
- Flexural strength tests—concrete beam subjected to flexure using symmetrical two-point loading until failure occurs. The theoretical maximum tensile stress reached in the bottom fibre of the test beam is known as the modulus of rupture. Tested on sides as the top is float-finished.
- Material test—to take representative samples
- Hardened concrete—compressive strength—cube test/cylinder test—test sample is finished with float so no stress concentration is possible—capping with high-strength gypsum plaster/molten sulphur mortar and also regulated set cement
- Weigh batcher should conform to the requirements of IS: 457 regarding calibration

Water

- pH meter to check that pH value is not less than 6 so that alkalinity is maintained.
- Vicat apparatus for checking initial setting time using available water and distilled water – samples made with water available for construction work should be compared with samples made with distilled water vide the clause 5.4.1.2 of IS: 456 – 2000.
- Compression testing machine for checking compressive strength of concrete using available water and distilled water – samples made with water available for construction work should be compared with samples made with distilled water using standard sand.
- Determination of the total quantity of solids is done by evaporating and drying a measured sample in an oven at 105°C for an hour, and measuring the residue.

7.24 NON-DESTRUCTIVE TESTING OF HARDENED CONCRETE

Rebound test

The rebound test is based on the principle that the rebound of an elastic mass depends on the hardness of the surface upon which it impinges.

This test is most suitable for concrete in the range of 20 – 60 N/mm^2 strength. In this test, a spring-controlled hammer mass weighing less than 2 kg slides on a plunger within a tubular housing. The plunger retracts against a spring when pressed against the concrete surface and this spring is released automatically when fully tensioned thereby forcing the hammer mass to impact against the concrete through the plunger. As the spring-controlled mass rebounds, it carries a rider which slides along a scale. This movement is visible through a small window in the side of the casing. The rider, by depressing the locking button, can be held in position on the scale. This simple equipment (Fig. 7.11) may be operated either horizontally or vertically. The plunger is pressed strongly and

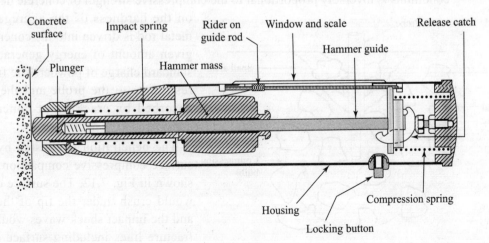

Fig. 7.11 Rebound hammer

steadily against the concrete at right angles to its surface, until the spring loaded mass is triggered from its locked position. Immediately after the impact, when the hammer is in the test position, the scale index is read. The locking button, alternatively, may be pressed to allow the reading to be retained. The results can also be automatically recorded by pen recorder. The scale reading is known as the rebound number, and is deemed as an arbitrary measure as it is dependent on the energy stored in the given spring and on the mass used.

This test necessitates taking of several readings so as to find their average as the reading is sensitive to local variation of concrete. The results give a measure of the relative hardness of the zone, and the same cannot have any relation with other property of the concrete.

The factors that influence test results are:
- Mix characteristics—cement type, cement content, coarse aggregate
- Member characteristics—mass, compaction, surface type, age, rate of hardening, curing method, surface carbonation, moisture condition, stress state and temperature

The equipment manufacturers provide general calibration curve relating rebound number to strength. The above factors influence test results so much that the calibration curve would not be of much practical value. Strength calibration must be based on the particular mix under investigation, and the mould surface, curing and laboratory specimens should correspond as closely as possible to the in-situ concrete. It is also necessary to check correct functioning of the rebound hammer by checking occasionally using a standard steel anvil of known mass because wear may change the spring and internal friction characteristics of the equipment. Calibrations prepared for a particular hammer may not be applicable for other hammers.

Windsor probe test

This test is based on the principle that the penetration depth by a metal rod under standard conditions is inversely proportional to the compressive strength of concrete depending on the hardness of the aggregate. The metal rod is driven into the concrete by a given amount of energy generated by a standard charge of powder. The frictional resistance to the probe and the energy absorbed by cracking of concrete are considered to be negligible.

The penetration is resisted by a sub-surface compressive compaction bulb as shown in Fig. 7.12. The surface concrete would crush under the tip of the probe, and the impact shock waves would cause fracture lines including surface spalling adjacent to the probe as it penetrates the

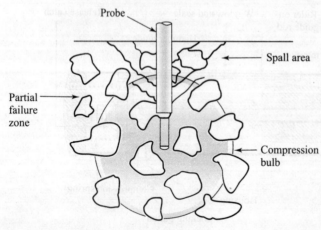

Fig. 7.12 Compaction bulb

concrete mass. The main advantage of this test is that the hardness is measured up to a certain depth, thus giving more realistic result compared to rebound hammer test.

This test basically measures hardness and cannot be indicative of reliable values of strength. The type of aggregate affects the penetration depth which requires separate type-wise penetration chart.

Ultrasonic pulse method

Refer ultrasonic inspection in Chapter 8, Section 8.4.

This method can be applied to in-situ structural concrete members or to laboratory specimens. The method consists of measuring the time for the onset of a longitudinal pulse of vibrations to travel between two crystals transducers placed in contact with the concrete on the opposite faces. This method cannot be applied when transducers cannot be placed on the opposite faces of the concrete. The ultrasonic pulse is generated by an electro-acoustical transducer. As the pulse is introduced into the concrete from the transducer, a complex system of three types of waves is developed. Surface waves having an elliptical particle displacement are the slowest while shear or transverse waves with particle displacement at right angles to the direction of travel are faster. Longitudinal waves with particle displacement in the direction of travel are the most important as these are the fastest and provide more useful information. Electro-acoustical transducers produce longitudinal waves detected by the receiving transducer in the fastest possible time.

The intent of ultrasonic pulse velocity test is to establish the homogeneity of concrete, changes in the structure of concrete occurring in course of time, elastic modulus of concrete, quality of one element of concrete relative to another, and also the presence of cracks, voids and other imperfections. The ultrasonic pulse method may not be indicative of any measure of strength directly.

The velocity of ultrasonic pulse through any material like concrete is important and depends on: (i) density (ii) modulus of elasticity (iii) presence of steel reinforcement (iv) Poison's ratio. This means that higher pulse velocity would be achieved in concrete having good quality in terms of density, homogeneity, and uniformity.

There are three basic ways in which the transducers may be arranged (vide Fig 7.13):

- Opposite faces (direct transmission)
- Adjacent faces (semi-direct transmission)
- Same face (indirect transmission)

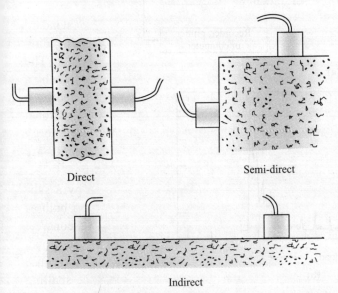

Fig. 7.13 Ultrasonic pulse velocity test – types of reading

The direct method is the most reliable in terms of transit time measurement. The semi-direct method can sometimes be used satisfactorily if the angle between the transducers is not too great and the path length is not large. The indirect method is the least satisfactory as the received signal amplitude may be less than 3% compared to that of direct transmission.

The ultrasonic method of testing has a very important application in determining when it is safe to release the tension in the prestressing wires of prestressed concrete. It is also of utmost importance for detecting whether the concrete in a structure is of the same strength as the test cubes.

Acoustic emission method

On loading of a material, localized points may be strained beyond their elastic limit resulting in crushing and micro-cracking. The kinetic energy released would propagate small amplitude elastic stress waves all through the specimen. These are, although not in audible range, acoustic emissions, which may be detected as small displacement by transducers positioned on the material surface. Acoustic emission transducers are basically high-frequency microphones and so fixed on the surface as to ensure best possible acoustic contact.

The signal detected by piezoelectric transducer is amplified, filtered, processed, and recorded in some convenient form as shown in Fig. 7.14(b). The application of acoustic emission methods has not yet been fully developed.

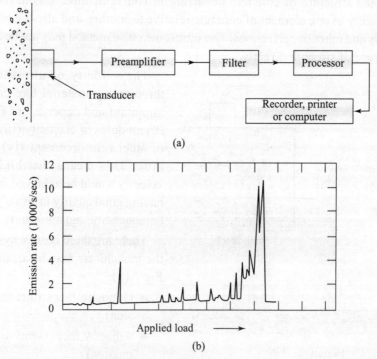

Fig. 7.14 Acoustic emission method

Pulse echo method

This method, which is based on the analysis of reflected pulse traces to detect defects or varying soil support conditions, has been developed for pile testing. A single hammer blow is applied to the pile top and the subsequent movements, including reflected shock waves, are detected by a hand-held accelerometer. A signal processor, which is connected to the accelerometer, integrates the signal. The signal and a trace of vertical displacement against time can be displayed on an oscilloscope. Figure 7.15 shows (a) basic equipment (b) typical trace.

The trace may be used for assessing pile length reliably and for detection of any defects and lack of uniformity of the pile, which would show as distortions to the trace.

These may be interpreted by a specialist to determine pile type and position. The equipment can record and store several hammer blows to check signal consistency, and the oscilloscope traces may be recorded by an instant camera attachment. This method, although simple and cheap, cannot be used to determine the cross sectional area of a pile and its bearing capacity.

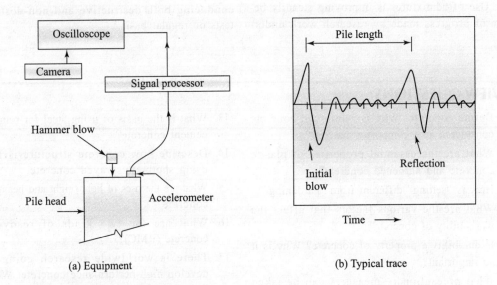

Fig. 7.15 Pulse echo method

SUMMARY

Concrete is the lead construction material that has been used in some form since the last century because of its high strength, durability, fire resistance and also because it can be easily produced using locally available aggregates as filler materials or moulded into any desired shape or size. Scientific and technological progress in construction is directly connected to the advanced development of both precast and cast-in-situ concrete. Concrete would continue to retain its lead position as construction material for years as the share of concrete in the construction projects is steadily growing. Concrete has found use in all types of construction from highways, bridges, dams, canal/tunnel linings to towering buildings. Concrete technology has progressed and evolved with the passing of time because of research work undertaken all over the world. The trend now is to use precast prestressed concrete to improve quality and save time. Another trend now is to use artificial aggregates alongside traditional crushed stone aggregates and gravels.

For a given aggregate, physical and chemical characteristics of the cement paste determine the most important properties of plastic concrete like homogeneity and workability as well as the most significant engineering characteristics of hardened concrete like strength, durability and dimensional stability. It has, however, certain weaknesses like brittleness and poor in tension.

Admixtures have become regular and important ingredient of concrete mix because of manifold benefits. Use of admixtures is improving steadily because of progress made by research work undertaken globally. Different types of concrete are included in this chapter and described above.

Plastic concrete is placed into the mould and compacted and it is set to the inner profile of the mould to conform to the dimensions shown in drawings. Shuttering comprises form proper, supporting scaffolding and fasteners.

Curing process, natural or artificial, assures that there is enough water present in the concrete for the hydration to continue for quality concrete production.

Concrete may be reinforced with either plain or deformed steel bars to resist tensile stresses to which a structural member is subjected while the concrete resists compressive stresses. The integrity of reinforced cement concrete depends to a large extent on the reinforcing steel.

Quality of concrete is to be doubly assured by conducing both destructive and non-destructive tests on regular basis.

REVIEW QUESTIONS

1. Define concrete. Why is reinforced concrete considered as a composite material?
2. What are the essential properties of plastic concrete and hardened concrete?
3. How is "setting" different from "hardening"?
4. What are the various factors that affect the workability of concrete?
5. Is durability a property of concrete? Why is it so important?
6. What precautionary measures can be taken against adverse alkali-silica reaction?
7. Explain what causes corrosion of steel in RCC.
8. Fineness of cement should be maintained at the optimum level. Why?
9. Explain the importance of water:cement ratio.
10. Although admixture is not an essential ingredient of concrete, it is commonly used. Why is it so?
11. Define a blended admixture. What are its uses?
12. What are the functions of formwork?
13. What is the basis of using steel for reinforcing cement concrete?
14. Describe how concrete structure is formed using shotcrete/sprayed concrete.
15. What are the uses of lightweight and heavyweight concrete?
16. What are the two kinds of ready-mixed concrete (RMC)?
17. There is worldwide research going on to develop high-performance concrete. What are the performance requirements?
18. What is meant by self-compacting concrete?
19. Briefly describe the difference between extreme weather concreting and conventional concreting.
20. What are the applications of fibre-reinforced concrete?
21. How does pre-stressing of reinforcing steel increase the load-bearing capacity of concrete?
22. Describe the measures that are to be taken for underwater concreting.

23. What are the features that distinguish polymer concrete from polymer-impregnated concrete and polymer-modified cement concrete?
24. Why does concrete need to be cured after hardening?
25. What are the destructive tests that need to be carried out for quality control of cement, aggregates, water, and concrete?
26. What are the common non-destructive tests carried out on hardened concrete?

8
Fabrication and Erection Work

INTRODUCTION

Fabrication is involved in steel structures where the use of steel is an economically viable proposition. Fabrication work is generally carried out at shops, but can be done at sites if suitable mechanized facilities could be mobilized, and experienced/qualified personnel assigned. Shop fabricated steelwork would be better quality-wise, but transportation of the same from shop to site could pose problems. Also, storage of site-fabricated steel may be eliminated by synchronizing construction of foundations/bases and fabrication of the corresponding structures for immediate erection.

Welding is joining of steel, rendered plastic, or liquid by raising the temperature by a heat source that would be localized along the joint line. Metal arc welding characterized by low voltage and high current is predominantly used in steel fabrication and is carried out by qualified welders. Regular inspection of welded structures is the best way to assure product quality. Large shop fabricated structures are transported in pieces and pre-assembled at the site before erection. For actual erection, the skilled crews need to refer to the fabrication drawings and erect fabricated items on priority basis. The final operation in steel erection is the permanent fastening/welding of the erected elements.

In case of precast concrete, maximum efficiency in erection is achieved by placing the elements in their final positions directly from the transport or storage area in single operation. Temporary bracings for columns, frames, and slabs are designed to take both tension and compression so that wind loads and unbalanced loads are taken care of before completion of the final aligning, leveling and plumbing of the erected elements.

Erection of single span steel or concrete bridge between abutments can easily be lifted by mobile crane positioned onshore. Single or multi-span pre-stressed precast bridge girders can be erected by pushing. In bridges with several spans, onshore or offshore cranes may be used to lift beams/trusses. For long bridges, erection of beams by using launching truss or gantry girder could be economical. In balanced cantilever method of construction, the bridge deck is formed by adding segments on either side of the pier asymmetrically maintaining balance. Cable stayed bridges are suitable up to a specific length. Suspension bridges comprise a main span with two side spans on both sides of the main span and bridge deck is suspended from the cables by hangers.

8.1 FABRICATION OF STRUCTURAL STEEL AT SHOPS AND SITES

Structural steel is used widely in the fabrication of structures for industrial or non-industrial buildings, towers, stacks, storage tanks, masts, bridges, and other structures where the use of steel is an economically viable proposition. The functional requirements of steel structures are: (i) strength and stability, (ii) durability and freedom from maintenance, and (iii) fire safety.

Time is the essence of all construction projects (vide Chapter 1, Section 1.6). Completion of all project activities within a tight time schedule would require simultaneous completion of different inter-related activities in proper logical sequence so that the required progress could be achieved without interruption. This simultaneous and parallel completion of different activities is the most significant concept in choosing steel structures erected over reinforced concrete foundations and bases. These apart, steel structures can be conveniently used to accommodate modifications or support additional features not contemplated earlier.

The construction of foundations, basements, and substructures may be a highly time-consuming process. In a site where foundations are built on rocks, excavation by blasting and mucking would be very long drawn process. An equally time-consuming process is piling work in poor soil or controlling ground water where water table is high or where construction is to be carried out off-shore. The steel structures can easily be fabricated at shops in parallel during this period or even at sites if suitable facilities could be developed and construction equipment and machineries could be mobilized and deployed. For quality assurance, experienced and qualified manpower also need to be assigned at sites.

The quality of fabrication work at a shop under totally controlled conditions would definitely be better, but transportation of shop-fabricated structures could pose problems. Adequate precautions would have to be taken to eliminate deformation and damage to end-plate connections, milled end surfaces and abutting edges during transportation.

However, site fabrication may eliminate the necessity of storage of structural members by synchronizing construction of foundations/bases and completion of fabrication of the corresponding structures for immediate erection. Normally, an executing agency is hired for both fabrication and erection of structural steel work. This contractor, if specialized in fabrication work, may hire a sub-contractor or two for executing the erection work only. Even the owner may split the contract into two parts and award the work of fabrication and erection to two separate executing agencies.

Structural steel members are fabricated by a contractor on the basis of working drawings approved by the design engineers. The contractor is provided with design drawings, technical specifications, and details by the owner or their engineering consultants. The contractor awarded with the contract for fabrication work would have the responsibility of preparation of the working drawings and bill of quantities (BOQ). BOQ contains the cutting lengths of each individual member. The fabrication work would be taken up only on approval of the working drawings by the owner or their consultants. BOQ forms part of a detailed drawing. Based on approved BOQ, welding procedure specifications (WPS) are prepared.

The fabrication of structural steel is taken up on the basis of approved time schedule, which should conform to the project time schedule. Obviously, the fabrication work would be executed in logical sequence—whatever is required first in sequence is to be fabricated first. The same logical sequence should be followed on procurement of steel materials—sections, shapes, and plates produced by mills as hot-rolled products. Otherwise, work continuity would be disrupted and more of the site space is used up for storage of fabricated materials.

One of the many problems involved in the design of structural steel work is designing connections. All individual components that comprise structural steel cannot function effectively if the connection of these components is in such a way that it does not permit all forces and moments to be transferred to each other as designed. Two basic types of connections are typically used in fabrication work, bolting and welding. A third type of connection, which is riveting, is virtually abandoned now although the Indian Railways still prefer riveted connections. Bolts in general are used for connections made during actual erection whereas shop connections are made by welding. High-strength bolts are generally used for bolted connections and these bolts are tightened in accordance with the contract specifications using torque wrenches.

Structural steel shapes and plates as received at a shop or a site may not be in the correct form for immediate use. Deformity must be rectified before the initiation of even marking and cutting. The rolling machines are used for straightening deformed plates. Deformity in rolled sections and shapes is rectified with the aid of a straightening machine.

Raw steel is to be marked as per the relevant approved drawings for cutting and/or drilling. The lines for cutting and centre lines for drilling should be punched on the raw steel. Flame cutting along the punch marks is executed manually. It is possible to mount a flame torch on rail for ensuring long straight-line cutting. If the rail is fixed correctly, curved lines can similarly be cut.

Raw steel can be cut by a shearing machine. The shearing operation has its own limitations. Shearing is effective in straight-line cutting of materials of limited thickness and relatively short length. Manual drilling of holes is also possible. But the process can be hastened by fitting jigging arrangement with the drill.

In order to avoid welding because of the requirement of the service conditions of the components under consideration, there would still be demand for structural components to be bolted. The requirements may be: (i) low temperature criteria; (ii) the need to avoid welding stresses; (iii) the requirement for the component to be taken apart during services like bolted-on crane rails. In case of lattice structures, the designer should specify bolting considering the effect of hole clearance around bolt shanks. High-strength friction grip (HSFG) bolts would not cause problems, but other bolts in clear holes could allow a 'shake-out' which may cause significant additional displacement at joints. A truss with bolted connections may lose its theoretical camber because of ill-fitting of bolts. Hole clearance in relation to bolt size causes this problem. The use of close-tolerance bolts in holes of the same diameter reduces this effect without incurring the cost of turned and fitted bolts in reamed holes.

Tack bolting is used in positioning the components prior to welding, combining the effective use of CNC (computerized numerically controlled) punching and drilling with the speed and convenience of welding. The tack-bolted assemblies are self-aligning and hence expensive jigging is not required. The tack bolts are sacrificed as they cannot be considered as contributing to the strength of a welded joint. Tack bolting is used for lattice work where angles, channels, or tees are used as the booms or lacings.

Large and complex assemblies which are to be bolted together on receipt at the site may be trial-assembled in the fabrication shop. This additional work would increase fabrication cost but, at the same time, ensure that steel delivered to the site would fit.

There are basically four types of bolts as follows.
- Structural bolts—also known as 'black bolts' which are now available with protective coating and in a range of tensile strength
- Friction-grip bolts (HSFG – high strength friction grip)
- Close-tolerance bolts
- Turned-barrel bolts

Increasing numbers of buildings and bridges are being built according to the precepts of good welded design. There are many reasons for using welded design and construction, but the most important reasons are as follows: (i) welded design offers most efficient use of materials; (ii) the speed of fabrication and erection of structural steel result in drastically reducing the project completion time. The progress made in welding equipment and electrodes as well as the economy and efficiency achieved in design based on welding result in a very efficient and cost-effective project implementation.

8.2 WELDING TECHNOLOGY

Mechanical properties like strength, elasticity, ductility, hardness, toughness, and fatigue determine the behaviour of metals under the action of applied loads. Metals are composed of tiny crystals whose submicroscopic building blocks are atoms arranged in a definite pattern within each crystal. The atoms exist within certain well-defined planes in the crystal structure. It is believed that if the metal is subjected to applied load, then the load would tend to alter the orderly arrangement of atoms by displacing them from their original stable positions within the crystals. The atomic planes of a crystal may be altered under load by elastic or plastic deformation that results from the relative movement of atomic planes parallel to each other, or by rupture when the planes are pulled apart. Thus, the mechanical properties of the metal depend on the degree of strength that binds all the atoms to each other.

Welding is the joining of pieces of metal at faces rendered plastic or liquid by raising the temperature at the joint or by forging or under pressure. Adequate strength can be produced by only inter-atomic bonding in a welded joint. The basic function of welding is to provide links between atoms at the interface of the joint. Two conditions that must be satisfied for these links to be formed are: (i) the surfaces must be in intimate contact and (ii) the surfaces must, from the metallurgical point of view, be clean. Any molecules of moisture, paint, grease, oxygen, and nitrogen present on the surface would prevent the metal atoms from bonding to each other. If a practical welding system is to be based on the principle of simple inter-atomic bonding, a method of bringing the surfaces into intimate contact and, simultaneously, dispensing the surface contaminants is essential.

Cold-pressure welding

In cold-pressure welding, intimate contact and movement of contaminants are achieved by forcing the surfaces together. The surfaces deform under cold pressure thus breaking up the contaminants thereby bringing areas of clean metal into intimate contact. The thickness of work pieces is reduced appreciably under pressure exerted to disperse the contaminants. Optimum joint strength is achieved at the threshold deformation level (percentage reduction in thickness). The softer the metal, the lower would be the deformation to start welding at room temperature. Anyway, cold-pressure welding has limited applications.

Hot-pressure welding

In hot-pressure welding, the wrought iron or steel bars that are to be joined are heated to about 1350°C at which temperature the molten oxides on the surface are melted and, when the components are hammered together, the molten oxides are squeezed out of the joint thus allowing bonding at low deformation levels. Of the various methods used to heat a joint for hot-pressure welding, three of the most successful are: (i) gas heating, (ii) resistance heating, and (iii) induction heating.

Friction welding

There is also friction welding which is not suited for fabrication work. The machine used for this type of welding looks like a large lathe fitted with two chucks. One of the two chucks is driven by a motor whereas the other one is fixed. The two parts to be joined are clamped in the chucks and one part is rotated. This poses problem of shape and size.

Structural steel fabrication involves steel sections, shapes, plates, and bars of large dimensions. In such cases, neither pressure nor friction welding would be of any use. This led to the development of good welding by the use of controlled fusion by heat application. As the solid metal reaches its melting point, oxide films which exist on the surface of the plate are disrupted and dissolved in the steel pool. The surfaces are thus clean from the metallurgical point of view. The localized bond made by melting just one point is not useful for fabrication. Bonding along a line is possible by moving the heat source along the joint line.

Fusion welding

Fusion welding for structural fabrication would be possible by a heat source that is localized along the joint line. The fusion should not be more than what is required to accomplish bonding. Overall heating should be restricted to a narrow band along each side of the weld.

The metallurgical properties of the joint and distortion are related to the cycle of heating and cooling in the area of joint. The heat source must have some salient features if it is to be used in actual welding in fabrication work which are as follows:

(i) The heat source must be small so that weld pool can be limited to manageable size.
(ii) The heat source must operate at a temperature distinctly higher than the melting point of the metal being welded. Otherwise, heat would spread to a wider area before fusion is realized.
(iii) The heating capacity should be adequate as the rate of heat required not only depends on the metal properties but also on the joint configuration and dimensions.
(iv) The heat source must be such as can be regulated.

> The most common heat sources used in industrial welding work are:
> - Flame welding that uses gaseous fuel and oxygen, for example, oxy-acetylene welding
> - An electric arc
> - Resistance heating at an interface
> - Resistance heating at a slag bath

Flame welding

Quite high temperature can be achieved by burning a gaseous fuel with oxygen. But such temperatures are high enough for welding only a few metals of low melting point such as lead, zinc, and tin. Acetylene gas, however, is an exception. When mixed in correct proportions, oxy-acetylene gases burn with a flame temperature of about 3100°C, which is adequate for many welding applications in structural fabrication.

Metal arc welding

Apart from electricity that passes through conductors such as wires, electric current can flow across a gap between a metallic or carbon rod and the surface of a plate generating an appreciable amount of heat that can be used to melt metal in a small area such as a metallic joint. An electric arc characterized by low voltage (20–40 V) and high current (30–100 A) between the end of a rod electrode and the flat surface of a plate constitutes a very effective heat source that forms the basis of the most common techniques for structural fabrication work.

Resistance heating

At the interface of two overlapping sheets, high current could be passed through interface. There would be spot weld over a limited area if the spot where melting has occurred is allowed to solidify. This is resistance heating at an interface.

Electr-slag welding

Mineral or ceramic slags are usually good insulators, but they conduct electricity when melted. Their resistance to the flow of electricity is, however, low because of which considerable heat would be generated in the body of the molten slag. The generated heat can be utilized to melt metal plates in electro-slag welding.

Metal arc welding is predominantly used in steel fabrication. An arc is the passage of electricity across a gap in the electrical circuit between an electrode and the parent metal. For this to occur, the atmosphere in the gap must be ionized. Here, ions are formed by subtracting electrons by strong electric fields in a gas, which remains positively charged. The free electrons travel from the cathode to the anode maintaining the flow of current. Some electrical energy is converted to heat in the arc column by the resistance of air or gas and is radiated into the surrounding air, but most of the energy is released at the end of the electrode and at the parent metal surface. Heat generated at the end of the electrode causes melting. The molten metal from the electrode is then transferred to the weld pool. Heating at the parent metal produces fusion of the joint faces. The factors governing heat input in arc welding are:

- Arc voltage
- Arc length—which is related to arc voltage
- Electrode feed rate
- Arc current—through a source that can be controlled (amperage and voltage, AC/DC)
- Travel speed—which is under the direct control of the welder

The parent metal could be connected to the positive or negative pole of the power supply source. Accordingly, the parent metal will become the anode or cathode. This choice of polarity is governed by a number of considerations such as arc stability, rate of electrode melting, transfer of molten metal from the electrode to the weld pool, and removal of oxides from the surface of the parent metal. The proportion of the total energy supplied to the arc that is used to melt the parent metal depends on the polarity, that is, anode or cathode.

There must be good bonding between the weld and parent metals. This is possible only when the parent metal has been melted before the filler/electrode metal is allowed to flow into the joint. If this sequence is not followed, a lack of fusion is observed at the boundary of the weld metal as the molten filler metal tends to be cooled down by the unmelted parent metal.

The quality of welding operation is also dependent on the heat input to the joint. Immediately on supply of heat to the area to be welded, either by an arc or a flame, the heat starts flowing into the metal on either side, as the parent metal is at a lower temperature. For quality welding, therefore, the rate at which heat is being supplied to the joint must be greater than the rate at which heat flows into the parent metal. Preheating the parent metal before welding, if done, would help in rapid attaining of the melting point. Thus, the thermal conductivity of parent metals is probably one of the most important considerations when choosing welding conditions.

Heat sources discussed thus far produce a fusion region having a temperature well in excess of the melting point of the metal being welded. If a joint is exposed to open atmosphere immediately after welding, it absorbs oxygen and/or nitrogen and/or hydrogen depending on the composition of the parent metal. A small percentage of gas either dissolved in or combined with the weld metal may not cause any harm to the joint properties. However, the presence of significant amounts of dissolved or combined gases is undesirable as it often results in the formation of gas pores or voids in the completed weld. Further, properties or integrity of the joint can be impaired. Dissolved hydrogen in high-strength steel weld metal can lead to cracks in the joint.

Prevention of atmospheric contamination must be inherent in the welding process. There are two basic preventive techniques. In the first of the two techniques, the completed weld is blanketed by molten flux forming a slag layer that is impervious in

nature for the gases. In the other alternative, the ambient air is replaced by a gas that does not react with the molten metal. The two aspects that need to be considered for each fusion welding process are as follows:
- A suitable source of heat
- Shielding to control the complex arc phenomenon and to improve the physical properties of the weld deposit

Some of the commonly used *fusion welding processes* (Fig. 8.1) are described below:
- *Manual metal arc (MMA) welding:* This process involves establishment of an arc between the end of the electrode and the parent metal at the joint line. The arc, which produces a temperature of about 3600°C at the tip of the electrode, melts both the parent metal and the electrode to form a common weld pool which is protected by the molten flux layer and gas generated by the flux that covers the electrode. As the areas solidify, the metals are joined into one solid homogeneous piece. The welder moves the electrode towards the weld pool to maintain the arc gap at a constant length. The current is controlled from the power source. The electrodes are normally 450 mm in length. As the electrode melts down to 50 mm, the arc is extinguished. The solidified slag or flux is removed, and the welding is continued with a fresh electrode.
- *Submerged arc welding (SAW):* Two separate consumables are used in submerged arc welding—the wire and the flux. The arc operates under a layer of granular flux because of which the process is referred to as submerged arc welding. Some of the flux melts to provide a protective blanket over the weld pool. The flux that is not melted is recovered and re-used. The process involves establishment of an arc between the end of a bare wire electrode and the parent metal. The current is controlled from the power source. As the electrode is melted, it is fed into the arc by a servo-controlled motor. The electrode feed rate should be controlled in such a way so as to keep the arc length constant. The electrode and drive assembly is moved along the joint line by a mechanized traverse system.
- *Metal inert gas (MIG) welding:* In this process, an arc is established between the end of the electrode and the parent metal at the joint line. The electrode is fed at a constant speed by a controlled motor—the electrode feed rate determines the current. The arc length is controlled by the power source. It is necessary that the welder keeps the nozzle at a fixed height. The arc area and weld metal are protected by a gas. This gas is selected to suit the metal being welded. The gases that are normally used are: (i) argon, (ii) carbon dioxide, and (iii) argon mixed with 5% oxygen or 20% carbon dioxide. This type of welding is used in joining plates or fabricating box-girders.
- *Tungsten inert gas (TIG) welding:* In this process, an arc is established between the end of a tungsten electrode and the parent metal at the joint line, but the electrode is not melted. The welder maintains a constant arc gap. The current is controlled at the power supply source. Filler metal, usually available in wires of

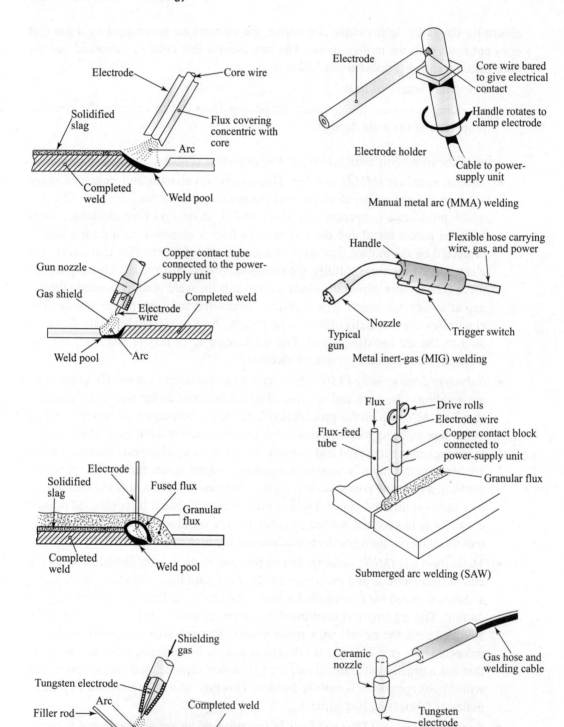

Fig. 8.1 Welding processes

length of 1 m, is added at the leading edge of the weld pool as required. The molten pool is shielded by an inert gas which replaces the air in the arc area. The most commonly used shielding gas is argon. TIG welding is a special type as the metal is not transferred across the arc.

Penetration

The penetration is the depth to which the parent metal has been fused, that is, the extent of melting into the parent metal as, for example, in case of fillet-welded or butt-welded joints. As both voltage and current contribute to heating in the arc, penetration would be affected by a change in either voltage or current. The width of the weld and the surface profile depend on the arc length, and arc length in turn is related closely to voltage. In general terms, high voltage results in long arc lengths, wide flat welds along with the risk of unstable arc. Oxygen and nitrogen may be drawn into the arc column simultaneously. Very low voltages (short arcs), in contrast, can result in narrow welds having a high profile with the possibility of lack of fusion at the edge of the weld. It is, therefore, necessary that the welder chooses a voltage and arc length that gives a stable arc and a satisfactory weld surface profile. The current becomes the main factor in the control of penetration as and when the voltage has been fixed.

The molten weld pool comprises a mixture of parent metal and electrode. The proportions in the mixture are determined largely by the heat input, which controls the rate of electrode melting as well as the penetration. The weld metal is considered as a metal deposited from the electrode or filler material diluted by mixing with the molten parent metal. Conventionally, dilution is expressed as the percentage of the molten parent metal in the weld and this can be used to predict individual alloy content in the molten pool.

Welding Positions

In arc welding, the position of welding is very important. In ideal position, the molten weld metal is held in place by gravity. This is possible when the position would be such as termed 'flat' (Fig. 8.2). This position also allows high currents to be used thereby resulting in faster welding. Most welding work is carried out in position. Due to this reason, flat welding is rarely possible.

Apart from flat welding, three main positions of welding are: 'horizontal', 'vertical', and 'overhead'. There is another position also known as 'horizontal–vertical', which is related to a T-joint where one of the members is vertical and another horizontal. In all these, there is gravitational disturbance which can be overcome by the welding technique and the welder's ingenuity. Heat input needs to be lowered, and weld pool should be small for quick solidification. Simultaneously, the direction of arc can be varied to position the weld pool to the best advantage. The arc force also helps to keep the weld pool in place.

Although 350 A of current can be readily used for joints in the flat position, not more than 160 A can be used in welding overhead. The sequence used to deposit weld of a given size differs from one position to another. The leg length of weld is the distance

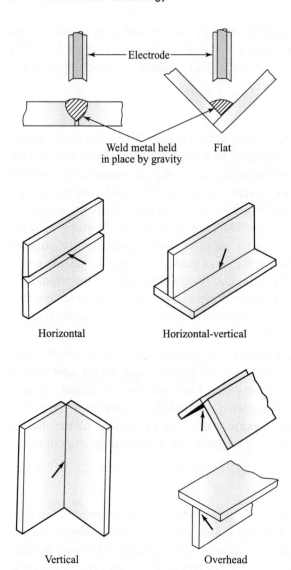

Fig. 8.2 Welding positions

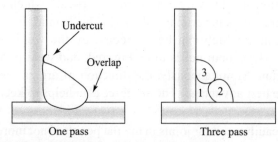

Fig. 8.3 Welding positions: Horizontal-vertical

from the root to the toe—one side of the triangle formed by the fillet. In the flat position, the weld can be deposited in one pass with a current of 300 A at the rate of approximately a metre in 10 min. The values of ampere and rate of welding as mentioned are indicative only. In practice, instructions obtained from the electrode or filler manufacturers should be followed.

In case of the horizontal–vertical joint (a subdivision of horizontal position), the weld is to be completed in three passes to ensure quality and proper shape (Fig. 8.3). The time involved in completing welding in this position is 50% longer compared with the flat position. In overhead position, four or five runs may be required to obtain the required size and the progress is at the rate of a metre in about 24 min.

In case of vertical joint, the welding is done in vertical-up technique. It is started at the bottom, and the arc is moved upwards. Welding is completed in two runs with 145 A of current. To prevent possible sagging, the current may be reduced to 120 A and the welding is completed in three passes. Alternatively, the vertical joint can be carried out using a vertical-down technique. The welding is started at the top of the joint. Here, the rate of travel is critical. The molten metal must not run down the joint ahead of the arc. In case it does, fusion of the parent metal may not take place. For this reason, 10-mm fillet weld may need five or six weld runs. However, the travel speed of the vertical-down technique is comparatively high. In this position, the weld can be deposited in one pass at the rate of approximately a metre in 17 min. The travel speed is critical as the molten metal must not run down the joint ahead of the arc. If it so happens, fusion of the parent metal may not occur.

Edge Preparation

The main factors to be considered for completing a welded joint economically are:
- Joint type, included angle, root opening, and root face (Fig. 8.4)
- Type and size of electrode
- Type of current, polarity, and quantum in amperes
- Arc length (arc voltage)
- Arc speed
- Welding position

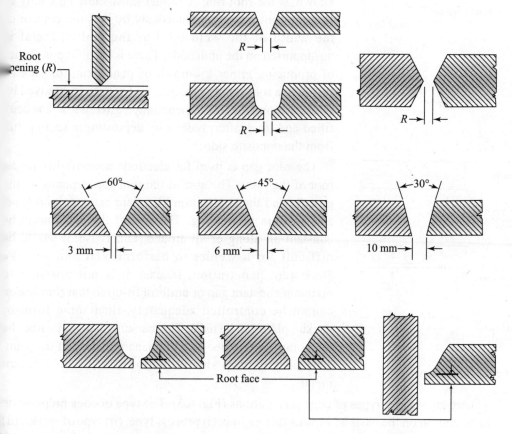

Fig. 8.4 Root opening and root face

Quite a good number of the factors mentioned can be determined by a trial sample joint. A fabricator can easily determine the electrode type/size, polarity, current, arc characteristics, and welding techniques.

Without any edge preparation, the thickness of the metal which can be butt-welded from one side by maintaining square edges is 3 mm if the welding is carried out from one side, and 5 mm if the welding is carried out from both sides—this is relevant in case of MMA and TIG processes only. In case of MIG welding, the same limitations are applicable with currents up to 200 A. But it would be possible to butt-weld joints up to 6 mm from one side with square edges in MIG welding with high currents such as 400–450 A.

The simplest edge preparation (Fig. 8.5) is a single bevel that forms a groove when assembled with root gap that separates the components to be joined. The groove is filled up by depositing a number of weld runs into it ensuring that each run is fused into both the surface of the preceding weld and side walls of the groove. This ensures that the joint is completely bonded. The first weld or pass, which requires considerable skill to execute, to be deposited is known as the root run. A welder must show ingenuity to fuse the root faces simultaneously by keeping control of the width of the weld bead as the molten metal is unsupported on the underside. There is harmful possibility of producing either too much of penetration or lack of fusion in a root run. In practice, the root run is removed by gouging along the line of penetration after the 'v' has been filled and, thereafter, redone by depositing a sealing run from the opposite side.

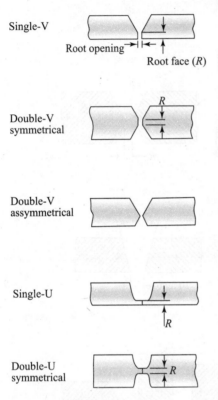

Fig. 8.5 Edge preparations

The root gap is used for electrode accessibility to the root of the joint. The size of the root gap depends on the process and the application like plate or pipe, joint type and position of welding. The size of the gap should be constant all along or all around. Otherwise, it would be difficult for a welder to perform well and achieve consistent penetration. In case it is not possible to maintain constant gap or uniform fit-up so that root fusion cannot be controlled adequately, then some form of backup plate can be used. This backup plate may either be temporary or retained as a permanent feature of the joint. This backup strip should conform to the base/parent metal.

There are various types of edge preparations (Fig. 8.5). The type of edge preparation can be chosen on the basis of various factors like: (i) process type, (ii) type of work, (iii) welding position, (iv) access for arc or electrode, (v) volume of deposited weld metal, (vi) dilution, (vii) cost of edge preparation, and (viii) shrinkage and distortion.

Electrodes

In manual metal arc welding of steel structures, four main groups of electrode coverings (Fig. 8.6) are used. The major constituents of the flux covering determine their major operating characteristics. The materials used include various organic materials, minerals, chemical compounds, clays, and ferroalloys.

Cellulose: The coverings of the cellulose type have large quantities of organic materials, which decompose in the arc to form hydrogen and carbon dioxide. The voluminous gas shield replaces air in the arc column and provides a protective gas shield to the weld metal. The penetration characteristics of the electrode are also increased by about 70% for a given current. As most of the flux is decomposed, the resultant slag layer is thin and the slag is easily detached. This type of electrode is not used in high-strength steel welding.

Acid: The medium and thick coverings of the acid type are composed mainly of oxides and silicates and have high oxygen content. This type of electrode produces iron oxide, manganese oxide, silica rich slag, etc. whose metallurgical characteristic is acidic. The slag generally has a porous or honeycomb structure and is easily detached. Despite having good ductility, the welds tend to be of less strength. Because of this, acid electrodes are not widely used. Acid electrodes are most suitable for welding in flat position.

Rutile: The coverings contain a large quantity of rutile or components derived from the oxide of titanium. This type has good slag-forming characteristics, and produces a stable arc which can be easily used with very little spatter. Rutile electrodes are widely used in welding as these electrodes result in a fine, smooth finish to the weld and fulfil a general purpose in the fabrication work. Slag detachability is good. By varying the addition of fluxing agents, the viscosity and surface tension can be adjusted to give electrodes that are suitable either for the flat position only or for use in all positions. Mechanical properties are sufficient for most structural steels although achievement of high tensile strength is not easy.

Basic: The coverings of this type comprise mainly calcium compounds such as calcium fluoride (fluorspar is the natural form of calcium fluoride), and calcium carbonate so that from metallurgical considerations, they are basic in character as flux. The basic types are low hydrogen electrodes. They are also sometimes called 'lime-coated'. The flux can be dried by baking at 480°C, and if the electrodes are stored at 150°C (or as directed by the manufacturers) till used, the hydrogen content of the weld metal can be reduced to between 10 and 15 ml/100 g at which level of controlled hydrogen, the possibility of cracking in high-strength steels is minimized. The mechanical properties are very good. The oxygen content of the slag is low, and surface profiles of weld deposits are convex. Low hydrogen electrodes help in maintaining a low gas content in the weld metal.

For manual metal arc welding, the diameters and lengths of core wires vary—diameters from 1.6–8.0 mm, and lengths from 150–450 mm. The entire welding current flows along a core wire as the electrical connection is made at the top of the electrode. Heat is generated with the flow of current as

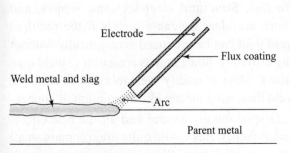

Fig. 8.6 Electrode coverings

the core wire has electrical resistance. If the temperature increase is too high, there could be the possibility of premature melting of the electrodes. Even flux can be damaged by heat before the premature melting of electrode. The moisture evaporated from the compounds in the flux causes it to flake off leaving lengths of the core wire uncovered. Oxidation of some of the alloying elements can simultaneously affect the composition of the resultant weld. Because of all these reasons, the electrode manufacturers stipulate the maximum current for each electrode core-wire diameter. There is also a lower limit of current below which the arc becomes unstable.

The following aspects need to be considered in choosing the correct electrode for the welding work for optimum economic performance:

- The required mechanical properties
- Possibility of weld metal cracking
- Composition of the metal to be melted
- Thickness of the parent metal
- Type of joint specified
- Power source
- Welding position
- Limitations on heat input, if any

Weldability

Most steel metals are weldable and can be economically arc-welded with satisfactory results achieving sound and strong welded joints. It is commercially possible to create a molten pool between two pieces of metal which solidifies to turn into a welded joint. The 'weldability' of a metal refers to the relative ease in producing a welded joint conforming to a given specification into a specific, suitably designed structure. The joint needs to be crack-free and sound so that the fabricated structure performs satisfactorily in the intended service. Sound refers to the condition that the joint has to be totally free of defects and discontinuities. The better the weldability, the greater is the ease with which these requirements are met. Structural steel sections, shapes, and plates are ideal weldable metals if the required weld joint can be produced economically without any difficulty. Some metals are easier to weld than others. Steel is readily weldable and is easier to weld than other metals like aluminium or copper.

Despite quality control and all the best efforts undertaken, welds would quite often contain small amounts of defects like porosity, slag, or oxide

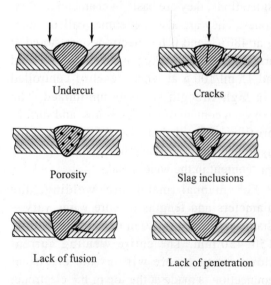

Fig. 8.7 Typical defects in welding

inclusions, and lack of fusion (Fig. 8.7). Such faults are related to the welding techniques adopted and a welder's performance and skill. Cracks are mostly related to the composition of the material being welded and the risk of their formation is an important element in the question of weldability.

Continual improvement in the metallurgy of steel making and welding techniques is evident all over the world. Automation is also being incorporated in the steel production process and also the welding techniques. There is thus sure indication of widening the range of 'weldability' on the basis of steel analysis.

Heat Input

In the case of welding arc, the total energy is converted to heat except for a small proportion that is converted into bright light and ultraviolet radiation. It is possible to determine power input in watts by measuring current and voltage using meters.

Input power in watts = arc voltage × arc current in amperes

With an electrode operating at 125 A and 20 V:

Input power = 20 V × 125 A = 2500 W = 2.5 kW

This value can be converted into energy as 2500 J/s. This is the heat energy input if the arc is stationary. In practice, however, a welder moves the arc along the joint line. The heat is, therefore, distributed, and the heat input at any particular point depends on the travel speed. Once the energy being supplied to the arc and travel speed are known, it would be possible to express the same as amount of heat per unit length of weld run in terms of J/mm.

The formula for calculating heat input in any arc process is:

$$\text{Heat input (J/mm)} = \frac{\text{welding current (A)} \times \text{arc voltage (V)} \times 60}{\text{travel speed (mm/min)}}$$

Heat-affected Zone (HAZ)

When the process of cooling of the molten weld begins, only a small amount of heat escapes through the weld pool. The bulk of the heat flows through the parent metal on either side of the joint. The parent metal, consequently, passes through a cycle of heating and cooling. The two vital aspects of this cycle are the highest temperature attained and the rate of cooling that follows. The peak temperature attained depends on the distance from the weld. The parent metal just outside the boundary of the fusion remains heated to a temperature just below the melting point. On moving away from the fusion boundary, the peak temperature falls progressively till the parent metal ceases to be heated. Before this point is reached, the temperature decreases below the lowest value at which any metallurgical change takes place. The HAZ is the area which has undergone metallurgical change as a result of the thermal cycle.

As far as structural steel is concerned, the parent metal in the HAZ remains either unaffected or hardened. The main concern with any hardened structure is the possibility of cracking which would severely affect the joint's performance in service. There would not be any concern for the tensile strength of joint if the HAZ remains free of cracks.

The heating and cooling cycle in the HAZ would have effects that could be illustrated by three types of metallurgical reaction: (i) softening of work-hardened metals, (ii) softening of precipitation-hardened alloys, and (iii) hardening of steels.

In the earlier stages of rolling of metal from ingots or thick plates, the metal is kept at high temperature to achieve large reduction in thickness. In subsequent rolling for achieving accurate thickness and better finish by rolling at room temperature (cold working), there would be increase in hardness and tensile strength. Work-hardening can also be achieved at fabrication work as a result of pressing, bending, or forming at room temperature. If the metal is heated to a critical temperature instead, softening occurs. This is known as re-crystallization. In the welding work, the metal which is raised to this temperature is re-crystallized. Consequently, the HAZ contains softened metal having lower tensile strength than the unaffected parent metal. The properties lost thus cannot be recovered.

The process of 'precipitation-hardened alloys' is related to alloy aluminium or alloy steel. This involves heating, cooling, re-heating, etc., that are not relevant in structural steel work. The treatment consists of heating the sheet or section to a high temperature depending on the composition of the alloy and cooling it rapidly. It is then re-heated to an intermediate temperature and held there for a prolonged period during which the hardness of the metal increases with the formation of the precipitates. The precipitates grow to a critical size a little while later and the hardness reaches a maximum value at which point the metal is cooled rapidly.

For heat-treatment, steel is hardened by heating to a temperature of about 1000°C and then cooled rapidly by quenching the metal into oil or water thereby changing the metallurgical structure of the steel – from ductile to hardened brittle form. The extent of hardening depends on the rate of cooling and the composition of steel. The main concern regarding the hard HAZ is that it creates conditions that tend to form cracks in the HAZ. There is a high risk of cracking if the following three conditions exist: (i) hardness above critical level, (ii) existence of tensile stress, and (iii) presence of hydrogen in the weld pool. Even if all these conditions exist, cracks may not be formed although the question of 'risk' would remain.

Steel used in the welded structural steel fabrication work contains a number of alloying elements apart from carbon. Some of these appear to play a little role in the formation of HAZ cracks. But some like manganese and chromium increase the ease with which a hardened structure can be formed. In assessing their significance, it is possible to simplify the relationship of all alloying elements in steel to the occurrence of underbead cracking. The simplification is expressed in a single formula known as the carbon equivalent (CE). In CE, the amount of each alloying element is factored according to its contribution to hardening. Accordingly, manganese is allocated a factor of one-sixth as it has been estimated that a manganese content of 0.6% produces the same effect on hardening as 0.1% carbon. There are many variations of the formula used to calculate the CE.

For a given analysis of steel, there is a maximum rate at which the weld and parent metals may be cooled without development of any underbead cracking. The higher the carbon equivalent, lower is the maximum cooling rate. In other words if the steel's CE is higher, the use of low hydrogen welding and preheating becomes more important in welding. At lower CE, faster cooling rate can be endured before there is any risk of cracking. Underbead cracking in the HAZ rarely occurs when the CE values are below 0.39%. At high values of CE like 0.48%, there is a high risk of underbead cracking with even slow rate of cooling.

Weld Cracks

The most vital requirement of any welded joint is that it must be crack-free. Cracking may occur either in the weld metal or heat-affected parent metal plates. A crack in a weld is never minor and should not be overlooked. Good design and appropriate welding procedure would prevent problems resulting in weld cracks: (i) weld cracks occurring during welding, (ii) cracking in the HAZ of the parent metal, and (iii) failing in service of the welded joints.

As long as the thickness of the steel plate remains an 'average', there would not be any reason for worry. However, with the increase of plate thickness (thick plates) as well as the carbon and alloying content, weld cracks and underbead cracks may become problems necessitating special precautions for their control.

The requirements for this are as follows:
- Good welding procedure particularly on bead shape (slightly convex) and control of admixture (carbon and alloy content).
- Reducing rigidity by intentional spacing of plates—a slight gap between plates assures crack-free fillet welds especially if the plates are thick.
- Using low-hydrogen welding materials to escape any possibility of forming cracks.
- Controlled cooling rate including control of welding current and travel speed.

Preheating

Preheating the parent metal before welding assists in attaining the melting point rapidly by reducing the temperature difference between the weld and parent metals. The thermal conductivity of the metal, simultaneously, falls as the temperature is raised. In other words, preheating reduces the rate of flow of heat. Preheating, though not necessary at all times, is used for one of the following reasons.
- Preventing excessive hardening and lowered ductility in both weld and heat-affected parent metal base plate by providing slower rate of cooling through the critical temperature range (980°–720°C).
- A slower rate of cooling through the aforementioned temperature range permits more time for any hydrogen that is present to diffuse away from the weld and parent metal to stop underbead cracking.

- Reducing shrinkage stresses in the weld and adjacent base metal; this is particularly important in highly restrained joints.
- Increasing the allowable critical rate of cooling below which there would not be any underbead cracking; the welding procedure remaining unchanged, preheating increases the safe rate of cooling while simultaneously reducing the actual rate of cooling thereby making heat input from the welding process less critical.
- Lowering the transition temperature of the weld and the adjacent parent metal.

Preheating, which increases cost, is normally not required as a preventive measure against underbead cracking. Preheating at high temperature, however, could be required for other reasons like a highly restrained joint between two highly thick plates.

Shrinkage and Distortion

In the welding process, the prime concern is the shrinkage in both parent metal and filler materials always caused by the heating and cooling cycle. This happens during cooling after the heat source has passed along the joint line. The shrinkage forces tend to cause a degree of distortion. Shrinkage during welding is made of: (i) contraction of liquid metal, (ii) volume change on solidification, and (iii) contraction of solid metal. With the increase in temperature, there is decrease in values of properties such as yield strength, modulus of elasticity and thermal conductivity, and increase of coefficient of thermal expansion and specific heat. Restraint from external clamping, internal restraint due to mass, and stiffness of the steel plate itself must also be considered. The sum of all these factors definitely influences the degree of movement. The period of time during which a specific condition is in effect controls the importance of that condition. In addition to all these variable conditions, there is influence of the welding process also – different welding procedures, type and size of electrode, welding current, travel speed, joint design, preheating, and cooling rates. A solution of shrinkage and distortion should be based on correcting the combined effect.

As melting and solidification in welding progress simultaneously, it would be difficult to separate the contraction of liquid metal and the change of volume on solidification. As the solidification process progresses upwards towards the centre line, the solid metal occupies a smaller volume compared with the replaced liquid metal, thus resulting in increase of density. The molten metal also contracts resulting in receding of the surface level below the original level. Simultaneously, further molten metal is being added into the area by melting the parent metal at the leading edge of the weld pool and by the addition of the melted electrode. Despite the occurrence of appreciable shrinkage during cooling and solidification from the liquid state, the surface of the weld does not normally show any evidence of this unless sufficient added metal is available to make up for the reduction in volume. As no further melting would take place at the termination of a weld run, a welder adds sufficient electrode or filler metal to build up the surface so that a crater is not formed at the end. In practice, a welder uses sufficient electrode for building up the surface of the terminal end gradually reducing the heat input simultaneously.

Until the entire joint is cooled to the room temperature, heat from the weld continues to flow into the parent metal. During this phase of cooling, the weld metal contracts longitudinally along the length of the joint and transversely at right angles to it. Thus, the weld metal should become smaller which does not happen because of the restraint exercised by the parent metal to which it is bonded.

The hot weld metal is not free to shrink longitudinally on cooling as the same is attached to the parent metal plates on either side. The plates should be deemed as rigid not allowing the weld metal to shrink. The weld metal is plastically deformed by being held to the original length. This means that tensile forces are being set up in the weld region that is being balanced by the compressive forces in the plates. The contraction along the length of the joint should be appreciable. In reality, however, the measured contraction is significantly low, as low as 1 mm per metre of weld. The compressive stresses induced on finishing of the cooling cycle are of considerable magnitude and are in excess of the compressive yield stress of the parent metal. Consequently, the plates are plastically deformed. This explains the reduction of the overall length of the welded joint and accounts for 1 mm per metre of the shrinkage mentioned earlier.

In case of shrinkage transverse to the weld, the aforementioned considerations should apply. In this case, the contracting weld metal tries to pull the parent metal plates towards the centre line of the joint. This results in the whole area being in transverse tension. Again a situation arises where, because the hot weld metal has lower yield stress compared to the cold parent metal plates, deformation first takes place in the weld. Subsequently at a later stage of cooling, when the relative yield stresses become rather equal, some yielding of the parent metal takes place as a result of which the overall width of the welded plates is reduced. As a general rule, for a given plate thickness, the overall reduction in width transverse to the joint at any point is related directly to the cross-sectional area of the weld. Thus, the total shrinkage increases with the thickness of the plate as the weld area is greater.

Apart from longitudinal and transverse distortion as mentioned, possible angular distortion and longitudinal bowing should also be taken into consideration mainly because: (i) shrinkage is not uniformly distributed along the neutral axis, and (ii) weld cools progressively, not all at one time. Due to edge preparation, the included angle in a butt-joint may form 60° included angle which would make the width at the top of the joint greater compared to the root. As the shrinkage is proportional to the length of cooling metal, the contraction would be greater at the top of the weld. In fabrication work, the parent metal plates are free to move and, therefore, would rotate with respect to each other resulting in such angular distortion (Fig. 8.8), as would not be acceptable for fabrication work. Remedial measures, therefore, should be devised. Clamps may be used to restrain the movement of plates or sheets making up the joints, but this is not always possible. Instead, a better way of dealing with such problems of distortion is to devise suitable sequential welding procedure to balance the amount of shrinkage about the neutral axis. There are two ways for performing this: (i) welding should be carried out on both sides of the joint; (ii) an edge preparation that gives more uniform width of weld through the thickness should be used. In the direction of welding, asymmetrical shrinkage shows up as longitudinal bowing. This is the sum total of building-up effects as the heating and cooling cycle makes progress along the joint. Some control against bowing can be achieved by welding short lengths on a sequential basis. Welding both sides of the joint could be a corrective measure against bowing, but this could be accompanied by local buckling. Angular distortion and longitudinal bowing are also observed in joints made with fillet welds. Distortion is readily visible in fillet welding in the form of reduction of angles between the plates and is mostly for the first run. The second weld run on the other side of the joint tends to pull the web plate back into line, but the angular rotation would be less this time. The solution to this problem lies in holding the web plate at an angle of more than 60° before the first weld and then ending up after the second weld with web and flange at right angles probably with slight warping at the flange plate.

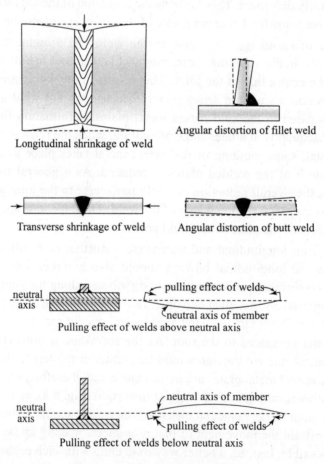

Fig. 8.8 Shrinkage and distortion

Residual Stresses

As mentioned earlier, the hot weld metal is not free to shrink longitudinally on cooling as the same is attached to the parent metal plates on either side rigidly. The weld metal is plastically deformed by being held to the original length. In other words, the tensile forces that are developed in the weld region are being balanced by the compressive forces in the parent metal plates. Regarding shrinkage transverse to the weld, the considerations mentioned before should apply. Here, the contracting weld metal tries to pull the parent metal plates towards the centre line of the joint. This results in transverse tension in the entire area. The stresses left in the joint on completion of welding are referred to as residual stresses. Regarding longitudinal stresses, the weld metal plus some of the plate that has been heated are at or near yield stress level. Moving out into the plate from the HAZ, the stresses first fall to zero. Beyond this, there is a region of compressive stress.

All fusion welds not subjected to post-weld treatments contain residual stresses. The procedures that are developed to counter or minimize distortion may change the distribution of the residual stresses without eliminating them or reducing their peak level. This should not be any cause of concern regarding welded joints for such fabrication work as that of building frames, storage tanks, low-pressure pipes, etc.

Residual stresses in certain cases are undesirable. In such cases, the designer should be apprised of the requirements so that appropriate materials may be selected on the basis of safe working stress. Some kind of structures like bridges, earthmoving equipment, and cranes are subjected to fluctuating loads resulting in the formation and propagation of fatigue cracks. The designer selects materials to conform to the requirement of the working stress range to obviate fatigue cracks. But there would be cases where residual stresses in the welded joint need to be reduced like in pressure vessels where stress-relieving is a statutory/insurance requirement as risk exists of disastrous failure of brittle fracture. The stress corrosion is the process in which the damage caused by stress and corrosion acting together greatly exceeds that produced when they act separately. Some metals corrode rapidly in certain environments in the presence of tensile stress causing stress corrosion. In such cases, the metal should be stress-relieved. In addition, in cases where the welded components are machined, removing metal in layers near the welded joint could disturb the balance between the tensile and compressive residual stresses resulting in further deformation or warping. In all such and similar cases, stress-relieving would be required.

Stress-relieving

Of all the methods in practice for reducing the level of residual stresses, the most common method is thermal stress relief based on controlled heating and cooling cycle. The other methods are overloading and vibratory treatment. The thermal stress relief method is based on the fact that the yield stresses of a metal decreases as its temperature is raised. If a welded joint is heated to a temperature of as much as 650°C, the residual tensile stress that was equivalent to the yield stress at room temperature would now be in excess of the yield stress of the metal at the raised temperature. Plastic deformation occurring locally reduces tensile stresses. Simultaneously, the compressive stresses that were in equilibrium with the tensile stresses are also reduced so as to maintain equilibrium. In stress-relieving, the temperature is raised until the yield stress has fallen to a low value at which residual stresses can no longer be supported. This, however, depends on the metal being treated as its alloy content critically influences the equation between temperature and yield stress.

In thermal stress relief method, there remains the possibility of development of new residual stresses if differential expansion and contraction occurs. A uniform temperature must be maintained throughout the component by carefully controlled uniform heating and cooling. There are special furnaces equipped with comprehensive temperature control systems wherein the entire component or fabricated item may be heated thereby eliminating the problem of temperature gradients. Although localized heating is not recommended for stress-relieving, local stress relief for pipe joints in-situ is allowed on condition that the temperature distribution would be controlled by specifying minimum temperatures at points remote from the weld. Localized heating for the purpose of stress-relieving especially in joints in flat plates is not recommended because of the risk of creating further stresses.

Mechanizing Arc Welding

A welder's skill and ingenuity can best be utilized to the best advantage in versatile manual arc welding in relatively short run, constant changes in the position of welding, movement across the shop floor from one fabrication job to another, and a mix of different types of work or joints. In addition, higher speeds of welding and greater disposition rates can be achieved by mechanizing welding processes rather than resorting to manual welding.

High welding speeds involve (equation formulated under 'Heat Input' in this chapter) more than just travelling faster along the joint. For a joint to be sound, the minimum heat input for fusion must always be in excess. As the travel speed is raised, the rate of heat input to the arc must also be increased. That is, it would be necessary to use high current to achieve high welding speed as making large changes in arc voltage is not possible in welding. The current must be increased to a value between 500 and 1000 A to ensure more rapid electrode melting to deposit significantly larger weld runs. Currents of such range are not easily manageable manually. In case of normal manual welding work, the current values vary between 100 and 350 A. For currents of magnitudes such as 500–1000 A, some form of mechanization is required for controlling the arc length, feeding the electrode, and traversing the joint.

TIG welding can be mechanized by a method of moving the TIG torch along the joint line at a fixed height above the work thereby maintaining a constant arc length assuming that the work is flat. The abutting edges very often lift and distort under the influence of heat from the arc. For mechanized welding, special jigs are used to clamp to prevent the lifting of edges. In case the arc length control is critical, the torch is to be held in a clamp that can be raised or lowered in response to commands from a servo-control system linked to the arc voltage.

A continuous electrode is used with a self-adjusting arc in MIG/MAG welding. Self-adjustment works only when fast electrode feed rates are maintained with small-diameter electrodes. The use of larger-diameter electrodes would be advantageous in comparison as higher currents and feed rates can be achieved at slower wire feed speeds. This simplifies the construction of the wire drive unit. On mechanization, the electrode is fed into the arc area by means of a motorized drive system, which is used in SAW. The arc operates hidden under a protective layer of granulated flux. The voltage, current, and travel speed are set on the control panel. The current from the power source is maintained at the predetermined value. Arc length control is achieved by monitoring the arc voltage and adjusting the wire feed speed to conform to the norm. If the arc length increases, the voltage rises. The control system restores the system to preset values.

SAW is the mechanized system generally in use for steel fabrication work like girders (box and I-sections), longitudinal butt joints of pipes, and many other products where long straight joints are required. The currents required range from mere 200 A to as high as 1000 A, but currents used in practice range from 450 to 850 A as SAW ceases to be advantageous over MIG/MAG below 450 A. As the electrode is melted, it is fed into the arc by a servo-controlled motor. There are a number of alternative traversing mechanisms capable of moving the complete welding head including wire head, electrode spool, and control box at a stable controllable speed along the joint line. The opposite of all the alternatives is to keep the welding head fixed and the joint line be moved under it. This technique is very much used when welding large-diameter pipes or shells.

Welding at about 1000 A leads to deeper penetration apart from the increase in welding speeds. This would have bearing on the dimensions and type of edge preparation for joints that are to be welded by SAW. A square edge can be used up to a thickness of 10 mm. The included angle in both single-V and double-V can be reduced to 40°. The double-V can be used up to a thickness of 30 mm. Beyond this thickness, it is better to switch to single- or double-U type edge preparation. Consumption of electrode should be kept in view while selecting the type of welding or type of edge preparation. It is necessary that the difficulty experienced in controlling root penetration when using high currents is not overlooked.

The advantages of mechanizing SAW may be summed up as: (i) use of high currents resulting in faster welding speeds; (ii) uninterrupted welding of longer lengths; (iii) controlled arc voltage, arc current, and wire-feed speed that are essential for traverse speed assure consistent weld quality; (iv) less shrinkage and distortion because of smaller heated bandwidths because of high welding speeds; (v) apart from larger deposition rate because of high currents, deeper penetration is seen in the welding of T joints. Disadvantages are related to holding of a larger weld pool in place because of high currents. If any problem arises, the current is to be adjusted. High currents also create problems in depositing root runs. This problem in butt-welding can be solved by welding from both sides or using backup plates.

The defects in welding are:
- Undercut
- Hydrogen cracking
- Porosity
- Lack of fusion
- Slag inclusion – can occur when multi-pass welds are used due to incomplete clearance of slag from earlier weld runs – treat like lack of sidewall fusion
- Root concavity
- Excessive penetration
- Excessive reinforcement
- Lack of penetration
- Lamellar tearing

The defects induced by poor process techniques are as follows:
- Undercut – weakens joints
- Overlap – weakens joints
- Porosity
- Lack of sidewall fusion – can be tolerated provided it occurs infrequently along the weld and is of limited size
- Incomplete penetration – poor technique plus poor weld design/both need to be corrected/back gouging root runs will often cure the problem and should be carried out and regarded as good welding practice
- Crater pipes
- Worm holes
- Lamellar tearing
- Hot cracking
- HAZ cracking (refer to page 275) – seriously undermines the structural strength of the joint similar to hot cracking and lamellar tearing

Critical highly stressed joints may require much higher standard of weld integrity than other joints. It is not economic to demand perfect results for all weld situations. Visual inspection of weld is important as the size and physical appearance of welds is some indication that the joint is sound.

8.3 QUALIFICATION OF WELDERS

The welding quality is supposed to be as good as a welder's skill and ingenuity. Hence, it is necessary that a welder is tested for evaluating his skill before he is deployed in actual welding work. Simultaneously, it would be a good idea to test the welder's understanding of safety, consumables, weld defects, approvals, and the welding process although such test is not mandatory. Welding in actual fabrication depends on the performance test of the welder on the basis of the Welding Procedure Specifications (WPS).

The following are the qualification tests for manual and semi-automatic welders as per the American Welding Society (AWS):
- Groove weld qualification test for plate of unlimited thickness
- Groove weld qualification test for plate of limited thickness
- Fillet weld qualification test for fillet welds only
- The pipe or tubing qualification tests for manual and semi-automatic welders: (i) groove weld qualification test for butt joints on pipe or square/rectangular tubing; (ii) groove weld qualification test for T, K, or Y connections on pipe and rectangular tubing; (iii) groove weld qualification test for butt joints on square or rectangular tubing tested on flat plate

According to the definitions of the AWS, a joint between two components lying approximately in the same plane is termed butt joint. A groove weld is made in the groove between the two components to be joined.

Position of Test Welds

There are four weld test positions for groove and fillet plate welding – 1G to 4G/1F to 4F. A welder who is qualified for the 1G (flat – weld metal deposited from the upper side on horizontal plates) position welding is also qualified for the 1F (flat – axis of weld horizontal with throat approximately vertical) and the 2F (horizontal – deposited on the upper side of the horizontal surface) positions fillet welding. Qualification in the 2G (horizontal – test piece approximately vertical with horizontal groove) position qualifies for 1G and 2G positions groove and 1F and 2F positions fillet welding. Qualification in the 3G (vertical – both test pieces and groove vertical) position qualifies for 1G, 2G, and 3G positions groove and 1F, 2F, and 3F (vertical – both test pieces and axis of weld vertical) positions fillet welding. Qualification in the 4G (overhead – test pieces approximately horizontal with weld deposited from the underside) position qualifies for 1G, 2G, 3G, and 4G positions groove and 1F, 2F, and 4F (overhead – underside of the horizontal surface against vertical surface) positions fillet welding.

The following are the weld test positions for groove pipe welding. A welder who is qualified for the 1G (pipe horizontal rolled – pipe axis horizontal and groove approximately vertical) position welding is also qualified for the 1G (pipe horizontal rolled) position butt and 1F (flat) and 2F (horizontal) positions fillet welding of pipe, tubing, or plate. Qualification in the 2G (pipe vertical with axis of weld horizontal) position qualifies for 1G (horizontal rolled), 2G (pipe vertical) position butt and 1F (flat) and 2F (horizontal) positions fillet welding of pipe, tubing, or plate. Qualification in the 5G (pipe horizontal fixed with groove approximately vertical – pipe is not rotated during welding) position qualifies for 1G (pipe horizontal rolled) and 4G (overhead) position butt and 1F (flat), 3F (vertical), and 4F (overhead) positions fillet welding of pipe, tubing, and plate. Qualification in the 6G (pipe inclined fixed – inclined at 45° with the horizontal and not rotated during welding) position qualifies for all position butt and all position fillet welding of pipe, tubing, and plate. Qualification for T, K, or Y connections in the 6G (inclined fixed) position qualifies for groove T, K, and Y connections, butt and fillet welding in all positions of pipe, tubing, and plate.

The following are the weld test positions for fillet weld test. A welder who is qualified for the 1F (flat) position is also qualified for the 1F (flat) position fillet welding of plate, pipe, and tubing. Qualification in the 2F (horizontal) position qualifies for 1F (flat) and 2F (horizontal) positions fillet welding of plate, pipe, and tubing. Qualification in the 3F (vertical) position qualifies for 1F (flat), 2F (horizontal), and 3F (vertical) positions fillet welding of plate, pipe, and tubing. Qualification in the 4F (overhead) position qualifies for 1F (flat), 2F (horizontal), and 4F (overhead) positions fillet welding of pipe, tubing, and plate.

Qualification on groove plate test weld in 1G (flat) or 2G (horizontal) positions also qualifies for butt welding pipe with a backing in the same position qualified. If no backing is used in the groove plate test weld; this would also qualify for butt welding pipe with or without backing in the same position qualified.

The joint details for the qualification test of a welder: 25 mm plate, 20° included angle, 16 mm root opening with backing. Backing must be at least 10 mm × 76 mm if radiography is used for testing without removal of the backing. Backing must be at least 10 mm × 38 mm for mechanical testing or if radiography is used after removing the backing. The minimum length of the welding groove should be 380 mm. This test qualifies a welder for groove or fillet welding with materials of unlimited thickness. The base material must conform to the WPS.

For mechanical testing, two side-bend specimens should be cast from the test joint about 100 mm apart at the mid-length of the weld and prepared for testing. Radiographic testing of the test joint may be made at the contractor's option instead of the guided bend test.

For side-bend test, the specimen is bent using a jig. The convex side of the specimen should be examined for the appearance of cracks and other open defects. Any specimen in which a crack or other open defect is present after the bending (exceeding 3 mm), when measured in any direction, is deemed to have failed. Cracks that occur in the corners of the specimen during testing need not be considered.

Defects detected by radiography are porosity, slag inclusions, cavities, and lack of penetration. Cracks and lack of fusion are also detected if oriented correctly with respect to the beam. The welder's qualification should be considered as remaining in effect indefinitely unless the welder is not engaged in the process of welding for which he is qualified for a period exceeding 6 months.

8.4 SUPERVISION OF WELDING WORK AND APPROVAL

Regular inspection of welded structures, and not only after welding is completed, is the best way to assure product quality. A better approach to quality control is to resort to preventive inspection by arranging constant check-ups as welding progresses. Early detection of welding defects costs less in rectification of the same.

The welding is deemed satisfactory if the welded structures in buildings, bridges, etc. continue to perform the intended job indefinitely. During inspection, the supervisory personnel look for weld defects. If defects are found, they are adjudged 'acceptable' or 'unacceptable' on the basis of the estimate of possible influence of such defects in service. What is of 'acceptable' quality is, however, difficult to define. Usually, the supervisory personnel attempt to relate this definition to the presence of defects such as porosity, undercut, lack of bonding between filler metal and parent metal, cracks and pieces of slag trapped in the weld. Each of these defects does not make any weld liable for rejection. A specification should spell the kind and size of defects that would make any weld 'defective' and, therefore liable for rejection.

For using welded structures with confidence, there should be some way of testing whether unacceptable defects are present in the weld. Testing should be non-destructive (NDT) so that welded structure, if test results are found satisfactory, must be deemed 'fit' for use. It is wrong to assume that NDT consists of only radiography or ultrasonic flaw detection. However, before conducting any tests, an experienced inspector should examine the joint visually to discover defects by the naked eyes for repair and rectification at the early stage.

Attention is paid to three aspects in visual inspection:
- Dimensions of the weld particularly in fillet welds
- Penetration in joints welded from one side
- Surface defects

Preventive inspection involves all concerned in sharing responsibility in systematic observation of welding practices and adherence to specifications before, during, and after welding so as to visually detect and stop occurrences that might affect the quality of welds.
- Proper included angle, root opening, root face, proper alignment and cleanliness of joint should be checked before starting the welding operation.
- Cleanliness, proper size and type of electrode, proper welding current and polarity, proper tack welds, good fusion, proper preheating and maintaining of appropriate temperature level, proper sequencing of passes, proper travel speed, absence of overlap, tilt of crater in vertical welding, filled craters, absence of excessive undercut, and absence of cracks during the welding operation.
- Cleanliness of joint, good fusion, absence of overlap, filled craters, absence of excessive undercut, full size on fillet welds, and absence of cracks after the welding operation.

Even with good level of illumination, it is difficult quite often to detect cracks by visual inspection. This problem can be overcome by highlighting the cracks by using either dye-penetration test or magnetic particle test.

Dye-penetration Test

If a liquid of low surface tension is mixed with dye and poured on to the surface of the welded joint, the liquid, apart from wetting the surface, would seep into cracks, if any. Wiping of the wet surface would leave only the liquid in the cracks. If any absorbent layer of chalk is deposited on the dry surface, chalk would suck the liquids from the cracks. The coloured stains on white chalk would show the locations of cracks.

Magnetic Particle Test

Magnetic particle test is an alternative method of locating surface cracks in welded joints by creating magnetic field in cracks so as to attract iron oxide particles. The lines of force within a magnet in the shape of a bar run from one end of the magnet to the other (south to north poles). A magnetic field exists around the magnet concurrently. Iron powder sprinkled on to the bar gathers around the poles where the lines of force are close together. When another similar bar magnet is placed in its proximity so that its north pole is opposite the south pole of the first magnet, the lines of force flow between the two magnets. If iron powder is sprinkled on to the new arrangement, it collects in and around the gap between the north and south poles. Iron or iron oxide particles deposited on the weld gather at the point where the magnetic field leaks thereby indicating the presence of a crack.

Radiography

Visual, dye-penetration and magnetic particle tests would provide us information about defects that appear only on the surface whereas radiography is used to locate defects in the body of welds. The detection of defects in welds by radiography is based on the ability of X-rays and gamma rays to penetrate materials that are opaque to ordinary white light. Both are electromagnetic rays having characteristic wave lengths: (i) X-rays: 2×10^{-12} m to 10^{-8} m; (ii) gamma rays: 10^{-13} m to 10^{-12} m. X-rays are produced easily in a suitable vacuum tube. Gamma rays are emitted by radioactive materials.

When transmitted through any material, X-rays are absorbed. If such material is homogeneous in nature, the amount of absorption is uniform across the area exposed to the X-ray beam. If the material, on the contrary, contains a pore of gas for example, a smaller amount of rays passing through this point is absorbed and there is variation in the intensity of the emergent beam. This can be photographed in a film placed on the side of the material opposite to the source of the radiation. On a negative film, this shows up as a dark spot of the same shape as the pore. Defects in welded joints generally contain gas, air, or slag, which have appreciably smaller absorption coefficients than the parent metal. Their presence is thus easily detected as there is a marked variation in density in the film. Tungsten inclusion is an exceptional case as this metal is denser than those normally welded and shows up as a bright spot on the negative film because of absorption of more X-rays. One of the advantages of radiography is that the film can be saved as a permanent record of the quality of weld.

Some elements having unstable atomic nuclei disintegrate with time. Such disintegration is accompanied by the emission of radiation a part of which is composed of high-energy gamma rays, which can penetrate metals like X-rays. Gamma rays, therefore, can be used for NDT in welds. 'Artificial' radioactive isotopes, which are obtained by fission or irradiation in a nuclear reactor, are source of gamma rays used for NDT weld inspection. Isotopes that are used are iridium-192, cobalt-60, caesium-137, and caesium-134. Isotopes are small and, unlike X-ray sources, do not need any power supply unit. They are portable and need to be handled with care and stored in lead-lined containers. Gamma rays radiate in all directions and, therefore, can be used in the inspection of pipe welding. The radiation source can be located in the bore with the film wrapped around the outside of the pipe.

As exposure to both the X-rays and gamma rays are extremely harmful to human beings, all possible safety measures should be taken if radiographic inspection of welds is to be carried out.

Ultrasonic Inspection

This is a highly sensitive method of detecting flaws inside a metal by observing the way in which high-frequency vibrations are transmitted through the metal. The vibrations are reflected if there is any flaw or defect in the material being tested. The existence of flaw can be indicated by the following: (i) noting the strength of the transmitted vibrations, or (ii) monitoring the reflections. This is typical of a wide range of vibration frequencies starting from as low as 16 Hz to above the normal range of hearing at or about 20 kHz depending on the individual person. Reflections from an interface still occur at higher frequencies such as the ultrasonic range, but it is not possible to recognize the effect with unaided ear. It is necessary to use some sort of sensor. In the ultrasonic testing of metals, frequencies used range between 5 and 10 MHz and the vibrations are measured by using the piezoelectric crystal properties that may be shaped and used as a resonant circuit element or transducer (a device that converts physical magnitude of one form of energy into another, generally on a one-to-one correspondence).

8.5 HANDLING AND TRANSPORTATION OF UNITS TO BE ERECTED

Large structures like heavy-duty columns, trusses, trestles, overhead bridges, girders, etc. as fabricated at shops are transported in pieces as 'shipping units', which are pre-assembled at the sites before erection. This is done as per the approved working drawings. Working drawings should be prepared keeping in view the resources available and the mode of transportation. The available transport should be large enough to accommodate the largest pieces and strong enough to carry the heaviest loads. Fabricated members or pieces should be stiffened and guarded so as to avoid deformation during handling, loading, transportation, and unloading. Even then, there would remain the risk of steel structures becoming warped, dented, or bent. Sheet constructions can additionally develop changes in the radius of curvature. All these must be rectified prior to pre-assembly or shifting for actual erection.

Details of the plants for transportation, movement, and handling of steel structures are given in the Section 13.5 of Chapter 13.

8.6 ERECTION OF FABRICATED STEEL STRUCTURES

The erection of structural steel can be divided into two separate operations: (i) planning in advance at design stage, and (ii) site operations.

A designer should have a clear idea about the safe erection of structural steel stage by stage within the least possible time as the success of site operations depends to a large extent on the care and thoroughness of advance planning. For this, the designer should be sufficiently knowledgeable so that the advance planning could be meaningful. Huge bridge and complicated towering building work would require preparation of many drawings and layouts to guide the field personnel of each step. The sequence of shipping of fabricated structural elements must follow logical patterns based, of course, on site conditions and erection schedule. The size and weight of each individual piece must be within the fabrication capacity, shipping limitations, and the capacity of construction equipment. Accordingly, the methodology of erection of fabricated structural steel should be contemplated in advance and communicated to the fabricators and erectors through specifications and notes in drawings. Such notes should indicate if any stiffening or additional temporary support would be necessary during transportation, handling, pre-assembly (if and when required), and erection. If the site conditions require deviation of the methodology contemplated, then there should be revised methodology agreed upon by all concerned and documented.

In case of building or equipment foundations, level pads are made on the foundations maintaining precise levels so that the question of levelling does not arise in erection of columns. However, proper attention is called for in selection of the type and size of lifting equipment based on the number of pieces to be lifted into place, their average height, weight, size, and final position of the maximum lift. It would be necessary to optimize the choice by arranging that every lift matches the capacity of the crane at that particular jib length and radius. A crane should be deemed as under-utilized if it lifts anything that weighs less than its rated capacity.

Considerable economy can be effected, time saved in erection, and safety of personnel working at heights may be assured if pre-assembling of small pieces could be arranged at the ground level before lifting to heights. This has to be considered at the design stage so that there would not be any problem in splicing of joints from one side without the necessity of turning the whole assembly. If there is appropriate platform or floor at heights, pre-assembly can be performed at that height also, but small pieces would then be lifted by under-utilizing the rated capacity of the crane unless the small pieces are bundled and lifted thereby reducing the number of such lifts.

When erection of steel structures is in progress, there would possibly be other erectors or contractors who would be working at the same locations. It would, therefore, be necessary that incomplete work is stable and would not jeopardize safety of the work or personnel of other contractors. Accordingly, if required, additional temporary works should be carried out to ensure that the incomplete erection work remains stable.

The work involved in the erection of steel structures comprise: (i) receiving and unloading, (ii) storage after sorting, (iii) shifting to the location for erection, (iv) erection, (v) lining, levelling, and plumbing, and (vi) fastening. For the details on the construction equipment required for the erection of steel structures, refer Section 13.5 of Chapter 13.

Receiving and Unloading

Fabricated structures or other components when unloaded at a site need to be stored at a storage yard till the time they are required for erection. Each fabricated component is identified by a mark number indicated in the corresponding fabrication drawings. Components/structures should be stored in such a way as to allow easy spotting and handling when required. Erection is possible when all the required components are available in the storage yard. As shipments can be made by truck, railways, or barge, the location of the site for receiving materials would depend on the mode of shipment. Generally, trucks and crawler cranes are deployed for unloading purposes. Derricks may also be used instead of mobile cranes.

Storage after Sorting

Fabricated steel structures that are shipped are stored in some orderly pattern on receipt at site after being sorted on the basis of erection priority. This kind of storage is possible only at the erection site and not in any fabrication shop. High-rise building columns are stored floor-wise; beams are also stored in bundles floor-wise. In case of bridges, the individual components are stored on the basis of erection sequence.

Shifting to the Location for Erection

Shifting fabricated components or structures from the storage yard to the actual location of erection can be effected using trailers and trucks. For large projects, railways with captive engines may be used. For bridge construction over water, materials are moved on rail tracks or on barges for directly moving the materials to the erection crew.

Erection

For actual erection, the skilled crews need to refer to the fabrication drawings and identify the part structures or components on the basis of fabrication drawings and then erect them according to the markings. Fabrication drawings display how each individual component fits into the structure. The crew at the lower level selects the correct piece for lifting and the crew at the top guides the structure/piece to the exact location and secures it in position.

Lining, Levelling, and Plumbing

Despite erection of column bases on level pads and careful positioning of columns on the basis of centre lines, the structure may still need adjustment during lining, levelling, and plumbing. This would call for corrective action that begins with checks for conformity with approved drawings as the structures are fabricated and erected true to dimensions and details as per the drawings issued for fabrication. There should not be any error made during erection. It would be necessary to check alignment and levels thoroughly before the bolts are tightened finally. The alignment is checked using piano wires stretched truly parallel to the column axes. The verticality of columns or the components is checked using heavy plumb bobs or theodolites/modern survey instruments. The wedges are used for adjusting levels and a sufficient number of wedges should be placed so that level is not disturbed. Optical or laser plumbing units are available for use instead of plumb bobs especially in case of high-rise buildings. The columns should be connected with bracings and beams at the earliest for stability.

Site welding is performed in prevailing weather conditions and quite often in difficult positions thus making site welding more difficult than shop welding. In a fabrication shop, the conditions may be made favourable for welding work. Site welding in most cases is positional and is performed by manual metal arc welding as its use is more flexible and enables the performance of good fillet welds in the vertical up and overhead positions also. Safe means of access and working platform should be provided for the welder and also for the necessary equipment and materials required by him.

Wherever possible, site splices should be designed as bolted connections. Bolting is comparatively an easier way of joining the components. It is not affected by adverse weather conditions and needs simple tools/plants for fixing and tightening. When small quantities of HSFG bolts are to be used, manual torque wrenches are used. Where a large number of HSFG bolts are to be tightened, impact torque wrenches are very much needed.

HSFG bolts are first tightened to bring the plates together and then the torque wrench is used to make sure that the required torque has been reached. Otherwise, load-indicating washers may be used to indicate that the bolt has been tightened to the requisite tension. In case of bridgework, it is usually necessary to place drift pins and fit bolts in a connection before the component can support it and the derrick is then released. For this reason, the fitting-up crews follow immediately behind the connectors placing pins and bolts.

Fastening

The final operation in steel erection is the placing of the permanent fasteners: rivets, pins, and bolts. Even in case of welded connections, the elements are to be placed in positions and then temporarily connected with bolts. Welding is performed only when the temporary bolted connection is found in order conforming to the requirements of drawings and specification. All deviations must be within specified tolerances.

8.7 ERECTION OF PRECAST REINFORCED CONCRETE STRUCTURES

Leaving aside the exceptional cases, procedures for structural precast concrete erection are similar to those followed for structural steel erection. Precast concrete elements are not designed for resisting tensile stresses that result from indiscriminate handling and are also vulnerable to damage from possible impact. Maximum efficiency in erection is achieved by placing the precast elements in their final positions directly from the transport vehicles/plants or storage area in single operation. For this, both design of the precast elements and detailed erection procedures should be planned in advance so that unforeseen problems do not crop up during erection resulting in delay in work progress. The items that need to be considered during planning are: (i) sizes of precast elements, (ii) weights of precast elements, (iii) lifting capacities of construction equipment, (iv) reaches of lifting equipment, and (v) availability of clear access required for movement and manoeuvring for erection without interfering with the erected structural members. This may require review of the time schedules of all project activities.

For efficient use of lifting equipment, temporary devices are so designed as to ensure automatic aligning and levelling of the lifted elements so that lifting equipment would not be required again for adjustment work.

Temporary bracings for columns, frames, and slabs are designed to take both tension and compression so that wind loads and unbalanced loads are taken care of before final aligning, levelling, and plumbing of the erected precast concrete elements are completed. 'Push–pull' braces, yokes, clamps, etc., are used to facilitate connection of temporary components and must be able to withstand wind and unbalanced loads.

Precast concrete construction is controlled operation executed under ideal conditions of formwork, casting, vibration, curing, and inspection. The design considerations in case of precast concrete involve analysis that is not required in conventional design practices. The design of precast components calls for design of connection details and an investigation of stresses involved in the sequence of erection, which influences the framing and design of the elements. The designer must be familiar with the various steps of construction and erection and should meticulously plan the complete and specific design details accordingly.

Inserts are embedded in precast elements for: (i) attaching lifting devices for handling, (ii) temporary bracing of precast elements during erection, (iii) making joint connections, and (iv) attaching hangers, brackets, etc. for mechanical and/or electrical installations. The inserts should be fixed at the correct locations very carefully so that they are not disturbed during concreting. Where holes are required for through bolts, pipe sleeves are used in formwork.

A rigid structure is formed by assembling individual floor, roof, wall, and supporting frame elements at the time of erection. The design of the connections should be simple and practical and should be explicitly detailed in the drawings. Wherever possible, the connection should be located at a point of minimum stress, limiting the joints to minor welding, bolting, or grouting. The connections comprise simple bearing, receiving pockets to restrain lateral movement, physical connections, or a combination of these methods. The majority of connections are physical. The use of welding plates, adjustable inserts, reinforcement splices, and stud bolts permit some tolerance and facilitate completion of the connection without much juggling of the castings.

The precast concrete erection cycle comprises: (i) slinging, (ii) lifting and delivering to a point of erection, (iii) guiding, (iv) positioning and placing it in design position, (v) temporary fastening of component, (vi) releasing sling/s, (ix) back to (i). The operations where deployment of any crane would not be necessary, a precast item in such cases would only be placed in design position and connected followed by adjustments by aligning, levelling, and plumbing.

The time required for slinging, lifting, shifting to the place of installation, and removal of slings is more or less constant depending on the slinging devices and operator's skill. Precast concrete structures or parts, packaged materials, and other loads are slung by various devices like flexible slings, spreader bars, and grabs.

Flexible slings (Fig. 8.9) are made of steel wire ropes (cables). The slings are used to lift light columns, beams/girders, wall panel slabs, roof planks/slabs, pipes, containers,

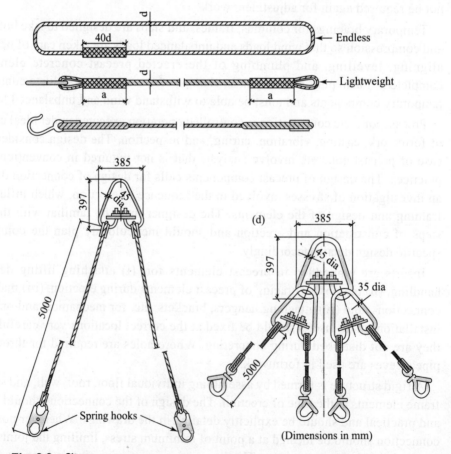

Fig. 8.9 Slings

buckets, hoppers with concrete, and other components. Slings may be of many designs with one to six legs. There are endless slings up to 15 m in length for direct binding of components. The lighter columns are erected using a standard endless sling.

The spreader bars (Fig. 8.10) are of two types – the beam and lattice types. The beam types are made of two back-to-back channels connected with cover plates and having pulleys at their ends with slings passed through them. This ensures uniform transfer of load to all the four points of attachment. Heavier columns are erected using spreader bars with double slings attached to erection eyes. Roof beams and trusses (>12 m spans) are lifted using box-type spreader bars of appropriate lengths. The spreader bars may carry various attachments like lightweight slings, clamps for lifting roof beams, T-crane girders, column tilters, etc.

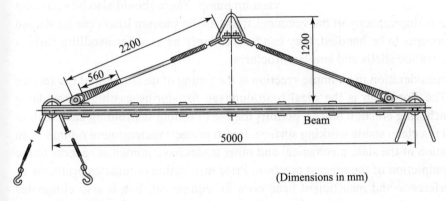

Fig. 8.10 Spreader bar

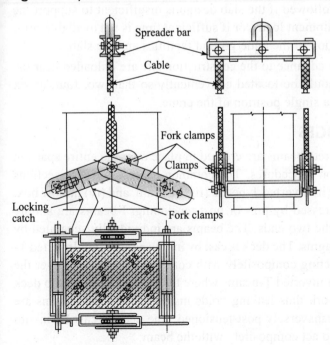

Fig. 8.11 Friction grips

Lattice spreader bars are made as metal triangular-welded trusses and used for handling long-line structures. Lightweight columns are handled with the aid of friction grips (Fig. 8.11). Heavy large wall slabs are handled with the use of lever grips which hold the slabs through friction forces. To ensure reliable gripping, the aggregate friction forces developed as jaws and forced against the load should be greater than the weight of load. Friction grips for handling slabs should have additional safety devices to hold the load in case of slippage.

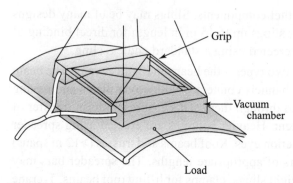

Fig. 8.12 Vacuum lifters

The vacuum lifter (Fig. 8.12) comprises vacuum chambers (cups), a hose, and a vacuum pump. The vacuum pump is suspended from a crane's hook whereas the vacuum pump and control instruments are accommodated in the operator's cabin. One or more vacuum chambers (cups) of various shapes may be fitted for lifting purposes. As the lifting depends on the working of the vacuum pump, standby safety devices should be provided for emergency like failure of the vacuum pump. There should also be sufficient margin in the lifting capacity of the vacuum lifters. As the vacuum lifters can be shaped to suit components to be handled, they can conveniently be used for handling various types of thin-walled shells and similar structures.

A major consideration in planning erection is the timing of installation of the service floor slab. The floor slab is the ideal base dunnage for storing castings, provides a means for anchoring erection braces, permits the use of rolling scaffolding and erection devices, and is a clean stable working surface. It also protects underground piping. With prior installation of the slab, mechanical and other trades have immediate access to the area upon completion of the precast erection. Prior installation eliminates problems of column interference and insufficient headroom for equipment; but, it also eliminates roof protection and exposes the slab to inclement weather and direct sun rays. This sequence obviously cannot be followed if the slab design is insufficient to support the weight of precast storage or equipment loads, or if sufficient time is not available prior to erection to complete installation of the underground facilities and the slab.

Lighter columns delivered in advance to the construction site are unloaded near the place of erection. The crane should be located conveniently so that two, four, or six columns could be erected from a single position of the crane.

8.8 ERECTION OF BRIDGES

Fabricated steel or precast concrete beams are erected in place over the entire spans in line with the bridge construction procedures. The common precast concrete sections generally in use are: (i) I-beam, (ii) inverted T-beam, (iii) M-beam, and (iv) hollow box. The beams in general are prestressed by pre- or post-tensioning. I-beams are simply supported on the abutments at the two ends. The beams are individually separated by evenly spaced transverse diaphragms. The deck is cast by in-situ concreting. Inverted T-beams are butted together for acting compositely with concrete between and over the units. M-beams are formed with inverted T-beams where the bottom slab and top deck are cast over permanent formwork thus leaving voids inside. Hollow box beams are butted against each other and transversely post-tensioned. The deck is cast by in-situ concreting and is not designed to act compositely with the beam.

Steel beams are used where simply supported spans are to be bridged. I-sections are used for bridging. The deck slab is designed to act compositely with the beams. Superstructure of a bridge can be fabricated using structural steel. Girders, trusses, arches are first fabricated and then erected. But decks in general are cast of reinforced concrete in composite construction.

The erection of single-span steel or concrete bridge between abutments can easily be lifted by mobile crane positioned on the shore or on the approach road of the bridge. This may require deployment of a crane of high capacity for just lifting beams for bridging a single span. This kind of erection is limited to the light beam bridges for cost-effectiveness.

Erection by pushing

Single- or multi-span prestressed precast, steel or box girder bridges can be erected by pushing (Fig. 8.13) using hydraulic rams or winch power if the alignment is nearly straight. In case of downward grade (4–5%), braking system should be provided. Bridge launching demands precise surveying and setting out with continuous checks made of deck deflections. Usually spans are limited to 50–60 m to avoid excessive deflection and cantilever stresses. Where the span exceeds 60 m, temporary support piers are installed. A light aluminium or steel launching nose forms the head of the deck to provide guidance over the pier. Special chrome-nickel or similar materials are used to reduce sliding friction to about 5% of the weight. Segmented girders are joined and prestressed before initiating launching operation.

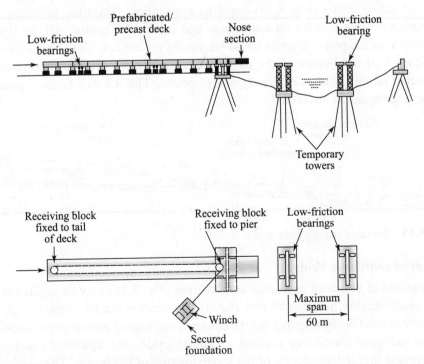

Fig. 8.13 Erection by pushing

In composite design, the deck takes care of the compression leaving the tension to be taken care of mostly by beams. The decks are cast in-situ and the beams, especially steel beams, are likely to deflect when loaded. As a preventive measure, camber is induced in the beams before concreting the top deck. Each beam is jacked up equally against a trussed frame for camber induction. The frame is supported on the piers/abutments.

Erection using crane/shear legs

In long bridges with several non-standard spans, truck-mounted strut-jib cranes are available with capacities up to 2000 tonnes. Truck-mounted hydraulic-jib cranes are more versatile and can lift more than 800 tonnes. Such cranes can lift beams/trusses in positions from pontoons located between piers. The erection of beams can also be done economically using shear legs mounted on pontoons (Fig. 8.14). But hostile currents and tides present severe problems in both launching and transporting operations. Erection, therefore, should be planned in fair weather.

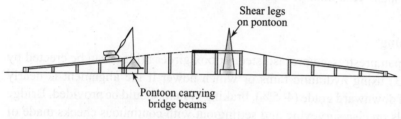

Fig. 8.14 Beam erection using crane/pontoon

Erection using gantry girder

For long bridges comprising a number of spans, erection of beams by using launching truss or gantry girder (Fig. 8.15) could be economical. The truss or gantry girder is supported on two trestles on either side and one at the centre. All the trestles are supported on the piers. Trestles could be moved on rails on piers across the deck for placing of all the beams of the particular span. Beams are transported on bogies supported on rails laid over already placed beams. This kind of erection procedure is suitable for spans of 30–60 m.

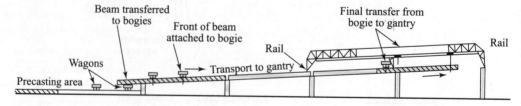

Fig. 8.15 Erection using gantry girder

Balanced cantilever bridge

The method of balanced cantilever construction (Fig. 8.16) may be applied to bridges with spans ranging between 50 and 150 m. The method can be applied to cast-in-situ concrete or steel bridges apart from the common segmented precast prestressed bridges. In the balanced cantilever method of construction, the abutments and piers are constructed first independently of the superstructure of the bridge. The segment just at the top of each pier may be cast-in-situ or erected as a precast unit. The bridge deck is formed thereafter by adding segments on either side of the pier asymmetrically maintaining balance. In practice, the segments on both the sides cannot be erected simultaneously resulting in imbalance for a while due to the weight of a segment, materials, construction plant, and wind. This is to be considered in design. The two individual cantilevers are linked at the centre by a key segment that is normally cast-in-situ. At the abutment end when the beam is shorter than the span, the overturning moment must be resisted by a downward reaction provided by anchor/support. The application of prestress requires the supporting column and the adjacent one or two segments to be temporarily supported and then post-tensioned. Continuity tendons are prestressed on completion of casting of the key section joining the cantilevers from the two sides.

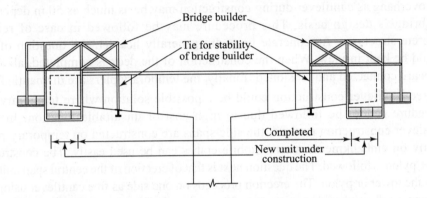

Fig. 8.16 Balanced cantilever bridge

Cable-stayed bridge

Cable-stayed bridges (Fig. 8.17) are suitable for spans up to about 800 m. Such span along with approach roads on either ends make it possible to bridge long spans, which is more than what can be covered economically by girder- or suspension-type bridges. Deck segments, which are made of precast or cast-in-situ concrete or steel box, are supported and stayed to a tower/pylon. This kind of arrangement provides compression in the deck by utilizing self-weight. The deck may be made up of individual segments in this way to be ultimately shaped up so as to act like a prestressed beam.

The pylon (tower) may be fabricated from structural steel, precast concrete elements, or may even be cast-in-situ concrete. Similarly, the deck may also be assembled using precast concrete elements or structural steel plates or girders or may even be cast-in-situ. The most common form being the box section which offers most torsional restraint. The cable material is similar to that used for prestressing work and comprises multi-strand cable made of cold drawn wires or single-strand cable consisting of parallel wires. Protection against corrosion can be avoided by using galvanized wires but a more time-tested practice is to cover the cable in steel or plastic duct/sheath and then finally injecting grout under pressure on positioning the cable in place.

The cable is generally connected to the pylon with pin-type joints or as shown in the approved drawings. The appropriate erection method is controlled by the stiffness of pylon, cable anchorage system, scope of installing temporary supports, maximum unsupported spans, ease of transporting materials, etc. The system's stability depends to a large extent on the transfer of the cable's horizontal component of the force through the stiffening girder; it is essential that girder continuity is maintained between each pair of stays. This may be achieved by different procedures: (i) erection on temporary staging, (ii) free cantilever with progressive placing, and (iii) balanced cantilever.

Temporary staging (piers) would be required for erection of cable-stayed bridge, if the pylon is so designed as to be stable with anchor cables fixed. Temporary piers are to be erected first on solid base so that it can support a derrick-type crane to lift and place deck units one by one. The crane is to be mounted on rails for mobility. The length of a fabricated steel deck unit varies between 5 and 15 m. The length of the deck units that can overhang as cantilever during construction may be as much as 50 m depending on the bridge's design basis. This procedure may be followed in case of reinforced concrete structures. As concrete units are generally heavier, the question of lengths should be kept in view. When the construction of the deck is completed, all the cable stays are connected and tensioned. Finally, the temporary piers are dismantled.

Free cantilever construction could be a possible solution where temporary staging procedure cannot be followed due to high cost or unsuitable location. In the free cantilever construction method, the side spans are constructed on temporary propping mostly on embankments where mobile cranes can be used easily. The construction of tower/pylon is followed. The question next is that of erection of the central span unit-by-unit from the tower or pylon. The erection proceeds on one side as free cantilever using derrick or crane. Each unit could be as much as 20 m in length. When a unit is joined, the permanent cable stays are fixed. The provision of temporary stays is relevant on erection of concrete units as the weight of a concrete unit can be as much as 300 tonnes. Temporary post-tensioning is done in concrete units to bring them together. When the permanent stays are fixed, temporary stays are removed. This free cantilever procedure is not viable in case of cast-in-situ concrete.

The balanced cantilever technique of bridge construction provides uninterrupted space under the deck for movement of transports including ships. The first few units on either side of the tower are erected on falsework till such time as cable stays are attached – the erection progresses simultaneously on both sides of the tower. There may be imbalances caused by construction equipment, variation of segments' dead weights, and tension in the cables. It is necessary that the tower be stiff and well-fixed with the pier. The cable stay connecting the tower with the deck should be adequately strong.

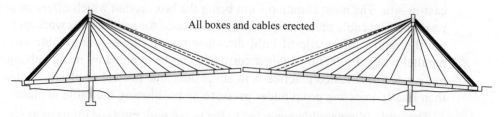

Fig. 8.17 Cable-stayed bridge

Suspension bridge

Suspension bridges comprise a main span with two side spans on either side of the main span. The bridge deck is suspended from the cables by suspended hangers because of which this kind of bridge is called suspension bridge. The cable ends are usually anchored ultimately into the ground. The main cable connecting the two towers form a catenary from which the hangers are suspended.

The tower may be fabricated of structural steel or made of precast concrete sections. It may also be constructed cast-in-situ by slip-forming. Ultimately, the required height and mode of erection would decide the most suitable tower for bridge construction.

The cable suitable for cable-stayed bridges could be used in suspension bridges also. Hanger cables are clamped directly to the suspension cable with tightly bolted bands. Steel box sections are fabricated at shops and then pre-assembled completely with footpaths and hand railings. The units can be connected together by bolting/welding followed by concrete surfacing.

The anchorage of the suspension cable is of vital importance and to an extent is similar to the anchoring of cables in cable-stayed bridges. Anchoring can be done following any of the three methods: (i) rock anchors, (ii) tunnel anchorage, and (iii) gravity anchorage. Rock anchors require drilling in rocks for grouting bolt-type anchors to which cable strands are subsequently attached. As and where solid rock is available, U-tunnel can be constructed so as to allow the two suspension cables to form a loop. The general practice, however, is to insert anchor bolts in heavy mass foundation. The cable strands are attached to the anchor bolts.

The erection of suspension bridges (Fig. 8.18) can be done by: (i) starting erection from the centre of the main span or (ii) starting erection from the towers. While erecting deck sections starting from the centre of the main span, the erected sections should remain without being connected till the last section is in place as significant displacement of the main cable and towers takes place with the progressive increase of dead load. The cable tension and shape, and distortion of the towers during erection should be estimated so that tolerances within permissible limits could be ensured.

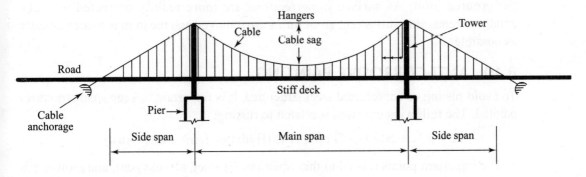

Fig. 8.18 Suspension bridge

8.9 GROUTING OF JOINTS OF PRECAST REINFORCED CONCRETE STRUCTURES

Grout is a mixture of cementitious material, fine aggregate, and sufficient water to produce pouring consistency without segregation of the constituents. The design of the precast reinforced concrete connections should be explicitly detailed in the drawings. If possible, the connection should be located at a point of minimum stress, limiting the joint to minor welding, bolting, or grouting.

The simplest method of connecting a precast column to the pier or footing is by grouting the base into a pocket in the base slab. The column is supported on the pier/footing and restricted from lateral movement by the base slab. This is a very simple form of connection requiring little manual efforts. Chamfering the column base provides more uniform bearing.

For vertical reinforcing steel to be continuous in the precast column, vertical dowels protrude from the footing. There should be holes in the precast column to accommodate the reinforcements. On erection of the column, the hole is filled up with grouting through a pour hole at the hole top. The connection can also be made by welding or bolting through plates of appropriate sizes instead of leaving dowels. Such connections also need to be protected by grouting. The question of erection, scaffolding, and grouting should be considered in planning and designing precast elements.

Continuity of reinforcing steel is required in rigid, continuous beams/girders, and other such precast elements. The continuity is maintained by welding reinforcing steel or by providing lap lengths or mechanical joints. After jointing, the reinforcing steel in all joints should be encased in the grout.

In general, floor and roof slabs are connected to the building frame. Grout keys are provided on the vertical edges to provide for load transfer between the slab elements. In joining the roof or floor slabs to the supporting frame, those connections designed for the completion work, mainly grouting, should better be carried out from the top of the erected slabs easily—the erected elements serve as a safe working platform. Wall precast elements are connected by bolting or welding with columns and floor/roof slabs.

A light wood frame covered with curing blanket or insulation-filled craft paper cures the grouted joint. As surface imperfections are more readily corrected with the condition remaining still green, grout forms are removed and the joint is treated as early as possible.

8.10 ANTI-CORROSIVE PAINTING

To avoid rusting of unprotected steel structures, it is important to keep steel structures painted. The following equation is related to rusting:

$$4Fe + 3O_2 + 2H_2O \rightarrow 2Fe_2O_3\, H_2O \text{ (Hydrated ferric oxide – rust)}$$

The important points related to this equation: (i) steel, air (oxygen), and moisture is needed for rusting; (ii) steel does not rust in dry air; (iii) under normal conditions, there is continuous supply of air and moisture; (iv) all corrosion occurs at the anode and no corrosion occurs at the cathode.

Although painting is done as an anti-corrosive measure, occasional problems arise with paint discolouring, flaking or blistering over welds, or in immediate adjacent area. Dust, smoke film, iron oxide film, grease, and such harmful materials on the surface of the weld and immediate adjacent area prevent the paint from coming in close contact with the steel surface and properly adhering to it. Steel/weld surface that has been rubbed smoothly with power wire brush could also prevent proper adhesion.

Occasional problems may arise from the welding fumes. The elements of fumes when deposited in the slag as a film on the steel surface may combine with moist air to produce an alkaline solution that reacts with paint resulting in discolouring and blistering. This problem deteriorates with increasing humidity. As the slag is generally removed in SAW, the process leaves no film of smoke or iron oxide on the adjacent plate and, therefore, submerged arc welds are relatively free of problems related to paint.

The remedial measure in overcoming paint-related problems is to be initiated with cleaning up by removing slag, spatter, smoke film, iron oxide film, and other similar materials. Apart from providing clean surface to which paint can adhere, other chemical substances that could react with paint need also to be removed. However, power brush should not be used for cleaning.

There are other measures if discolouration or blister prevails after normal cleaning. A wash in any mild acid (such as boric acid) solution followed by good rinse with clear water would neutralize the alkaline solution so that it would not affect the paint. Otherwise, highly alkaline-resistant paints like vinyl, epoxy, or chlorinated rubber base may be used.

Painting should be done as quickly as possible after welding. This would prevent the chemicals in the deposited film from absorbing more moisture. As a result, less alkaline solution would be formed to attack the paint. Two coats of paints, including an alkaline-resistant primer at the earliest, is usually better than a single coat.

The inside of closed-in hollow box structural sections could be left unpainted because, it is believed, that any slight oxidation of the steel would soon come to equilibrium as there would be no continual supply of air and moisture. The assumption that condensation in hollow steel sections is very slight is largely substantiated. Inaccessible or difficult-to-reach sections should always be welded airtight. Manholes, if any, should be closed with rubber gaskets. With these precautions, corrosion protection of the inner parts becomes unnecessary.

Surface Preparation

The removal of mill-scale that forms on the surface of the steel plates and sections during the hot rolling process is the essential element in a quality anti-corrosion treatment. The modern high-grade successful protective systems are much better than the traditional method of wire brushing. Structural steel is a product hot-rolled at about

1000°C. As the products cool after rolling, the steel surfaces react with the atmospheric oxygen to produce mill-scale, a complex oxide that appears as a blue-grey tenacious scale completely covering the surface of the as-rolled steel section. Mill-scale, however, is unstable and rusting of the steel surface occurs on weathering as moisture penetrates into the fissures in the scale. Mill-scales, should, therefore, be removed before application of protective coatings. The surface preparation of fabricated steel is basically concerned with removal of both mill-scale as well as rust. The methods used for surface preparation of structural steelwork are delineated hereunder.

Manual preparation

Although not very effective, the manual preparation is often done for economic reasons. This method involves chipping, scraping, and brushing with hand-held implements.

Mechanical preparation

This method is similar to the manual preparation except for the fact that power-driven tools like rotary wire brushes etc., are used.

Flame cleaning

In this method, an oxy-acetylene flame is applied rapidly to the surface locally by heating at about 95–150°C. Differential thermal expansion rates of the steel and scale results in the loosening of the mill-scale and simultaneously facilitates the removal of rust. The surface is then cleaned by mechanical scraping and immediately given a priming coat of paint preferably before cooling of the steel. Not all the tightly adherent scale is removed by this method. If heating is done on thin plates (<6 mm), the flame heat may cause buckling and distortion.

Acid pickling

This method is the established means of surface preparation for hot-dip galvanizing, but has not been widely used for surface preparation for painting of the fabricated structural steel. Acid cleaning, however, gives results comparable with blast cleaning.

The steel is immersed in a bath with cold dilute hydrochloric acid or hot dilute sulphuric acid for as long as required (from a few minutes to a few hours depending on the acid concentration) for the ferric oxides to be removed by chemical action. Individual components up to 12 m may be treated. The inhibitors are added to the acid to slow down attack on the raw steel once the scale and rust have been removed.

Blast cleaning

In this method, hard abrasive materials are projected towards the surface at high speed. Abrasive particles, always metallic, comprise either spherical particles termed 'shot' or angular particles termed 'grit'. The particles impinge on the steel surface, removing scale and dust, producing a rough, clean surface.

The abrasive materials are projected towards the surface either in a jet of compressed air or by centrifugal impeller wheel. The air is projected through chilled iron or ceramic nozzles. The particles impinge on the steel surface so as to remove scale and rust thereby producing a rough and clean surface. The size and shape of the abrasive materials used cause equivalent size and shape of the surface roughness—angular grits produce angular surface profiles and round shots produce rounded profile. The rates of cleaning depend on the shape and size of the steelwork and the degree of cleaning required.

Grit blast abrasive materials can be either metallic like chilled iron grit or non-metallic like slag grit, which is termed 'expendable' as the same can be used only once. They are used at project sites only. Metallic grits are expensive and can be used where the same can be recycled. Grit blast abrasive materials are used for some paint coatings particularly on site and for primers where adhesion may be a problem. It is always used for metal-sprayed coatings, where adhesion is partly dependent on mechanical keying.

Shot blast abrasive materials are always metallic, usually cast steel shot, and are used particularly on shot-blast plants, utilizing impeller wheels and abrasive recycling. They are the preferred abrasive materials for paints, particularly for thin film coatings.

In wet blasting, which is a variation of the dry blasting, a small amount of water is entrained in the abrasive/compressed air stream. Atmospheric pollutants like chlorides and sulphates form soluble iron salts during weathering. Wet blasting washes soluble salts from the surfaces. Such salts are often located in corrosion pits on the steel surface and cannot be removed by ordinary dry blasting. Wet blasting is useful especially for off-shore structures. There are two criteria for assessing the quality of surface preparation: (i) surface cleanliness and (ii) surface roughness.

Paint coatings

Painting is the main process of protecting structural steel from corrosion. Paints comprise pigments, binders, and solvents. Paints are applied to steel surfaces by many methods but ultimately produce wet film irrespective of the method used. As the solvent evaporates, the binder and pigments are left to form a 'dry film' on the surface. The corrosion protection afforded by paint film depends directly on the thickness of its dry film.

The most common methods of classifying paints are either by their pigmentation or the binder type. Primers for steel are usually classified according to the main corrosion-inhibitive pigments used in their formulation like zinc phosphate, zinc chromate, red-lead, and metallic zinc. Each of these inhibitive pigments can be incorporated into a range of binder resins, as for example, zinc-phosphate alkyd primers, zinc-phosphate epoxy primers, and zinc-phosphate chlorinated rubber primers. Intermediate coats and finishing coats are usually classified according to their binders like vinyl finishes and urethane finishes.

Paints are usually applied as one coat on top of the other. Application of each coat is meant for a specific purpose/function. The primer is applied on the prepared surfaces so as to wet the surfaces and provide good adhesion for coats applied subsequently. Primers meant for steel surfaces are required for corrosion inhibition. The intermediate coats (could be several coats) are meant to build the total film thickness of the system. The finishing coats provide frontline defence against the environmental hazards and, of course, determine the final gloss, colour, etc. All the paint coats starting from the primers should be compatible with one another.

The standard methods used for application of paints to structural steelwork are: (i) brush, (ii) roller, (iii) conventional air-spray, and (iv) airless spray. The use of brush is the simplest and slowest way of execution of painting work; it can be done in the restricted spaces also. With rollers, painting is done more quickly depending on the rheological properties of the paints. Spraying of paints atomized at the gun-nozzle by jets of compressed air accomplishes a faster painting compared with painting using rollers at the cost of wastage. The spraying of paints atomized at the gun-nozzle by very high hydraulic pressures accomplishes painting faster compared to air-spray with comparatively less wastage. Brush and roller applications are more common at the construction sites.

Hot dip galvanizing is one of the oldest and still one of the most effective methods of protecting steel against corrosion. The galvanizing process involves the following.

- The surface preparation of steel is first thoroughly accomplished by blast cleaning mill-scale and roughening of the surface followed invariably by pickling in hydrochloric acid producing ferric chloride on the steel surface which acts as flux for the hot-dip galvanizing process.
- The cleaned and fluxed steel is dipped into a bath of molten zinc heated to a temperature of about 450°C at which temperature the steel reacts with the molten zinc to form a series of iron–zinc alloys on its surface. That portion of zinc surface through which the material to be coated enters the zinc bath must be kept covered with a flux; ammonium chloride and zinc chloride are widely used for this purpose. Exposed structural steelwork like towers is generally zinc-coated by this process.
- As the structural steel is removed from the bath, an additional thin layer of pure zinc is added.

The weight of coating deposited on the article is influenced by various factors like the bath temperature, time of immersion, quality of steel surface, withdrawal speed as well as the shape and size of the article.

SUMMARY

Completion time schedule of a project can be reduced significantly by incorporating fabricated steel superstructures on reinforced concrete foundations/bases by simultaneous completion of different inter-related activities in logical sequence so that the progress could be achieved uninterruptedly. Fabrication of steel structures is generally carried out at shops, but can be done at sites for which suitable infrastructures would have to be built and necessary equipment and machineries mobilized. Besides, highly skilled personnel need to be assigned for assured quality. For obvious reasons, quality of fabrication work at a shop under totally controlled conditions would be better, but transportation of the same from shop to site could pose problems. On the other hand, site fabrication may eliminate the storage of fabricated structures by construction of foundations/bases and completion of fabrication of the corresponding steel structures for immediate erection.

Welding technology has progressed and evolved with the passing of time because of research and development work undertaken worldwide. More and more structures are being built according to the precepts of good welded design. An electric arc characterized by low voltage and high current between the end of an electrode and the parent metal constitutes a very effective heat source that forms the basis of the most common techniques for structural fabrication work. Welding quality can be assured by incorporating automation in the welding process.

The sequence of shipping of shop fabricated elements must follow logical patterns based on site conditions and erection schedule. Large structures fabricated at shops are transported in pieces as 'shipping units', which are pre-assembled at sites before erection. A designer should foresee how structural steel would be erected safely stage by stage within the least possible time as the success of site operations depends to a large extent on the care and thoroughness of the advance planning. Considerable economy can be effected, time saved on erection and safety of personnel working at heights may be ensured if pre-assembly of small elements could be arranged at the ground level before lifting for erection. Site welding has to be carried out in prevailing weather conditions and quite often in difficult positions thus making site welding more difficult than shop welding.

Where steel structures are not planned, precast concrete structures could be used instead. Precast concrete elements are not designed for resisting tensile stresses resulting from indiscriminate handling, and are also vulnerable to damage from possible impact. Maximum efficiency in erection is achieved by placing the precast elements in their final positions directly from the transport or storage area in single operation. For efficient use of lifting equipment, temporary devices are so designed as to ensure automatic aligning and levelling of the lifted elements so that lifting equipment would not be required again for any more adjustment work.

Balanced cantilever, cable stayed, and suspension bridges are planned keeping in view erection of precast prestressed deck segments. Erection of these bridges can be planned and executed efficiently and timely. Erection of conventional bridges is also being carried out by pursuing traditional methods of using cranes from shores or pontoons. For long bridges comprising a number of spans, erection of beams by using launching truss or gantry girder could be done efficiently. Single or multi-span prestressed precast bridge girders can be erected by pushing using hydraulic rams or winch power if the alignment is nearly straight.

REVIEW QUESTIONS

1. How does fabrication of steel superstructures reduce project implementation time?
2. Describe briefly the four basic types of bolts.
3. What are the two important reasons for using welded design and construction?
4. What is the basic function of weld? What are the two conditions that aid in the formation of inter-atomic links?
5. What are the salient features any heat source must have to be usable in actual welding involved in steel fabrication work?
6. What is arc? Describe the factors which govern heat input in arc welding.
7. How is submerged arc welding (SAW) different from manual metal arc welding (MMA)?
8. Compare metal inert gas (MIG) welding with tungsten inert gas (TIG) welding.
9. Welding position wise, which technique yields better results – vertical-up or vertical-down?
10. State the various factors on which the selection of the type of edge preparation depends.
11. Why do electrode manufacturers stipulate maximum current for each electrode core-wire diameter?
12. How is heat input measured in arc welding?

13. Define heat-affected zone. What is carbon equivalent? What precautions are required for ensuring crack-free welds?
14. Submerged arc welding (SAW) is the mechanized system generally in use for steel fabrication work. What are the advantages of mechanizing SAW?
15. What are the poor process technique-induced defects in welding?
16. Name the defects that need to be checked by supervisory personnel during welding supervision.
17. Describe briefly the non-destructive tests carried out on the fabricated steel structures.
18. What are the different aspects of structural steel erection starting from receiving and unloading of fabricated steel structures?
19. What are the different steps involved in the erection cycle of precast concrete structures?
20. Describe in detail the erection procedure of a balanced cantilever bridge.
21. Why is surface preparation of steel structures performed?

9
Cladding and Wall

INTRODUCTION

A building is designed as steel or reinforced cement concrete (RCC) skeleton with solid brick masonry used for cladding and raising partition walls. Load-bearing walls are used for minor buildings.

Cladding refers to the external envelope of framed buildings made of brick/stone masonry, concrete, or precast concrete panels. Concrete cladding can be constructed along with RCC columns by slip-forming in high-rise buildings. The spaces within the framed building are generally enclosed by constructing masonry walls. Bricks or stones are joined together by mortar to make monolithic and strong masonry walls.

Walls serve many purposes including protection from fire, environment, and vagaries of nature. The term 'facings' is used to describe the materials which are used as non-structural, thin, and decorative external finish to enhance architectural features. Refractory masonry is intended for insulating machineries and structures, such as furnaces, boilers, and chimneys from heat. Refractory mortars are also made of insulating materials. Temporary scaffoldings are required for masonry work construction. Quality of masonry work is assured by continuous supervision during execution.

9.1 MASONRY MATERIALS

In India, the term 'masonry' is used for both stone masonry as well as brickwork. Strictly speaking, masonry refers to the parts of a building that are made of stone.

Most industrial and high-rise buildings are of framed construction. The frames are made of either steel or RCC or composite frames of steel and RCC. In some countries, frames are made of timber.

Cladding refers to the external envelope of framed buildings. Cladding is supported by the skeleton or structural frames. Cladding can also be done with framed glasses. Generally cladding is done with either masonry walls made of brick, stone, concrete, or precast concrete panels. Concrete cladding can be constructed with columns by slip-forming, especially in high-rise buildings.

The spaces within the framed building are enclosed by constructing walls around them. The walls are generally built of masonry work although a wide variety of walling materials are available to meet the changing needs of use, economy, and architectural trend.

Walls and partitions are classified as load-bearing and non-load-bearing. Different design criteria are applied to the two types. Both enveloping and enclosing walls differ in design as load-bearing walls are different from the walls supported on framed building structures. Load-bearing walls transfer the building's load to the soil below the ground level. As regards the functions of walls of any type, load-bearing or otherwise, the requirements are:

- Strength—must be strong enough to support self weight between floors/frames or the total weight
- Environmental protection against wind pressure—must withstand wind thrust
- Environmental protection against ingress of rain or falling water—must protect the inside by absorbing rainwater that evaporates during dry periods
- Environmental protection against outside heat—must protect the inside from cold weather during the winter and heat during the summer—heat transfer to and from outside should be favourable
- Environmental protection against outside noise—must be an ective barrier to airborne sound—thicker cladding wall would be more effective in this respect
- Fire safety—must not allow the spreading of fire and the damage caused due to radiant heat—fire is damaging and may destroy a building.
- Glazed walls allow daylight entry.
- Walls add to architectural excellence.
- Durability and freedom from maintenance—walls of brick and stone facings require very little maintenance over the expected life of buildings in general—deterioration that becomes apparent over a period of time is due to weathering

Compared to heavy construction activities, construction/erection of cladding walls is relatively easy. There are five different forms of cladding, as follows:

- Brick/block masonry – most common in India
- Precast concrete panels
- Steel/asbestos sheeting
- Glazed curtain walling—most expensive
- Timber

This chapter mainly deals with masonry work. The term 'facings' is used to describe materials used as a non-structural, thin, decorative, external finish such as the natural stone facings applied to brick or concrete backing to enhance architectural features.

Masonry is built of non-metallic, incombustible materials, such as brick, stone, concrete block, and a number of such materials. Brick is the most popular material for masonry work and is used extensively. Unit masonry consists of pieces of such materials. The units are bonded together with mortar or other cementitious materials. Both natural and artificial materials are used for masonry construction. Limestone, dolomite, sandstone, and other natural rocks weighing up to 50 kg are used for stone masonry depending on the nature of work. The most popular artificial materials like solid (clay), porous, hollow, porous-and-hollow, pulverized fuel ash bricks, and lightweight concrete stones are used for masonry work.

Stones or bricks are joined together by mortar to make monolithic and strong masonry. Bricks are classified as follows:
- First class—well-burnt, sound, hard and even vitrified if of good shape
- Second class—not sufficiently burnt for exposure to running water, but good enough for exposure to weather, that is, bricks used for walls other than the enveloping walls
- Third class bricks—totally under burnt bricks

Bricks delivered to a site should show the manufacturers' identification marks visibly marked on the frogs (vide definitions in Section 9.3). The bricks should show no efflorescence when soaked in water followed by drying in shade. Dimensions of bricks vary between $250 \times 125 \times 75$ (generally in the Eastern India) and $230 \times 115 \times 75$. Modular bricks have dimensions of $200 \times 100 \times 100$. On delivery at site, instead of dumping at random, bricks should be stacked in regular tiers so as to minimize breakage and defacement of bricks. Stacking should be planned in order of priority for actual use in construction. Stones are dumped at random taking care not to break the stones into pieces.

9.2 MASONRY BONDING

In masonry work, bricks or masonry stone pieces are joined together by mortar, which makes the masonry monolithic and strong. Mortar comprises fine aggregates (sand) and a binder, which usually is cement. Additives like admixtures may be used with cement to improve workability. Lime and cement, or simply lime may also be used as binder. The present trend is to use cement: sand mortars in various proportions, depending on the nature of work and design strength. The factors controlling the strength of any particular mix are the ratio of binder to aggregate (sand) plus the water:cement ratio.

The purposes of brick bonding are as follows:
- To achieve maximum strength during transfer of loads mainly through the wall, and also through column or pier
- To resist side thrusts for ensuring lateral stability
- To have an aesthetically acceptable appearance

Masonry is built in courses, each of which consists of separate stones or bricks. The gaps between adjacent stones or bricks, termed joints, are filled with mortars. Such joints could be horizontal or vertical. In stone masonry, stones are mostly laid flat, and less frequently on edges or on ends.

Masonry should be laid in courses parallel to each other and perpendicular to the direction of loading as masonry materials can resist compression, bending, and shear. Brick or stone faces should be laid in the plane of a course and should rest entirely upon the underlying course. The mortar bed exercises a beneficial effect in this respect.

When stones are not fitted firmly to each other, oblique acting loads may induce bending moments and shearing forces in masonry, thus overstressing the masonry causing its failure. This kind of development can be taken care of as is done in case of construction of arches.

Within each course, brickwork should be divided by a system of joints mutually perpendicular and also perpendicular to the bed—one type of joint being at right angle to the outside face of brickwork, and the other parallel to it. If the joints are arbitrarily done, the result may be harmful for proper bonding. Proper bonding makes the brickwork monolithic and ensures uniform distribution of loads on underlying layers.

9.3 STONE MASONRY

Stone masonry is different from brickwork. In brickwork, even-sized homogeneous materials are used. In case of stone masonry, sizes vary beyond limits. Stone masonry is, therefore, capable of infinite variety because of the wide variation in quality and its method of use. Listed below is a list of frequently used terms with their respective definitions.

- Ashlar masonry—masonry composed of rectangular units, usually larger in size than brick and properly bonded having sawed, dressed, or squared beds – laid in mortar
- Arris—arris is the exterior edge in masonry and should never be less than 90^0
- Abutment—the outer support of an arch
- Backing—the stone that forms the back of the wall
- Bats—bits of bricks are called bats
- Batter—uniform slope from the top to the bottom
- Bed—the horizontal surface upon which a course of bricks or stones is laid in mortar
- Bond—a term used to describe the principle or method on which the bricks are laid in horizontal courses to form vertical walls of various thickness—the object is to prevent the vertical joints in any course from falling directly over those in the course below
- Buttress—bonded masonry column built as an integral part of the wall and decreasing in thickness from base to top – though never thinner than the wall
- Ceramic veneer—hard-burnt, non-load-bearing, clay building units, glazed or unglazed, plain or ornamental
- Chase—a continuous recess in a wall to receive pipes, ducts, and conduits
- Closer—the reduced bricks (a quarter or three quarter) to create bond and to form vertical ends or openings
- Coping—a cap or finish on top of a wall, pier, chimney, or pilaster to prevent penetration of water to the masonry below
- Cramp—to draw the abutting stones together

- Corbel—stones or bricks projecting from a wall to form a support for the load – may also be an architectural feature
- Concrete block – machine-formed masonry building unit composed of Portland cement, aggregate, and water
- Course—a continuous horizontal layer of masonry units bonded together
- Dowels—small blocks of stone, slate, or metal (other than iron) to prevent easy fracture
- Drip—a groove in the projecting underside of a stone wider than the wall to prevent water from reaching the wall – also called gorge or throat for the same purpose
- Facing—the stone that forms the front of the wall
- Frog—an indent is formed on the top surface when clay is moulded into bricks
- Grout—a mixture of cementitious material, fine aggregate, and sufficient water to produce consistency in pouring without segregation of the constituents
- Grouted masonry—masonry in which the interior joints are filled by pouring grout into them as the work progresses
- Header—a whole brick laid with its length across or at right angles with the face of the wall to bond two wythes
- Jambs—the sides of the piers and abutments
- Lintels—large flat stones to cover openings in a wall
- Load-bearing wall—that supports any vertical load over and above its own weight
- Quoins—the corner stones of masonry buildings
- Masonry—a built-up construction or combination of masonry units bonded together with mortar or other cementitious material
- Mortar—a plastic mixture of cementitious materials, fine aggregates, and water
- Mullions—long vertical stones to subdivide a window opening
- Partitions—the dividing wall, one story or less in height, used to subdivide the interior space
- Pier—an isolated column of masonry – a bearing wall not bonded at the sides into associated masonry is considered a pier when its horizontal dimension measured at right angles to the thickness does not exceed four times its thickness
- Pilaster—a bonded or keyed column of masonry built as a part of the wall, but thicker than the wall, and of uniform thickness throughout its height – it serves as vertical beam, column, or both
- Quoin—an exterior angle of a building, especially the one formed of large squared corner-stones projecting beyond the general faces of the meeting wall surfaces

- Solid masonry wall—a wall built of solid masonry units laid continuously, with joints between units filled with mortar or grout
- Stretcher—a masonry unit laid with its length horizontal and parallel with the wall face
- Transverse joint—a joint that crosses the bed in a continuous line is a transverse joint
- Veneer—a wythe securely attached to a wall but not considered as sharing load or adding strength to it
- Wall—vertical or near-vertical construction, with its length exceeding three times the thickness, for enclosing space or retaining earth or stored materials
- Weathering—sloped dressing over the stone masonry for draining of rainwater
- Wythe—each continuous vertical section of a wall that is one masonry unit in thickness

Stone masonry is divided into two categories: (i) soft stone masonry, (ii) hard stone masonry. Soft stone can be easily modified, but design with soft stone should conform to the quality of stone as available. The quality should be ascertained by testing samples.

There are three types of stone masonry: (i) rubble, (ii) ashlar, and (iii) block-in-course. Among these, there are many gradations particularly in rubble masonry, such as: (i) random squared coursed rubble, (ii) random coursed rubble, (iii) random un-coursed rubble, and (iv) dry rubble.

Irrespective of the stone used, the same should be hard, durable, and tough. All stones should be soaked in water for at least 4 hours before use so that there is no loss of water from the mortar.

Rubble masonry

Stones that are rough and irregular in shape are used for rubble masonry. Stones are used for construction of foundations, basement walls, small height buildings, relieving/supporting walls, and other installations in areas where natural stones are available in plenty. Course rubble is laid conveniently in courses varying in height on the basis of availability which should not be less than 115 mm and not more than 230 mm. The courses may be of equal or varying heights. In case of varying heights, higher courses should be laid at the bottom of the structure. All beds and joints need to be perfectly true all through, both horizontally and vertically, and should not be more than 12 mm in thickness. The line of each course needs to be perfectly levelled and no joint should overlie another by less than 110 mm measured on the face of the wall. Stones that are to be used in masonry should not be less than 0.014 m^3 in size, and its bed should not be less than 1.5 times its height. Headers should constitute one-fifth of the face of the wall. For walls having thickness up to a metre, headers should be all through stones.

Random square coursed rubble—Every course would be made up of perfectly dressed square stones of unequal sizes of possible dimensions. Two stones may have their joints immediately over another, but the third course should always overlap by at least 90 mm. One fifth of the face of the wall should consist of through headers. All other stones should be in half bond or should overlap each other by at least one-third of the width of the wall. All beds should be perfectly horizontal, and all joints vertical.

Random coursed rubble—Each stone should be cut into such number of sides as could be conveniently dressed followed by fitting into the wall so that the thickness of joints remain 12 mm throughout. The vertical joints of each course must break the joints at least by 75 mm with those of the courses above and below it and no face stone, on any account, is to be narrower or shorter than its height. If it is of irregular shape, its length across the wall face must be at least 1.5 times its height. Random coursed rubble masonry should be supplied with equal or unequal quoins and should be coursed every 450 mm, one-fifth of the face should be through headers, and no stone should be less in depth than 1.5 times its height, every stone being well flushed in mortar. All the stones which are not headers should half bond or overlap with each other, at least one-third the width of the wall. There must be two quoins to every 450 mm of wall in height.

Random un-coursed rubble—The stones would be laid at random without being brought up to any level courses, each stone would be laid in its quarry bed in more than adequate mortar supply and would be wedged or pinned strongly into its position in the wall. The joints should be as small as possible.

Dry rubble—All the above rubble masonry can be executed dry without using mortar and can be used as retaining walls. For dry retaining walls, the front batter should not be less than 1: 4 and not more than 1: 2 if there is base room. The back batter is about 1: 6. The coursing would always be normal to the wall. The footing at the base should be broad. The top thickness of the wall must not be less than 600 mm. Should there be any possibility of surcharge, the thickness of wall should be increased to about 900 mm. The height of dry walls should not exceed 3.65 m.

Ashlar masonry

The dimensions of the stones should exactly conform to the details shown in the drawings. The beds and joints are to be dressed to be square and true in every respect. All joints should not only be within 6 mm in thickness, but all joints and beds should also be truly vertical and horizontal respectively. Each stone should be well set in the mortar as specified. Ashlar masonry should never be laid in courses of less than 250 mm in height. Volume of no stone should be less than 0.0354 m^3. One-fifth of the face should be headers, and width of bed of no stone should be less than 1.5 times its height. All stones should, if possible, break joints by as much as 1 to 1.5 times the depth of the course. In heavy work, stones should be mechanically lifted and lowered into their exact positions. Manual handling would be alright in light work. It is to be ensured during handling and placing that the freshly laid masonry is not disturbed. All the stones need to be washed and cleaned before placing. The outer face of stones may be rock-faced, finely chisel-dressed, rock-faced with chisel margin, rock-faced with chisel margin and chamfered edge.

Block-in-course masonry

This is a class of masonry that occupies an intermediate place between ashlar and rubble. The stones are of large size, so that they may be procured in blocks and not as rubble. The joints and beds are only roughly dressed, so the work cannot be designated as ashlar work. Block-in-course masonry is mainly used in large engineering works and rarely in small building works. The height of the courses should not be less than 180 mm. The work should be carried out in the same way as ashlar masonry.

9.4 SOLID BRICKWORK

Clay bricks are generally used in India for construction work. Dimensions of bricks are mentioned under Section 9.1 above. Bricks are bonded together to form brickwork. Bricks arranged in a manner such that the bricks at the upper course cover the joints of the lower course. Based on the method of staggering the joints, there are several systems of bonding. Mortar is used as the binding material to join the bricks. Bond of different kinds is designated below.

Heading bond—When each brick is laid as a header, the bond is designated as a heading bond. This kind of bond is useful for curved brickwork – convex or concave. This bond is also useful for the construction of corbels requiring projections.

Stretching bond—When each brick is laid as a stretcher, the bond is designated as a stretching bond. Every vertical joint is located at the centre of stretchers, above and below. This kind of bond is useful for constructing 125 mm or 75 mm partition walls. This bond is also useful for lining or facing of hollow walls for countering dampness – ties may be required in such cases.

English bond—This kind of bonding (Fig. 9.1) is ideal for construction of brick walls—heading and stretching courses always alternate in this kind of bond. No stretchers are laid barring those seen on the face of the wall. Bricks in the same course should not break joints with each other. In providing vertical joints, it is to be ensured that a brick overlaps at least a quarter of the brick in the lower course and, by the same principle, upper course should be having similar overlaps. In all the walls having thickness of whole bricks, every course would have headers or stretchers on both the faces. English bond is not suitable where the wall thickness is of one brick because variation in sizes of bricks cannot be taken care of maintaining aesthetic quality with even surfaces.

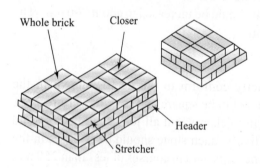

Fig. 9.1 English bond

Flemish bond – This bond (Fig. 9.2) comprises of both headers and stretchers in every course. A header is placed at the centre of the stretcher of the course, both above and below it. Each header, unlike in English bond, is separated from the next in the course by a stretcher. Flemish bond is as strong as English bond because the same kind

Cladding and Wall 317

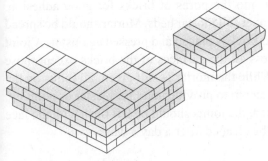

Fig. 9.2 Flemish bond

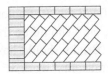

Fig. 9.3 Raking bond   Fig. 9.4 Herringbone bond

of overlapping with the upper and lower courses is maintained here as well. The drawback of this bond is the necessity of the use of bats in the wall. For one brick wall, smoother appearance can be achieved on both sides of the wall by using Flemish bond. Single Flemish bond is a combination with English bond for the body of the wall and Flemish bond for the face work. Double Flemish bond has Flemish bond on both faces.

Raking bond – In case of walls of exceptional thickness, raking bond (Fig. 9.3) is used so that vertical pressure could be radially distributed through the joints. The courses travel diagonally across the interior of the wall. The face work is made up with triangular bats. Raking or diagonal bond is useful in the courses of high wall footings.

Herringbone bond—This bond (Fig. 9.4) has the advantage of making effective bond in the middle. From a centreline in the middle of the wall, this is an arrangement of raking bond in both directions and is suitable for pavings and thick walls comprising four or more bricks.

The junction of walls is formed at right angles. The walls at the junction are joined by the alternate insertion of headers and stretchers of each wall.

Bricklaying

Ideally, the gaps to be covered by solid brickwork should be multiple of half bricks as the mason engaged in bricklaying has to build in brick sizes. Vertical heights are also to be controlled in a similar way by brick sizes. In the horizontal course, any odd dimension is covered by a closer at a centrally located gap where a bond cannot be maintained. In the western countries, brick sizes are maintained true to dimensions, and building dimensions are worked out on the basis of brick sizes.

A mason should check the brick face alignment by stretching a string along the edges, and verticality of walls by a plumb bob. He must have straight edge for maintaining alignment. Level of courses should be checked with spirit level. This is necessary for both good workmanship as well as enhancing the architectural features.

Bricks are soaked in clean water for at least two hours before put to use. The surfaces of bricks, however, should be dry. Bricklaying requires steady supply of soaked bricks and mortar for continuing the work without interruption. A mason spreads plastic mortar over a course of bricks with trowel so as to prepare a bed for laying bricks. The strength and quality of brickwork are governed by the care taken in spreading and levelling the bed of mortar. Bricks are pressed on the mortar bed by tapping with mallet in such a manner as to force the mortar into the pores of bricks for good adhesion. Bricks should be laid with frogs upwards over full mortar beds. Mortar should be spread on the inside faces of brick before the next brick is placed and pressed against it. A joint is not properly filled between bricks unless all the surfaces of the bricks except those exposed are fully buttered with mortar. While the mortar is still green, all joints on the face should be raked to minimum 10 mm depth to provide bond for plaster or pointing. Where plastering or pointing is not required, the joints should be left flush with the face without raking. All extra mortars should be cleaned after a day's work.

9.5 REFRACTORY MASONRY

Refractory masonry is intended for insulating machineries and structures like furnaces, boilers, chimneys, etc., from the heat generated by high temperature.

Materials for refractory masonry should conform to the specifications. Refractory items and brick (fireclay, magnesite, dolomite, alumina, chromite, etc.) should be free of extensive mechanical damage, cracks, voids, and other flaws. Bricks should have adequate mechanical strength. Quality refractory materials for masonry work with joints as thin as 1 mm are first selected according to dimensions. When laying pillars, corners, intersections, abutments and curbed sections, specially shaped shop manufactured bricks and bats are to be used.

Size tolerance of refractory bricks should be ±1% or ±1 mm whichever is greater, or as specified.

Refractory mortars are made as dry mixtures at manufacturing shops from the same materials as masonry items. Mortars are delivered in two ways: (i) powder in bag and binder in container, (ii) single component system—water is to be added to the powder. Generally, water is added at site to such mixes as: fireclay-clay on fireclay-alumina or soluble glass to fireclay-bauxite, fireclay-alumina or others. The size of aggregates for refractory mortars should be half-the-thickness of the refractory. Depending on the joint thickness; liquid, semi-thick, or thick mortars are used. Mortar should be used within half-an-hour of mixing.

Mallet should be used for compacting bricks in position. Greased plywood should be used for expansion joints for easy removal on setting of bricks. Expansion joints should be tight-packed with ceramic fibre materials after cleaning the expansion gap of dust, debris, or foreign matter—all these to be carried out on completion of the refractory lining.

As a function of service, refractory masonry is divided into four standard categories with thickness of joints as 1, 2, 3, and more than 3 mm. In an extra special category, the requirement of joint thickness is 0.5 mm

9.6 ENABLING WORK FOR CLADDING

There is a limit to the safe working height at which a worker can access the masonry work. Temporary structures like scaffolding should be designed to suit their purpose. Scaffolding is required for providing temporary, safe, and convenient working platform above ground level. Scaffolding and associated temporary work enable carrying out work also on existing buildings, including repair and maintenance work. Different types of scaffolding systems are available to suit different conditions prevailing at different construction sites to facilitate construction, alteration or demolition of structures or to allow hoisting, lowering or standing/supporting of men and materials above ground level.

In construction work, scaffolding may be deemed as temporary 'enabling work' as scaffolding support would be necessary till the permanent structure becomes self-supporting. The basic functional requirements for scaffolding are to:
- Provide safe and accessible horizontal working platform
- Provide safe horizontal and vertical access to buildings

Scaffolding is generally erected of steel tubes clipped or coupled together and adequately laced and braced to make it strong, stable, and safe to work on the platforms made of timber planks. Platforms at different levels can be accessed using timber or metal ladders. In India, scaffolding is also made of timber or bamboo poles. Bamboo is largely used in small building construction. Bamboos are tied with steel wires or non-metallic ropes.

Scaffolding is erected on either side of a wall, generally on level ground. The upright poles called 'standards' are first erected at about 3 m centres and about 450 mm from the wall. It is to be ensured that the bottom of any upright is not disturbed at any time during its use. Pressure on soil is distributed evenly by placing each upright tube/pole on base plates over sole plates so that the load is distributed evenly on the ground. The top of the upright is fitted with head or cap plates for supporting formwork for slab or platform. A single row of upright scaffold is erected, partly supported by the structure under construction for stability. When two rows of uprights are erected, the scaffolding becomes independent and self-standing. Two rows of uprights can be supported on rails for mobility. How the scaffolding would be erected would depend on the planning of construction activities. If the materials are to be manoeuvered on the scaffold, a clear passage of 600 mm should be maintained at all times. The uprights are supported horizontally with the wall at about 1200 mm intervals. When the scaffold is high or exposed, it is stiffened by long diagonal braces.

Care should be taken not to overload the scaffolding. Toe boards must be provided at the end of platforms so that materials and tools do not fall off the scaffold. All temporary structures are to be checked properly before they are put into use, more so in extreme weather conditions. Scaffoldings are discussed in detail in Chapter 13, Section 13.7.

9.7 SUPERVISION AND APPROVAL OF EXECUTED CLADDING AND WALL

Compressive strength of masonry largely depends on workmanship and the completeness with which stones, bricks, or refractory are bedded.

Systematic quality control of masonry work involves supervision of all stages of operation, workmanship, and identification of defects. The checklist includes quality of bonding, thickness and filling of joints, horizontal and vertical conformance to the requirement, finished surfaces, and the corner angles. Corrective action should be taken for defects identified. Quality of executed cladding and wall would be approved on satisfactory completion of corrective action on the defects and departures/deviations from the specifications.

The quality of work must comply with the following acceptance criteria before the cladding and wall is approved.

Bricks

Strength—minimum average compressive strength must conform to that specified for various classes as per the Clause No. 3.1 of IS: 1077—latest edition or relevant technical specification. However, compressive strength of any individual brick should not be less than 80 percent of the specified average minimum compressive strength.

Water absorption—The bricks when tested after immersion in cold water for 24 hours, the average water absorption should not be more than 15 percent by weight when dry.

Efflorescence—The bricks when tested in accordance with IS: 3495, the rating of efflorescence should not be more than 'slight'.

Water

Water used for brickwork, plaster, curing, and other construction-related purposes should be free from injurious amounts of oil, acid, alkali, salts, sugar, organic, or any other materials that may be harmful for masonry, concrete, and steel. The pH value of water should not be less than 6.

Cement

Cement should be checked to ensure its conformance to specified grade.

Mortar

Mortar should be checked to ensure its conformance to specified mix. Lime, if mixed, should conform to IS: 712 and sand should conform to IS: 383.

Stones

Only non-reactive stone should be used for stone masonry.

Workmanship

Workmanship should conform to the technical specifications.

SUMMARY

Cladding refers to the external envelope of framed buildings. Walls are constructed to enclose or separate spaces by raising partitions. Generally, cladding is done with masonry walls made of bricks or stone pieces. Masonry construction is labour-intensive and, therefore, time-consuming work. Concrete cladding can be constructed along with RCC columns by slip-forming, especially in high-rise buildings. Cladding can also be done with framed glasses if it is not included in slip-forming work. The spaces within a framed building can be enclosed by erecting precast concrete panels. At present the walls are built of masonry work, although a large variety of walling materials are available. Stones or bricks are joined together by mortar to make monolithic and strong masonry to serve many purposes including protection from fire, environment, and vagaries of nature. The term 'facings' is used to describe the materials used as non-structural, thin, decorative, external finish such as the natural stone facings applied to brick or concrete backing to enhance architectural features. Brick sizes should be standardized and manufactured true to dimensions. Stone masonry is different from brickwork because stone sizes vary considerably. Refractory masonry is carried out maintaining quality at all stages. Similar quality control is necessary in masonry work involved in cladding and raising walls.

REVIEW QUESTIONS

1. What is meant by masonry wall?
2. What is meant by cladding?
3. What are the functional requirements of walls?
4. What are the different forms of cladding? Which is the common form in India?
5. Describe what kind of materials are used in 'facings' and for what purposes.
6. Bricks are classified into three types. Explain the basis of the classification.
7. Bricks are to be stacked in tiers at site. Why?
8. Bricks are generally bonded with cement mortar. What are the purposes of brick bonding?
9. State how bricks or stone pieces are to be laid in masonry work.
10. Of the three types of stone masonry, state how ashlar masonry work is carried out.
11. How is English bond different from Flemish bond?
12. State how the quality of bricklaying is maintained by checking alignment, verticality, and level.
13. How is refractory masonry laying different from brick masonry laying?
14. What are the basic functional requirements of scaffolding in masonry work?

10
Roof and Roofing

INTRODUCTION

A roof is constructed on top of a building to protect it from heat, rain, wind, solar radiation, and other environmental agents. Roofing, however, protects the roof from damage and deterioration. Roofs are of two types—flat and pitched. Flat roof, which is the most common type, may not be suitable for quick drain-off of rainwater. Pitched roofs supported on triangular-shaped steel/timber trusses are covered with waterproof sheets.

Precast reinforced concrete roof slabs are supported on steel purlins or crossbeams. Roofing for protection of roofs comprises grading underbed, insulation, and waterproofing. A shell roof may be defined as a thin structural curved skin covering an area in the given plan shape. The term 'shell' emphasizes the considerable strength and rigidity of thin, natural, and curved forms such as the shell of an egg.

10.1 CAST-IN-SITU REINFORCED CONCRETE ROOFS

Roofs protect buildings from the vagaries of weather – heat, precipitation, wind, solar radiation and other environmental agents. All these should be taken into consideration in appropriately designing and constructing roofs to be durable so as to reduce maintenance problem to minimum. Depending upon the construction materials used and the geometrical features, roofs may be classified into the following two main types:

- Flat roofs of different kinds
- Pitched roofs with different covering materials

Flat roofs have about 1 percent slope depending on the span and intensity of rain. Flat roof is the most common type of roof built in this country. However, flat roof may not be suitable in places where heavy rainfall occurs. It is necessary that the rainwater drains off as quickly as possible.

Roofs may have any of a variety of shapes. More often, roofs are made of one or more plane surfaces. There are two basic types of flat roof slab design:

- Beam/girder construction – slabs supported by and acting together with beams/girders running in one or both directions ultimately transferring the load to the frames or walls
- Flat-slab construction – loads are carried directly by columns

A girder is always a primary member which is used to support beams. Girders are designed to span between columns, walls, or other girders. Floor girders may be either single-span or multi-span in construction.

Beam/girder construction refers to floors formed by slabs acting together with beams/girders spanning in one or both directions ultimately transferring the loads to walls and/or columns. Flat slabs are supported directly on columns, which are flared at the top. The flared portion is called *capital*.

Reinforced concrete roof slabs may be cast in-situ using:
- Temporary forms
- Permanent forms

Temporary formwork is included in Chapter 7, Section 7.7. Permanent formwork, unlike the formwork used normally, is left in place for the life of the roof slab it is supporting. This system, apart from roof slab, has been successfully used in many areas of concrete construction. There are, however, two types of permanent formwork used in composite decks:
- Structurally participating—designed to provide the temporary support for the plastic concrete and loads caused due to the construction activities ultimately becoming part of the permanent roof slab contributing to its strength
- Structurally non-participating—designed solely to support the plastic concrete and loads caused due to the construction activities

The use of permanent formwork has the following benefits:
- Eliminates the necessity of installation of falsework
- Reduces the need for skilled labour for carpentry
- Increases the potential for standardization related to roof construction among other possibilities
- speeds up fixing time and eliminates stripping of forms on hardening of concrete

There are two systems which are generally used for non-participating formwork:

(i) Glass fibre reinforced plastic (GRP) – composite of durable, thermosetting polyester resin, reinforced with glass fibre, that is used as a thin laminate with high strength, low density, good corrosion and weather resistance but a low modulus of elasticity

(ii) Glass fibre reinforced cement (GRC) – made from cellulose and polymeric fibres, cement and water, and pressed into a range of profiles.

GRC panels are available in the form of flat sheets for spans up to approximately 800 mm.

Panels made from GRP have metal bar stiffeners encapsulated in the material to improve the stiffness and are designed to cater for spans up to 4 m. Advice and information regarding the maximum spans and the predicted deflections is provided by every manufacturer. The length and width of the unit is limited by weight so that two men can handle a panel into position. The joints between panels and main beams should be sealed in accordance with the manufacturer's instructions as recommendations can vary depending on the type of main beam material used - concrete or steel.

For concreting work of cast-in-situ roof, vide Chapter 13, Section 13.6. Pitched roofs are supported on triangular-shaped framed structures/trusses made of structural steel or timber. The pitch (depth to span ratio) should conform to the specifications. Roofs are covered with sheets, vide Section 10.3 below. Small span timber trusses may be used in the construction of residential buildings.

10.2 PRECAST REINFORCED CONCRETE ROOFS

In building construction, roof as well as floor slabs are designed as slabs supported by girders or beams. Several design alternatives are usually worked out and compared, and the least expensive one having the least weight is adopted for construction. Width of precast slabs should be within the lifting capacity of the available lifting equipment, vide Chapter 13, Section 13.5.

As steel is readily available all over India, building frames at construction sites are generally fabricated out of structural steel. Fabrication work is carried out either at fabrication shops or even at the sites. For site fabrication work, some infrastructural facilities are built at the site so that there would not be any problem of carrying out fabrication work simultaneously with the construction of foundations and substructures. Parallel implementation of reinforced concrete construction and structural steel fabrication drastically reduces the project implementation time schedule.

A precast roof element is designed as a single span beam in bending. The effective span is taken as the center-to-centre distance between its supports.

If the span of a building is large as in the turbine bay of a powerhouse or steel fabrication shop where electrical overhead travelling (EOT) crane is required, then the columns are connected using trusses instead of girders. Precast roof slabs are placed over purlins, which are supported on trusses. For erection of precast slabs, vide Chapter 8, Section 8.7.

Roofing is built of the following materials:
- Grading underbed
- Insulation
- Waterproofing

The joints of the erected precast roof slabs are checked and sealed with mortar before the underbed is laid. The roof surface which is to receive the underbed should be thoroughly cleaned of oil, grease, and other slippery material and roughened up with wire brush after cleaning the surface with compressed air. The surface should be soaked with water and all excess water should be removed before laying of underbed of cement-sand screed as specified. The underbed should be laid maintaining gradient for drainage (1: 100) with a thickness of 25mm at the drainage end. The underbed should, therefore, be of screed concrete of nominal mix as specified, and compacted.

Considering the thickness, placing of underbed should be avoided under hot sun, and the underbed as placed should be cured for 7 days.

10.3 ROOFS COVERED WITH SHEETS

Suitable materials for covering roofs are:
- Hot deep galvanized corrugated sheets
- Aluminium profiled sheets
- Asbestos cement sheets—not used any more on the ground of health and safety—replaced by fibre cement sheets (see the item below)
- Asbestos free profiled sheets—products based on a mixture of Portland cement, mineral fibres, and density modifiers

The functional requirements of roof covering are:
- Strength and stability
- Resistance to weather
- Durability and freedom from maintenance
- Safe access during maintenance
- Fire safety
- Resistance to passage of heat—thermal insulation
- Resistance to passage of sound
- Security
- Aesthetics

The strength of roof depends on the properties of materials used and their ability to sustain self-weight and imposed loads. Roof sheets should be of low self-weight to be cost-effective.

The thin sheets of steel and aluminium derive strength and stability mainly from the depth and spacing of profiles—shallow depth of profile for small spans to deep trapezoidal profiles and standing seams for medium-to-large spans between supports. Compared to such sheets, thick corrugated and profiled fibre cement cladding sheets have adequate strength for assumed loads, and rigidity in the material to resist distortion and loss of stability over moderate spans between supports. The thickness of the sheets, the strength of the fasteners used to fix the sheets and the ability of the sheets to resist the tearing effect of the fasteners fixed through the sheets would determine the stability of the frames. Manufacturers furnish data on the size and thickness of sheets, minimum end lap, maximum purlin and rail centres, and maximum unsupported overhang of the sheets as well as guidance on the type and spacing of fixing to match the site exposure.

The lowest allowable pitch or slope of the roof is dictated by the end lap of the sheets to keep water out. Thermal and structural movement is accommodated by the profiles, the end lap, and the designed tolerances at the fixings. Perforations on roof sheets are sealed by providing integral steel and neoprene washers on the screw heads. Top fixing is preferable to bottom fixing as perforations would be less exposed to water. Fibre cement sheets resist water penetration by material density, roof slope, and end laps. Fibre sheets take care of moisture, thermal and structural movement through end/side laps and rather large fixing holes.

Durability of coated/galvanized sheets depends, to some extent, on careful handling and fixing as the sheets are easily damaged. Damage to protective coatings, even around fixing holes, can cause corrosion. Roof sheets on buildings located in polluted locality or marine environment deteriorate more rapidly compared to those in sheltered and less polluted areas. Sheets with light-coloured coatings are less susceptible to ultraviolet damages. Fibre cement sheets do not deteriorate easily if a steep slope is provided for quick shedding of water. This material is, however, brittle and susceptible to damage even during maintenance work. Reinforced fibre cement sheets have higher impact strength, but because of their coarse texture, the fibre cement sheets become dirty very quickly.

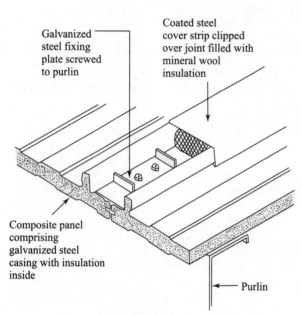

Fig. 10.1 Composite roofing panel

There is safety hazard on carrying out maintenance work on sloped or pitched roof. The question of safe maintenance should be kept in view during both design and construction of such roofs. The question of fire hazard should also receive utmost attention, particularly in selecting both roof sheets and supporting materials.

Heat insulation on flat roof is the content of the next Section in this Chapter. However, in case of sloped roof, insulation can be an integral part of the roof sheets in composite panels (Fig. 10.1). False ceiling under the truss bottom may also insulate the inside from heat apart from enhancing the aesthetics of the interior.

As regards sound insulation, fibre cement sheets provide much better airborne and impact sound insulation. As far as security is concerned, roof sheets are susceptible to vandalism. Curved sheeting with colour is available.

10.4 THERMAL INSULATION OVER ROOFS

The rate of heat flow under steady-state conditions from a unit area from the air on one side to the air on the other side of a material is the coefficient of thermal transmittance, which is represented by 'U'. Insulation material having high resistance, R, to flow of heat is judiciously used as thermal insulation in roof construction.

Any material with high resistance to flow of heat is called an insulator. There are many insulation materials available in the market including proprietary materials that are available with instructions for use. The insulating materials generally used at construction sites are:

- Foam concrete
- Expanded polystyrene blocks

Foam concrete

Lightweight concrete is superior to stone concrete for heat and sound insulation. As the weight of lightweight concrete increases, its insulating value decreases proportionately. Lightweight insulating concrete is produced at the site by combining insulating aggregates like vermiculite or perlite with Portland cement and water. Another variation of this type is referred to as 'foam concrete' which is produced with foaming agent that creates small air cells within the matrix. The compressive strength and thermal resistance of lightweight insulating concrete depends on the mix design and composition. Lightweight foam concrete is cast atop roof slabs. The thickness of foam concrete should be 50 mm or as specified and density should be within 300 to 500 kg/m^3. Foam concrete is placed in-situ or precast blocks may be laid—the option depends on the site conditions or specification. Insulation work with foam concrete can actually be taken up only on satisfactory testing of samples of foam concrete to be laid.

Expanded polystyrene blocks

Polystyrene made into rigid plastic foam has two distinctly different forms. For enhanced thermal resistance, moulded expanded polystyrene (EPS) may be incorporated into lightweight insulating concrete blocks. EPS has air-filled cells and, therefore, is not subject to thermal aging. EPS is available with a variety of densities, and its R (thermal resistance) value is a function of the density. Extruded polystyrene is blown with CFC (chlorofluorocarbon) or HCFC (hydro chlorofluorocarbon), thus it has higher R value compared to EPS. Extruded polystyrene is resistant to water and water vapour and is available in very high compressive strength. The blocks must be strong enough to withstand expected workload on the roof without undergoing any deformation. The work on the roof insulation should be carried out by highly skilled workers under the direct supervision of the manufacturer's specialist representative.

Roof insulation, apart from insulating the roof, protects the built-up roof from the harmful effects of thermal cycling, ultraviolet degradation, weathering, and roof traffic, if any.

10.5 WATERPROOFING OVER ROOFS

Many roof failures have been caused by excessive water accumulation. Waterproofing provides protection against penetration of rainwater or leaking of accumulated water through the roof. Water may penetrate the roof concrete via interconnected voids within the cement paste matrix or at the paste-aggregate interface. Ignoring defects such as gross voids, cracks or open joints, the interconnected porosity is of two kinds: (i) micro pores in C-S-H gel, and (ii) capillary pores between cement hydration products.

Waterproofing of roof is carried out using membrane. The traditional low-slope membrane roof covering is composed of bitumen (asphalt or coal tar pitch), usually applied hot; felts (hessian, glass fibre, or polyester); and a surfacing, such as aggregate, coating, or cap sheet. This is a kind of built-up roof.

The membrane should be of one of the following types:

- Two plies of hessian/polyester base felt and three mopping of bitumen—(i) mopping of hot bitumen on the roof surface (ii) polyester/hessian base felt (iii) mopping of hot bitumen on the felt (iv) polyester/hessian base felt (v) mopping of hot bitumen on the felt (vi) finishing with pea-sized gravel on bitumen
- Three plies of hessian base felt and four mopping of bitumen—(i) mopping of hot bitumen on the roof surface (ii) hessian base felt (iii) mopping of hot bitumen on the felt (iv) hessian base felt (v) mopping of hot bitumen on the felt (vi) hessian base felt (vii) mopping of hot bitumen on the felt (viii) finishing with pea-sized gravel on bitumen (the same thing can be done replacing hessian base felt with glass fibre felt)
- Four plies of fibre base felt and five mopping of bitumen—(i) mopping of hot bitumen on the roof surface (ii) fibre base felt (iii) mopping of hot bitumen on the felt (iv) fibre base felt (v) mopping of hot bitumen on the felt (vi) fibre base felt (vii) mopping of hot bitumen on the felt (viii) fibre base felt (ix) mopping of hot bitumen on the felt (x) finishing with pea sized gravel on bitumen (the same thing can be done replacing fibre base felt with hessian base felt)

In special cases, however, more courses or combination of hessian base felts and fibre base felts may be used. The number of moppings done exceeds the number of plies by one.

Cotton fabric, which should be high-grade cotton cloth, saturated thoroughly and uniformly with bitumen, is stronger and is more extensible than bituminous saturated felt. However, bituminous saturated cotton fabric is more expensive and more difficult to lay. At least one or two of the plies in a membrane should be of saturated cotton fabric to strengthen and provide ductility and extensibility to the membrane.

The roof surface that receives waterproofing treatment must be cleaned and dried satisfactorily. Concrete surfaces should be well cured before waterproofing is done. Membrane waterproofing should not be carried out in wet weather or when the temperature is as low as $10\,°C$.

Polymer is used for modification of asphalt. The most used polymers for asphalt modification are atactic polypropylene (APP) and styrene-butadiene-styrene (SBS). These prefabricated sheets are generally installed over a base sheet, which may not be composed of modified bitumen.

Waterproofing membranes laid on the flat roofs harden as time passes and do not retain sufficient elasticity or tensile strength to resist deterioration due to thermal movements resulting from considerable variation of temperature between day and night and also seasonal temperature variations.

10.6 SHELL ROOFS

Shell roofs are skin structures. Skin construction may be formed either by joining plates, as in cellular and hipped plate construction, or by curving the surfaces as in single- or double-curved shells. A shell roof may be defined as a thin, structural, curved skin covering an area in the given plan shape. Roofs utilizing skin construction are formed by folding flat plates into corrugated shapes or by shaping the skin into single- or double-curved shells. A structural membrane or shell structure is a curved surface structure which is capable of transmitting loads to the supports in more than two directions. Structurally, the shell structure is highly efficient. It is shaped, proportioned, and supported in such a manner that the load transfer upon bending or twisting is minimized while the axial forces (compressive or tensile) in the shell or membrane are maximized. The term 'shell' emphasizes the considerable strength and rigidity of thin, natural, and curved forms such as the shell of an egg.

The basic materials which can be used in the formation of a shell roof are concrete, timber, and steel. Concrete shell roofs consist of a thin, curved, reinforced membrane cast in-situ over timber formwork, whereas timber shells are usually formed of carefully designed laminated timber, and steel shells are generally formed of a single layer grid.

Concrete shell roofs, although popular, are very costly to construct, since the formwork required is usually purpose-made from timber, which in itself is a shell roof and has little chance of being reused to enable the cost of the formwork to be apportioned over several construction contracts. In other words, the main disadvantages of shell roofs are related to their cost, and also poor thermal insulation properties.

Concrete is the most suitable material for the construction of shell structure. Wet concrete is highly plastic material and can take up any shape inside the formwork. Reinforcing steel bars of small diameters can be bent to follow the curvature of shells. Wet concrete is placed and spread over the formwork and around the reinforcing steel bars. The stiff concrete mix and reinforcement prevent the wet concrete from running down the curvature. Concrete should be compacted to the required thickness. On hardening of concrete, the reinforced concrete membrane or slab acts as a strong and rigid shell that serves both as a structure and a covering to the building. A shell roof offers considerable material saving over conventional beam slab construction. Cost of the curved shuttering may, however, offset the cost saving achieved in material consumption.

Shell structures can be classified as: (i) single curvature and (ii) double curvature shells. Single curvature shell structures are curved on one linear axis and form part of a cylinder in the form of a barrel vault or conoid shell. Double curvature shells are either part of a sphere as a dome or hyperboloid of revolution. Double curvature of a shell enhances its stiffness, adds to the resistance to deformation under load, and reduces the need for restraint against deformation

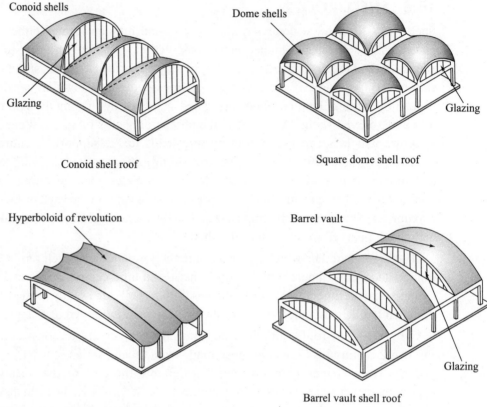

Fig. 10.2 Shell roof forms (typical)

Traditional formwork is the term that describes the adequately supported temporary formwork that takes the loads of curved reinforced concrete shell structure during and after casting. The formwork for a single curvature barrel vault is less complex compared to the kind of formwork required for a dome that is curved from a centre point. Advances in software and use of IT have made the design of shell structures and setting out of formwork easy. Nevertheless, formwork and associated supporting falsework is labour-intensive work. Besides, there would be more cutting and wastage of materials as the shape of shell structure becomes more and more complex. The simplest and most economical of all shell structures is the barrel vault.

Barrel vault

Barrel vaults (Fig. 10.3) comprise a thin membrane of reinforced concrete, positively curved in one direction so that the vault acts as both roof and supporting structure. In other words, a barrel vault is essentially a cut cylinder that must be restrained at both ends to resist the tendency of flattening. A barrel vault acts as a beam at right angles to the curvature and the beam's span is equal to the length of the roof.

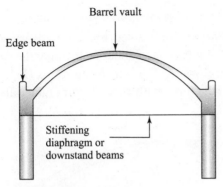
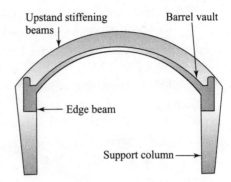

Fig. 10.3 Barrel vault

Long-span barrel vaults are vaults having longer span compared to their width or chord length. Similarly, short-span barrel vaults are vaults having shorter span compared to their width or chord length. Typical spans range from 12 to 30 m with the width being about half the span and the rise being about one-fifth of the width. Large areas may be utilized by constructing multi-span, multi-bay barrel vault roofs. The thickness of the shell should be sufficient so that the reinforcing steel is protected against damage by fire and corrosion.

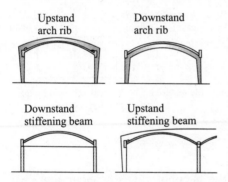

Fig. 10.4 Stiffening of barrel vaults

The thin shell of the barrel vault would show a tendency to distort and lose shape; and if such distortion is of sufficient magnitude, the shell could ultimately collapse. To counter such possibility, stiffening beams and arches (Fig. 10.4) are cast integrally with the shell. The general practice is to provide a stiffening member between the columns supporting the shell. In every long-span barrel vault, expansion joints should be provided whereby a series of weather-sealed abutting barrel vaults would be created.

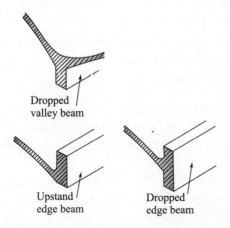

Fig. 10.5 Upstand and dropped beams

To prevent the thin shells from flattening their curvature due to self-weight and imposed loads, reinforced concrete edge beams are cast between the columns as an integral part of the shell. The edge beams (Fig. 10.5) may be cast as dropped beams, upstand beams, or partly upstand and partly dropped beams. Between multi-bay vaults, the loads on the vaults are largely transmitted to adjacent shells and then to the edge beams, thus allowing the use of comparatively slender feather-edge beams.

Roof-lights are fixed to the upstand curb, cast integrally with the shell.

To counter the expansion and contraction caused by temperature variation, continuous expansion joints are formed at intervals of about 30 m along the span across the width of multi-bay, multi-span barrel roofs. The expansion joints (Fig. 10.6) are formed by constructing separate shell structures with flexible joint materials between adjoining shell structures.

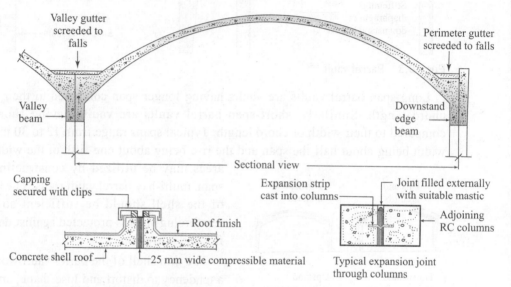

Fig. 10.6 Typical expansion joints

Conoid shells

Conoids are double-curvature barrel vaults as opposed to single curvature barrel vaults or as an alternative to barrel vaults (Fig. 10.7). The two basic geometrical forms that have been conceived are:

- Shape generated by moving a straight line along a curved line of, say, a *barrel vault* at one end and a different curved line at the other end
- Shape generated by moving a straight line along a curved line of, say, a *barrel vault* at one end and a straight line at the other end

Hyperbolic paraboloid shells

The name 'hyperbolic paraboloid shell' is derived from the geometry of the shape. The surface is generated by moving a straight line over two skew straight lines. Thus, the formwork can be formed using straight planks over two supporting skew beams of the shuttering system.

In order to obtain a more practical shape than the true saddle (the usual shape is that of a warped parallelogram), a straight line limited hyperbolic paraboloid is formed by raising or lowering one corner out of the plane containing the other three corners of a square/rectangle as shown in Fig. 10.8. By virtue of its shape, this form of shell roof has a greater resistance to buckling than dome shapes.

Roof and Roofing 333

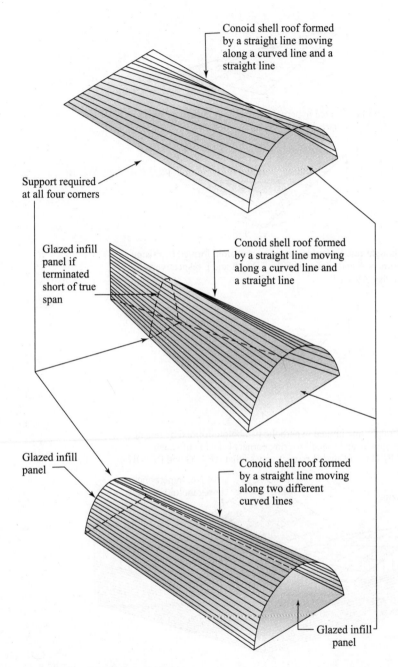

Fig. 10.7 Conoid shell roofs

Hyperbolic paraboloid shells can be used alone or in conjunction with one another to cover a particular planned shape or size. If the rise – the difference between the high and low points of the roof – is small the result will be a hyperbolic paraboloid of low curvature acting structurally like a plate which will have to be relatively thick to provide the necessary resistance to deflection. To obtain full advantage of the inbuilt strength of

334 Construction Technology

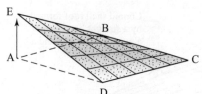

Straight line limited hyperbolic paraboloid formed by raising corner A of square ABCD to E

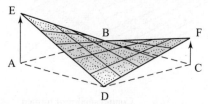

Straight line limited hyperbolic paraboloid formed by raising corners A and C of square ABCD to E and F respectively so that AE ≠ CF

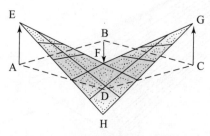

Straight line limited hyperbolic paraboloid formed by raising corners A and C and lowering corners B and D of square ABCD to E F G and H respectively so that AE=CG and BF=DH

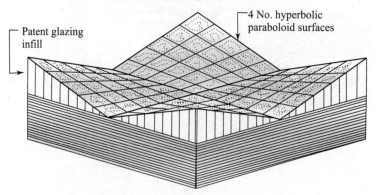

Hyperbolic paraboloids combined to form single roof

Fig. 10.8 Hyperbolic paraboloid surface

the shape, the rise to diagonal span ratio should not be less than 1:15; indeed higher the rise, greater will be the strength and the shell can be thinner.

By adopting a suitable rise-to-span ratio, it is possible to construct concrete shells with diagonal spans of up to 35 m with a shell thickness of only 50 mm. Timber hyperbolic paraboloid roofs can also be constructed using laminated edge beams with three layers of 20-mm tongued and grooved boards. The top and bottom layers of boards are laid parallel to the edges but at right angles to one another and the middle layer is laid diagonally. This is to overcome the problem of having to twist the boards across their width and at the same time bend them in their length.

Domes

Construction of domes using individually shaped wedge blocks or traditional timber roof construction techniques is in practice over the centuries. But, the method of construction of domes and the materials used have changed over the years. Domes in their simplest form comprise half-spheres. But, it is possible to construct domes having elliptical, paraboloid, and hyperboloid shapes (Fig. 10.9). Domes are double-curvature shells and are formed by a curved line rotating about a vertical axis (surface of revolution). Or, they can be translational domes that are formed by a curved line moving over another curved line. Pendentive domes are formed by inscribing a polygon within the base circle and cutting vertical planes through the true hemispherical dome.

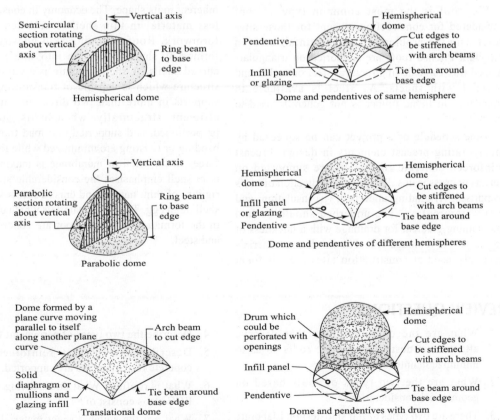

Fig. 10.9 Typical dome roof shapes

A dome shell roof would have the tendency to flatten due to the loadings and this tendency must be resisted by stiffening beams or something similar to all the cut edges. A dome which rises in excess of one-sixth of its diameter would generally require a ring beam.

Timber domes, like steel domes, are constructed on a single-layer grid system and covered with a suitable thin skin membrane.

SUMMARY

Roof design and construction is as important as design and construction of foundations. The roof is constructed to protect a building from heat, rain, wind, solar radiation, and other hostile environmental agents. Extensive variation in the design and construction of roof is possible similar to the variation done in the design and construction of foundations. A roof is a kind of shelter for a building or a structure, whereas roofing protects the roof from damage and deterioration.

Flat roof is the most common type of roof considered for construction, except for those sites beset with the problem of heavy rainfall or even snowfall. Pitched roofs are supported on triangular-shaped trusses made of structural steel or timber and covered with sheets. As steel is available in abundance in India, timber is not generally used to support roofs.

Time schedule of a project can be squeezed by incorporating precast elements in design. Precast reinforced concrete roof slabs are supported on purlins or crossbeams. Roofing for protection of roofs comprises: (i) grading underbed, (ii) insulation, and (iii) waterproofing. The underbed should be laid maintaining gradient for drainage with a thickness of 25 mm at the drainage end. The insulating materials generally used at construction sites are: (i) foam concrete, and (ii) expanded polystyrene blocks. Waterproofing provides protection against possible leaking of rainwater or accumulated water through the roof. Sheets used for covering pitched roofs are: (i) corrugated galvanized iron sheets, (ii) aluminium profiled sheets, and (iii) asbestos-free profiled sheets. The strength of the sheet roof depends on the properties of materials used and their ability to sustain self-weight and imposed loads.

A shell roof is a structural element with its strength inherent in its shape. The economy in consumption of less material in thin shell is nullified by costly formwork. Roofs utilizing skin construction are formed by shaping the skin into single or double curved shells. A shell structure is a curved surface structure which is capable of transmitting loads to supports in more than two directions. It is highly efficient structurally when it is so shaped, proportioned, and supported that load transfer when bending or twisting are minimized while that of axial forces in the shell or membrane is maximized. The term shell emphasises the considerable strength and rigidity of thin, natural and curved forms such as the shell of an egg. The basic materials which can be used in the formation of a shell roof are concrete, timber and steel.

REVIEW QUESTIONS

1. What are the considerations that deserve attention in designing durable roofs so as to minimize maintenance?
2. What are the two types of roofs based on geometrical considerations?
3. There are two types of flat slab design. Elaborate the two types.
4. Describe the two types of permanent formwork.
5. Describe how precast reinforced cement concrete roof elements are supported.
6. What is roofing? What does roofing on precast RCC slabs consist of?
7. What are the suitable sheet materials used for covering a roof?

8. What are the functional requirements related to roof covering sheets?
9. How is heat insulation provided in sheets used for covering roof?
10. What are the materials used for thermal insulation over roofs?
11. Describe the three types of membrane used for roof waterproofing.
12. What is the definition of shell roof? What are the basic materials that can be used to form shell roofs?
13. Describe how material saving in thin shells is offset by formwork.
14. How would you define a barrel vault? How can large areas be utilized by using barrel vaults?
15. Explain the two basic geometrical forms that have been conceived to form conoid shells.
16. State how a straight line limited hyperbolic paraboloid can be formed.
17. How can the flattening tendency of a dome shell roof be resisted?

11
Finishing Work

INTRODUCTION

Finishing work includes plastering, facing, glazing, flooring, and painting. Finishing work is executed for protection against environmental deterioration as well as for architectural attractiveness. Plastering is required because plastered surfaces are hard, abrasion resistant, rigid, incombustible, and provide a monolithic surface even at corners. In plasterwork, mortar is the most important material.

The term 'facing' is used to describe materials used as non-structural, thin, decorative, external finish like natural stone facings applied to brick or concrete backing to enhance architectural features. Facing is applied to the outside of basement walls, building fronts, and similar places. Facing work is carried out with both natural and artificially made products. Glazing is the act of fixing glass, and the main factors to be considered in choosing glasses are resistance to wind load, clear vision, privacy, security, fire resistance, and aesthetic appeal. Paint is a fluid material which when spread on a surface dry up and harden to form a continuous, adhering, and cohesive film to provide a protective surface as well as a visually attractive finish.

11.1 PLASTERING

Plastering, a thin plastic covering of different compositions applied using trowel to generally on walls and ceilings, is a protective coating on structures, buildings, and other installations and has architectural, decorative, and protective functions. Plastering is classified into ordinary, decorative, and special based on the kind of surfaces to be plastered. Plaster finishes are desirable finishing work as they are hard, abrasion-resistant, rigid, incombustible, and provide a monolithic surface, even at corners. Plastering should be properly applied so as to avoid cracking when movements due to drying, shrinkage, or thermal changes are restrained.

Ordinary plastering is applied mainly for coating internal premises and fronts of industrial, residential, and civil buildings in one or several layers. A single-layer plastering is done on flat brick masonry, concrete, and other similar surfaces. A multilayer plastering work comprises a preliminary coat, one or more floating coats, and a final setting coat. The preliminary coat, a tough fluid mortar, is meant to bond the plaster with the surface that is to be protected. Floating coats are intended to remove unevenness of plastered surfaces. The requirement of one or more floating coats would depend on the existing surface conditions. The thickness of each coat should not exceed 5 mm in case of cement mortars. The final setting coat, after application and smoothening with trowel should not be more than 2 mm in thickness.

Decorative plastering is applied to meet architectural and decorative requirements. This type of plasterwork comprises one or more floating coats, and one or several setting coats to provide the desired architectural or ornamental effect. Special plasterwork is intended for protection against harmful effects of the environment.

Plastering involves: (i) preparation of mortar in a mixing plant; (ii) transportation of mortar to the actual location of plasterwork; (iii) delivery to the personnel for actual application; (iv) erection of scaffolding for facilitating actual plasterwork; (v) preparation of surfaces for receiving plaster; and (vi) application of plaster coats in sequence.

In plastering, mortar is the key material. Mortar comprises fine aggregates (sand) and a binder, which usually is cement. Additives like admixtures may be used with cement to improve workability. Lime and cement, and simply lime may also be used as binder. The present trend is to use cement:sand mortars in various proportions depending on the nature of work and design strength. Properties of the mortar vary considerably on the basis of the properties of the cementitious materials used, ratio of the cementitious material to sand, characteristics and grading of the sand, water:cement ratio, and ratio of water to solids.

Aggregates for mortar could be of two types—heavyweight and lightweight. Heavyweight aggregates include river/sea sands, wastes resulting from grinding of heavy rocks like marble, granite, limestone, and so on. Lightweight aggregates are the wastes from grinding of lightweight rocks or artificial materials like pumice, tuffs, and so on.

Additions to plaster materials are intended to improve their quality, impart specific properties, and economize on the use of binders. The other effects of such additions are: plasticizing, damp-proofing, anti-freezing, decorative, and others.

Mortars with one binder like lime, cement, etc., are called plain binders. A combination of binders would result when more than one mortar is used. The kind and composition of binders would depend on the surface to be plastered, application and thickness of plaster, and the specifications as required. Properties of the plaster mainly include strength, density, plasticity, water-retaining capacity, porosity, heat conductivity, water absorption, water impermeability, shrinkage, resistance to harmful environmental effects, and so on.

The activities that need to be completed before initiating plastering are: (i) installing and securing door and window frames; (ii) filling and sealing gaps between walls and frames; (iii) covering all orifices in walls; (iv) fixing inserts for plumbing, electrical, and piping work; (v) embedding pipes/tubes that should not be exposed into walls; and (vi) other miscellaneous work.

Brick, concrete, and other stone surfaces to be plastered should be roughened and cleaned of grease, stains/paints, oil, bitumen, dust, and other dirty objects. After cleaning, all the surfaces to be plastered are flushed with water. Swelling and projections of bedding mortars should be chipped off. If bedding and side mortar joints are flushed with masonry, they must be raked to a depth of 10–20 mm or the surface should be scratched. If for any reason the thickness of the plaster exceeds 20 mm, wire mesh should be nailed on the wall to keep the plaster intact.

Every mason must possess a plumb bob so as to ensure that plastering work is executed true to the plumb. Mortar datum marks/patches are made on the walls to facilitate plastering in plumb and also to ensure even thickness and true surface. Patches of plaster, 100 or 150 mm square of the required thickness, should be fixed vertically at 2–3 m apart to act as gauges. All drips, grooves, mouldings, and cornices as shown on drawings should be done with special care to maintain true lines, levels, and profiles.

Plastering is carried out in sequence as follows: (i) ceiling and top parts of walls; (ii) cornices, grooves, and other moulding work; (iii) ceiling internal angles; (iv) ceilings and top parts of walls – coated and floated; (v) top parts of window openings and bottom parts of walls and also door/window opening sides; (vi) internal and external angles; and (vii) bottom parts of walls and openings – covered with mortar and finished. Plastering is carried out in this country manually by highly skilled masons who exhibit ingenuity in carrying out excellent work in their own ways. In course of time, however, mechanization would be incorporated in the plastering process also to achieve more progress in less time.

For imparting strength and resisting cracking, a metal base is sometimes used in plastering. The plaster sticks to metal lath by mechanical bond between the initial coat of plaster and the metal lath. It is, therefore, very important that the plaster completely surrounds and embeds the metal. Basic types of lath generally used are expanded metal, punched sheet metal, and paper-backed welded wire. The plastering should have a minimum thickness of 20 mm when measured from the back of the lath and the metal lath should conform to the specification or as shown in the relevant drawings.

Plastering Surface Finishes with Ordinary Mortars

- For standard finish, the interior plaster should be finished rough if punning is indicated. If not, the interior plaster should be finished smooth.
- For neat cement finish, the entire area should be uniformly treated with neat cement paste and rubbed smooth with a trowel as soon as a true plastered surface is achieved with the help of wooden straight edge.
- For coloured plaster finish, coloured cement should be used instead of ordinary cement. The plaster of thickness exceeding 12 mm should be applied in two coats. When plastering is to be done in two coats, coloured cement should be used in the top coat only. Instead of coloured cement, ordinary cement can also be used for this purpose with colour pigments.

- For pebble-dash finish, small pebbles or crushed stones of sizes ranging from 10 mm to 20 mm should be thrown on the plastic plastered surface made of cement:sand mortar (1:4 by volume). The pebble or aggregate pieces should be lightly tapped into the mortar with wooden float or mallet so that the pebbles or aggregate pieces remain secured to the plaster.
- For scraped finish, ordinary plaster is levelled and allowed to stiffen for a few hours. The stiffened surface is then scraped with a straight edge to remove the surface skin for a scraped finish.
- For textured finish, mortar of mix of cement:sand (1:3 by volume) should be applied as usual. Decorative treatments in the form of horizontal or vertical rib texture, fan texture, etc., should be applied using suitable tools to the freshly applied plastered surface.
- For grooves in plaster finish, grooves should have straight edges conforming to the pattern shown in drawings. Grooves in plaster should be done when the mortar is still in the plastic stage.
- For rough-coat finish, a wet plastic mix of coloured cement:sand:aggregate (3:6:4 by volume) should be thrown on to the wall using trowel and left in the rough condition. Stone aggregate should be of 6–12 mm in size.
- For sponge finish, the plaster is applied as usual using mortar having ratio of cement:sand (1:6 by volume). Cement slurry of moderate consistency should be sprinkled when the plastering mortar is still green. The slurry is sprinkled with a brush and evened with a wooden float till laitance is formed on the surface. Sponge of sizes such as 250 mm × 250 mm × 75 mm should be soaked in fresh water and softly pressed on the laitance to give it a rough textured finish.

All plastered surfaces should be cured for at least 7 days after completion of the process and should also be protected from excessive heat and sunlight.

11.2 FACING

Facing is used to describe materials used as a non-structural, thin, decorative, external finish such as natural stone facings applied to brick or concrete backing to enhance architectural beauty. Facing is intended for:

- Making surfaces attractive in appearance—for decorative purposes
- Protecting building from harmful environmental effects—engineering protection

Facing may be external or internal. Facing is applied to the outside of basement walls, building fronts, and internally in kitchens, toilets, industrial/chemical/utility plants, hospitals, subways, underground stations, and so on.

The background wall or frame is required for supporting the whole of facing up to a storey height or at 3-m intervals vertically by means of corbels or angles. Apart from supporting the facing materials, the question of resisting wind pressure and suction forces should also be considered.

Facing work is carried out with both natural and artificially made products. Slabs of natural stones like marble, granite, sandstone, limestone, and others are used for external facing. Artificial products like ceramic materials, glass, polymer-coated cement:sand tiles, polystyrene and amino plastic resin-based tiles, wood laminate plastics, etc., are also used.

Natural stone facings as well as reconstructed stone facings are to be fastened to the face of buildings to simulate the effect of solidity and permanence traditionally associated with solid masonry. Because of considerable cost involved in preparation and fixing, natural and reconstructed stone facings are in most cases used in prestigious buildings such as banks, city centres, and places of tourist interests. Granite is the natural stone which is very much in use as facing slabs for the hard durable finish. Polished granite slabs are used for the fine gloss surface that is maintained throughout the useful life of a building. Granite may be tooled to provide more rugged finish.

Marble is not chosen much as external facing materials because of polished marble finishes tend to lose their shiny appearance soon. Coarser surfaces like honed or eggshell finishes would maintain their finish as and when white or travertine marble is used.

Sandstone is chosen for use as facing slabs for the colour and grain of the natural material. The colour of standstone would gradually change over some years of exposure. The coarse grain of the material causes irregular runoff of water down the face and hence may result in stain. Some care and experiences are necessary for selection of sandstone as the quality, and therefore, durability, of the stone may vary between stones taken from the same quarry.

Limestone is used as facing, by and large, to resemble solid Ashlar masonry work (vide Chapter 9, Section 9.3), the slabs having a smooth finish to reveal the grain and texture of the material. These comparatively soft stones undergo a gradual change of colour over the years and this weathering is deemed as an attractive feature of these stones. Hard limestone is used as facing materials due to their hardness and durability. This type of stone is generally used as flat, level-finished, facing slabs.

As indicated in the relevant drawings, the stone slabs should be cut and finished to the required sizes and shapes. The stone should be fastened to the wall with suitable non-corrodible anchorage made of stainless steel or non-ferrous materials. The holes drilled in the stones for the bolts/anchors are then filled with pellets of stone to match the stone of the slabs. Joints between the stones are filled with mastic sealant to provide a weather-tight joint to accommodate differential thermal and structural movement. After the completion of curing, the exposed surfaces should be cleaned and pointing, if specified, should be carried out.

Ceramic tiles are universal facing materials. They are produced of ordinary and light-coloured clays by pressing and burning. Outside facing is generally carried out with ceramic tiles of ordinary clays with or without glazing of the face, which may be smooth or patterned. For better bonding with mortar, the back of each tile is grooved. The tiles are available in a variety of dimensions having such shapes. Facing work in the interiors is carried out with glazed ceramic tiles produced of white clays. Glass tiles made of glass waste are waterproof and acid-resistant, but not strong. Cement:sand tiles and cement:sand tiles with polymer coats are also used for facing work. The tiles can be cut and drilled as required.

Before starting facing work, the surfaces should initially be cleaned off unevenness, dust, dirt, grease, bitumen stains, and others. The surfaces to be covered with facings should be vertical and devoid of defects. If required, repair work should be undertaken and completed before starting facing.

11.3 GLAZING

Windows, sashes, doors, skylights, shop windows, and other light-transmitting enclosures are glazed with the following kinds of glass.

- Heat-absorbing glass—reduces heat, glare, and to a large extent ultraviolet rays; only resilient glazing material should be used; edges should be ground smooth and rounded off.
- Clear window glass—flat drawn sheet glass; the most widely used type for windows in all classes of buildings; used for doors also; available in thickness ranges from 3 mm to 25 mm.
- Plate and float glass—superior quality, more expensive, and better performance with no distortion of vision at any angle; ideal for showcase or quality windows.
- Obscure glass—process and rolled figured sheet; polished surface on one side with pattern on the other side for obscurity; is relatively popular.
- Obscure wired glass—have resistance against fire and breakage.
- Polished wired glass—more expensive compared to obscure wired glass; used in schools and institutions; may be clear or coloured; nominal thickness of 7 mm.
- Corrugated glass, wired glass, and plastic panels—used for decorative treatments, diffusing light, or as translucent structural panels with colour.
- Laminated glass—two or more layers of glass laminated together by one or more coatings of transparent plastic to add strength; some types provide security to an extent, sound insulation, heat absorption, and glare reduction; fade-proof opaque colours may provide privacy; lamination minimizes flying splinters.
- Bullet-resistant glass—made of three or four layers of glass laminated under heat and pressure; thickness varies from 20 mm to 75 mm; transparent plastics are also bullet-resistant materials.
- Tempered glass—this is produced by a process of reheating and sudden cooling thereby increasing strength substantially; all cutting and fabricating must be done before tempering.
- Tinted and coated glasses—for light and heat reflection, lower light transmission, greater safety, noise reduction, reduced glare, and increased privacy.

- Transparent mirror glasses—appears as a mirror when viewed from brightly illuminated side, and transparent to a viewer from the darker opposite side; available as laminate, plate, or float, tinted, and in tempered quality.
- Plastic window glazing—plastic glazing, made of plastic or polycarbonate is used for school buildings or where vandalism is apprehended; have considerably higher impact strength compared with glass or tempered glass.

Definitions Related to Glazing

- Glazing—the act of fixing glass
- Bar—either a vertical or horizontal component that extends to the full height or width of the glass opening
- Frame—a group of wood or metal parts machined and assembled so as to form an enclosure and support for a window or sash; plastics may be considered in special cases; aluminium, bronze, stainless steel, galvanized steel, steel are the metals used especially for fire resistance
- Jamb—side jamb is the upright component forming the vertical side of the frame whereas head jamb is the horizontal component forming the top of the frame
- Rabbet—rectangular groove along one or both edges to receive a window or sash
- Stiles—upright, or vertical, outside pieces of a sash or screen
- Rails—cross, or horizontal, pieces of the framework of a sash or screen
- Sash—a single assembly of stiles and rails made into a frame for holding glass, with or without dividing bars. It may be supplied either open or glazed
- Sill—the horizontal component forming the bottom of the frame
- Window—one or more single sash made to fill a given opening. It may be supplied either open or glazed

The principal factors to be considered in choosing glasses are: (i) resistance to wind load, (ii) clear vision, (iii) privacy, (iv) security, (v) fire resistance, and (vi) aesthetics. Glazing involves the following processes and operations:

- Preparation of putty
- Cutting of glass
- Handling of materials
- Preparation of sashes for glazing
- Glazing

The following kinds of putties are generally used: (i) chalk putties (ground chalk in natural linseed oil) for wooden sashes, (ii) quality putty (screened ground chalk, dry white lead, natural linseed oil) for critical applications, (iii) minimum putty (ground chalk, lead or iron minimum, natural linseed oil) for external metallic sashes, (iv) bitumen putty for skylights and external windows of industrial buildings, (v) coal tar putty (coal tar pitch, ground chalk, gasoline), and others. Glazing putty is required to be plastic, free from segregation, adhering to sashes and glass, hardening in not more than 15 days, and resistant to washing out by water.

Apart from putty, glazing compound, rubber, plastic strips, and metal or wood moulds are used for holding glass in place in sash. Metal clips for metal sash and glazing points for wood sash are also used for this purpose. Wooden sashes for residential houses are generally delivered to the construction sites glazed and primed with a coat of paint. When glazing is done in design position at a site, sashes should be properly fitted and primed with one coat of paint. Each glass pane should be held in position by special glazing metal clips of the approved type.

Glazing should conform to the specifications and design details. Putty should be free of cracks and should not peel off glass and rabbets. Pins, cramp irons, metal clips, and other means to secure glass panes should be fully coated with putty. Putty should terminate on the glass in a neat line, parallel to the rabbet. The external faces of glass panes should be clean and free of traces of putty, grease, and others.

Scaffolding, trestles, and stairs used in glazing work should be adequately strong and stable.

11.4 FLOORING

Flooring is floor covering that would meet the maximum standards at reasonable cost. Flooring, or floor covering, covers the floor. The functions of flooring are as follows:

- Appearance—should be aesthetically attractive still having reasonable wearing properties
- High resistance—should have wearing and impact resistance properties for busy areas
- Hygiene—should have reasonable aesthetic appeal with impervious and easy-to-clean surface

The ground floor of a residential building consists of hardcore, damp-proof membrane, and concrete slab. The removal of top soil and bad soil is essential for replacing the same with the specified soil, which is compacted to form the hardcore. Damp-proof membrane is an impervious layer like polythene sheet to prevent moisture passing into the interior through the floor. Concrete bed is the ground floor cast-in-situ concrete slab for taking loads of men and materials, and also to receive the desired floor finishes. Large cast-in-situ ground floors are designed to take medium-to-heavy loads as those used in factories, warehouses, shops, garages, and similar buildings. Large floors are cast in panels with expansion joints wherever required to take care of thermal expansion/contraction.

Suspended floors are cast in-situ in this country, although there is scope of precasting the slab elements and then erecting them thereby saving construction time. Terrazzo is composed of two parts of marble/granite chips to one part of Portland cement or as specified. Colour pigments may be added to make surfaces attractive. Terrazzo can be cast in-situ over structural concrete slab in three ways: (i) sand cushion, (ii) bonded, and (iii) monolithic.

Sand cushioning is provided where the possibility of structural movement exists due to settlement, expansion, contraction, or vibration. The underlying concrete slab is first covered with dry sand bed of 6–12 mm in thickness. Then, a wire-fabric reinforcing membrane is laid over this. The terrazzo underbed is built up to 16 mm (or as specified) below the finished floor level. Terrazzo topping is placed between divider strips of aluminium or glass.

Bonded terrazzo is placed over the structural concrete slab after flushing the floor slab with water for cleaning. The slab surface is then covered with neat Portland cement to ensure good bonding with the terrazzo as terrazzo is placed on the underbed after installing divider strips of aluminium or glass.

Monolithic terrazzo is constructed as 16 mm topping (or as specified) monolithic with the green concrete floor slab. Using epoxy resin adhesive, monolithic terrazzo has been successfully bonded with a topping thickness of only 10 mm. In-situ terrazzo work should be completed by grinding and polishing.

For Indian Patent Stone (IPS) finish, underbed would be the same as that of terrazzo finish but would be laid in alternate panels of 3 m^2 each. The topping would comprise cement:sand (1:1) slurry laid on the underbed when the same is not fully set. The topping should be adequately pressed for proper bonding followed by spreading, pressing, and rolling for neat cement finish.

For Ironite floor finish, one part of Ironite should be mixed dry with four parts of cement by weight or as specified. One part of this mixture should be mixed with two parts of sand by volume. The topping should be adequately pressed for proper bond followed by spreading, pressing, and rolling ensuring that cement:Ironite powder is at the top.

Terrazzo in precast form is used in treads, risers, platforms, and stringers on stairs. As it is possible to attain a large variety of colours and textures with terrazzo, it is used extensively as interior and exterior decorative flooring. For laying terrazzo tiles, underbed mortar should be evenly spread to bring it to proper grade and then compact it to a uniform surface. Before the underbed sets, cement slurry is applied and tiles are immediately placed and firmly pressed on to the underbed until the desired level of the tile is achieved. Tiles may be saw-cut as and when required. Tiles that are fixed should be cured for at least 7 days.

Ceramic tile is preferred in places like shower or swimming pool because of its attractive and water-resistant surface. Ceramic tile is a surfacing unit – relatively thin with respect to surface area. The ceramic tile has body of clay, or a mixture of clay and other ceramic materials that has been fired above red heat. It is referred to as the glazed tile if it is given a surface finish composed of ceramic materials that have been fused into the body of the tile and that may be impervious, vitreous, semi-vitreous, or non-vitreous. Impervious tile is required to have water absorption of less than 0.5% when tested compared to vitreous tile having water absorption between 0.5 and 3%. Ceramic mosaic tile is made of porcelain or natural clay and may be glazed or unglazed having a nominal thickness of 6 mm.

For setting ceramic tile, Portland cement mortar is generally specified for thick beds. Chemical-resistant tiles should be resistant to chemicals described in the drawings and specifications and should not absorb water more than 2% by weight when soaked in water for 24 h and should have a least compression strength of 700 kg/cm^2. The surface should be abrasion-resistant and durable. The tiles should have straight edges, uniform thickness, plain surface, uniform non-fading colour, and texture. Glazed tiles if permitted to act as chemical-resistant tile should be considered as mentioned earlier.

The mortar used for setting or for underbed for the chemical-resistant tiles should be durable and strong as required. The grout which should be up to the entire depth of the tile should have similar chemical-resistant properties as that of the tiles. The joints should be properly caulked and pointed if required.

11.5 PAINTING

Paint is a fluid material which when spread on a surface dries up and hardens to form a continuous, adhering, and cohesive film. Painting in buildings and structures is a finishing operation intended for protection and decoration purposes. There are also paints for special purposes like electrical insulation, fire retarding, anti-fouling, and signalling. The term 'paint' covers a wide range of paints, wood stains, varnishes, and oil for internal and external uses. The function of paints etc. is twofold: (i) to provide protective surface to materials like timber, steel, etc. and (ii) to provide visually attractive finish to the material. Colour and surface texture make a difference to our mood and can be used to create a strong or subtle visual effect.

As regards protective function, painting safeguards structures against attack by the environmental agents like moisture, acids, alkalis, and others. Paint is composed of a number of ingredients. Each ingredient is added with a specific purpose. Paint is composed of: (i) binder, (ii) pigment, (iii) solvent, and (iv) additive.

Binder is the medium which dries and solidifies after application to develop into a protective surface film. Alkyd resins and vinyl or acrylic resins have replaced linseed oil as binders for the majority of paints. The addition of a drier induces polymerization of the binder to ensure rapid drying. The main function of a paint medium is to provide a means of spreading the paint over the surface and act as a binder to the pigment at the same time.

Pigments are meant to provide colour to painting films. Unlike dyes, they are insoluble in water and solvents. By origin, pigments may be mineral or organic, but according to the manufacturing process, they are either natural or artificial. Pigment provides the body, colour, durability, and corrosion protection properties of the paint. White lead pigments are highly durable and moisture-resistant but toxic. Their use is restricted to priming and undercoating paints. The lead pigment content must be declared on the container. The general pigment used in paint is titanium dioxide, which is not toxic and which obliterates the undercoats. Pigments give opaque and translucent coloured films.

The function of solvent is to lower the viscosity of drying oils, oils, and to dissolve resins. The solvent, either water or an organic material, helps to create fluidity to the paint to facilitate painting with a brush and/or roller or spraying. Water-based paints, compared to paints made from organic solvents, are more favourably considered due to environmental, health, and safety reasons. Colour is added with organic or inorganic dyes and pigments.

The additives are driers, wetting agents, flow promoters, anti-skinning agents, anti-settling ingredients, and mar-proofing items. There is a wide range of paints which can be divided into water-based paints and oil-based paints.

Water-based paints are called emulsion paints. The additives like alkyd resin and polyvinyl acetate when mixed with water provide various finishes like matt, eggshell, semi-gloss, and gloss. Emulsion paints are easily applied, quickly dried, can be obtained with washable finish, and are suitable for many applications.

Oil-based paints are available in priming, undercoat, and finishing coats. The undercoat and finishing coats are available in a wide range of colours and finishes like matt, semi-matt, eggshell, satin, gloss, and enamel. Polyurethane paints have good hardness and resistance to water and cleaning. Oil-based paints are suitable for most applications if used in conjunction with correct primer and undercoat.

The finished paint system is made up of a series of 'coats' – each coat performing a specific task. For emulsion painting of the interiors, it is a routine to apply a 'mist coat' to the plastered wall, substrate. This is an undercoat of emulsion thinned with water. This undercoat is followed by two finishing coats of the emulsion. For gloss painting of both the interiors as well as the exteriors, the substrate requires a coat of primer followed by an undercoat and one or two finishing coats.

Paint manufacturers extend all kinds of guidance on the different types of paints that they manufacture and provide typical specifications for application for a wide range of materials.

Surface preparation: The substrate is the surface of the material to be painted. Preparation of the surface is an important criterion in achieving durable paint finish. The soundness and readiness of the surface to be painted must conform to the requirement.

- Timber: The moisture content should be less than 18% for adhesion of paint film. The timber should be finished smooth using abrasive paper. All surfaces should be free from dirt and loose or peeling paints. Careful treatment of knots is essential.
- Metals: All metal surfaces should be totally clean, dry, and free from wax, grease, etc. Further, all iron and steel surfaces should be free from rust. All galvanized steel surfaces should be pretreated with a compatible primer at the manufacturer's direction.
- Plaster: This should be perfectly dry, smooth, and free of defects. The plaster which contains lime should be treated with alkali-resistant primer.

Primers: The substrate must be clean and free of all loose materials so that the primer can adhere to it so as to inhibit corrosion or deterioration and offer a good base for the undercoat. The timber substrate should be dry with very low moisture content. Different primers are available for different materials and circumstances.

Undercoats: The primary function of the undercoat is to provide an opaque cover and build up necessary base for the application of the finishing coats. Undercoats are normally based on acrylic emulsions or alkyd resins.

Finishing coats: These are applied in one or more coats to impart the required protective and decorative surface finish.

Application of paints and stains is possible using brush, spray, roller, and application pads. For special effects, cotton rags and sponges may be used. All paints should be applied following the manufacturers' instructions. The internal rooms should be well ventilated to avoid accumulation of fumes. The temperature of the surface to be painted is also an important factor in achieving durable finish.

There are different kinds of paints. The most common ones are:

- Masonry paints—whitewashing, dry distemper, oil-bound washable distemper, waterproof cement paint, acrylic emulsion paint, and so on.
- Fungicide paint—applied in kitchens and bathrooms; this acrylic paint contains a mix of fungicides; the fungicide is released gradually over a long period.
- Water-repellent paint—silicone-based paint may be applied to porous surfaces to prevent water penetrating the wall and allowing evaporation of the water from the masonry.
- Waterproofing paints—epoxy waterproofing paints provide an impervious surface; bituminous paints serve a similar function.
- Heat-resisting paints—aluminium paint is resistant to temperatures of up to 250°C.
- Flame-retardant paints—a flame-retardant paint would release non-combustible gases when under fire thus retarding the spread of flame.
- Intumescent coatings—applied to give fire protection at 30, 60, and 120 min. Fire protection of structural steel is achieved through a thin coat (1–2 mm) of intumescent paint sprayed at the fabrication shop to control thickness. Intumescent emulsion paints and clear varnishes are also available for timber.
- Varnishes for timber—polyurethane varnishes are either water- or solvent-based systems available in gloss, matt, or satin finishes. The solvent-based varnishes are more durable coatings. Two coats of varnish provide adequate protection for most applications.

SUMMARY

Environmental factors that may adversely affect the integrity, behaviour, and service life of the structural elements of buildings and structures are related to corrosion, erosion, moisture, solar radiation and hot/cold weather. The trend now is to neutralize the ill effects of environment by improving traditional finishing work and introducing new products/processes. Finishing work includes plastering, facing, glazing, flooring and painting. Plastering is desirable finishing work as plastered surfaces are hard, abrasion resistant, rigid, incombustible, and provide a monolithic surface. Possibility of cracking when movements due to drying shrinkage or variation due to thermal changes are restrained in plasterwork by carrying out plastering correctly conforming to specification. The term 'facing' is used to describe materials used as non-structural, thin, decorative, external finish like natural stone facings applied to brick or concrete backing to add to architectural attractiveness. Facing is applied to outside of basement walls, building fronts etc and internally in kitchens, toilets etc. Facing work is carried out with both natural and artificially made products. Regarding glazing, the main factors to be considered in choosing glasses are resistance to wind load, clear vision, privacy, security, fire resistance and aesthetic appeal. Most flooring is to fulfill a particular function or functions like withstanding loading/unloading of materials and vehicular movement. Paints can dry up and harden in room temperature and are widely used as protective coatings with visually attractive finish. Colour and surface texture can influence our mood and can be used to create a strong or subtle visual effect.

REVIEW QUESTIONS

1. Describe how multilayer plastering is carried out.
2. What are the activities involved in plastering?
3. What are the useful properties of plaster?
4. What is the reason behind using a metal base in plastering? What type of laths are used as metal base?
5. What is the purpose of providing facing?
6. What are the natural or artificial materials used for facing?
7. Granite is preferred over marble for external facing work. Why?
8. Define glazing.
9. What are the different activities involved in glazing?
10. What are the different kinds of glasses used for glazing purposes?
11. What function or functions can be served by flooring?
12. Terrazzo can be cast in-situ over structural concrete slab in three ways. What are they?
13. When is ceramic tile referred to as glazed tile?
14. Paint is composed of four ingredients. Name them and provide brief details of each ingredient.
15. What are the requirements for surface preparation of timber, metal, and plaster?
16. There are different kinds of paints serving specific purposes. Provide details of five such paints.

12
External Work

INTRODUCTION

Construction of roads and drainage system is involved virtually in all project sites. Roads/highways also connect towns, cities, districts, states, and countries. Similarly, drainage is an external work, not confined to any boundary. Routes along any proposed road are surveyed in stages for working out an accurate layout.

Earthwork for road construction involves both excavation and filling. Where the water table is high, the drainage system should be completed well before the commencement of road excavation work. Excavation in rock is done by deploying construction equipment. Approved materials are used for forming road embankments including sub-grades, earthen shoulders, etc. Soil compaction is done by deploying construction equipment. During formation, shaping of the fill is checked continuously to ensure accurate embankment profile. The sub-base, apart from protecting the sub-grade, provides a foundation on which to lay the base, kerb, etc.

Bituminous flexible pavement courses are made by mixing coarse and fine aggregates with bituminous binder in a hot mix plant. Bituminous concrete should be used in wearing and profile corrective courses. Rigid concrete pavement comprises a concrete slab which is laid over the dry lean concrete sub-base. In road and highway work, the requirements of drainage concerns both sub-surface and surface water drainage.

The objective of designing a drainage system includes conducting storm-water from the carriageway to the outfalls, maintaining the water table well below the carriageway, blocking the passage of both ground and surface water flowing towards the road, and controlling drainage across the road alignment by constructing culverts and bridges.

12.1 ROADS

The main purpose of the road structure is to provide a means of reducing the stress or pressure to a level at which the ground beneath can support without deformation. Simultaneously, the pavement right up to the sub-grade needs to be strong enough to withstand the stresses and strains to which each layer is subjected to. Most roads are built to promote development by providing facilities for the transport of people and materials.

For both project onsite roads or offsite roads and highways, detailed survey is needed for determining and finalizing the layout. Route surveys of offsite roads and highways are executed in the following three stages.

- Reconnaissance survey—includes study of topographic survey sheets, revenue map, geological map, aerial photographs, and others.
- Preliminary survey—the purpose is to enable preparation of plans with respect to an accurate base line. The survey consists of plotting a traverse comprising a series of straight lines and their included angles measured very carefully. The accuracy of the final centreline depends to a great extent on the accuracy of the preliminary survey.
- Final location survey—on the basis of the preliminary survey that shows profile, cross-sections and contours, the best one is selected from the alternatives – the one that is best on technical, aesthetic, environmental, and economic viability considerations.

Survey

The modern precision theodolite which is capable of directly measuring both horizontal and vertical angles very accurately right up to 0.5–0.1 second is used in road surveys. Traverse survey by theodolite is the system of direct measurement of distances between two consecutive stations of observation and the angle made at each station by the two adjacent stations on either sides of it. In a closed traverse survey, the instrument is set up at each station, that is, the converging point of the two lines for measuring the included angles and the lengths of the lines. An alternative to this is to have an open traverse. In another alternative, the instrument is placed at a central point and the lengths of the radiating lines are measured. The 'closed traverse' method is generally followed at the construction sites (vide Chapter 4, Section 4.3). Open traverse survey is relevant in road construction work.

There are two basic types of theodolites: (i) optical and (ii) digital. Both the types are good for precise measurement of angles. The new technology of electronic reading instruments which are more accurate and more easily read are in common use now. In the latest models of electronic instruments, the facilities incorporated could be automatic data recording and processing, storage for downloading information when required, computer-aided design, mapping and global positioning systems (GPS). Some are even linked to laptops to enable more effective setting with accurate co-ordinates.

Electromagnetic distance measuring (EDM) equipment has radically changed the surveying procedures as distances can now be measured easily, quickly, and very accurately irrespective of the conditions of the terrain. For most engineering surveys, 'total stations' may be used. This requires a fully integrated instrument that captures all the data necessary for a three-dimensional position fix and displays it on digital readout systems, recorded at the press of a button. The common survey instruments are now 'total stations' combined with electronic data loggers. They are virtually standard equipment onsite particularly for road construction projects.

It is possible to transform basic theodolites into total stations by adding top-mounted EDM modules. A standard measurement can now be taken within 3 seconds. The principle that governs the working of every EDM is the same involving emission of a carrier wave that is usually infrared in the electromagnetic spectrum. This wave is reflected back to the instrument, usually from a corner cube prism. The distance travelled by the beam is computed by varying the modulation frequencies of the carrier wave and comparing the resulting phase shift. The speed of the wave may be affected by meteorological conditions and corrections for this may be made either manually or automatically. The accuracy of the readings within 3 mm at distances of up to 2.5 km can be achieved for a single prism reflector or up to 3.5 km for a prism cluster. The instrument is being upgraded continuously. In the remote positioning unit (RPU) version, the operator can be at the reflector end remotely controlling the instrument from that position.

Procedures setting with the EDM instruments or total stations vary as there are a variety of instruments available in the market. The lengths given in drawings are horizontal distances as seen on plans, but the modern survey instruments measure lengths between the instrument and the reflecting prism. This would require slope correction.

Using a total station, computer programme can be utilized to rotate the instrument forward/backward or right/left to the appropriate direction to set out the stored co-ordinates. The GPS is space-based microwave technology. GPS, apart from strategic military use, is now being used in engineering surveying also. There is a constellation of satellites in orbit around the earth. The GPS would eventually dominate the survey equipment field. The distances in GPS are called 'ranges' and are measured to satellites orbiting nominally at 20,183 km above the earth instead of control points on the earth. This system requires four distances, each from a different satellite, to achieve a positive fix and this is then complimented by a pseudo-random noise code (PRN), which enables the achievement of point accuracy. The use of the GPS for survey requires that the line of sight of the satellites remain unobstructed.

Automatic and semi-automatic levels are replacing dumpy level types which need to be adjusted before readings are taken. All the rotating beam laser levels incorporate automatic self-levelling systems and can generate horizontal planes by infrared beams produced by lasers. The levelling operation forms part of the total stations.

Layout

Roads are planned and staked on the basis of both horizontal and vertical alignments, and cross-sections. The alignment details should be enough for a surveyor to lay the road exactly as depicted in the drawings. The curves are laid out as arcs of circles – each point on such an arc is equidistant from the centre of the arc. A sharp curve would have higher degree of curve than a gradual one.

Relative to the curve, straight portions that are tangential to the connecting curves are called tangents. Tangents vary in length from a fraction of a metre to many kilometres.

For the exact survey of a road, a baseline should first be laid, which may be totally or partially outside of the right of way. This line is often made before the exact location of the road is decided.

The centreline is the basic location reference of the road itself. The setting is done on the basis of angles and distances from the baseline points. The profile of a road represents its rise and fall without an indication of a straight or curved route. The cross-sections show details of pavement width and thickness, shoulder and gutter width, crown or side slope, and other construction details.

Normally wooden or metal stakes are driven into the ground to guide the executing agency/contractor and the employees deployed for execution to follow the work involved. To start with, stakes are driven along the centreline showing depth of cut and height of fill, and slope stakes that show the outer limits of the area to be cleared, grubbed, and graded, and usually the cut and fill information also. When heavy cuts or fills are required, most of the work is performed with guidance of only slope stakes that are driven as the work progresses. When the working levels approach the sub-grade, additional stakes are needed. Centreline stakes are then restored.

Centreline stakes are driven at 30 m intervals, and sometimes as closely as 7.5 m in narrow, winding roads or in finishing operations. Slope stakes are set where the outer slopes of the cuts and fills meet the original grade usually at 30 m intervals along the road, and also at other points where the ground slope changes or other structures affect the slope. Slope stakes are usually set with a transit theodolite, dumpy level, and 30 m steel tape, if the total station is not used for this purpose.

Stakes would be dug away in cutting and buried in filling. In shallow cuts, the stakes could be retained temporarily on islands. It is, therefore, necessary to drive one or more reference stake/s well outside the work lines for resetting of the work stakes.

The distance of the ground surface above or below a desired elevation or grade is shown by the grade stakes. Vertical distances to the grade are marked on the stakes in metric system preceded by the letter C for cutting and F for filling. The extent of cuts or fills may be figured out from the base of the stake, from its top, or from the marking on the stake. The construction equipment operator should be able to view the stakes and markings, if any, on them. The fine grading or the final operation is guided by stakes driven until the tops are at sub-grade level.

Site Clearance and Preparation

This is basically the preparation of the entire site in advance so that the subsequent work may be continued smoothly without interruption. The problems are different in urban and rural areas. In rural areas, trees may have to be felled, uprooted, and removed from the site. Hedges may need to be grubbed up (grubbing is the removal of objects to a nominal depth below the surface) and burnt on site. But accumulation of a tangle of roots and top soil is not advisable owing to the difficulty involved in disposing the accumulated rubbish at a later stage in the work. Site clearance, therefore, comprises work involving cutting, removing, and disposing of all materials like trees, hedges, stumps, roots, grass, weeds, top soil, rubbish, and so on. The extent of stripping of the top soil would depend on the height of embankment to be formed. For a high embankment of 2 m or more, stripping of the top soil may not be necessary, depending, however, on the quality of the top soil. When the top soil is stripped, it should be salvaged and stored in heaps for subsequent reuse after completion of the roadwork particularly for covering embankment slopes, cut slopes, or other disturbed areas where re-vegetation is desired. Deployment of construction equipment for earthwork including land clearing is discussed in Section 13.4 of Chapter 13.

Temporary fences may need to be erected and access points provided to re-route the public past the construction site. The existing overhead or underground facilities or adjacent properties should not be disturbed during both site clearing and subsequent construction work.

Dismantling of existing structures

The existing structures and facilities/utilities like culverts, bridges, pavements, kerbs, guardrails, fences, manholes, catch basins, inlets, and others should be dismantled to the extent required in accordance with the decisions made and indicated in the relevant drawings and documents. This dismantling work should be performed prior to the beginning of the actual road construction. Structures above the ground level should be dismantled completely. Substructures may not be dismantled unless they cause problem or interference for the new construction. Dismantled materials should be disposed off as per the contractual provisions.

Earthwork

Earthwork is involved in both excavation and embankment formation. Excavation again involves both soil and rock and excavation in rocks involves both soft and hard rocks. In hard rocks, excavation is performed by normal and/or controlled blasting. In the case of soil, excavation is carried out in different kinds of soil. Earthwork in excavation involves excavation, removal, and disposal of the excavated soil satisfactorily. The excavated soil needs to be hauled for either dumping or use in formation of embankment and/or sub-grade. For details on soil classification, refer to Section 4.1 of Chapter 4 and Section 5.4 of Chapter 5 for information on rock drilling.

Excavation

Excavation may be defined as the process of loosening and removing soil or rock from its original position and transporting the same for filling embankments or disposal as waste. This is often categorized into three divisions: (i) road and drainage excavation, (ii) excavation for structures, and (iii) excavation resulting in borrow pits.

For details on the construction equipment to be deployed for earthwork, refer to Sections 4.4 and 13.4 of Chapters 4 and 13, respectively. These sections provide enough details in helping an engineer in selecting the appropriate equipment for road construction.

On completion of site clearance, the limits of excavation should be set out true to lines, curves, slopes, grades, and sections as per the contract specifications and drawings in order to begin excavation. The executing agency (contractor) should maintain the bench marks and reference points properly till the completion of all related work.

Excavation should be carried out immediately after the removal of the top soil so as to retain the optimum moisture content (of a sample taken at a depth of 1 m below the surface) of the soil. If the excavated soil is used for forming embankments, then the moisture content should be an important criterion to be considered as a high moisture content would render the soil plastic. Dry soil may crack and crumble. A handful of soil would stick together in a cohesive lump at optimum moisture content.

It is important that all excavated cuts are kept well drained during the entire excavation operation to prevent wetting and softening of the soil. It would, therefore, be necessary to construct a drainage system prior to the commencement of road excavation. In addition, cuttings are normally left with the final 300 mm of excavation undisturbed above the formation level until the final moment, as earth is its own best protector. In cutting, the excavation to formation level may well be lower than the water table at that location. In such cases, roadside drains should be laid before starting excavation. The provision of such drainage would be most useful in lowering the water table so as to avoid disruption of road construction.

The side slopes should be carefully maintained in cutting as backfilling to maintain slopes is not advisable except in cases where boulders or soft materials are encountered. In such cases, boulders and soft materials are removed and replaced with the approved materials to depths in line with the provisions in the specifications and design drawings. The filled up materials should be well compacted in the approved manner.

On completion of excavation, the side slopes should be trimmed so as to avoid either erosion or ponding, thereby ensuring natural drainage. Similarly, the side slopes should be trimmed in case of rock excavation also.

Cuttings through sound rock can stand close to vertical. In road excavation, the encountered rocks should be removed up to the formation level or such levels as indicated in the relevant drawings and specification. Where, however, unstable or unsuitable rock materials are encountered at the formation level, excavation should normally be continued to the extent of 500 mm below the formation level or in line with the provisions in the specification. Rock in no case should protrude above the specified levels. Rocks and large boulders that may cause problems of drainage and differential settlement should be removed and corrective action taken as per the specification.

In case of marshy and swampy land, the entire area along the road alignment should be excavated up to the firm level to the extent practicable before starting backfilling. In case of widening of existing pavements, care must be taken to ensure that flooding of the existing carriageway does not become a hazard to road uses. Excavation for removal of road shoulders should be executed in such a way as not to disturb the designated widths of the pavement. If disturbed anyway, corrective action as specified should be taken. Excavation should be done in proper sequence for surface or subsurface drains.

Excavation by blasting is covered in detail in Chapter 5, and the necessary safety measures related to blasting are discussed in Section 15.12 of Chapter 15 whereas smooth blasting by both conventional methods and pre-splitting has been discussed in Section 5.10 of Chapter 5. Pre-splitting is done to avoid overbreaking.

Pre-splitting is a technique of smooth blasting for reducing excessive overbreak and ground vibrations especially in soft rock by controlled use of explosives in properly aligned and spaced blast holes of depth not exceeding 9 m. Pre-splitting involves drilling a row of closely spaced, about 60–75 mm in diameter, full-depth blast holes along the design limit/periphery for forming rock excavation slopes at locations that are marked on the drawings. The beginning and end of excavations concerning pre-splitting should be carefully marked so that all overburden soil and weathered rock along the top of excavation for a distance of 5–15 m beyond the drilling limits, or to the end of excavation, must be removed before initiating drilling of pre-splitting blast holes. The blast holes should neither deviate from the plane of the planned slope by more than 300 mm nor should any hole deviate from being parallel to an adjacent hole by more than two-thirds of the spacing between the holes.

Specifications for Road and Bridge Works of the Indian Road Congress (IRC) contain details of Blasting Operations including pre-splitting in Clause 302. Structural excavation includes the excavation of materials to permit the construction of culverts, foundations for bridges, retaining walls, headwalls, cut-off walls, pipe culverts, and so on. Structural excavation would involve building of temporary structures for facilitating the actual construction work. The methods of construction of such temporary structures like cofferdams, sheet piles, shoring, bracing, pumping, etc. have been explained in detail in Sections 4.4 and 4.5 of Chapter 4. Section 4.6 would also be relevant in cases that require relocation of the existing facilities. Suitable materials obtained from structural excavations are used either in backfilling around the completed structure or other parts of the construction site. Clause 304 of the Specifications for Road and Bridge Works of the IRC contain details on 'Excavation'.

Embankments

Use of lasers and software in grading is discussed in Section 4.7 of Chapter 4. Embankments are formed in road construction when the alignment of the road necessitates level elevation of the existing ground to conform to the design requirements so as to prevent damage from surface or groundwater. Design heights of many embankments are only 0.5–1.5 m, but such heights could be 5 m or more on major highways.

High embankments impose heavy load on the underlying soil foundation resulting in settlement in some soils. If the soil foundation is very weak, a slip failure may occur. The residual soils are not compressible.

Approved materials obtained from road and drain excavation, borrow pits, or other sources are used in forming embankments including sub-grades, earthen shoulders, and other backfilling work. The materials used in formation are soil, moorum, gravel, a mixture of these, or any other approved materials.

The materials considered unsuitable for embankment formation are:
- Those originating from swamps, marshes and bogs
- Perishable materials like peat, log stump, etc.
- Materials susceptible to spontaneous combustion
- Materials in a frozen condition
- Clay having a liquid limit exceeding 70 and plasticity index exceeding 45
- Materials with salts resulting in leaching in the embankment

The materials for embankment formation should be obtained from approved sources preferably using materials from cutting for filling. Clause 305.2 (Materials and General Requirement) of the IRC Specifications for Road and Bridge Works furnish details of materials that are suitable to be used. The sizes of materials in the earth mixture should normally be not more than 75 mm in the embankment formation or 50 mm in the sub-grade. In any case, the maximum size should not be more than two-thirds of the compacted layer. As regards density, the maximum laboratory dry unit weight when tested as per IS:2720 (Part 8) should be (not applicable for lightweight filling materials like cinder, fly ash, etc.) as follows:

- 15.2 kN/m^3—embankments up to 3 m in height not subject to extensive flooding
- 16.0 kN/m^3—embankments exceeding 3 m in height or embankments of any height subject to long periods of inundation
- 17.5 kN/m^3—sub-grade and earthen shoulders/verges/backfill

The execution of filling should be planned in such a way as to ensure saving the best materials for use in filling the sub-grade and the portions immediately beneath the sub-grade. Haulage of materials of approved quality to be obtained from borrow pits for filling embankments or other areas should proceed only on switching on the spreading and compaction equipment in operation. It is to be ensured that the compaction yields the designed California-bearing ratio (CBR) value of the sub-grade. Relative compaction as percentage of maximum laboratory dry density as per IS:2720 (Part 8) is as follows:

- Not less than 97—sub-grade and earthen shoulders
- Not less than 95—embankment formation
- Not allowed—expansive clays in sub-grade and 500 mm below the sub-grade
- Not less than 90—expansive clays in the remaining portion of the embankment

On completion of site clearance, the limits of embankment/sub-grade need to be marked by fixing batter pegs on both sides at regular intervals as guides before earthwork in filling is started. The embankment or sub-grade should be sufficiently wider than the design requirement as this allows trimming of the surplus material ensuring that the remaining material could achieve the desired density and the alignment is maintained with side slopes as specified.

Where the base of the embankment is in a waterlogged area, water should, if possible, be removed by bailing or pumping making sure that the discharged water does not damage works, crops, or properties. Ordinary pipes or porous concrete pipes may be laid to drain the water away from the area.

Where filling is to be executed under water, only granular materials or rocks of approved quality should generally be used. Granular materials comprising graded, hard, and durable particles of maximum 75 mm in size could be approved for filling under water. The material should be non-plastic. The material should be deposited in open water by end-tipping without compaction.

Compaction

Compaction brings the soil particles into more intimate contact with each other by expelling air from the voids. This is the cheapest and simplest method for improving the shearing resistance of soil and minimizing future settlements.

The original ground, if deemed necessary, should be levelled to facilitate placement of the first embankment layer. It is necessary that the fill materials used to form embankments are compacted to the maximum dry density and to accurate surface profile. This achievement would depend on the following factors:

- Characteristics of fill materials
- Moisture content of fill materials
- Type of compaction equipment deployed
- The weight of the compaction equipment relative to roller width or base plate
- Energy applied
- Thickness of the layer to be compacted
- Number of passes for the required compaction

Clause 305 (Embankment Construction) of IRC Specifications for Road and Bridge Works provides detailed guidance on use of the range of materials for embankment construction. Construction equipment for achieving best possible compaction should be selected very carefully (Chapter 13). The plants for earth filling compaction consist of:

- Rollers of various types (Section 13.4 of Chapter 13)
- Impact compactors of two types: (i) vibratory plate type and (ii) impact type—in areas where rollers are unsuitable due to space restriction (Section 13.4 of Chapter 13).

The compaction should be done with 8–10 tonne vibratory roller of adequate capacity capable of achieving the required compaction. The final compaction is specified by either (i) method—based on type of fill, thickness of layer, type of equipment, number of passes, roller speed, and moisture content of soil, or (ii) performance requirement—such as specified dry density to be checked by laboratory tests. The performance tests can be arranged only at large and important road projects. In smaller projects, compaction of thin layers of filling should be continued till the compaction marks disappear. During the formation of embankment, the fill shaping should be checked continuously during the course of filling for the surface to develop an accurate profile.

Clause 306 of the IRC Specifications for Road and Bridge Works deals with Soil Erosion and Sedimentation Control. Soil erosion, sedimentation, and water pollution control is implemented through the use of berm, dykes, sediment basins, fibre mats, mulches, grass, slope drains, and other devices.

Lightweight and flexible synthetic geotextiles (also referred to as 'filter fabrics' and encompasses the entire range of fabrics, from woven to non-woven) made of high-strength filaments encased in polyolefin sheath in flat form are used for soil stabilization. Plastic mesh structures, which allow free passage of water without getting clogged, can be used to stabilize and improve the load-bearing capacity of soft ground. The mesh helps to produce high-density, shear-resistant layer of soil that distributes vertical loads and restrains soft ground in the lower layers of the sub-grade.

Geotextiles act as a separating material in certain types of pervious soils (i) for reducing the possibility of water seepage through the pavement into the sub-grade; (ii) for preventing intermixing of the sub-grade and sub-base during construction; (iii) for continuing work in inclement weather; and (iv) for executing work over such ground conditions as are unsuitable for traditional methods of road construction. Jute geotextiles have been successfully inducted in a road construction project over weak soils.

Geotextiles in the form of plastic mesh are used on the side slopes and between layers to reinforce the embankment formed. They are also used for erosion control. Gabions are made of galvanized wire mesh filled with durable stones. The mesh size should be suitable for filling stones of 125–250 mm in size. To be effective, the gabions are tied together and anchored to form a solid structure for stabilization purposes.

Stabilization is a process by which a soil is improved and made more stable. The addition of cement or lime can give increased strength for soil stabilization purposes. The most widely used binders for soil stabilization are cement and hydrated lime. A cement- or lime-treated soil is an intimate mixture of pulverized soil, cement or lime, and water. Content above 2% cement or 2.5% lime is recommended with sufficient water for hydration or slaking. The exact cement or lime content is determined by laboratory tests. Mixing is carried out in place using stabilizing machine in layers of 130–250 mm thickness. The layers should be compacted immediately on placing.

The sub-grade is the soil acting as a foundation for the pavement. The sub-grade is the result of the earthwork, and it may consist of the undisturbed, local soil or material excavated elsewhere and placed as fill. The surface of the sub-grade is referred to as formation and its strength is assessed in terms of the CBR value.

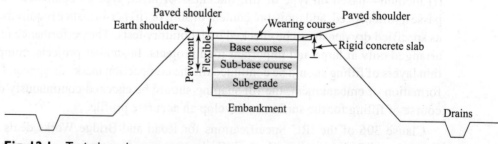

Fig. 12.1 Typical road cross-section

A capping layer is a strengthening layer that may be used on very weak sub-grades. It is normally defined as part of the sub-grade and is made from imported, selected fill or from sub-grade material stabilized with lime.

Sub-bases

Clause 400 (Sub-Bases, Non-Bituminous Bases, and Shoulders) of the IRC Specifications for Road and Bridge Works deals with sub-bases and bases above sub-grades. The functions of the sub-bases are:

- To assist in spreading the design load—a secondary load-spreading layer
- To provide some protection to the sub-grade as soon as it is exposed to both rain (soggy/slushy) or sun (dry/cracking)
- To provide a foundation on which to lay the base, and kerb, bed, and backing

The material to be used for the work could be natural sand, moorum, gravel, crushed stone, or combination thereof depending on the grading required. Materials like crushed slag, crushed concrete, brick metal, and kankar may also be used. All materials should be free of organic matter and deleterious materials. It is to be ensured prior to actual execution that the material to be used in the sub-base satisfies the requirements of CBR and other physical requirements when compacted and finished.

Lime-stabilized soil should be used for improved sub-grade or sub-base—especially for soils containing a large percentage of clay and silty clay (plasticity index > 8). Cement-stabilized soil should be used for improved sub-base/base on prepared sub-grade/sub-base.

Water bound macadam (WBM) sub-base/base comprises clean, crushed aggregates mechanically interlocked by rolling and bonding together with screening, binding material where necessary and water laid on a properly prepared sub-grade/sub-base/base or existing pavement. Coarse aggregates are either crushed or broken stone, crushed slag (air-cooled), over-burnt brick aggregates, or any other naturally occurring aggregates such as kankar and laterite of suitable quality. Crushed cement concrete sub-base/base work consists of breaking and crushing the damaged cement concrete slabs and consolidating the same again in one or more layers as sub-base and base.

Wet mix macadam (WMM)

This work consists of laying and consolidating clean, crushed, graded aggregate and granular material, premixed with water, to a dense mass on a prepared sub-grade/sub-base/base or existing pavement as the case may be. The materials should be laid in one or more layers as necessary to lines, grades, and cross-sections as shown in approved drawings. The thickness of a single compact WMM layer should be not less than 75 mm. However, if compaction is done by vibration or similar method, the compact depth of the sub-base course may be increased to 200 mm. WMM would be mixed, placed, and consolidated as per Clause 406 of the IRC Specifications of Road and Bridge Works.

Shoulders, islands, and median

This work consists of constructing shoulder (hard/paved/earthen with brick or stone block edging) on either side of the pavement, median in the road dividing the carriageway into separate lanes and islands for directing the traffic at junctions in accordance with details shown on drawings. The selected earth/granular/paved materials should conform to the requirements of Clauses 305/401 of the IRC Specifications of Road and Bridge Works.

Cement concrete kerb and kerb with channel

This work consists of constructing cement concrete kerbs and kerbs with channel in the central median and/or along the footpaths or separators as per the drawings.

Base and surface courses (bituminous)

Bitumen is composed principally of high-molecular weight hydrocarbons. It is a thermoplastic material that gradually softens and eventually liquefies on heating. Bitumen is characterized by its consistency at certain temperatures. The bituminous binder material should conform to IS:73. The binder must be sufficiently viscous not to drain off from the road surface and strong enough to retain the aggregates when the road is opened to traffic.

Bituminous flexible pavement courses are made by mixing coarse and fine aggregates with the bituminous binder. Coarse aggregates are crushed rock or gravel or other material retained on the 2.36-mm sieve whereas fine aggregates pass through the 2.36-mm sieve and are retained on the 75-micron sieve. The aggregates should be clean, hard, durable, dry and free from dust, soft or friable or organic or other deleterious materials; the coarse aggregates should be of cubical shape. All materials should be tested and approved before commencing actual construction.

Paving bituminous materials including bituminous macadam, dense bituminous macadam, semi-dense bituminous concrete, and dense bituminous concrete should be premixed in a hot mix plant of adequate capacity and capable of yielding a mix of appropriate and uniform quality with thoroughly coated aggregates. The hot mix should be transported and placed in the shortest possible time taking appropriate measures so that the quality of construction work is not affected even by adverse weather. The bituminous materials should be spread, levelled, and tamped by an approved self-propelled paving machine comprising a tractor and screed unit. The materials should be supplied continuously in a receiving hopper and laid without delay. The bituminous material should be kept clean and uncontaminated. The only traffic permitted to run on bituminous material to be overlaid should be that engaged in laying and consolidating the next course or, where a binder course is to be sealed or surface-dressed, that engaged on such surface treatment.

The binder course material should not remain uncovered for more than three consecutive days after being laid by either the wearing course or surface treatment, whichever is specified in the contract. This period is extended in case of unfavourable weather conditions.

The bituminous materials, which are laid in layers, should be compacted to achieve optimum density of the mix and provide a smooth riding surface. The compaction is commenced immediately after laying without any delay and it should be completed as much as possible before the temperature falls below the specified minimum rolling temperature/s. Rolling of the longitudinal joints is done immediately after the paving operation. Rolling thereafter should commence at the edges and progress towards the centre longitudinally except that on super elevated and unidirectional cambered portions, it should progress from the lower to the upper edges parallel to the centreline of the pavement. The rolling is continued till all the roller marks are erased. The initial or breakdown rolling is done with 8–10 tonnes weight of smooth-wheeled rollers and the intermediate rolling with 8–10 tonnes of deadweight or vibratory roller or with pneumatic wheel-typed roller of 12–15 tonnes in weight having nine wheels, with a tyre pressure of at least 5.6 kg/cm^2. The finish rolling is done with a weight of 6–8 tonnes of smooth-wheeled tandem roller. More details on the construction of bituminous pavement are included in Clauses 501.6 to 501.8 of the IRC Specifications of Road and Bridge Works.

Prime coat comprises a single coat of low viscosity liquid bituminous material over a granular base surface preparatory to the superimposition of bituminous treatment or mix. The prime coat distributor should be self-propelled or towed bitumen pressure sprayer equipped for spraying the material uniformly at the specified rates and temperature.

Tack coat comprises a single coat of low viscosity liquid bituminous material to an existing bituminous road surface to promote adhesion between the old and new superimposed bituminous mix, if specified or instructed by the engineer. The tack coat distributor should be self-propelled or towed bitumen pressure sprayer equipped for spraying the material uniformly at a specified rate (0.4–0.6 l/m^2) and temperature.

Bituminous macadam

The work involved in bituminous macadam comprises construction in single course having 50–100 mm thickness or in multiple courses of compacted crushed aggregates premixed with a bituminous binder (paving bitumen) on a previously prepared base. Bituminous macadam is more open graded than the dense-graded bituminous materials described next. The detailed specification of bituminous macadam is furnished in Clause 504 of the IRC Specifications of Road and Bridge Works.

In bituminous penetration macadam, one or more layers of compacted crushed course of aggregates are constructed with alternate applications of the bituminous binder (paving bitumen of specified penetration grade) and key aggregates in accordance with the requirements of the IRC Specifications to be used as base course on roads, subject to the requirements of the overall pavement design, in conformity with the lines, grades, and cross-sections shown on the drawings or as directed by the engineer. The thickness of an individual course should be 50 mm or 75 mm or as specified. After the spreading of coarse aggregates using self-propelled or tipper tail-mounted aggregate spreader capable of spreading aggregate uniformly at a specified rate over the required widths, dry rolling should be carried out with a 8–10 tonne smooth steel roller. After initial dry rolling, the surface should be checked with a crown template and a 3 m straight edge. The surface should not vary more than 10 mm from the template or straight edge. All surface irregularities exceeding this limit should be corrected by removing or adding aggregates as required. After the coarse aggregate has been rolled and checked, the bituminous binder should be applied at a specified rate and at specified/approved temperature. Immediately after the first application of bitumen, the key aggregates, which should be clean, dry, and dust-free, are spread uniformly over the surface by means of an approved mechanical spreader. The entire surface should then be rolled with 8–10 tonne smooth steel wheel roller. The surface finish of the completed construction should conform to the requirements of Clause 902 of the IRC Specifications of Road and Bridge Works. The penetration macadam should be provided with surfacing (binder/wearing course) within a maximum of 48 hours.

The built-up spray grout consists of a two-layer composite construction of compacted crushed coarse aggregates with application of bituminous binder after each layer, and with key aggregates placed on top of the second layer, in accordance with the requirements of the IRC Specifications, to serve as a base course and in conformity with the lines, grades, and cross-sections shown on the drawings or as directed by the engineer. The thickness of the course should be 75 mm.

Immediately after spreading of the aggregates, the entire surface should be rolled with 8–10 tonne smooth wheel steel roller. The rolling should commence at the edges and progress towards the centre except in super-elevated or unidirectional cambered portions where it should proceed from the lower to the higher edges. Each pass of the roller should uniformly overlap not less than one-third of the track made in the preceding pass. Care should be taken in not overcompacting the layer.

Dense graded bituminous macadam

Built-up spray grout should be used in a single course in a pavement structure. Dense graded Bituminous Macadam (DBM) is used chiefly, but not exclusively, in base/binder and profile-corrective courses. DBM is also intended for use as road base material. This work involves construction in a single or multiple layers of DBM on a previously prepared base or sub-base. The thickness of a single layer should be 50–100 mm. The bitumen should be paving bitumen of penetration grade conforming to IS:73. The detailed specification of the materials including binder, mixture design, laying trials, and construction operations are included in Clause 507 of the IRC Specifications of Road and Bridge Works.

Semi-dense bituminous concrete

Semi-dense bituminous concrete is used in wearing/binder and profile-corrective courses. This work comprises construction in single or multiple layers of semi-dense bituminous concrete on a previously prepared bituminous bound surface. A single layer should be 25–100 mm in thickness. The detailed specification of the materials including binder, mixture design, laying trials, and construction operations are included in Clause 508 of the IRC Specifications of Road and Bridge Works.

Bituminous concrete

Bituminous concrete is used in wearing and profile-corrective courses. This work comprises construction in single or multiple layers of bituminous concrete on a previously prepared bituminous bound surface. A single layer should be 25–100 mm in thickness. The detailed specification of the materials including binder, mixture design, laying trials, and construction operations are included in Clause 509 of the IRC Specifications of Road and Bridge Works.

Surface dressing

The uses of surface dressing in maintenance work are:
- Waterproofing the road surface
- Arresting disintegration
- Providing a non-skid surface

Surface dressing involves application of one or two coats of a layer of bituminous binder sprayed on a previously prepared base. This is followed by a cover of stone chips rolled in to form a wearing course. Refer to the Manual for Construction and Supervision of Bituminous Works for information on surface-dressing design. The detailed specification of the materials including binder, mixture design, laying trials, and construction operations are included in Clause 510 of the IRC Specifications of Road and Bridge Works.

Open-graded premix surfacing comprises preparation, laying, and compaction of surfacing material of 20 mm thickness composed of small-sized aggregates premixed with a bituminous binder on a previously prepared base to serve as a wearing course. In case of close-graded premix surfacing/mixed seal surfacing, however, surfacing material of 20 mm thickness composed of graded aggregates premixed with a bituminous binder are used. The detailed specification of the materials including binder, proportioning of materials, preparation of premix, spreading and rolling, seal coat, etc. are included in Clauses 511 and 512 of the IRC Specifications of Road and Bridge Works.

The seal coat involves application of a coat for sealing the voids in a bituminous surface laid to the specified levels, grade, and cross-fall (camber). The seal coat can be of the following two types:

(i) Liquid seal coat comprises an application of a layer of bituminous binder followed by a cover of stone chips.

(ii) Premixed seal coat comprises a thin application of fine aggregates premixed with bituminous binder.

The detailed specification of the materials including binders for the aforementioned types, construction operations, and opening to traffic are included in Clause 513 of the IRC Specifications of Road and Bridge Works.

Slurry seals are mixtures of fine aggregate, cement filler, bitumen emulsion, and additional water. When freshly mixed, they have a thick consistency and can be spread to a thickness of 1.5–5 mm. The mixture may be used to seal cracks, arrest fretting, and fill voids and minor depressions to provide a more even riding surface or a base for further treatment; the mixture may also be used above a single-coat surface dressing. It is usual to use additives like cement, hydrated lime, or others to control consistency, mix segregation, and setting rate. The proportion of the additive should not normally exceed 2% by weight of dry aggregates.

Recycling the existing flexible pavement materials

Recycling the existing bituminous pavement materials from a paved road that has served for the period it was built for has emerged as a good cost-effective method in upgrading an existing bituminous pavement. Two key issues that need to be addressed are: (i) the quality of materials to be recycled and (ii) type of binder to be added to make the rejuvenated bitumen mix work. The work involved comprises pavement removal, stockpiling of the materials from the old pavement, addition of new bitumen and untreated aggregate in the requisite proportions, mixing, spreading, and compaction of the blended materials. Recycling could be:

- *In-situ* recycling—where processing takes place onsite
- Central plant recycling—where reclaimed materials are processed offsite; it allows for multilayered construction and greater control of the material

The processes can be further subdivided into:

- Cold process
- Hot process—IRC specifications deal with hot process only

The removal of the pavement materials to the required depth should be accomplished either at ambient temperature (cold process) or at an elevated temperature (hot process) as specified. In the cold process, the work of ripping and crushing is carried out using scarifiers, grid rollers, or rippers or as specified. The materials thus obtained are loaded and hauled for crushing to the specified size. On removal of the pavement material, drainage deficiencies, if any, need to be corrected. Then, the base/sub-base as the case may be should be cut, graded, and compacted to the required profile and density.

In the hot removal process, the road surface should be heated by suitable means before scarification. A self-propelled plant removes the heated soft bituminous layer. During the heating process, the surface temperature of the road should not exceed 200°C for more than 5 min. Then the recycled mixture is spread on the preheated surface and tamped and compacted to the required profile.

Rigid concrete pavement

The concrete pavement comprises a concrete slab which is laid over the dry lean concrete sub-base. Compared to a flexible pavement, this concrete slab is the equivalent of the surface course, binder course, base and sub-base combined. It is termed rigid as the concrete slab does not deflect within itself under traffic load and is designed for about four decades before needing major reconstruction. The work is carried out in accordance with the lines, grades, and cross-sections shown on the approved design drawings. The design parameters like width, thickness, grade of concrete, details of joints, etc., of dry lean concrete sub-base should also conform to the approved design drawings.

Dry lean cement concrete sub-base

The sub-grade should conform to the grades and cross-sections depicted in the drawings and should be uniformly compacted to the specified strength. The lean concrete sub-base should not be laid on any sub-grade softened by rain after its final preparation. Surface trenches and soft spots, if any, must be properly backfilled and compacted so as not to have any weak or soft spot. The construction traffic should, as far as possible, be avoided on the prepared sub-grade. In order to stabilize the loose surface, water should be sprayed on the sub-grade surface and the sub-grade should be rolled with one or two passes of a smooth-wheeled roller after a lapse of 2–3 hours. This is performed one day before placing of the sub-base.

The pace and programme of the lean concrete sub-base construction should be synchronized with the programme of laying cement concrete pavement over it. The overlaying of cement concrete pavement follows only after 7 days of the sub-base construction. The design parameters of dry lean concrete sub-base like width, thickness, grade of concrete, details of joints if any, etc., should conform to the details in the approved drawings.

Cement conforming to IS:269 or IS:455 or IS:1489 (Part 1) or IS:12330 may be used. Aggregates for lean concrete should conform to IS:383 and properly graded. The aggregates should not be alkali-reactive. Water should meet the requirements of IS:456.

The lean concrete mix is proportioned with a maximum aggregate:cement ratio of 15:1. The correct amount of water should be provided for the lean concrete so as to ensure complete compaction under rolling and should be decided at the time of trial rolling. High water content would cause the concrete to compress and heave in front of the roller. The cement content in the mix should not be less than 150 kg/m^3 of concrete. The average compressive strength of lean concrete should not be less than 10 MPa at 7 days whereas the minimum compressive strength of any sample should not be less than 7.5 MPa at 7 days.

The design features of the batching plant feeding the paving mixers should be such that the shifting operations of the plant do not consume more time when shifting is required during work progress. For details on batching plant and paving mixers refer to Section 13.6 of Chapter 13. Contraction and longitudinal joints should also be provided if shown on the drawings.

Curing of sub-base lean concrete is carried out in one of the following two methods.
- The initial curing is done by spraying liquid-curing compound. The curing compound is to be white-pigmented or transparent type with a water retention index of 90% when tested in accordance with BS:7542. The curing compound is sprayed immediately after rolling is complete. As soon as the curing compound has lost its tackiness, the surface should be covered with wet hessian for 3 days.
- Curing is done by covering the surface with gunny bags/hessian, which should be kept continuously moist for 7 days by sprinkling water.

Cement concrete pavement

The cement concrete pavement may be constructed in any one of the following methods as approved:

- Un-reinforced concrete
- Dowel-jointed concrete
- Continuously reinforced concrete pavement

The work generally comprises construction of un-reinforced, dowel-jointed, plain cement concrete pavement in accordance with the lines, grades, and cross-sections shown on the drawings. The design parameters like thickness of pavement slab, grade of concrete, joint details, etc., should be as per the approved drawings.

Designing of concrete mixes for rigid pavement follows SP:23 – *Handbook on Concrete Mixes of the Bureau of Indian Standards*. For the details on the ingredients and properties of concrete, refer to Chapter 7. Admixtures may be added to improve workability without having any adverse effect on the properties of concrete on strength, volume change, and durability. The admixture should also not have any damaging effect on reinforcing steel. The satisfactory performance of the admixtures should be proved both on trial mixes and in trial paving work. The aggregates should conform to IS:383 but with LA abrasion test result not more than 35%. They must be free of chert, flint, chalcedony, or other silica in a form that can react with the alkalis in cement. In addition, the total chlorides content expressed as chloride ion content should not exceed 0.06% by weight and the total sulphate content expressed as sulphuric anhydride (SO_3) should not exceed 0.25% by weight. The maximum size of coarse aggregates should not exceed 25 mm for pavement concrete and they should be tested for soundness in accordance with IS:2386 (Part 5). After five cycles of testing, the loss should not exceed 12% if sodium sulphate solution is used or 18% if magnesium sulphate solution is used. Fine aggregates should be free from soft particles, clay, shale, loam, cemented particles, mica, organic, and other foreign matter. They can contain deleterious substances in the following ratios only: (i) clay lumps – 4%; (ii) coal and lignite – 1%; and (iii) minerals that can pass through the 75-micron sieve.

Pre-moulded firm compressible joint filler board for expansion joints that are proposed for use only at some abutting structures like bridges and culverts should be 20–25 mm in thickness with a tolerance of ±1.5 mm (IS:1838). The filler's thickness should be 25 mm less than the thickness of the slab within a tolerance of ±3 mm provided to the full width of the side forms.

The joint sealing compound should be hot-poured, elastomeric, or the cold polysulphide type having flexibility, resistance to age hardening, and durability. The cement content should be within 350–425 kg/m^3 of concrete. The strength of concrete needs to be increased to provide greater durability in order to offset wear due to increasing volume and weight of traffic.

The cement concrete pavement would be laid over the sub-base in accordance with the design drawings and relevant specifications. The sub-base should be repaired if found damaged at some places before the separation membrane is placed between the concrete slab and the sub-base. The membrane should be impermeable plastic sheeting 125-microns thick laid flat without creases. Overlapping of the membrane, where necessary, should be 300 mm. This waterproof membrane serves two purposes: (i) it prevents the wet concrete from losing some of its moisture into the sub-base below so that the concrete can attain full strength on completion of total hydration; (ii) it allows the concrete slab, when set and hardened, to move reasonably freely over the sub-base to relieve thermal stresses.

The location and type of joints is provided as per the drawings. The joints are constructed depending on their functional requirements as follows.

- Transverse joints are contraction and expansion joints (Figs 12.2 to 12.4). These joints, which are transverse to the longitudinal axis, are provided at the spacing specified in the drawings. The contraction joints should be cut as soon as the concrete undergoing initial hardening is hard enough to bear the load of sawing machine without causing damage to the slab. The expansion joints comprise filler board and dowel bars as detailed in the drawings with joint sealant at the top. These joints are not provided at junctions and roundabouts. Any reinforcement in the deck must be discontinued at the joint.

- Transverse construction joints (Fig. 12.5) are placed whenever concreting is completed after a day's work or is suspended for more than 30 minutes. These joints are provided at the regular location of contraction joints using dowel bars and they are made as butt type joints. At all construction joints, steel bulk heads are used to retain the concrete while the surface is finished.

- In order to prevent differential vertical movement between adjacent slabs, dowel bars are provided, set at mid-depth of the slab and parallel to the longitudinal axis of the road. One end of the dowel bar is de-bonded to keep it free for dowel's movement. The other end is cast into the concrete of the adjacent slab.

- Longitudinal joints should be saw cut as per the details in the drawing. The groove may be cut after the final set of the concrete. Joints should be sawn to at least one-third of the depth of the slab ±5 mm as indicated in the drawing. Longitudinal differential vertical movement between the slabs is prevented by fixing mild steel

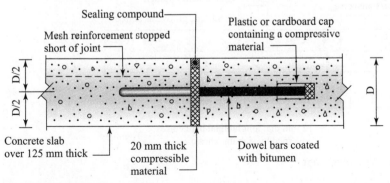

Fig. 12.2 Expansion joint

tie bars of generally 12 mm in diameter. Tie bars prevent the joint from opening by more than a fraction of 1 mm thus maintaining the interlocking of aggregate particles on the two sides of the track.

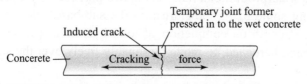

Fig. 12.3 Contraction joint for plain concrete

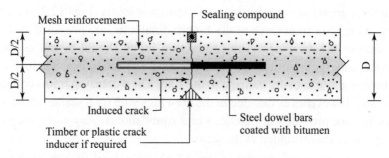

Fig. 12.4 Contraction joint for reinforced concrete

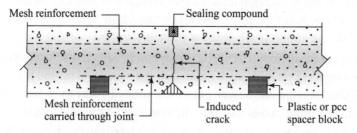

Fig. 12.5 Construction joint

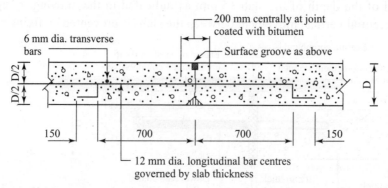

Fig. 12.6 Warping joint

Timber forms are not used any more. All side forms are of mild steel of depth equal to the thickness of pavement or slightly less to accommodate the surface regularity of the sub-base. The forms can be placed on a series of steel packing plates or shims to avoid unevenness/irregularity of sub-base. They should be sufficiently robust and rigid to support the weight and pressure caused by the paving equipment. Side forms incorporate metal rails and are firmly fixed at a height below the top of the forms to be used with the wheeled paving machine. The forms and rails should be firmly secured in place by at least three stakes per pin (must be firmly driven to withstand lateral thrust) per each 3-m length so as to prevent movement in any direction. The forms and rails should be straight within a tolerance of 3 mm in 3 m and when in place should not settle in excess of 1.5 mm in 3 m while paving is being done. The forms should be bedded on a continuous bed of low moisture content lean cement mortar or concrete and set to the line and levels shown on the drawings within tolerances of ±10 mm and ±3 mm, respectively.

Mechanized construction of concrete pavement is now possible using two main types of construction equipments:

- Slip-form pavers of models—(i) conforming plate and (ii) oscillating beam models
- Concreting trains

The conforming plate of slip-form paver is built around the parallel side forms which are connected together by a horizontal top plate. These shape the concrete pavement as the construction equipment moves ahead. The concrete is fed to the front of the conforming plate through a hopper.

Ready-mixed concrete is fed into the hopper at the front of the equipment by trucks or transit mixers. These trucks can approach by reversing along the prepared sub-base. The waterproof membrane is laid over the sub-base from rolls immediately under the front of the hopper.

The weight of the concrete in the hopper forces the material under the conforming plate as the paving machine moves forward. The compaction of concrete is effected by a vibrating beam across the hopper and a row of poker vibrators. In the larger slip-form pavers, a spreader is incorporated to facilitate distribution of the concrete across the entire width of the hopper. The paving machine can readily be adjusted to lay various widths of pavements up to a maximum of 13 m.

The insertion of dowel bar (Fig. 12.7) is precisely controlled by a plunger mechanism linked to the forward movement of the paving machine. Dowel bars are ejected from the feed tube simultaneously. The feed tubes are spaced at intervals of 300 mm across the paver.

In the oscillating beam slip-form pavers, the oscillating beams compact and shape up the top of the road slab whereas the side

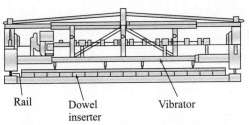

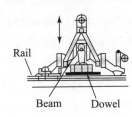

Fig. 12.7 Dowel bar insertion machine

forms give the shape to the edge of the road slab. Initial compaction may be effected by vibrators in the hopper. Figure 12.8 shows this arrangement.

Longitudinal joints are cut mechanically as the paving machine moves ahead. The joints are generally cut centrally in the slab being laid. The slot is filled by a temporary plastic insert fed from a drum on the paver into the slot as it is cut – just behind the conforming plate or the oscillating beam.

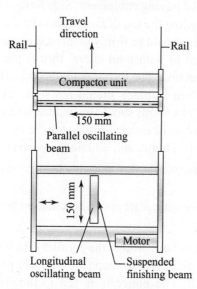

Fig. 12.8 Oscillating finishing beams

For facilitating performance of a slip-form paver, two wires one on each of the slab being laid are stretched at a constant height above and parallel to one edge of the true pavement surface. In addition, one of the wires must be at a constant horizontal distance from the slab edge. For more details on this, refer to the IRC Specifications for Road and Bridge Works.

The paving machine is equipped with an electronic sensing system, which picks up signals from contact with the guide wire and these signals initiate alterations at the controls, either raising or lowering the level of the slab surface, or causing the machine to veer to the left or right. Accurate fixing of the guide wires is to be ensured as the same has a bearing on the construction of the slab to the correct line and level.

A concreting train having a length of up to 600 m comprises a series of machines travelling in sequence along prepared side forms (Fig. 12.9). There are different types of side forms including rails. The question of the use of a concreting train would arise only if the job involved is of a gigantic nature. Prior to initiating the concreting job, the rails meant for supporting the train must be fixed over a length of several kilometres. The underlying membrane has to be laid and crack inducers should be fixed at the planned transverse joint positions. Before the commencement of concreting, longitudinal crack inducers together with tie bars that are made up into prefabricating units on support cradles need to be fixed in position. The sequential units of the train are:

- First spreading and compacting unit
- Dowel bar insertion unit (vibratory)
- Second spreading and compacting unit
- Transverse joint forming unit
- Diagonal beam final shaping unit
- Surface texture unit
- Curing membrane spray unit
- Mobile cover unit

External Work   373

The capacity of the paving equipment, batching plant, and all the ancillary equipment should be adequate for a paving rate of at least 300 m in a day.

Batching and mixing of the concrete should be done at a central batching plant with automatic controls located at a convenient location taking into account sufficient space for stockpiling of cement, aggregates, and stationary water tanks. This should be, however, situated at an approved distance, duly considering the properties of the mix and the mode of transportation of concrete available with the contractor. The capacity of the batching and mixing plant should be at least 25% higher than the proposed capacity of the placing/paving equipment.

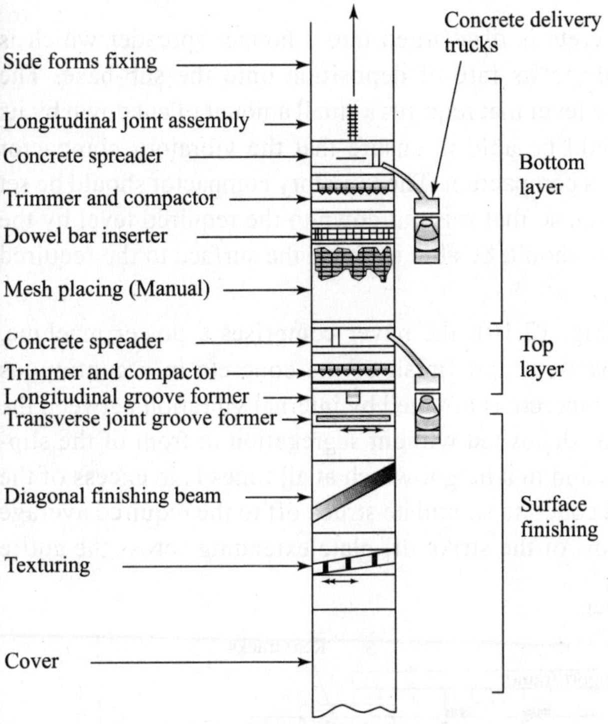

The batching plant (Section 13.6 of Chapter 13) should comprise at least four storage bins, weighing hoppers, and scales for the fine aggregates and each size of the coarse aggregate. If the cement is used in bulk, a separate scale for cement should be included. Mixers should be of the pan type, reversible type, or any other type capable of combining the aggregates, cement, and water into a thoroughly mixed and uniform mass within the specific mixing period, and of discharging the mixture without segregation.

Fig. 12.9 Paving train equipment

The concrete should be placed with an approved fixed form or slip-form paver with independent units designed to: (i) spread, (ii) consolidate, screed and float-finish, (iii) texture and cure the freshly placed concrete in one complete pass in such a manner that a minimum hand finish would be required so as to provide a dense and homogeneous pavement conforming to the plans and specifications. The paver should be equipped with electronic controls to control/sensor line and grade from either or both sides of the machine.

There should be an adequate number of concrete saws with sufficient number of diamond-edged saw blades. The saw machine should be electric or petrol-/diesel-driven type. A water tank with flexible hoses and pumps should be made available in this activity on priority basis. The contractor should have at least one standby saw in good working condition and the concreting work should not commence if the saws are not in working condition.

The concrete should be transported from the central batching and mixing plant to the paver site by means of trucks/tippers of adequate capacity and approved design in sufficient number so as to ensure uninterrupted supply of concrete. Covers should be used for concrete protection against the weather. The trucks/tippers should be capable of maintaining the mixed concrete in a homogeneous state and discharging the same without segregation and loss of cement slurry. The feeding to the paver is to be regulated in such a way that the paving is done in an uninterrupted manner with uniform speed throughout the day's work.

In a fixed-form paver, the concrete is discharged into a hopper spreader which is equipped with means for controlling its rate of deposition onto the sub-base. The spreader strikes off concrete up to a level that requires a small amount of cut down by its distributor. The spread layer should be able to ensure that the vibratory compactor thoroughly accomplishes the layer's compaction. The vibratory compactor should be set to strike off the surface slightly high so that it is cut down to the required level by the oscillating beam. The final finisher should be able to finish the surface to the required level and smoothness as specified.

In slip-forming construction (Fig. 12.10), the paver comprises a power machine, which executes spreading, compaction, and finish of the concrete in a continuous operation. The compaction of the concrete is effected by internal vibration between the side forms. The concrete should be deposited without segregation in front of the slip-form paver across the entire width and to a height which at all times is in excess of the required surcharge. The deposited concrete should be struck off to the required average and differential surcharge by means of the strike off plate extending across the entire

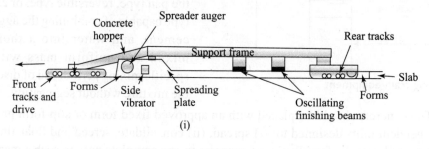

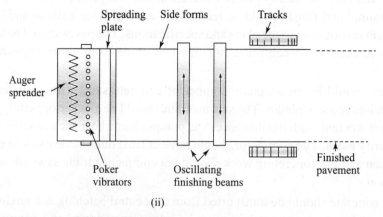

Fig. 12.10 Slip-form paving machine

width of the slab. The equipment that strikes off the concrete should be capable of being rapidly adjusted for changes of the average and differential surcharge necessitated by change in slab thickness or cross-fall.

The level of the conforming plate and finishing beams are controlled by the guide wires as mentioned earlier. The alignment of the paver is controlled automatically from the guide wire by at least one set of sensors attached to the paver. The alignment and level of the ancillary machines for finishing, texturing, and curing of concrete are automatically controlled relative to the guide wire or to the surface and edge of the slab.

The slip-form paving machine should have vibrators of variable output, with a maximum energy output of not less than 2.5 kW per metre of width and 300 mm depth of slab for a laying speed of up to 1.5 m per minute or for higher speeds. The machines should be of sufficient mass to provide adequate reaction during spreading and paving operations on the traction units to maintain forward movements when placing concrete in all situations. Before the application of curing membrane, the surface of the concrete slab should be brush-textured in a direction at right angles to the longitudinal axis of the carriageway.

On completion of surface texturing, the surface and sides of the slab should be cured by the application of approved resin-based aluminized reflective curing compound, which hardens into an impervious film or membrane with the help of a mechanical sprayer. The curing compound, which should be thoroughly agitated in its container immediately before application, should not react chemically with the concrete and the film or membrane should not crack, peel, or disintegrate within three weeks after application.

12.2 DRAINAGE

Wastewater for drainage includes domestic, industrial, and stormwater. Wastewater drainage is classified into two categories:
- Stormwater drainage to prevent flooding caused by rainfall
- Foul water drainage to dispose off domestic and industrial wastewater without causing any environmental pollution

Rainwater has a tendency to drain downwards due to gravity. The rate at which this drainage occurs depends mainly on the soil type and its properties. The downward percolation of soil continues until a depth is reached below which the soil pores are saturated with water. The water in this zone is known as groundwater and the upper level of the groundwater is known as the water table. This water table can create considerable problems for roads and highways. If the soil contained only gravitational water, then the soil above the water table would always be dry except for the time when gravitational water was flowing through. It is, however, found that a zone exists in a section through the soil where the pores are wholly or partially wet with moisture. This zone is often termed as 'capillary fringe'.

Roads constructed for the movement of people and materials should be protected from both surface water as well as groundwater. If water is allowed to enter the structure of the road, the strength and deformation resistance of the pavement and the sub-grade would be weakened making the road susceptible to damage by the traffic. Water can enter the road as a result of rain penetrating the surface, or as a result of the infiltration of the groundwater. The objective of designing road drainage system is to deal with most severe storm conditions and conduct the water away from the road by gravity, eventually discharging to a natural watercourse. It is to be ensured that the level of the watercourse is low enough for one-way flow. As regards groundwater, the water table should be maintained sufficiently below the carriageway.

The basic functions of the drainage system are as follows:
- Conducting stormwater from the carriageway to the outfalls
- Maintaining water table in the sub-grade (~1 m) below the carriageway
- Blocking the passage of both ground and surface water flowing towards the road
- Controlled draining across the road alignment

The fourth item is about cross-drainage whereas the top three functions concern longitudinal drainage along the alignment of a road or highway. Structures like bridges, culverts, trenches, etc., are built for cross-drainage.

In road and highway work, drainage requirements fall into two categories.
- Subsurface water drainage—to ensure that the water table is maintained at a level low enough not to allow saturation of the road structure and its surrounding soil because the excess water can cause the structure and ground to become plastic and incapable of withstanding the traffic loads.
- Surface water drainage—the surface of the road must be kept free of standing water so as not to endanger traffic flow; to achieve this, the roads should be cambered when straight, laid to cross falls on bends, and provided sufficiently with gullies or grips to dispose off rain water that falls on the carriageway.

Subsurface water drainage

In many cases, drainage of groundwater as well as surface water can be combined. Subsurface water drainage assumes great importance in maintaining the water table below the road formation level. This is particularly true for wet low lying areas where the ground could be sufficiently drained by constructing a system of drains, collector drains, and a long outfall pipe. Sand drains are used to drain subsurface water in impervious soil so that the subsurface water cannot affect embankment formation (Chapter 4, Fig. 4.15). The structural work of the road is started only when the drainage system is found effective. Springs, if encountered in a position close to or under the proposed road structure, should be connected to the drainage system using pipes.

Surface water drainage

As construction of most roadwork is in progress outside the urban areas, both cutting and filling are involved. The surplus from cutting could be utilized in embankment formation. It would be a good idea to lay French drains along the limits of the roadwork. On completion of earthwork, similar drains should be provided in the median area and the verges of the road surface. This is necessary to keep the area of the road under construction free of surface water. The surface water should be discharged to a natural watercourse. In urban areas, surface water is connected to the existing storm sewers.

All the pipeline work is done maintaining such gradients that would allow surface water to move at self-cleansing velocity so that solid particles remain in suspension while flowing without settling down. A simple way of working out the gradients required for self-cleansing velocity is to divide the pipe diameter by 2. Thus a 100-mm diameter pipe should have a gradient of 1 in 50 for self-cleansing flow. Likewise, a 300-mm diameter pipe should have a gradient of 1 in 150. These gradients are much steeper than what is required. The gradients can be properly worked out if all the design inputs are available. In practice, there remains a gap between the required and the actual quality of materials and workmanship. With steeper gradients, self-cleansing velocity may be assured for temporary drainage to facilitate construction work.

All sewer pipelines must be laid in straight lengths. Wastewater lines are called sewers. Inside any building, however, wastewater lines are called plumbing or process piping. Each straight length of a sewer pipe must be laid at constant gradient. A manhole (Fig. 12.11) is required at every change of: (i) direction, (ii) gradient, (iii) pipe size, and (iv) pipe type.

Manhole

A manhole (Fig. 12.11) or inspection chamber should be built where several lengths of sewer are connected. In straight sewer pipelines, manhole spacing should not be more than 100 m. A manhole can be used to attend blocked sewer. There are many variations in the details regarding the construction of manholes, but the key factors are as follows:

- The manhole must have a strong cover and frame fitted on the top
- The access shaft should preferably be 500 mm wide or more
- Iron rungs should be fitted at 300 mm spacing for climbing in and out of the manhole pit
- There should be sufficient headroom in a manhole for workers to work safely inside
- The benching must have reasonable cross-fall so that a worker can stand on it
- The manhole should be watertight

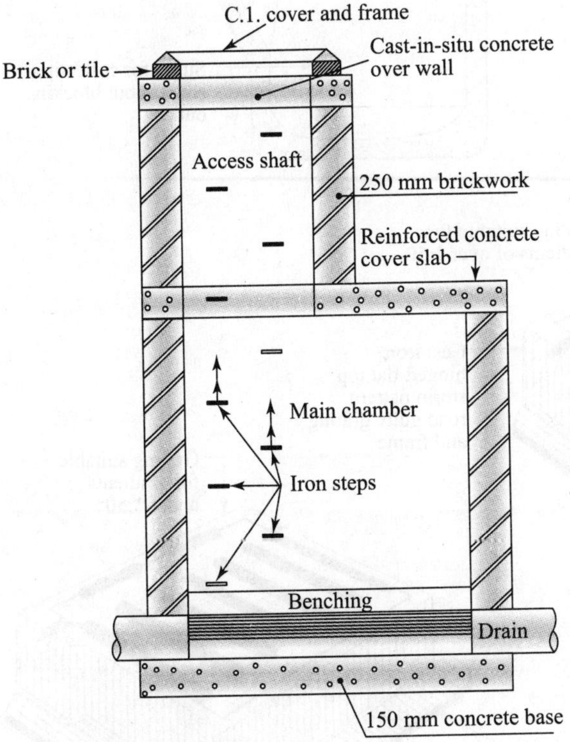

Fig. 12.11 Manhole

Gullies

Gullies (Fig. 12.12) are fitted below the road channel to collect surface water from the carriageway and trap any silt carried by the water. The surface water is thus conducted away from the road by connecting the gully to a surface water sewer or open watercourse using pipe lengths. The water is ultimately led to a river or stream via a ditch or sewerage system. There should be adequate number of gullies to prevent water standing on the carriageway or crossing it. The positions and number of gullies in a carriageway would depend on the area that is being drained, the gradient of the road, and the type of the surface. Gullies should be set on concrete bed and surrounded by 150 mm of concrete.

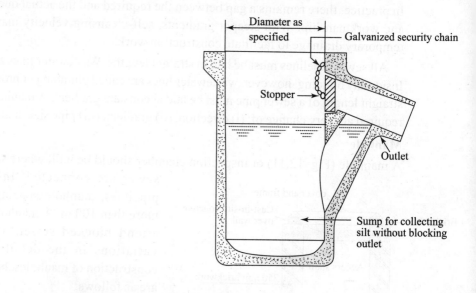

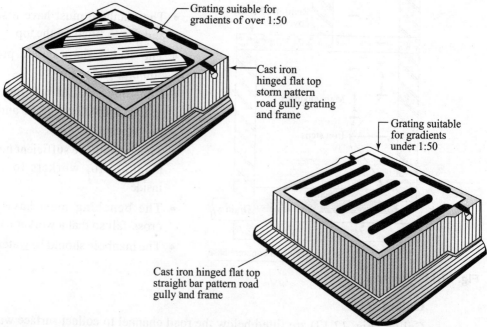

Fig. 12.12 Gullies

Gullies are made of:
- Glazed stoneware (vitrified clay)—heavy and easily broken/damaged
- Concrete—heavy and difficult to handle
- Glass fibre—light, easy to handle, not easily damaged, but subject to distortion
- Plastic—light, easy to handle, not easily damaged, but subject to distortion

Procedures for setting out, earthwork in excavation, and laying of pipelines are discussed next. Reference pegs are driven strategically as follows after carrying out site survey using the laser beam of a total station:
- At both the ends of each pipeline, additional pegs if pipeline length exceeds 50 m
- At the centre of each manhole, inspection chamber, etc.
- At each change of direction or gradient even if manholes are not provided

The centreline that connects two pegs is used for aligning the construction equipment to be deployed for excavation of trenches. If the equipment is adjusted so as to remain horizontal, then the trench sides would be vertical. A nylon line can be stretched alongside the trench to be excavated and parallel to the pipe centreline, and then the excavation can be executed with respect to the nylon line. Otherwise, the outline of the trench can be marked or painted on the ground for executing excavation work on that basis.

Once excavation is completed, the laser may also be used in pipe laying. The laser is installed on the centreline of the trench and then oriented in the correct direction. It should be angled at the required gradient of the pipe. By this, a pipe can be laid in one operation up to a distance of 150 m. Otherwise, nylon line or steel wire can be stretched along the trench and then pipelines are laid with reference to that.

Unless the sides of trenches are sloped at an angle less than the angle of repose of the soil, there remains the possibility of collapse of the sides. Be it soil or rock, a vertical face can collapse under certain conditions. Trench sides should be supported by timber or steel shoring against any possibility of collapse. Detailed side-supporting information are provided in Section 4.5 of Chapter 4. Trenchless 'no dig' methods of laying pipes are discussed in Section 4.6 of Chapter 4.

Pipes are of two types: (i) rigid – made of cast iron, concrete, asbestos cement, clay, etc.; (ii) flexible – made of plastic materials, steel, and so on. Pipe joints fall into the following two categories: (i) rigid if jointed by welding or bolting (metal flanges) or caulking socket/spigot connections and (ii) flexible if jointed using sleeves or collars or deformable rings/gaskets in socket/spigot connections.

If the soil encountered during trench excavation is found to be of unstable nature, extra excavation to the extent of about 150 mm is carried out and filled with concrete for the supporting pipes. Additional support may be provided by placing concrete surrounding the pipe to the extent desired.

The points that need to be considered in pipe laying are as follows:
- Overexcavation below the invert levels should be avoided by constant supervision
- The excavated trench bottom should be firm and stable for laying pipes

- Pipe laying should be accurately true to line and gradient
- Pipes should be jointed properly
- Packfilling of laid pipes should be executed without dislodging or damaging the pipes

Culverts

Culverts (Fig. 12.13) are used to conduct a watercourse through the road embankment from one side to the other. There is a basic difference between culverts and bridges. Whereas bridge decks form part of the pavement, culverts are placed in the embankments under the road pavement. Usually, bridges have larger spans compared to that of the culverts. A culvert usually comprises steel or concrete pipe or reinforced concrete box. Corrugated galvanized steel pipes are available in large diameters and are lighter and easier to handle. The ends of steel culverts are cut at an angle to match the embankment side slope so that no part of the pipe protrudes from the sloped embankment. A culvert should be kept clean from accumulated silt or debris. Corrosion may occur to metal culverts in certain instances. A flat top culvert may be constructed on a concrete bed with masonry walls and reinforced concrete cover. Standard culverts or box culverts comprise precast concrete units having rectangular cross-sections. These units can be designed to suit the site conditions and to withstand all possible loads. Most culverts have an upstream head wall and terminate downstream with an end wall. Both protect the embankment from erosion by flood water.

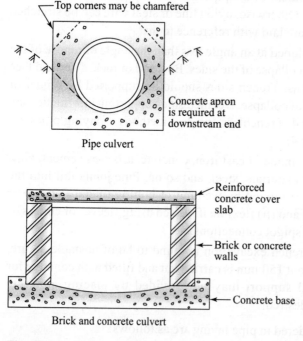

Fig. 12.13 Culverts

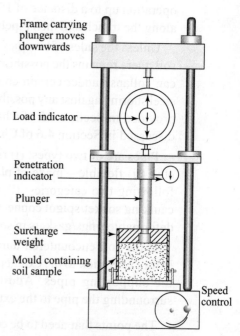

Fig. 12.14 California-bearing ratio apparatus

Bridges are required for crossing streams and rivers where culverts are deemed inadequate. The erection of bridges is discussed in Section 8.8 of Chapter 8.

12.3 CONSTRUCTION—ACCOMMODATION OF SERVICES AND IMPACT

Road construction may be simple, but it becomes complicated if the following services are to be accommodated under paved areas:

- Sewers
- Buried cables for power
- Gas mains in selected urban areas
- Water mains
- Telephone cables
- Television relay cables

Overhead electrical power lines are drawn by maintaining margins for construction of roads and highways because of which problems are not anticipated. In order to accommodate the services under the pavements, co-ordination with the concerned public or private enterprises would be necessary. Such services should be accommodated preferably under the footpaths or verges so that fewer disturbances are caused at the time of repair/maintenance work. The question of possible repair/ maintenance work must be kept in view when designing and working out details.

The environmental impact of construction of roads/highways could be very serious and disrupt lives of the people of the affected areas. The proposed alignment of the roads/highways may pass through such areas as inhabited by the local people. Such people would have to be rehabilitated so that construction of roads/highways may be initiated. If the arrangements made for rehabilitation are not satisfactory, then there would be discontent, which may either stop or slow down the process of execution of roads/highways. Satisfactory rehabilitation should be accorded the highest priority.

If the alignment of the road passes through fertile agricultural land then the livelihood of people living on agriculture would be seriously affected. Such disruption of livelihood would require resettlement of the affected people and making alternative arrangements for their living. Resettlement may lead to development of new areas which may attract resourceful people from other areas of different background. This would affect social compatibility resulting in unrest and disturbances. All these should be taken into account while planning rehabilitation prioritizing the upliftment of the affected people in general.

The alignment of the road may require uprooting and felling of trees in a large scale thus creating environmental imbalance. Such imbalance can be mitigated by equally large-scale or larger-scale afforestation. Roads and highways would make ways for movement of all kinds of vehicles and movement of a large number of people in such vehicles. This would create noise pollution. Hence, resettlement of people should be planned considering such noise.

Vehicles moving on roads/highways burn fuels and hence discharge pollutants into the atmosphere. The authorities empowered to control air pollution must control the movement of such vehicles that discharge pollutants more than what could be discharged under the laws.

SUMMARY

Surface connectivity across a country is essential for progress and prosperity. Construction of road network is, therefore, not only necessary but need to be implemented in least possible time. Route survey can be carried out using modern survey instruments in double quick time to finalize road alignments and cross-sections. Details of alignment should be detailed enough for a surveyor to lay out the road correctly. Earthwork is involved in both excavation and embankment formation. To prevent wetting and softening of the soil, drainage system should be ready before road excavation work is started. Where water table is high, roadside drains should be laid before initiating excavation work. Excavation in rock is carried out by blasting. Approved materials obtained from road and drain excavation, borrow pits or other sources are used on forming embankments including sub-grades, earthen shoulders, etc. The compaction should be done using construction equipment of adequate capacity.

During the formation of embankment, shaping of the fill should be checked continuously so that, in the course of filling, the surface develops an accurate profile. Flexible pavement is founded on water-bound macadam sub-base and base. Wet mix macadam is laid and consolidated on sub-grade/sub-base/base as the case may be. Bituminous flexible pavement courses are made by mixing coarse and fine aggregates with bituminous binder in a hot mix plant. Bituminous concrete should be used in wearing and profile corrective courses. Rigid concrete pavement comprises a concrete slab which is laid over the dry lean concrete sub-base. This concrete slab, compared to a flexible pavement, is the equivalent of the surface course, binder course, base and sub-base combined.

Drainage of wastewater means drainage of domestic, industrial and storm water. Foul water in wastewater must not cause any environmental pollution. Rainwater has a tendency to drain downwards due to gravity depending mainly upon the soil type and its properties. The downward percolation continues until groundwater table is reached. Above this water table, a zone exists where the pores are wholly or partially wet with moisture held by capillary forces. Water can enter the road as a result of rain penetrating the surface, or as a result of the infiltration of the groundwater. If water is allowed to enter the structure of the road, the strength and deformation resistance of the pavement and the sub-grade would be weakened and make the road susceptible to damage by the traffic. In road and highway work, both sub-surface and surface water drainage is meticulously designed and implemented.

Physical connectivity across the country is directly related to its development and economic prosperity. As India is flourishing economically, road and highway construction for physical connectivity is at the top of the country's priorty list. Consequently, the trend now is to incorporate the latest technology in building roads and highways fast so that the economic disparity between the urban and rural areas is reduced to minimal level.

REVIEW QUESTIONS

1. What is EDM? Describe the principle that governs the working of every EDM.
2. What is 'total station'? How can a basic theodolite be transformed into a total station?
3. What is GPS? What are the requirements for GPS to be used in engineering survey?
4. How are stakes used in laying out roads involving both cutting and filling?
5. What kind of site clearance is involved in road construction work?
6. Define excavation involved in roadwork. What are the three categories of road excavation?
7. Explain why earthwork in excavation should be carried out up to 300 mm above the formation level.
8. What is meant by structural excavation in roadwork?
9. Why should the best filling materials be saved for filling the sub-grade and the portions immediately below the sub-grade in embankment formation?
10. Name the construction equipment/plants that are used in compacting earth in embankment formation.
11. Why are geotextiles used in embankment formation?
12. What are the functions of sub-base? What kinds of materials are used in sub-base?
13. How is wet mix macadam (WMM) produced? What are the uses of WMM?
14. How are paving bituminous materials produced, transported, placed, and compacted? Explain in detail how compaction is achieved.
15. On what basis is dry lean cement concrete proportioned?
16. How rigid concrete pavement is cast using slip-forms?
17. What are the basic functions of the drainage system?
18. How can drainage of subsurface water be accomplished for keeping the water table below the pavement?
19. What is the basic difference between culverts and bridges built across road embankments for cross-drainage?

13
Mechanized Construction

INTRODUCTION

A project comprises various activities of diverse nature. Execution of each activity should be planned meticulously to ensure sequential execution in time. There is no alternative to adopting mechanized construction methods to achieve more in less time. Once the decision of deployment of construction equipment is taken, required construction equipment need to be identified. Knowledge of the basic elements that constitute a machine would help in such selection.

Earthmoving construction equipment fall under two basic categories:
- *Static*—for example, shovels, backhoes, draglines, clamshells, etc.
- *Mobile*—for example, bulldozers, loaders, scrapers, rollers, graders, trenchers, etc.

Drives of these earthmovers are of two types: wheel-mounted and track-mounted. Handling and lifting equipment are also of two types: static and mobile. Static equipment include winches, derricks, hoists, tower cranes, etc.; and mobile equipment include cranes, hoists, forklift trucks, aerial cableways, belt conveyors, monorails, etc.

For achieving proper gradation of aggregates for concrete work, stones are crushed in different types of crushers. In concreting work, construction equipment are available for batching, mixing, transporting/conveying, and placing. Scaffolding is temporary erection of timber or steel supports to facilitate construction, alteration, demolition or hoisting.

13.1 GENERAL CONSIDERATIONS

A construction project may be of two types:
- Linear project (involving concept, design, procurement, and construction)—traditional approach
- Fast-tracked project—design and build

On a fast-tracked job, design and construction begin almost simultaneously. Construction work is started even when the design engineers are working on the foundations of the structures. This happens when the completion time is considered as sacrosanct and competitive bidding is not essential. Design and construction work are set on parallel tracks.

The disadvantages of fast-tracked jobs are: (a) input for designing may not be available in time as a result of which reconstruction may be necessary to comply with the input when received, and (b) less control over costs.

In both cases, the intent of planning and implementation should be:

- Cost of construction should not exceed the budget
- Quality of executed work has to conform to the specifications, drawings, codes, etc
- Time of execution has to remain within the agreed construction time schedule

In other words, the aim of construction is to continue and complete a project at a reasonable cost within an agreed time schedule with assured quality.

Deployment of construction equipment

The cost of construction is a major factor in all projects. The factors that influence construction costs mainly are materials, labour, construction equipment, overhead and profit. The costs of construction equipment for civil engineering construction projects range from 25% to 40% of the total project cost. Deployment of construction equipment is done for the reasons as mentioned below:

- Larger output
- Cost-effective implementation
- For execution of work that is not feasible by manual efforts or when deployment of construction equipment may help in doing the work more cost efficiently
- To reduce the amount of heavy manual work/s which would cause fatigue, thereby allowing manual efforts to be more productive
- Large output can be maintained even if there is a shortage of skilled and semi-skilled manpower (as operators of construction equipment are of different category)
- Precision in implementation is required by sophisticated design engineering—such precision can be maintained by using modern construction equipment equipped with software controls

Construction equipment and machineries of very high capacities are available now. With very large outputs possible due to mechanization, adhering to construction schedules would not be difficult. General construction practices dependent on manual efforts result in time overrun and consequent cost escalation, and even production loss in case of industrial or utility projects.

Once the decision of mechanization of construction is finalized, identification and selection of construction equipment for different activities are to be made. For that, productivity analysis to select the optimum size of each construction equipment is to be done. This is necessary to determine the number of units and the size of equipment that would permit the construction agencies to complete each activity in time, resulting in execution at the lowest possible cost.

A project comprises various activities of diverse nature. Execution of each activity should be planned properly to ensure its timely completion at the lowest cost.

Equipment on lease

Mechanization based on rented construction equipment is cost effective. Construction equipment, when rented, can be selected exactly to match the requirement. For rented equipment, time to make the equipment ready for operation is important—like truck-mounted telescopic jib crane can be deployed for construction work without losing much time, but truck-mounted strut-boom crane needs time for assembling jib for heavy-duty work. The truck-mounted strut-boom crane is selected quite often on account of its ability to provide very specific heavy lifting capacity cost efficiently over a short period of time. To ensure effective and efficient performance, the crane should ideally be made ready for deployment immediately on arrival at the construction site. Unfortunately, the strut-boom cannot be transported in this ready form, and must be folded and securely held during transportation. Because of this, a strut-boom needs to be assembled and made ready for use only on arrival at the site.

Telescopic jib cranes are better in this respect as very short period of time is required to prepare the crane for deployment on arrival at the site. This type of cranes need not be rented for as long as strut-boom cranes need to be.

The main advantage of a tower crane is that its jib is supported at the top of a tall tower, which may be set at a sufficient height to clear all possible obstructions.

Modern construction equipment and machineries are manufactured with inbuilt arrangements for performing work maintaining specified quality. Such arrangements can also be incorporated in older models.

Equipment selection

Selecting the appropriate equipment for the job ideally forms part of the construction planning process and should be chosen for performing any particular task only after analysis of many interrelated factors. The important points for consideration are:

- *Function to be performed*: The choice of equipment in many instances would be dictated by the need to cope with the combinations of horizontal and vertical movements as well as travelling. A crane can provide both horizontal and vertical movement but is rather poor at manoeuvring, whereas most earthmoving equipment and machineries are mobile and would transport materials very efficiently over long distances, but cannot be used for lifting.
- *Capacity of the equipment*: The magnitude of work involved in moving materials relative to the time available as per the construction schedule is important. For cranes and other lifting equipment, the practical weight and the size of the units as designed are the most important criteria.
- *Method of operation*: The distance and direction of travel, speed and frequency of movement, sequence of movement, condition of the ground, etc., must all be taken into account.
- *Limitations of the method*: For many earthmoving operations, the choice of method may be limited by the cost of temporary work facilities. For example, haul

roads may not be permitted in a particular area; the establishing and dismantling costs of a crane or conveyor system may not be economic over the length of time required for the task. Every project is burdened with such problems.

- *Costs of the method*: The most appropriate method from feasibility point of view may involve expensive and uneconomic equipment/units, which also may not be available at the desired time without incurring high cost.
- *Cost comparison with other methods*: It is essential that all the alternatives be considered in the plan and their relative costs compared.
- *Possible modifications to the design of the project under consideration, in an attempt to accommodate the materials handling method*: For example, a single large lifting in a tower block could perhaps be reduced in size. A smaller crane then might be installed. Although this procedure may appear obvious and straightforward, because of the one-off nature of most designs and construction projects, the task of analysis is seldom simple. The choice of method is rarely clear-cut—a compromise is the more usual result.

Rent or purchase

Once the strategy of implementation is finalized and construction equipment and machineries are selected, every executing agency has the clear option of purchasing or renting or a combination of partly purchasing and partly renting the selected construction equipment and machineries. The owners may also purchase and rent out the same to the executing agencies for the sake of progress of implementation. This is particularly necessary if some equipment, e.g., a very heavy-duty tower crane is necessary for just lifting a one-off very heavy item. Part of such expenditure may be recovered by renting out the crane or any other equipment to the executing agencies (contractors).

For an executing agency, purchasing of construction equipment and machineries (resources) would be advantageous for the reasons as follows:

- Construction equipment and machineries would be available at all times for deployment
- These resources could be used in other projects as required according to the conceived programme
- Cost of such resources could be apportioned among different contracts

However, renting of construction equipment and machineries would be advantageous in some other ways for the reasons as mentioned below:

- Construction equipment and machineries could be rented as and when required for a calculated period of time
- Hiring agencies are responsible for repair, replacements, and even operation depending on the contractual terms
- On completion of the hiring period, the executing agency would be liability-free of the rented items

Rented items of construction equipment should not be left idle for prolonged periods of time so as to maintain a contract economically viable.

13.2 FUNDAMENTALS OF MECHANIZATION

How the construction machines work can be explained by referring to some basic laws of physics starting with the first law of the conservation of energy according to which energy can neither be created nor destroyed. In machines, speed and force are applied by the power generated by the engine. Speed, according to the above law, can be converted to power except for the loss, for example, due to the heat generated by friction.

Levers

The lever is one of the simplest machines and simplest energy converter. It may be considered as a beam or rod or link pivoted at one point called the fulcrum, a load being applied at one point in the beam and an effort, sufficient to balance the load, at another. It must be rigid for accurate distance-force ratio. These are worked out in straight lines from the point where effort is made to the fulcrum, and from the fulcrum to the load. The lever itself may be straight, angled, curved, or offset. A lever is used to achieve mechanical advantage. Three classes of lever may be distinguished as follows:

- Fulcrum between effort and load—class I
- Load between fulcrum and effort—class II
- Effort between fulcrum and load—class III

Class I lever is used in the beam of a balance, handle of a hand pump, scissor, pliers, rocker arm in internal combustion engine. For more advantage, effort arm should be greater than the load arm.

Class II lever is used in nut cracker, wheelbarrow, etc.

Wheel and axle is a special type of class III lever where the force (effort) is applied between the fulcrum and work (load) so that the leverage can be used only to increase speed and distance. In case of wheel, the centre of rotation is the fulcrum, the axle-surface is the effort point, and the wheel perimeter is the load. If power is applied to the axle and work is done at the outer edge of the wheel, speed is increased while the effort is decreased.

Springs

Springs are mechanical elastic devices that deflect under load to store energy. Springs recover their original form when the applied pressure is released. The essential functions of springs are:

- To absorb shock due to jolt as in car springs, railway rolling stock suspensions, elevator buffer springs, etc.
- To dampen vibration as in spring supports or elastic motor drives, coil spring coupling, etc
- To control motion by maintaining contact between two elements such as a cam and its follower by restoring a machine part to its normal position when the unsettling force is neutralized, in a governor or valve of an internal combustion engine or by producing the necessary pressure in a friction device as in a brake, and so on

- To store energy as a source of power as in clock motors, circuit breakers, starters, etc.
- To measure forces as in spring balances, gauges, etc.

Springs are often required to provide several of these functions simultaneously, especially when vibration and shock characteristics are important.

There are different types of springs, e.g., plate springs, cylindrical helical springs, helical conical springs, spiral springs, and special springs.

Helical-spring is formed by slender wire wound into a helix along the surface of a cylinder, sometimes erroneously termed as a spiral spring.

Spiral-spring is formed by coiling a steel ribbon into an elongated spiral or helix of increasing diameter. When compressed completely, it forms a true spiral.

Carriage-spring is used for the suspension of railway rolling stock and other vehicles. It comprises a number of steel strips of varying length, curved to semi-elliptic form, held together so as to be capable of acting independently, and loaded as a beam.

Gears

Gears are the most rugged and durable means of transmitting torque, rotary motion, and power from one shaft to another. A gear is a disc or a cylindrical wheel with teeth notched around its circumferential edges, either on the outside (external gear) or on the inside (internal gear). The velocities of two meshing gears must always be the same for quiet vibration-free operation.

Gears are used for providing a positive means of power transmission. This transmission is effected by the teeth on one gear meshing with the teeth on another gear or rack (that is, straight-line gear). Meshing teeth formed with special cutters provides a much more compact drive having unlimited capacity to change power-speed relation between the input (driving) to output (driven) shaft. Their operation is simple, efficiency is high, and service is reliable. Gears provide a much more compact drive than either belts or chain drives and can operate at much higher speeds and power.

Gears transform the torque applied to the driving wheels into forward/reverse motion of the tractor or truck. It is also the most common means of transmitting power in, for example, tiny watches to large turntables of the railways. They form vital elements in many machines, such as metal cutting machine tools, rolling mills, hoisting equipment, transmitting and transporting machinery, marine engines, etc.

If the axle or shaft is fastened to two wheels of different diameters, it would generate different proportions of power and force to their circumferences because of the different leverage ratios without any slippage. There are many applications where many pairs of gears are in mesh. Such a system is generally called a gear train. A typical example is a mechanical clock.

There are different types of gears, such as spur gear, helical gear, herringbone gear, meshed gears, bevel gear set, etc. However, spur gear is the one which is most commonly used. A variety of gears have evolved directly from the spur gear, but many of these gears are complex in design and manufacture.

Shafts

A shaft, generally a slender member of round cross-section, rotates and transmits power and motion. A shaft can also be non-rotating. An axle is a non-rotating member that supports rotating members. A shaft may also have non-circular cross-section. The general perception of a shaft, depending on its geometrical shape, is a straight shaft uniform or stepped cross section and crankshaft. A crankshaft is used to convert reciprocating motion into rotary motion and vice versa. A shaft not only supports rotating members but also transmits torque. Therefore, a shaft is subjected to both bending and torsion with various degrees of stress concentration.

The crankshaft takes the downward thrusts of the pistons and connecting rods when the fuel air mixture is burned in the cylinders and changes these thrusts into torque which is transferred to the drive line of the tractor or truck.

A shaft may be classified according to its use, for example, prime mover shaft, machine shaft, power transmission shaft, and so on.

Gears need to be connected to the power source and the work point. That is done by means of shafts or axles. A shaft simply holds a gear or gears in position. A rigid connection may be made to the shaft by means of keys, pins, snap rings, clamp collars and splines. When axial movement between the shaft and hub is required, relative rotation is prevented by means of splines machined on the shaft and into the hub. Universal joints or flexible couplings are placed in shafts connecting separate units to allow adjustment of misalignment. A universal joint is a kinematics linkage used to connect two shafts that have permanent intersecting axes.

Apart from gears, shafts support pulleys, flywheels, etc. A shaft or axle only supports a revolving part and, with respect to their supports, may be static or revolve with the parts mounted on them.

To reduce the deflections to minimum, materials used in shafts are cold-drawn or machined from hot-rolled, plain carbon steel. Alloy steels are used, where toughness, shock resistance, and greater strength are required.

Couplings

Couplings are required to connect two shafts or to ensure resilient connection between the drive shaft and the driven shaft compensating for minor misalignment between them. The couplings may be grouped into two classes:

- Rigid couplings—a rigid coupling locks the two shafts together without any allowance for either axial or radial relative motion among them, although some axial adjustment is possible at assembly.
- Flexible couplings—shafts are often susceptible to radial, angular, and axial misalignment in which cases flexible couplings must be used; severe misalignment, however, should be corrected.

The types of rigid couplings considered (from a large variety of such couplings) in this chapter are: flanged coupling, split coupling, and compression coupling. Both flanged and split couplings need the use of keys and keyways. A key enables the transmission of torque from the shaft to the hub. The grooves in the shaft and hub into which the key fits form the keyways or key seats. Splines are essentially built-in keys. Compression couplings are used when keys and keyways cannot be used.

Flanged couplings are used where there is free access to both shafts. The two shafts are bolted together to form a solid connection. Keys and split circular key rings or dowels are incorporated in the couplings to enable transmission of substantial torque.

Split or clamped couplings are basically sleeves that are split horizontally and clamp around both the shafts by means of bolts to form solid connections so as to transmit torque. Here also, keys and key-rings are incorporated in the couplings to eliminate frictional dependency for torque transmission.

Compression couplings comprise three pieces—a compressible core and two encompassing couplings that compress the core. The core is comprised of a slotted bushing that has been machine bored to fit both ends of the shafts. Compression couplings transmit torque by only frictional force between the shafts. The core has also been machined taper on its external diameter from the centre outward to both ends. The coupling halves are finish bored to fit the taper. When the coupling halves are bolted together, the core is compressed down on the shaft by the two halves and the resulting frictional grip eliminates the need of keys and keyways.

Flexible couplings, which are classified as material flexing, mechanical flexing, or combination of the two allow the coupled shafts to slide or move relative to each other. Flexible couplings are required to connect shafts that are beset with the problem of some small amount of misalignment. Substantial misalignment causes a whipping movement of the shaft, adds thrust to the shaft and bearings, precipitates axial vibrations, and results in premature wear or failure of the equipment.

One class of flexible couplings contains a flexing material inserts, such as laminated disk ring, bellows, flexible shaft, diaphragm, and elastomeric couplings. The insert cushions the effect of shock and impact loads that could be transferred between shafts. A shear type of rubber-inserted coupling can be used for higher speeds and horsepower.

Mechanical flexing couplings provide a flexible connection by permitting the coupling components to move or slide relative to each other. In order to permit such movement, clearance must be provided within permissible limits. It is important to keep cross-loading on the connected shafts at the minimum. This is accomplished by providing adequate lubrication to reduce wear on the coupling components. The most popular of the mechanical-flexing type are the chain and gear couplings, which are dealt separately.

Fluid coupling

A fluid coupling comprises an oil chamber that contains a centrifugal pump and a turbine. Oil is normally used in commercial units as the working fluid because of its lubricating properties, stability, and availability.

The principle of the fluid coupling can easily be demonstrated by the classic analogy of the two ordinary electric fans that are set facing each other. One fan connected to an electrical power source is turned on and put into motion. As its blades rotate, the air current developed in the process turns the blades of the other fan that is not connected to any power source. The rotation of the pump vanes by the input shaft driven by the engine puts the oil into motion, and the oil rotates the turbine in the same direction as the pump after sufficient energy or head has been developed. Since there are no torque-reacting elements in the fluid coupling besides the impeller and the runner; under steady operating conditions, the output torque would always equal the input torque, hence the term fluid coupling.

The fluid moves through a closed path. The speed of the drive shaft always exceeds that of the driven shaft when a torque is transmitted through the coupling. At the beginning of operation when the engine has just started, the drive shaft rotates while the driven shaft does not; the slip is 100%. At normal or rated speeds, the slip may be reduced to about 1–4%. At normal speeds and loads, the efficiency of a fluid coupling may be as high as 96–99%.

The fluid coupling is hardly used in construction equipment. Fluid torque converters are used instead.

Fluid torque converter

Unlike fluid coupling, a fluid torque converter has a stator and a disc with fan-like blades connected to the transmission via a fixed shaft with a one-way clutch that allows it to rotate only in the opposite direction of the fluid's radial motion.

Without the stator, fluid leaving the turbine would strike the impeller with a radial motion opposite its rotation, causing a braking effect. With the stator, the returning fluid strikes the stator blades, which reverses the radial direction of the fluid's motion so that it is moving in the same direction as the impeller when it re-enters the impeller chambers. This reversal of direction greatly increases the efficiency of the impeller, and the force of the fluid striking the stator blades also exerts torque on the turbine output shaft, providing additional torque. Thus, the turbine torque does not equal the pump torque, hence the name torque converter.

The converter provides for smooth starting of the load and absorbs torsional vibrations and shocks. In general, the efficiency of a torque converter is high at low speeds.

Bushings and bearings

Bushings are sleeves or hollow cylinders installed between a shaft and its supports, or between a gear, pulley or wheel and the shaft on which it turns. A bushing is usually made of some metal alloy (like bronze, brass, etc.) that is softer than steel, can retain an oil film that will prevent metal-to-metal friction due to contact, and easily worked.

Bushings absorb most of the wear caused by rotation of the shaft or wheel. A bushing may be replaced several times before the more expensive shaft has to be repaired. In addition, there are hard steel bushings that are meant to protect the support from damage by the shaft. Bushings in pillow blocks are cut lengthwise in two pieces to facilitate installation.

Bearings are machine parts that provide relative positioning and rotational freedom while transmitting the load between two parts, e.g., a shaft and its housing. They are used at most rotary friction points where exact alignment and low friction are required. Because of the relative motion between the bearing and the moving shaft or element, there would be loss of energy due to friction. There may be wear and tear if their surfaces touch each other. Bearings are provided against frictional energy loss and possible wear and tear. The primary objective of lubrication is to reduce friction, wear, and heating between the two surfaces moving relative to each other.

The two principal types of bearings are of rolling contact type bearings, and they are classified as ball bearings and roller bearings.

Ball bearings comprise four elements:
- Inner ring or race
- Outer ring or race
- Rolling element, i.e., balls
- Separator for keeping the balls isolated from each other

Of the two rings of hard steel, one of which fits closely on the shaft and has its outer surface machined into the smooth track; the other fits closely on the casing or hub and has a machined inner surface. Between the inner and outer rings or races, a circle of hard spherical steel balls is held in place by a light cage as antifriction bearings.

In roller bearings, the balls are replaced by straight, tapered or contoured cylindrical steel rollers, which roll similarly between the two races. Size-wise, roller bearings have much higher static and dynamic/shock load capacities than ball bearings. They are suitable for moderate speed, heavy-duty applications.

Both bushings and bearings reduce friction, provide lubrication, and are easily replaceable when worn at much less expenses.

Lubricants

Lubrication is a basic necessity of machine operation. The lubricants are used:
- To reduce the coefficient of friction and provide a slippery film between the surfaces rubbing, turning, or scraping each other thereby reducing wear
- To reduce the damages caused by friction—waste of power and heating
- To resist corrosion

Lubricants may be grouped into the following three types:
- Fluids

- Greases
- Solid film lubricants

Petroleum-based or synthetic oils are characterized by their viscosity. Oils for vehicle engines are classified by their viscosity as well as by the presence of additives for various service conditions. Other characteristics that may affect any lubricant are: acidity, resistance to oxidation, antifoaming, pour, flash, and fire hazard. These characteristics are related to oil quality for any particular operation; therefore, oils are named after the operation for which they are appropriate.

Synthetic lubricants are mainly silicones having high temperature stability, low temperature fluidity, and high internal resistance. Because of high cost, synthetic lubricants have limited application.

Greases are generally semi-solid or solid mixtures of mineral oil with special soaps or fillers that give the combination the qualities of adhesiveness, pressure endurance, water resistance, and melting point on the basis of which greases are selected. Unlike oils, greases cannot circulate, but they are expected to accomplish all the functions of fluid lubricants.

Solid lubricants are either brushed or sprayed directly on the bearing surfaces and are mixed with adhesives for retention. Basically, they are of two types: graphite and powdered metal. Solid lubricants are ideal for operating at high temperatures. Other varieties include Teflon and some chemical coatings.

Chain drives

A chain drive comprises an endless chain running over two sprocket gears. One of the sprockets is connected to a power unit while the other one is driven. Sprocket gears with teeth of special profile are used to drive through the roller chains. Unlike gear drive, the chain drive does not slip while in operation.

According to their performance, chain drives may be classified as:

- Driving chains
- Crane chains
- Tractive or pulling chains

Driving chains are relevant and are classified further as:

- Roller chains
- Bush chains
- Silent chains

Roller chains are used in revolving shovels in the track drive, crowd mechanism, and deck machinery; in ditching machines in track and digging drives; in grader final drives, and so on. Roller chain comprises two rows of outer and inner plates. The outer row of plates is known as *pin* or *coupling link*, and the inner row of plates is known as *roller link*. The chain rollers are mounted on the bushings and they roll over the sprocket teeth.

Bush chains are similar to roller chains except the fact that they do not have any rollers.

A silent chain is not really silent, but is quieter and has less vibration than roller chain and can be run at higher speed. Its links are small in proportion to its strength so that it can be used on smaller-diameter sprockets. It tends to cushion shocks and even out irregularities. It is more expensive and needs more careful maintenance.

Alignment—There are two types of misalignment in both roller and silent chain drives:
- Shafts are not parallel
- Sprockets are not in line

The angle between the shafts is important if the shafts are not parallel irrespective of the distance between them. This condition is common in construction equipment. This condition may result from: poor installation, sprocket's shifting on the shaft, or end play of the shaft. Misalignment should be rectified.

Regarding wearing with time and use, both roller and silent chains stretch as they wear, resulting in perceptible slapping. Then, the chain length should be adjusted. Otherwise, the chain should be replaced.

Roller chain's lubrication depends on chain speed. If the speed is less and the environment is dusty, then the roller chain should be run dry and occasionally cleaned in kerosene or fuel oil. Silent chains are run in covered cases with lubricant. Engine and transmission drives are also run in covered cases and are lubricated by plunging partially in oil bath.

A crawler track is a specialized type of roller chain. Crawler tractors and some other machines have separate shoes bolted on roller chains.

Belt drives

A belt drive transfers power between two parallel shafts by using a belt and connecting pulleys on the shafts. Belt drives can absorb shocks. In addition, they are cheaper and easier to install and service; but belt drives would not last as long or carry as heavy loads as chain drives. Belt drive cannot be used where precise timing is essential. A belt drive comprises:
- Endless flexible belt moving over two pulleys—a driving belt and a driven belt
- Belt drive/s

There are two separate arrangements of supplying driving power:
- A group drive using several motors/engines
- A single large motor/engine

A belt drive can be made to provide several speed ratios by mounting several pulleys side by side on the same shaft, and by shifting the belt from one set to another as required.

Basically there are four types of belts as follows:
- Flat—produce very little noise and absorb more vibration due to torsion

- Round
- V-shaped
- Timing

Of the four types, flat and V-shaped belts are used widely. Flat and round belts are made of urethane or rubber-impregnated fabric reinforced with steel or nylon cords take tension. Surfaces may have friction surface coatings. A V-belt is rubber cover with impregnated fabric and reinforced with nylon, Dacron, rayon, glass, or steel tensi cords. A timing (also known as toothed) belt is made of rubberized fabric and steel wi and has evenly spaced teeth on its inside circumference.

Flat belts are of two types:
- Endless that is made in closed circle
- Two ends are fastened together

Compared to V-belts, flat belts require more tension to prevent slippage. Flat bel are also cheaper and lasts more as their shape reduces internal friction. A flat belt driv has an efficiency of 98% whereas the efficiency of a V-belt drive varies between 70 96%.

V–belts are of endless type. V–belts are kept small in cross-section to reduce interna friction and heating to a minimum. To suit more capacity requirement, two or more V belts are used in parallel grooves of the same pulleys. This arrangement may result i tighter belts taking more loads causing unequal wear and tear. If a belt breaks, all th belts are to be replaced.

V–belts are made of several side slopes. If the belt's cross-section does not match th grooves in the pulley, the result may be disastrous. For example, if a belt flares out mor sharply than its pulley groove, there would be contact only at the upper belt corners. As a result, there would be damaging distortion of the belt, excessive wear at the corners and also damage to the pulley groove. If the belt does not flare out as much as the pulley groove, there would be damages at the bottom with similar damaging consequences. There may be other damaging mismatch of the belt sections and pulley grooves. Pulleys are generally made of cast iron or formed steel.

Inclined planes

The force that is required to lift loads can be reduced if the loads are pushed or pulled on a ramp. Ramps are inclined planes. Lifting would be quicker, but more force would be required in lifting. Friction would be an important criterion in using a ramp. The mechanical advantage of an inclined plane is calculated by dividing its length by the height of lift ignoring friction. If the mechanical advantage is 4, a quarter of the force required to lift the load would be required to apply along the ramp.

The principle of inclined plane is utilized in lifting loads by jacks along spiral threads, which fit along similar threads in the jack body. Bolts and nuts are examples of turning rotary motion into straight push or pull. This principle of inclined plane is also utilized in wedges, ratchet and pawl, etc. The cutting edges of most excavating machines are wedge-shaped.

Clutches

The function of the clutch is to alternately engage and disengage the driving and driven shafts turning on the same axis for which its torque potential must be adequate to overcome the torque of the load and inertia. In short, its function is to control mechanical energy within a machine as follows:

- Connecting and disconnecting the shafts as and when required
- Reduction of shocks transmitted between the machine shafts
- Engaging or disengaging a machine or rotating element without stopping the drive
- Restricting the torque to a limit
- Continuing constant speed, torque and power
- Effecting automatic disconnection, fast start and stop, gradual starts, etc

A jaw clutch comprises two toothed rings called jaws with the teeth facing each other. One jaw is keyed to the drive shaft while the other one is moved to be splined to the driven shaft. Jaw clutches are cheap to produce, occupies small space especially as one jaw might comprise just tooth sockets cut in the hub of a gear serving different purposes. Among its disadvantages are rough and inconvenient in action, difficult to engage at any time, and impossible to disengage under load. If engaged while in motion, it may give damaging shocks to shafts and gears. In stationary state, a tooth may strike against another tooth. It is not in wide use.

A single plate (disc) clutch is a friction clutch where friction is used to transmit power. Modern tractors, automobiles, and many other machines use single plate clutches. The clutch driving disc is connected to the engine flywheel, which turns with the engine crankshaft. At the centre of the flywheel is the pilot bearing in which jackshaft rests. The purpose of the clutch is to connect or disconnect these two shafts with each other. The clutch plate is tightly clamped between the pressure plate and the flywheel by a series of (coil) pressure springs held between the clutch cover and pressure plate. Owing to the friction developed between the friction surfaces of the flywheel, driven disc and pressure plate, torque now can be transmitted from the engine to the transmission input shaft. In this position, clutch is engaged. The clutch is disengaged by pulling the pressure plate back against the springs by means of three levers, which are pivoted on rear extension of the flywheel.

A double-plate clutch has two discs splined to the jackshaft to obtain extra friction without increasing diameter. The performance of the multi-plate clutch is limited.

In cars and trucks, a clutch may be operated by a foot pedal by releasing the clutch by foot pressure. The clutch is re-engaged by pressure plate springs. Clutches may also be operated by hand levers as in shovels and crawler tractors. Pressure plate springs are relatively weak in such cases.

Brakes

A brake is a device intended to slow down, stop and hold a tractor, an automobile or a machine at rest. A friction brake performs all the three functions, but a tooth or jaw brake is intended only to hold. A distinction is made between the service brake system that serves to slow down the running speed of the vehicle and stop it smoothly, the parking brake system intended to hold the vehicle stationary when parked on an incline, and the auxiliary or steering brake system that is used on tractors to help the operator make sharp turns. In doing all these, it would be necessary to absorb kinetic energy of the moving parts or potential energy of loads being handled and convert these energies into internal energy of the brakes and dissipate it in the form of heat.

Friction brakes comprise metal bands or shoes, composition lining, linkage and a drum. The composition lining can be pressed against a smooth, narrow and cylindrical drum using the linkage. Bands are ribbons of slightly flexible steel; shoes are rigid pieces shaped to a drum. Brakes that squeeze inward from outside the drum are called external contracting; those that push outward from inside are internal expanding.

Disc or plate brakes are similar to plate clutches described above, though half the plates do not rotate.

The drum-type hydraulic wheel brakes find application on automobiles.

Some tractors and automobiles are fitted with pneumatic brake systems. Such a system comprises wheel brake arrangements and pneumatic brake controls.

Pneumatic motor

Compressed air may be used to carry out work in controlled manner. The control of pneumatic power is realized using valves and other control devices that are connected in a well-designed circuit.

Pneumatic motors are used in wagon drill feeds, hoists and many other machines where compressed air is available. This is particularly necessary where a light motor developing high torque is required as well as where exhaust fumes could pose problem.

Pneumatic motors can be of both piston and vane types. In a four-cylinder motor, all cylinders operate off one throw of a counter balanced crankshaft. A rotary valve releases compressed air into each cylinder as its piston passes dead centre on the upstroke. The air expands, driving the piston down, and is exhausted on the next upstroke.

The vane motor has off-centre rotating cylindrical hub in an inside casing cylinder. Flat vanes are set in longitudinal slots in the hub. Air pressure in the hub and centrifugal force move them outward into light contact with the cylinder wall.

The curved space lying between the rotor and the more distant section of the casing has an inlet at one end and an outlet at the other. Compressed air entering through the inlet pushes against the nearest vane, moving it and rotating the hub until the air can escape through the outlet. The five vanes provide for a smooth continuous rotation.

Pneumatic motors, in certain specific cases, are provided as a back up to electric motors.

Electric drive

Motors are electric drives used for steering and other applications in construction equipment. Electric motors show very favourable torque characteristics. Power supplied to motors may be AC or DC. AC motors are run at set speed and are susceptible to damage because of low voltage or lugging down under load. DC motors are effective at wide range of speeds, subject to voltage and load, and are not easily damaged by lugging down.

Power required to turn the motors is obtained from petrol or diesel engines or local area electricity supplier. Electric motors may also be battery operated, as in camera.

Hydraulics

The extensive use of hydraulics for transmitting power is being made because of the fact that properly designed and constructed fluid power system posses a number of advantageous characteristics. They eliminate the reliance on complicated systems of gears, cams, and levers. Motion can be transmitted without the slack or mechanical looseness associated with the use of solid machine parts. The fluids used are not subject to breakage as are mechanical parts and the mechanisms are not subjected to great wear. Hydraulic systems of dynamic or flowing type are of great significance in construction equipment. Such systems are based on the fact that liquids cannot be compressed except under extreme laboratory conditions. Consequently, pressure exerted to the end of a column of confined liquid would be exerted by the liquid everywhere on the confining surfaces, if weight of water is ignored. The potential pressure at a working point is the same as that at the pump, as long as the line between them is open and adequate.

Hydraulic fluid for flowing systems is a petroleum-base oil, similar to but not identical with lubricating oil for engines. Viscosity needs to be checked as too thin oil tends to leak whereas too thick oil consumes more power and makes operation sluggish. Hydraulic systems, if properly designed and in good condition, may be highly efficient but are subject to many variables.

A hydraulic jack comprises a rigid casing that includes a large cylinder containing a piston that raises load and a small cylinder containing a piston, reciprocating pump and a reservoir. As and when the pump handle is raised drawing the small handle to the top of its cylinder, oil from the reservoir start flowing under it. When the piston is pushed down, oil flow from the reservoir is shut off. The oil received from the reservoir is pushed through the check valve into the large cylinder. The check valve comprises a ball spring held against a machined seat. The pump pressure forces the ball against the spring and oil is forced into the large cylinder. When the handle is raised, the spring and pressure from the large cylinder force the ball back into its seat for 'non-return' of oil. Each stroke of the pump thus increases oil content and boosts pressure in the large cylinder for pushing load up.

A reciprocating pump comprises a cylinder in which a piston is moved backwards and forwards by power from an external source. In the case of jack, the movement is upwards and downwards.

In most cases like in excavators, pumps are driven by rotary shafts. There are three principal types of rotary pumps known as gear, vane and piston pumps.

The displacement is the quantum of liquid a pump moves in a revolution or cycle. The displacement may further be classified as:

- Fixed displacement—can be varied by changing speed of revolution
- Variable displacement—hydrostatic piston pump is an example

Hydraulic gear pump comprises two accurately meshed spur gears between side plates, turning in a chamber so shaped as to allow close contact with housing, for about half the circumference of each gear. Gravity and/or atmospheric pressure force the oil in. The oil passes through the hollows of the gear teeth and is carried to the other side of the housing. With the teeth meshed in the centre, oil is forced into the outlet passage causing continuous flow. Very close clearances are required to prevent back leakage. This is the simplest and most economical of the rotary pumps.

Hydraulic vane pumps are complicated and more expensive than the hydraulic gear pumps. Vane pumps are of two types – unbalanced and balanced. In the unbalanced type, a cylindrical rotor has radial slots where sliding type flat vanes may be fitted. The rotor is placed off-centre in a circular ring or case. When the rotor turns, the vanes are pushed outward by centrifugal force until they reach the ring against which they fit closely. The chamber between the inlet and outlet expands and contracts as the rotor rotate. The chamber starts expanding at the inlet point of fluid creating suction and is discharged into the outlet where the vanes start contracting. The chamber beyond the outlet is such that the vanes fit tightly so that air pressure cannot follow the vanes back to the inlet. This vane pump does most of its work on one side resulting in rapid wear. For this reason, balanced type pumps with two port sets on opposite sides are used. In balanced pumps, the rotor is on-centre. The ring is oval-shaped instead of circular.

Hydrostatic piston pump comprises five or more pistons on a rotating cylinder block. The pistons are parallel to the axis of revolution. Their heads are in direct contact with non-rotating movable swash plate. The casing has two ports that carry low pressure fluid from a pump. The fluid goes through check valve, an open tube and an outlet to the working parts. In a particular pumping position, the charging pressure pushed the pistons deep into the cylinders till their ends press against the swash plate. The inlet port thereby closes forcing the fluid through the outlet valve. The oil is thus pumped around the closed circuit and in turn converted to torque to drive the shaft in the motor. Altering the tilt of the swash plate, oil flow may be reduced or increased. As the torque requirement increases, the oil flow may be slowed down and the oil pressure increased to deliver the required torque. This 'hydrostatic' system is always in equilibrium.

Hydraulic motors

The working of a hydraulic pump is made possible by rotating a shaft using hydraulic power. The hydraulic motor uses that power to rotate a shaft.

In hydraulic motors also, gears or pistons or vanes are used to utilize the hydraulic power. However, a pump is not interchangeable as a motor unless specially designed for the dual purposes.

Power sources

The power transmitted by a rotating machine component such as a shaft, flywheel, gear, pulley, or clutch is of interest in the study of construction equipment and machineries. The necessary torque is availed from steam engines, internal combustion engines and electric motors.

The steam engines along with steam generators are relatively cumbersome and expensive and are, therefore, hardly used now even though high torque is available from a steam engine.

The majority of construction equipment used for construction work like excavation, handling, hauling etc use diesel or petrol engines for energy. Diesel oil is generally used as it is less expensive. These engines are referred to as internal combustion engines because the fuel is burned inside the same unit that moves the shaft or combustion occurs within a cylinder. In a diesel engine, plain air is drawn on the intake stroke. This air is so highly compressed that it becomes very hot. If diesel oil is injected, it is ignited just by the contact of heated air. The diesel engine, therefore, needs no carburetor or ignition system. However, it requires a method of metering the fuel correctly and injecting it at just the right time as the air-diesel mixture occurs in a very short period of time. To make the mixture capable of burning rapidly and completely, it is essential that the fuel be broken into as fine particles as possible and that the amount of air surrounding each fuel particle be sufficient for its complete combustion. For this purpose, diesel is injected into the cylinder by a fuel injector.

Electric motors show very favourable torque characteristics, but electricity distribution circuit is expensive and susceptible to safety hazards. Stationary construction equipment like a tower crane is powered by electric motors. Electricity may be supplied by diesel generators without elaborate distribution cable network, but such arrangement would be expensive.

13.3 PLANTS AND TOOLS

Compressors

The volume of a gas (V), according to the Boyle's law, varies inversely as its pressure (P) when the temperature remains constant. This law may be represented as follows:

$$P_1 \times V_1 = P_2 \times V_2$$

Compressors used in construction work are of three types as follows:

- Reciprocating type
- Rotary vane type
- Rotary screw type

In the reciprocating type, air inflows into the cylinder through cleaner when the inlet valve is open on the downward suction stroke. During the upward stroke, the inlet valve is closed for the air to be compressed. At the full working pressure during a compression stroke, the discharge check valve opens into a line to storage tank (air receiver) from which compressed air is distributed to the tools. In the two stage compressor, air is compressed to a lower pressure to be compressed to the full working pressure in the second stage.

In a rotary vane type compressor, a cylindrical rotor is eccentrically mounted inside a larger cylindrical casing. Radially adjusting blades are fitted into lengthwise slots in the rotor. Because of the rotor being eccentric, the vanes are very small on one side of the casing, and large on the opposite side. As the vanes expand, air on entering is trapped in the space between the vanes and the cylinder wall. Air is compressed as the vanes contract and air volume decreases till the discharge port to the air receiver is reached. Lubrication of the vanes is required. Compressed air, therefore, contains oil contamination.

A rotary screw type compressor comprises a pair of counter-rotating screw rotor with matching screws to compress the air in single operation. There are no vanes in screw type compressors and no metal to metal contact. Lubrication, therefore, is not required. The rotary compressors have been developed quite recently.

Pumps

Water is incompressible and its movement is the effect of either impellers or pistons pushing the water body. Atmospheric pressure is used in the working of the lift pumps.

Pumps generally used in construction work are of three types:
- Centrifugal type
- Displacement type
- Submersible type

The centrifugal pump throws water outward from the centre by the rapid centrifugal rotation of vanes. The vacuum thus created at the centre by the discharged water is constantly filled up by water forced through the inlet passage by the atmospheric pressure. In case of a turbine, the vanes are moved by the pressure of water or steam. In centrifugal pumps, the vanes throw out water by the reverse turbine action. Water enters the pump through an inlet hose via a strainer. The hose is fastened to the inlet. A foot valve in the hose would prevent draining out of water when the pump is shut down. If lift of water is within the height permitted by the atmospheric pressure, water would then be drawn inside the pump casing imparting additional kinetic energy to it. Because of the casing's spiral shape, however, water velocity decreases because of the increasing area of flow. On reaching the delivery pipe, the velocity would be relatively low while the pressure would be high. Water, thus, can be pumped against high delivery heads. The pump body is a rigid casing that serves as a support for the pumping mechanism and as a tank to supply priming water. Centrifugal pumps need priming with water at each starting of pumping operation.

A displacement pump can either be a reciprocating pump or a diaphragm pump. A reciprocating pump comprises a rigid cylinder, a piston and valves for intake and discharge. Water is drawn in by the action of the piston in one stroke and forced out by the return stroke in single-acting operation. This would result in pulsating delivery. By incorporating additional valve, the pump may take in more water behind the piston as the water from the first stroke is discharged. This water behind the piston is discharged as the first stroke is repeated in the opposite direction. This back and forth double-acting operation handles double the amount of water compared to the single acting operation. Reciprocating pump is called simplex with a single cylinder, duplex with two cylinders and triplex with three cylinders. The quantity of water pumped is almost equal to the quantity of water displaced by the piston as 3 to 5% slip of water occurs as the valves open or close.

A diaphragm pump, which consists mainly of a movable diaphragm in a closed chamber between two check valves, can handle water containing solids while a reciprocating pump cannot. A diaphragm pump is a positive displacement type of pump and has flexibility in its main operating part, the circular disc or diaphragm, which moves vertically. On the upward movement, it causes a suction that makes the inlet valve open and sucks the contents of the inlet pipe. On the downward movement, the inlet valve is closed and the outlet valve is forced open, discharging part of the contents of the chamber into delivery pipe. The suction effect makes the pump self-priming. The flexible disc's alternate upward and downward movement is carried out by means of a rod connected to the rotating shaft of the pump's engine. Diaphragm pumps are available with two cylinders and two diaphragms for enhanced output and efficiency.

A submersible pump is powered by a waterproof electric motor having a common housing with the pump. The pump-motor set is suspended in the water to be pumped. A waterproof cable connects the motor to the source of power on the surface or above the water level. A submersible pump can operate totally or partially submerged. The pump can also run dry for a period of time. Because of this, the pump need not be moved with lowering of water level. As there is no suction pipe, there is no problem of priming also. It is also free of noise and fumes problem. The pump is to be submerged and it would be ready to perform. The disadvantage associated with submersible pump is the impeller blades being affected by granular particles in water. There would be loss of capacity due to wearing of impellers.

Apart from the construction equipment deployed at construction sites, small power tools and plants are essential for continuing with construction activities. Most of these tools are hand-held and powered by either electricity or compressed air. Electricity is used for rotating a tool and compressed air is used for percussion.

In the changing scenario, when construction work is being mechanized largely to complete projects in time, small power tools and plants also should be used so that the construction equipment deployed for producing larger volumes of work do not remain idle for preparatory or associated work being carried out manually instead of using tools and plants powered by electrical, pneumatic or hydraulic energy.

Electric tools

Electric hand tools include:
- Drills

- Hammers
- Screw drivers
- Circular saws and jigsaws for woodwork

There are other electric tools also not mentioned above. Electric drill is the most common hand-held tool used for making holes through timber, masonry, rocks, concrete and metals by rotary motion or percussion or combination of both rotary and percussion. For drilling into rock or concrete, special hardened metal bits, vide Chapter 5, Section 5.5, should be used at the tips. The success of a drilling operation depends on the ability of the bit to remain sharp under the impact of the drill. Regular bits are not suitable for drilling in some types of rocks that are so abrasive that steel bits need to be replaced after drilling only a short length of hole. The increased cost of bits and the time lost in changing bits are so high that it would generally be cost-efficient to use carbide-insert bits consisting of a very hard metal, tungsten carbide or silicon carbide, which is embedded in steel.

A pneumatic percussion drill hammers very rapidly. An electric hammer delivers powerful blows at slower rates.

An electric screw-driver operates only when the bit is in touch with the screw head and slips when the screw has been driven in at the pre-determined tension.

Other electric tools and plants like saws, jigsaws, planes etc may be used to reduce time of execution. Manually carrying out such work takes long time.

Pneumatic tools

The most common and visible pneumatic tool is breaker used for breaking hard surfaces. Compressed air required for the breakers is produced at the construction sites. Compressors that are operated at the construction sites are powered by either metered electricity or electrical power generated using diesel or petrol generators. Compressed air is stored in air receivers. Breakers are used for breaking concrete structures, roads and other hard surfaces. Chipping hammers are similar to breakers for lighter work of breaking. Pneumatic vibrators are used for compacting green concrete. Sheet piles can also be driven by pneumatic hammers. Other pneumatic tools and plants used in construction are for spraying concrete or paint and also include pneumatic drills, saws and grinders.

Cartridge hammers

Cartridges are fired by guns to penetrate into surfaces like concrete, rocks etc for fixing purposes. The cartridges contain hardened steel pins with threaded heads. Compressed gases in the cartridges provide power for firing. Otherwise, the driving force is provided by a piston in the gun. There are different types of cartridge guns for different purposes. A gun should be selected properly and carefully for the required purpose only. Cartridge guns are not used much in India.

Green concrete needs to be compacted after placement by using vibrators. This is essential as otherwise there would be voids in concrete, and the concrete would not attain the required strength on hardening. Concrete of small volume of minor nature is consolidated by tamping manually. The power for operating the vibrators could be metered electricity, diesel or petrol engines or compressed air. There are three types of vibrators:

- Poker vibrators—by immersing the vibrator in mass green concrete, consolidation is achieved by high vibration
- Tamping vibrators—are small vibrating engines on tamping boards for consolidating concrete slabs or concrete pavings
- Clamp vibrators—are attached to the outside of forms, which should be supported sufficiently strongly to take the load of concrete as well as the impact of vibrations.

13.4 PLANTS FOR EARTHWORK

Earthmoving operation involves:

- Loosening the materials to turn it, if required, into workable state
- Digging the materials to initiate earthmoving from the original location
- Moving the materials from the original location to dump location
- Dumping the materials
- Working on the materials to finish the same in desired shape and compaction

Earthmoving construction equipment fall under two basic types:

- Fixed position type
- Moving type

The selection of the type of construction equipment to be mobilized and deployed depends on the particular project size, its topography, earth materials, earthwork involved etc. Fixed position construction equipment are deployed for specific type of work. Moving construction equipment are deployed for ground levelling and bulk earthmoving. Shovels, backhoes, draglines, clamshells, etc. belong to the fixed position type; whereas bulldozers, loaders, scrapers, graders, trenchers, etc. belong to the moving type.

Construction-site tractors or prime movers fall into two basic classes:

- Wheeler or wheel-mounted—single or two axle units with two or four wheel drive—another single/more axle equipment or trailer balances a single axle tractor.
- Crawler or track-mounted

Wheel tyres

Tyres (Fig. 13.1) are required for:

- Spreading the weight of the vehicle/construction equipment on to the ground
- Converting rimpull at the point of contact with the ground into traction
- Improving steering potential of the construction equipment

The two principal methods of tyre construction are:
- Cross-ply rubber tyre comprises
  (a) Tread—the important part of the tyre in contact with the ground/road providing traction, cushioning qualities and eliminating resistance—the rate of wear and friction coefficient are influenced by the quality of the rubber and tread pattern
  (b) Beads—the tyre beads are bands of strong steel wire integrated with the body that prevent any change of shape that would interfere with fitting on the wheel rim
  (c) Sidewalls—layer of rubber intended to protect the sides of the tyre between the tread and the bead from damage and moisture
  (d) Belts—these provide an intermediate layer of protection confined to tread area to prevent separation of the tread from the carcass
  (e) Body plies—the plies are the individual fabric layers in the cord body and are usually woven of cotton, rayon, or nylon cords, or steel wires surrounded by rubber forming main carcass to provide tyre strength
  (f) Inner liner, which seals in the air
- Radial-ply rubber tyres—the beads are made from a single bundle of steel wires, the carcass is a single layer of steel cables run through the shortest distance radially from bead to bead to form an arch shape that results in wider bodied tyre.

A tubeless tyre has an airtight inner liner that seals in the air pressure. The rim must also be airtight. The basic advantages of such tyres are: greater reliability and better safety in case of an accident. The tubeless tyres deflate slowly in case of puncture. Other tyres have soft inner tubes to contain pressurized air.

Larger size tyres permit an increase in the hauling capacity of equipment or vehicle if the speed can be maintained with heavier load. In other words, the output of earthmoving equipment may be increased if site conditions permit the use of larger tyres.

Large-diameter tyres require lower tyre pressures. Rolling resistance is reduced because of larger contact area on the ground. This reduces tyre slippage and wear/tear. However, increased tyre diameter reduces the rimpull.

Decreasing tyre pressure increases the floatation and tractive efficiency. The increase in tractive efficiency results

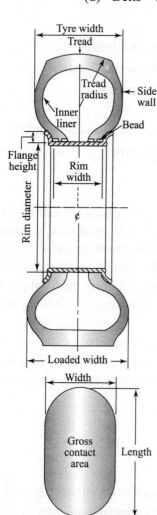

Fig. 13.1 Cross-section of a tyre

due to lower rolling resistance especially in hauling over soft surfaces. Rolling resistance is reduced due to lower penetration caused by the lower tyre pressure. However, low pressure increases the flexing of side walls resulting in heat generation that reduces the tyre's life.

The recent trend is to use radial tyres because of relatively lower heat build up and rolling resistance. The tyre life is affected by:
- Tyre type and size
- Heat build up
- Travel surface
- Tyre pressure and load
- Operator's/driver's skill
- Maintenance

Advantages of wheel-mounted tractor or power unit:
- Working conditions: (a) firm clay/soil, concrete; (b) abrasive soils but no sharp edges; (c) level or downhill work; and (d) suitable for fair and dry weather
- Effect on ground: (a) good compaction, and (b) variable with counterweight and ballast
- Application: (a) long distance; (b) loose soils; (c) fast return speed: 12–20 km/hr; (d) moderate blade loads, and long, thin cuts when dozing; and (e) highly mobile

When speed and mobility of the power unit are of major importance, wheel-mounted construction equipment is the obvious choice. A serious disadvantage of wheel-mounted equipment is that it can easily bog down in soil.

Tracks

The crawler track (Fig. 13.2) is a specialized type of roller chain. It is made up of a number of identical shoes cut and drilled at their ends so that they can be fastened together by pins. There are many types of track shoes. The standard construction is a flat plate with a single high cleat or grouser across it.

The track-mounted crawler travel unit is made up of a turntable welded to a base frame comprising two axles that connect two heavy truck frames. The axles are made from steel sections. The heavy frames rest on track wheels. The track wheels are surrounded by the track.

The tracks are driven by the bull wheels, which are mounted on the low rear of the frame. Tracks are classified on the basis of number of rollers by which they are supported. The multi-roller tracks are meant for deployment on soft and medium ground conditions. They feature a comparatively large number of small diameter track rollers whose axles are secured on a tough frame. The track shoes move on these rollers. Because of many number of small rollers, there is not enough gaps for the tracks to be pushed up between the rollers. This results in uniform load distribution over the ground, but affects negotiation of obstacles over the ground. Multi-roller tracks are used in power shovels.

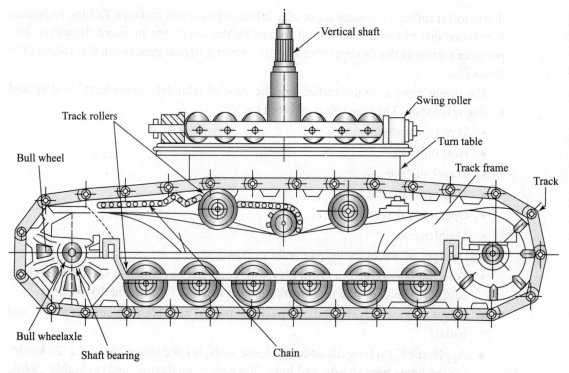

Fig. 13.2 Tracks and lower frame

In the other variation, only a few large diameter track rollers are used. Because of larger diameters, unsupported track lengths between the rollers would be more as a result of which load distribution on the ground would not be uniform. More unsupported lengths would make it easier for the tracks to negotiate obstacles over irregular hard and rocky ground conditions.

More wide and extra long tracks are used on weak swampy soils. Such tracks with wider shoes are heavier and more expensive.

There are different types of track shoes. The standard construction is a flat plate with a single high cleat or grouser across it. Advantages of track-mounted tractor/power unit:

- Working conditions: (a) variety of soils; (b) sharp edges not so damaging; (c) any terrain including uphill climbing; and (d) good in all weather
- Effect on ground: (a) good floatation and (b) low ground pressure with different shoe sizes and types because of large ground contact area
- Application: (a) short distances; (b) tight soils; (c) slow return speeds: 6–12 km/hr; (d) large blade loads and short, heavy cuts when dozing; and (e) restricted mobility

When traction requirements are high with less priority for speed and mobility, track-mounted construction equipment is the obvious choice. If track-mounted equipment bogs down in soil, pulling it out turns out to be a serious problem.

Self-propelled equipment is powered by a prime mover, which may be an engine or a motor. Otherwise, a tractor may be needed to propel the equipment. In both cases, the power unit must deliver force to the surface on which the equipment is operating.

The power output or force applied at the driving wheels or tracks is known as tractive effort. This force must be enough to overcome the combined resistance to motion. Traction is the amount of friction/grip between the supporting surface and the drive tyres or tracks on it.

Rolling resistance

Rolling resistance is the resistance encountered by a vehicle or construction equipment while moving over the ground surface or road. This resistance varies considerably depending on the condition of the surface over which a vehicle or construction equipment moves. Soft soil surface provides more rolling resistance than hard surface like concrete pavement. The maximum speed attainable by a vehicle or construction equipment depends on the rolling resistance of the ground or road. The traction between the ground surface or road and the wheels or tracks must be adequate to prevent slippage and, therefore, must exceed the supplied rimpull for the vehicle or construction equipment to move. In case of tyre-mounted vehicles and construction equipment, the rolling resistance depends on size, pressure and treads design whereas this resistance in case of track-mounted vehicles and construction equipment depends on ground condition or road surface. Rolling resistance is likely to vary under different climatic conditions or divergent ground conditions along the road.

Rolling resistance is expressed in pounds of tractive pull required to move each gross ton over a level surface of the specified type or condition. The rolling resistance of a haul road can be worked out by towing a truck or other vehicle whose gross weight is known along a level section of the haul road at a uniform speed. Appropriate instrument should be fitted to determine the average tension in the tow cable. This tension is the total rolling resistance of the gross weight of the truck.

The rolling resistance in pounds per gross ton will be:

$$R = \frac{P}{W}$$

where R = Rolling resistance, lb per ton
P = Total tension in tow cable, lb
W = Gross weight of truck, tons

Tractor

As already mentioned above, a tractor is a wheel-mounted (wheeler), Fig. 13.3(b), or track-mounted (crawler), Fig. 13.3(a), self-propelled vehicle used as a power unit for moving construction, agricultural or other equipment fitted with special attachments/tools. A tractor is also used for towing trailers. Wheel-mounted tractors may be single or two axle units with two or four wheel drive.

A tractor comprises:
- Engine
- Clutch
- Flexible coupling
- Transmission
- Rear axle mechanism

Engine

The engine converts thermal energy into mechanical power. Heat engines used are internal combustion engines as combustion takes place within the cylinder. The engine is usually diesel powered.

The driveline comprises a set of mechanisms that transmit the torque developed by the engine to the driving wheels or tracks and change the driving torque both in magnitude and direction. The driveline comprises the clutch, flexible coupling, transmission (gearbox or hydraulic), and rear axle.

Clutch

The clutch serves to disconnect the engine shaft from the transmission for a short time when the driver/operator is shifting gears and also to connect smoothly the flow of power from the engine to the driving wheels or tracks when starting the tractor from rest.

Flexible coupling and transmission

The flexible coupling incorporates elastic elements that allow connecting the clutch shaft and the transmission drive shaft with a slight misalignment. The transmission makes it possible to change the driving torque and the running speed of the tractor by engaging different pairs of gears. A manual gear change requires a driver's full concentration in selecting the right gear to prevent the engine from stalling in case of varying load conditions.

Rear axle mechanism

The rear axle mechanism increases the driving torque and transmits it to the driving wheels or tracks at right angles to the drive shaft. In most tractors, the rear axle also includes brakes. The steering mechanism serves to change the direction of the tractor by turning its front wheels (in wheel mounted tractors) or by varying the speed of one of the tracks (in track-mounted tractors). The working attachments of the tractor are used to utilize the useful power of the tractor engine for various tasks. The attachments include the power takeoff shaft, drawbar, implement attaching system and belt pulley.

Two-wheel and four-wheel drive

In two-wheel drive, the rear or drive wheels and tyres are very much larger than the front ones, which are provided for only supporting and steering. While heavy-duty rear tyres are required in almost all types of construction work, heavy front tyres are deployed only in case of front loader equipment. The main handicap of two-wheel drive is that it has poor traction on soft and slippery surfaces. Although the four-wheel drive has enough traction for all kinds of construction work, its traction is still less than that of

Mechanized Construction   411

Tracks   (a) Track-mounted

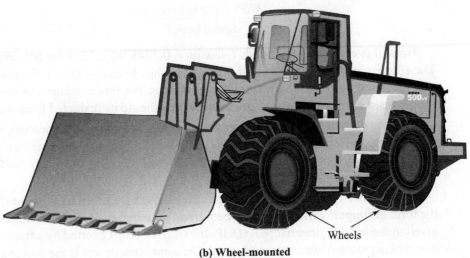

Wheels
(b) Wheel-mounted

Fig. 13.3 Wheel- and track-mounted tractors

track-mounted drives. Four-wheel drive equipment needs more power and weight than a crawler tractor to perform the same type of work although it has more advantage in terms of speed and working on highways apart from being more stable and comfortable on rough and hard terrain.

Drawbar

Drawbar is provided for attaching fully mounted or semi-mounted implements to the tractor. The drawbar comprises heavy steel drawbar with the necessary devices for fastening in position with pins or bolts. It is fastened under the centre of the tractor and can be extended backward across a support bracket. It projects to the rear and can swing horizontally. It is held in the desired position by a pin through the bracket or by a pair of bolts in a clamp.

In the hydraulic transmission system, the engine drives a hydraulic pump via a series of speed regulating gears, which in turn forces oil around a closed system to drive hydraulic displacement motors attached to each of the tracks. A special rotary coupling, centred in the slewing ring, is needed to overcome the problem that the power source is mounted on the superstructure, which must be able to rotate independent of the tracks and sub-frame, while maintaining the closed oil circuit.

Rimpull

A tractor is used as the prime mover by applying its power at the rim of its drive wheel as a tangential force known as rimpull. The rimpull is the tractive effort (TE) available between the tyres of the drive wheels and the ground surface to propel the vehicle or construction equipment without causing any slippage. Available power is a function of horsepower of the engine and operating speed. The relationship of the power to rimpull for a rubber-tyred tractor is influenced by the efficiency of transmission and can be expressed by the following equation:

$$\text{Rimpull} = \frac{375 \times \text{HP} \times \text{Efficiency}}{\text{Speed (mph)}}$$

Rimpull is expressed in pounds (Newton = 0.2248 lb). 375 is the conversion factor. The efficiency of most tractors and trucks will range from 80–85%. For working out the pull exerted by a tractor on a towed load, the tractive force required to overcome the rolling resistance as well as any grade resistance should be deducted from the rimpull.

The term gradability refers to the maximum slope, expressed as percent, up which a tyre-mounted or crawler-mounted empty or loaded vehicle or construction equipment may move at uniform speed. Obviously, the gradability of a tractor only would be more than a tractor with load attached to it.

While rubber-tyred equipment performance curves are given by the manufacturers in the form of rimpull versus ground speed, track-mounted tractor performance curves are given in the form of drawbar pull (DBP–the tractive effort exerted by a tractor on a load that is being towed) versus ground speed. Drawbar horsepower is the power available at the drawbar hitch (Fig. 13.4) for moving the tractor and its towed load forward on level ground. The drawbar hitch is mounted on the tractor for attaching drawn implements. The engine has to pull both the tractor as well as the load.

$$\text{Drawbar horsepower} = \frac{\text{DBP (lb)} \times \text{Speed (mph)}}{375}$$

For computation of the pull that a tractor can exert on a towed load, the tractive force required to overcome the rolling resistance as well as any grade resistance for the tractor need to be deducted from the rimpull of the tractor. Wheel-mounted tractor differs from the track-mounted tractor in this respect.

The forward motion of a tractor or prime mover is restricted by the following:
- Engine's power available as drawbar pull or rimpull
- Rolling resistance of the haul road
- Gross weight of the tractor and its load
- Grade to be negotiated—favourable or unfavourable. Unfavourable or adverse grade adds to the resistance while favourable grade reduces the resistance.

Production measurement

Distances—may be measured using tapes, distance metres, speedometers, etc.

Bank—a bank of earth or other materials may be measured before or during the time it is being excavated and loaded. Bank measurement is generally the basis for payment.

Swell—When soil or rock is excavated or blasted out of its original position, it breaks up into loose particles or chunks thereby creating voids. The increase from the bank volume to loose volume is called swell, which is expressed as a percentage of the bulk in the bank.

$$\text{Swell factor} = \frac{1}{1 + \text{percentage of swell}}$$

$$= \frac{1}{1 + 0.25} = 0.80 \text{ or } 80\%$$

Shrinkage—Soil shrinks on compaction. The extent of shrinkage would depend on the soil characteristics, its structure in the bank, fill thickness and type/weight of construction equipment deployed.

$$\text{Shrinkage factor} = \frac{\text{volume in fill}}{\text{volume in bank}}$$

$$= \frac{\text{volume in fill}}{\text{loose volume} \times \text{swell factor}}$$

Shrinkage percent = 1 − shrinkage factor

Capacity of blade or bowl or bucket—Manufacturers' rated capacity may be accepted. If the rated capacity is not known, the capacity has to be worked out from the blade/bowl/bucket dimensions making allowance for curves and irregularities.

Scraper bowl's capacity is measured as heaped with 1:1 slopes above the sides. Shovel's capacity is measured as struck volume. Struck as well as heaped capacity is explained in Chapter 3, Section 3.6. A loader's bucket capacity is also measured in struck volume.

A truck should haul materials on highways at struck capacity as heaped materials may spill over from the sides.

The ratio of a container's capacity to its actual loose loads is termed as the efficiency factor of the bucket, bowl or body. The container's efficiency factor (CEF) is expressed as follows:

$$\text{CEF} = \frac{\text{Material in container}}{\text{Container's rated capacity}}$$

The volume of the materials in container may be measured as bank volume or loose volume. CEF would vary accordingly.

Output of a construction equipment with a cycle $= \dfrac{Q \times K \times E \times 60 \times f}{C_m}$

where Q = Capacity in m^3
K = Efficiency factor of the bucket
E = Efficiency of construction equipment
f = Soil conversion factor
C_m = Cycle in minutes
60 = Number of minutes in an hour

Bulldozer

A bulldozer (Fig. 13.4) is in effect a short-range tractor equipped with a front dozing blade, which can be raised or lowered by hydraulic/mechanical control and is used for digging and pushing. It is necessary that tractor and dozer blades should match for best results. This matching is done on the basis of materials to be moved—particle size, particle shape, voids and moisture content. The weight and horsepower of the tractor determine its capacity to move materials. A bulldozer is versatile construction equipment that may be used from the start to the finish. It often grades both the cut and fill-materials in the same operation.

The front pusher blade of a bulldozer is a massive structure that has a rectangular base and back with two push arms. The back of the blade is reinforced with welded ribs and box sections to make it rigid both longitudinally and cross-wise. The leading edge of the base is a flat blade or knives of hardened steel which projects ahead of and below the rest of the blade. The front of the blade is called mouldboard and is concave and sloped back. As the blade is pushed into the ground, the knife cuts and breaks up the soil that is pushed up the curve of the mouldboard until falling forward. The weight of the soil first helps penetration and then, as the blade is kept more or less in rotary movement that tends to even up the load, thereby reducing power consumption in excavation – which means less or no cutting. A nose plate is welded to the blade top to prevent material being pushed from spilling over the blade.

The push arms are heavy, box-type hollow beams extending from a fork (hinged) connection with the tractor to the bottom of the blade to which they are welded. The fork is furnished with thermally treated adapter block that serve as bearing or the trunnion of the transverse beam of the bulldozer.

Bulldozers are used for:
- Cutting bushes, trees etc for land clearing
- Stripping of topsoil that is unsuitable as fill material or stable sub-grade
- Opening up pilot roads through hilly and rocky terrain
- Shallow excavation
- Grading
- Pushing scrapers to assist in their loading
- Spreading
- Backfilling trenches and pits
- Compacting fill
- Maintaining haul roads
- Ripping

In earthmoving, the three operations of a bulldozer are:
- Cutting in layers
- Moving up to a haul distance of 100 m
- Placing/levelling off

Based on the above, bulldozers may be deployed for:
- Construction of banks, dams and dykes
- Excavation of channels
- Backfilling temporary channels and trenches
- Development of sand and gravel quarries
- Piling bulk products

Track- and wheel-mounted bulldozer

Bulldozers are primarily tractors equipped with crawler tracks for heavy-duty work or wheels for lighter work fitted with front pusher blades, which can be raised or lowered by hydraulic or cable control and which are used for stripping and excavating up to a depth of 400 mm. The loosened materials may be pushed ahead of the blades up to a distance not exceeding 100 m. The blades can be set at an angle either in the vertical or horizontal plane for angle dozers. Power is transferred to the blades by means of two strong push arm connections on the two sides of the dozer. The output of a track-mounted (crawler) or wheel-mounted (wheeler) dozer varies with the soil type and prevailing conditions at the locations of operation. The control of the blades on most dozers is hydraulic. Otherwise, winch and cable (wire-rope) control system is used. The weight and power of the tractor determines its ability to push—higher value of kW per metre of the blade means the blade's more cutting power. A dozer of such huge power as 746 kW and 45 m^3 blade capacity is available in the market.

The advantages of winch and cable (mechanical) control are:
- Simplicity of installation and operation
- Simplicity of repairing the controls
- Less danger of damaging a machine as the blade can move up and ride over a rigid obstruction like a boulder

The advantages of hydraulic control are:
- Capability of producing high downward pressure on the blade, in addition to its dead weight, to force it into the ground
- Ability for more accurate positioning of the blade

The blade may be set at an angle to deposit the soil materials at one side along the push line. This is a very useful facility in case of opening up pilot roads in hilly terrain or backfilling trenches. Angled blades are meant for special purposes.

The track-mounted crawler dozers have large ground contact area and, therefore, have good track adhesion. They are capable of operating in rocky formations, on slopes and abruptly varying grades. The crawler dozers can also travel over muddy surfaces and operate under adverse service conditions. A dozer with gripped tracks can climb a 1:2 slope or even 1:1.5 provided the slope material gives adequate grip and is not composed of loose rounded cobbles. It is dangerous to operate a dozer (or any kind of a tractor) on ground slanting to one side, particularly if the ground is soft. The wheel-mounted dozers can travel at higher speeds on the job and have high mobility but their track adhesion is less. For off-road earthwork, functioning of wheel-mounted construction equipment is totally dependent on tyres. Better tyres mean better performance.

Track-mounted dozers cannot move over paved roads because of damage this would cause on the roads. They should not be allowed to move over finished formation surfaces also.

Four-wheel drive is required for wheel-mounted dozer to provide sufficient traction to perform the tasks of a crawler bulldozer and correspondingly larger engine than that needed for track-mounted dozer is necessary. The increased self-weight in turn raises the bearing pressure at the contact point between tyres and ground thus causing compaction of the soil. Such compaction is not desirable as pushing of loose materials is more effective. As the two axles are very close to one another, there may be stability problems on uneven ground. Wheel-mounted dozers are generally not in use for these reasons.

The big powerful dozers are costly to operate and maintain, so it is not worth to keep a bigger one at site for occasional use. A dozer's principal full time job is for cutting, or spreading fill for earthwork in the specified layer thickness and compacting and bonding it to the previously compacted layer. It's the weight and vibration of the dozer that achieves compaction continually. Heavier model would thus be better. The dozer cannot shift material very far; it can only spread it locally.

A bulldozer is important land-clearing equipment with standard blade. Its performance can be extended and improved by replacing the standard blades and connectors with special blades and devices. The stumper, when fitted in place of standard or angled blade, is used for pushing over trees and stumps, driving under stumps to push them out and digging around them. It is also useful in digging out boulders, clearing soil off loose stumps, digging up railways, ripping up shale and old paved areas and cutting shallow ditches. The V-cutter is designed to shear off all vegetation, large or small, at ground level. The tree shear is capable of cutting hardwood trunk of diameter up to 500 mm and softwood trunk of bigger diameters. There are more devices like root cutter, root rake, towed chain, disc harrow, rolling chopper, rotary mixer, rotary mower, brush chipper, tree chipper, stump chipper, log movers and so on.

Ripper

A dozer can be fitted with rippers at the rear for excavation work in rocky or hard soil for mechanical loosening of rock. The ripping action requires:

- Penetration of a strong ripper into the hard soil or rock by down pressure at the tip
- A tractor with sufficient horsepower to advance the ripper through the hard soil/rock (minimum 150 hp, maximum 400 hp or more) to be ripped
- A heavy tractor to generate sufficient traction
- A heavy-duty robust tractor to take the strain

A ripper is a metal (high tensile steel) blade known as a shank fitted with replaceable point or tip. It is used for ploughing soil up. The shank used may be straight or curved. The straight ones are used for massive or block formations, and curved ones are for bedded or laminated rock or road pavements. There are different types of rippers attached to dozers—rippers that are manually lowered at the required locations, rippers that are lowered hydraulically or rippers that are adjusted hydraulically to the required angle of attack. The cutting force of a ripper depends on the ripping angle. For weathered hard rock, the optional ripping angle is 30°– 45°. An increase of the ripping angle from 45° to 60° doubles the resistance of ripping. A decrease of the ripping angle to less than 30° would also raise the resistance of ripping. The number of shanks to be attached could be one to five or more depending on the nature of existing soil and required excavation depth. In most cases, the ripping operation is started with one shank. If the material is easily penetrated and fractured into pieces, a second shank then can be used. If one tractor proves ineffective, a second tractor may be deployed in tandem. The depth of ripping done with trailed rippers is normally up to 0.4 to 0.5 m, and with tractor mounted rippers up to 1.5–2.0 m. Ripping is possible in:

- Stratified rocks and soils
- Rocks with fractures, faults, and planes of weakness—would depend on the extent of weakening
- Large-grained rock of brittle texture
- Most materials like shale, slate and mudstone
- Soft rock—would depend on the extent of softening and weakening by weathering

Stratification and cleavage planes are necessary for ripping. Hard igneous rocks like granites, basalts, diorite etc do not have such stratification and cleavage. Sedimentary rocks like sandstone, limestone, shale, caliche, and conglomerate rocks are most easily ripped solid materials. Massive metamorphic rocks like gneiss, quartzite and marble are usually not rippable, but slate and thinly bedded schist slate can be ripped.

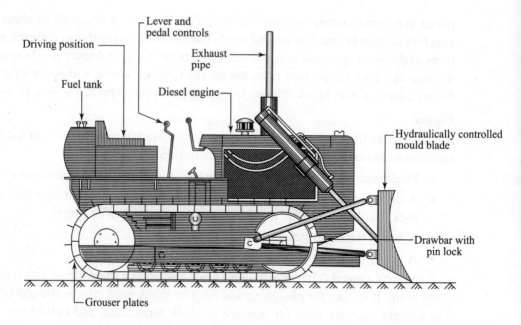

Fig. 13.4 Bulldozer

Example

Output of a construction equipment with a cycle = $\dfrac{Q \times K \times E \times 60 \times f}{C_m}$

where Q = Capacity in m^3

K = Efficiency factor of the bucket

E = Efficiency of construction equipment

f = Soil conversion factor

C_m = Cycle in minutes

60 = Number of minutes in an hour

If efficiency is approximately taken as $\dfrac{4}{5}$, then 50 minutes is considered in an hour

Output = $\dfrac{Q \times 50}{C_m}$ of loose earth in m^3/hr as $f = 1$ in case of loose earth

Working out the requirement of bulldozers for specific conditions

BEML model No. D–155A–1 is considered

Blade—4.13 m × 1.59 m, Blade weight—5.78 tonnes

Tractor weight—26.92 tonnes, Operating weight—32.51 tonnes

Capacity of moving earth = $\dfrac{1.732 \times 1.59 \times 1.59}{2} \times 4.13 \simeq 9.0$ m^3

Cycle time:
- Pushing 30 m @ 2.5 km/hr = 0.72 minutes
- Returning 30 m @ 5.0 km/hr = 0.36 minutes
- Fixed time, loading and changing gears = 0.42 minutes
- 1.50 minutes

Output = $\dfrac{9 \times 50}{1.5}$ = 300 m³/hr (loose)

Equipment time available = 1200 hrs/year

Annual output = 300 × 1200 = 360000 m³ (loose)
= 288000 m³ (bank)

For 350000 m³ of earthwork involved, number of bulldozers required = $\dfrac{350000}{288000} \simeq 2$

Scraper

Scrapers are generally cost-efficient earthmoving equipment when the haul distance is too long for bulldozers and yet too short for trucks. This distance generally ranges from 100–3000 m. Scrapers are of vital importance in earthmoving, and are standard equipment for altering cuts and fills under a wide range of conditions.

Scrapers are used for:
- Surface stripping
- Excavating
- Loading generally with the help of pushers
- Hauling
- Dumping
- Spreading/contour grading/levelling

Scrapers may be deployed, especially where large volumes are involved, for:
- Cutting of channels
- Formation of dams and embankments with smooth and accurate level
- Stripping overburden in pit operations

Points that need to be taken care of for better performance of scrapers are:
- Hard ground should be loosened by ripping or blasting, and, if necessary, pushing is to be arranged for cutting. Bulldozers are deployed for pushing.
- Scraper's movement is to be planned so that they pick up materials on a downgrade, their weight assisting in loading the bowls taking full advantage of gravity.
- Scrapers need wide areas to cut and fill and only gentle gradient on haul roads to pick up speed. A scraper needs a lot of area to manoeuvre. It also needs properly

maintained track called *haul road* along which to travel. A grader can be deployed to maintain haul road economically. A scraper is good for hauling materials up to a distance of about 3 km at 55 km/hr or so speed.

- Haul roads should be wide enough to allow overtaking. Haul road gradient should be less than 5%. The gradient should not exceed 12% in any case.
- High volume tyres of scrapers are expensive. Pushing reduces wearing out of tyres. Tyres should be inflated to recommended pressures so that scrapers would move without more resistance.
- Where rolling resistance is low for fast run on jobs, single engine scrapers are used. On more rugged sites, twin-engine scrapers produce extra power, speed and acceleration to overcome high rolling resistance and steep grades. The single engine pulls the scraper bowl from the front. The second engine, when deployed, is placed behind for pushing.
- For the same weight and scraper bowl size, a single axle tractor would provide more traction than double-axle tractor. Also a single-axle tractor is more maneuverable, but its operation is difficult and dangerous.
- Cuttings—this operation should better be done in combination with a grader or bulldozer maintaining a convex formation of the area to be scraped so that cutting may be done downward. Cutting should be started along the periphery of the contemplated cutting area and work downward towards the fill area/s. Cut areas should be properly graded and loose materials should be pushed into scraper bowls for removal.
- Fillings—this operation is to be done maintaining a convex formation so that the scrapers lean inwards thus reducing the possibility of scrapers slipping over the edges. Embankments are formed in layers working inwards from the periphery.
- The supplied rimpull, as already mentioned earlier, is the force applied between the tyres of the drive wheels and ground surface, measured in newtons and is determined by the power of engine/s and influenced by the efficiency of the transmission, the gearing ratios and wheel/track arrangements, and is further affected by the influence of altitude and air temperature on the engine power. The usable rimpull depends upon the co-efficient of traction for rubber tyres or tracks moving on particular soils. The co-efficient of traction is defined as the factor by which the load/downward force on the drive wheels or tracks may be multiplied before slipping occurs. This value is the maximum usable rimpull, which clearly must be greater than the supplied rimpull or the wheels would spin.

A scraper is used for continuous operation of cutting/loading, hauling and dumping/filling. It comprises:

- A power unit such as a tractor—the prime mover
- A large open scraper bowl

In front of the bowl, a gooseneck or yoke has a vertical swivel connection with the tractor, which is usually in two parts with two pivots, upper and lower. It permits turns of 85° to 90° to each side of the centre. The bowl, which is basically a box with rigid sides, with the apron forming a movable front and the ejector a movable back. The bowl is supported at the back, by the rear or scraper axle; at the centre, by trunnions on the ends of draft arms; and at the front, by a pair of hydraulic cylinders suspended from the gooseneck. The pull of the tractor is applied through the gooseneck. The apron forms the forward side and a variable amount of the scraper assembly.

In closed position, it rests against the bowl at the cutting edge. When lifted, it moves up far enough to leave the entire bowl open. A scraper has cutting edge of the same width as the bowl at the bottom that can be lowered to cut the top surface of soil up to a depth of 300 mm or so. As the bowl moves forward at 5–35 km/hr speed, the loosened soil/rock mass is forced into the bowl and collected there. About 50–100 m of travel is required to fill the bowl. And when the bowl is full, the cutting edge is raised to seal the bowl. For filling the bowl, a bulldozer is often used as a pusher to add force to a scraper's cutting edge. The bowl is emptied by just dumping or semi-forced dumping or forced dumping. Just dumping is done in case of small scrapers by raising the front apron. Semi-forced dumping in case of medium scraper is done by turning bottom and back of bowl. Forced dumping in case of a medium or large scraper is done by forcing the back of bowl, called the ejector (or tailgate), forward. A scraper is classified on the basis of bowl size—small (up to 3 m^3), medium (up from 3–10 m^3) and large (above

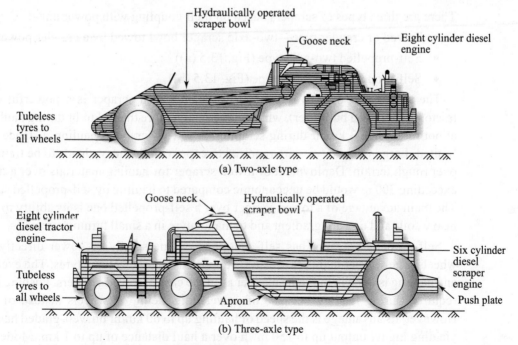

Fig. 13.5 Two axle and three axle scrapers

10 m³). A scraper's operation would be cost-effective if it is used to move large quantities of soil quickly over considerable distances.

The bowl may be filled by means of self-loading scraper elevator also (see sketch). In such scraper, the apron is replaced by an elevator made up of two bicycle-type chains carrying a number of cross bars called flights bolted to it. Its foot is near the bowl cutting edge, it slopes back 40°–45° and to an extent higher than the sides of bowl. The elevator conveys the excavated spoil into the bowl. This elevator is rotated by a power source different from that of the scraper.

The cycle time of a scraper is:
- Loading
- Hauling the load to the dumping area
- Dumping the load
- Returning to the loading area
- Getting in position for loading

Time required for loading as well as time required for dumping are uniform. Travel time varies depending on the hauling distance. Time required for loading, however, can be reduced by pre-ripping tight and hard soils, pre-wetting the soils and planning scraping operation downgrade. A scraper is the result of a compromise between the best loading and the best hauling construction equipment.

Types of scraper

There are three types of scrapers based on their coupling with power units:
- Towed or crawler type—two-axle scraper bowl towed by a crawler power unit
- Self-propelled two-axle type (Fig. 13.5 (a))
- Self-propelled three-axle type (Fig. 13.5 (b))

The power unit of a two-axle four-wheeled towed scraper is a powerful crawler tractor (generally a bulldozer), which moves at not more than 8 km/hr during hauling and at not more than 3 km/hr during scraping. A towed scraper's hauling distance is very limited. Large volume tyres are required to cope with the heavy loads to be transported over rough terrain. Deployment of towed scraper for hauling materials over a distance exceeding 300 m would be uneconomic compared to hauling by self-propelled scrapers. The main advantage of a towed scraper over a self-propelled one is its ability to load in heavy soils and upward gradient and to manoeuvre in a small turning circle.

Self-propelled scraper has self-contained engine supplying power directly to the wheels. The scraper in this case also is supported on large volume tyres. The excavation and hauling is done in the same manner as in the case of towed scrapers. Pushing is often required during loading because of power units being on wheels instead of tracks. However, its advantage is the speed of hauling up to 60 km/hr on well-graded haul roads yielding higher output up to 150 m³/h over a haul distance of up to 1 km. Modern self-propelled scrapers are large earth moving construction equipment having capacity up to 50 m³ with single or double engines.

The standard single-engine self-propelled scraper comprises a scraper bowl mounted on a single rear axle. The front end is attached to the drive axle. The bowl height can be controlled. A bulldozer is required to push the standard scraper to provide the necessary traction for loading the bowl. This type has more manoeuvrability over the three-axle type.

Three-axle self-propelled type is similar to the above standard scraper in both its construction and operation. The double-engine power unit is mounted on two-axles and can be used for other purposes also. This type can also use its top speed more frequently. The rimpull required on soft soils or rough uneven surfaces with high resistance or high upward gradient is best achieved with four-wheel drive to utilize the total available grip/traction. Large volume tyres are needed to cope with the heavy loads to be transported over rough uneven and climbing surfaces.

Scrapers are the principal bulk excavation and earth-placing equipment used where the soil is soft enough for them to work. They cannot excavate hard bands or rock unless made loose by ripping or by blasting. They cannot cut near-vertical sided excavations. They are deployed where, as already mentioned, haul distance is too long for bulldozers. Their movement needs to be planned so that they pick up material on a downgrade, their weight assisting in loading; if this is not possible or the soil is tough, they may need a dozer for pushing at the time of bowl-loading. This way, deployment of a more expensive higher-powered scraper could be avoided. A scraper is cost-efficient in terms of cost of excavation per cubic metre compared to similar equipment, but it needs a wide area to excavate or fill. In wide area, its movement should be in spiral, elliptical or digit eight (8) pattern.

The demerits of scrapers:
- Comparatively short service life
- Drop in output with increasing haulage distance, size of lumps and flooding of excavated rocks

Example

Working out the requirement of scrapers for specific conditions

Caterpillar model No. 435F made by Tractor India Ltd is considered

Bowl capacity— 13.8 m^3 (heaped), 10.7 m^3 (struck)

Cycle time = Fixed time + Variable travel time = FT + VT = C_m

FT = LT + ADB + TT + DT

$$\text{LT} = \text{Loading time} = \frac{W_L \times 100}{T_{EL} + E_L} = \frac{100}{67} = 1.5 \text{ minutes}$$

W_L = Weight of materials to be loaded

T_{EL} = Tractive effort available for loading

E_L = Loading efficiency factor—67 % for loose gravel

T_{EL} = W_L assumed as T_{EL} is not known

ADB = Acceleration – Deceleration – Braking time

= 0.5 minute for crawler tractor

TT = Time to allow for turns = 0.25 minute @ 0.25 minute/180° turn

DT = Time for dumping load = 0.5 + 0.25 including 180° turn for return

FT = 1.5 + 0.5 + 0.25 + 0.75 = 3.0 minutes

VT = Round trip, about 500 m (average) each way = 6 minutes @ 10 km/hr

Cycle time = 3 + 6 = 9 minutes

$$\text{Output} = \frac{Q \times 50}{C_m} = \frac{13.8 \times 50}{9} = 77 \text{ m}^3/\text{hr}$$

Annual output = 77 × 1200 m³ (loose)

= 77000 m³ (bank) considering 20 % swell

(For obtaining bank measure volume, the average loose volume should be divided by 1 plus swell, expressed as a fraction, that is, 1.2)

As earth is continuously forced into a scraper thus effecting little compaction, swell usually is less and hence 20% considered.

Quantity involved = 350000 m³

$$\text{Number of scrapers required} = \frac{350000}{77000} \approx 5$$

Face shovel

A face shovel (Fig. 13.6) is characterized by considerable digging force generated by its momentum. It is basically a power unit like a crawler or wheel-mounted tractor with a cable (wire-rope) or hydraulically controlled bucket (with/without teeth) mounted in front of it. It is earthmoving equipment used primarily to excavate earth face; manoeuvre the excavated materials to a point of disposal for dumping into spoil heap or transport. Its major advantage is its high output and ability to excavate without loosening in all classes of soils except solid rock. Its digging action is cyclic, that is, discontinuous.

Shovels are used for:
- Stripping top soil
- Cutting/scooping up
- Moving/manoeuvring
- Loading (dumping/discharging) into transport/conveyor
- Tree clearing/stump uprooting

Shovels are deployed for:
- Industrial projects
- Road construction
- Land reclamation
- Mining

If a transport is used for disposal of earth spoil, a shovel's capacity should be such as to fill up the transport in very few cycles. A basic shovel comprises:

- Heavy steel deck to carry engine, pumps, attachments, controls and cabin. The upper unit is supported by rollers or balls on a circular track or turntable, and is rotated by a vertical pinion gear that meshes with bull gear teeth in the turntable. The centre of rotation is forward of the centre and balanced by other parts plus counterweight. Deck-mounted drums are fitted with spools for cables or hydraulic pumps and hoses to rotate and stop the deck by clutches and brakes. The gear is controlled by reversing clutch, which can turn it in either direction.
- Cabin—a set of equipment for bucket control–in-and-out and up-and-down movement for excavating and dumping
- Boom/jib—short and sturdy
- Handle connecting the bucket—for excavating and manoeuvring for disposal
- Bucket – with/without teeth—sometimes called dipper
- Hoist lines—or one or more pumps may be used in hydraulic system
- Power unit like diesel engine or electric motor to drive all the mechanisms and propel —the standard one is a diesel engine

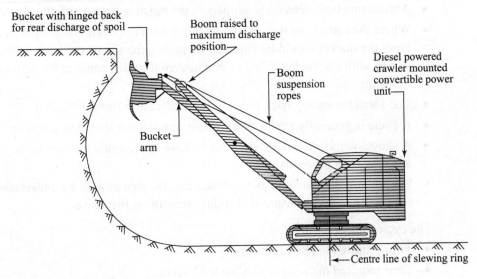

Fig. 13.6 Face shovel

Hydraulically operated face shovels have many advantages over mechanically (cable/winch) operated ones. Hydraulically operated shovels have swivel buckets, shorter arm action, constant ripping power, light arrangement, no discharging problem whereas mechanically operated shovels have fixed buckets, longer arm action with less output, full ripping power possible only at the top of the reaches and difficulties in discharging cohesive soils. However, cable operated machines may have bigger buckets ($\simeq 40$–50 m^3) than hydraulically operated buckets ($\simeq 25$ m^3).

Shovels are classified as full revolving (360°) and partially revolving (up to 270°), but average swing angle does not exceed 90°. Shovels are differentiated by types of application, working tools, bucket sizes, power units, angles of deck swing, controls and undercarriages. A shovel comes in all sizes, from small to giant; but for typical major excavation jobs, it would have a relatively large bucket. The size adopted depends on what rate of excavation must be achieved, the capacity of dump truck/dumper it feeds to move excavated spoil, and the haul distance to dispose or earthwork formation to be constructed.

Shovels are versatile construction equipment deployed to perform more than half of earthmoving jobs. The working cycle of a shovel is as follows:

- A shovel is driven to the correct position, near the face of the earth to be excavated using the main hoist power mechanism
- Its bucket is lowered and positioned at the toe with the teeth pointing into the excavation face although it can work a limited height of excavation face
- It must stand on firm level ground when working
- Its momentum is used to force the bucket bite into the earth heap for filling the bucket using the secondary hoist and this action can be carried out simultaneously with the main hoist for powerful digging
- At the same time, tension is applied on the hoisting line to pull the bucket up
- Where the earth heap is of sufficient depth with respect to the bucket size and soil type, the bucket would be filled as it reaches the top of the heap - the shovel is swung with the bucket filled to manoeuvre it to discharge at the point of dump/transport
- The shovel is swung back to bring the bucket into further digging
- A cycle is generally performed without moving the shovel as a whole
- A shovel works on one location for as long as required, moving its position only as excavation proceeds
- For moving the shovel to another location, the move has to be undertaken slowly, care being taken to ensure that it does remain on firm base

The cycle time of a shovel:
- Time required to load the bucket
- Time required to swing with a loaded bucket
- Time to dump the bucket
- Time to swing back with an empty bucket

The face shovel is the most efficient excavator where bulk excavation of large quantities is involved. The height of the face depends on the size of the equipment, but its deployment would not be cost-efficient for face heights lower than 1 m or so. For deep foundation excavation, face shovel has to stand on the excavation level to feed into trucks standing on the same level. This would require extra space for a ramp for moving in and out of the excavation pit. For this reason, a face shovel is unlikely to be deployed for deep excavations in confined areas where forming a ramp would not be possible.

A shovel's production can be worked out as follows:

$$q_s = \frac{3600 \times B_c \times E \times (A:D)}{CT_s}$$

where q_s = maximum production, yard³/hour (1 yd³ = 0.765 m³)

B_c = bucket capacity in yard³

E = bucket efficiency in percentage of capacity

$A:D$ = combined factor for the shovel's angle of swing and depth of cut

CT_s = cycle time in seconds for standard shovel operation of 90° swing and optimum cut

Backhoe

Backhoe (Fig. 13.7) is a variation of the face shovel. It is also referred to as hoe, pull shovel, back shovel, and ditching shovel. The backhoe moves its bucket down and towards the operator to carry out excavation below the equipment's mounting. In other words, a shovel may be converted into a backhoe by replacing the bucket along with the fasteners and attachments. A backhoe is often equipped with goose neck boom/jib to increase its digging depth. A backhoe is designed to dig much deeper than face shovel.

Backhoes are used generally for excavating below the ground level on which they are positioned. They are used for:
- Earthwork in excavation in trenches, foundations and basements requiring precise control of depths
- Excavation in hard and firm materials because direct pull is exerted on the bucket
- Deep excavation in confined areas
- Efficient excavation maintaining correct profile
- Direct dumping on trucks at close range
- Handling duties in pipe-laying, installing trench sheets etc

A backhoe comprises:
- Heavy steel deck to carry engine, pumps, attachment, controls and cabin. The deck rests and revolves on a turntable. The centre of rotation is forward of the centre and balanced by other parts plus counterweight
- Cabin—a set of equipment for bucket control—in-and-out and up-and-down movement

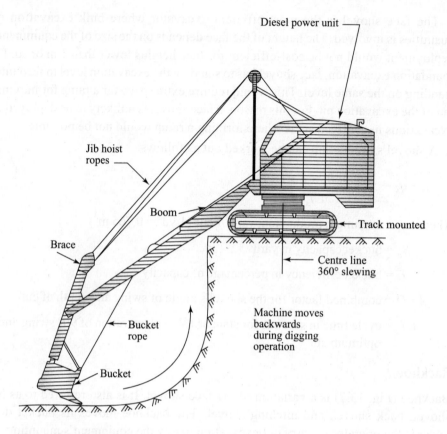

Fig. 13.7 Backhoe

- Boom/jib—short and sturdy—the boom angle varies between 30°–60°
- Handle connecting the bucket—for excavating and manoeuvring for disposal
- Bucket—with/without teeth—sometimes called dipper
- Hoist lines—or one or more pumps may be used in hydraulic system
- Power unit like diesel engine or electric motor to drive all the mechanisms—the standard one is a diesel engine

A backhoe is as versatile construction equipment as a face shovel. The working cycle of a backhoe is as follows:

- Detailed planning would be necessary before starting excavation work of large and deep pits covering disposal procedure of the excavated spoil
- Provision of temporary ramps should be contemplated so that trucks and dumpers may be used for spoil disposal
- Excavation of small pits for construction of column foundations should be carried out with utmost care resulting in reduced output

- In drain trenches especially where the ground is water-bearing, excavation should be carried out starting from the deeper towards the shallower depths—the sides of excavation should be sloped to avoid their collapse
- If a sump is located at the deeper section, water can be pumped out from there and possibility of sides collapsing could be avoided
- For engaging backhoes for lifting duties, utmost care should be exercised in estimating the lifting capacity of the equipment

Dragline

A dragline (Fig. 13.8) is in effect a crane fitted with long jib for greater reach and drag bucket for excavating in loose and soft soils below its standing level and loading the same into hauling units or depositing in areas earmarked for the purpose. A dragline can be positioned on firm ground while excavating materials from a pit, canal or ditch. If there is any possibility of depositing the excavated materials along a canal or ditch or nearby pit, then it would be possible to carry out both excavation and disposal in single operation if boom length is sufficiently long. Depending on operating restrictions, outputs of dragline excavators would vary from 30 to 80 bucket loads per hour.

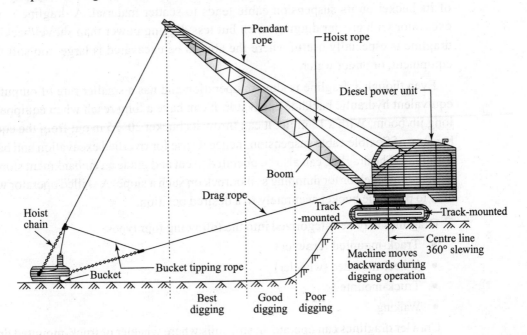

Fig. 13.8 Dragline

Draglines are used for:
- Cutting (open cuts)
- Grading embankment
- Dumping soil or rocks on embankment slopes or on transporting units
- River dredging

Draglines are deployed for:
- Cutting ditches with sloping sides
- Excavating drainage and irrigation canals
- Dredging silted up river/canal beds
- Sand and gravel pit production
- Strip mining

As already mentioned, a dragline is basically an excavator as well as a crane with a long lattice jib to which is attached a drag bucket for bulk excavation. This bucket, which is a bit similar to a scoop whose front edge has teeth, is generally used for excavating in loose and soft soils below the level of the equipment. Its sides protrude forward and are reinforced at the top with an arch to make the bucket front stiff. It is suspended from a hoist cable. A power shovel can be converted into a dragline by replacing its jib/boom with that of a crane and suspending a bucket from the crane's hoist cable.

A dragline stands on firm ground and its bucket digs below its level even under water. For dumping/discharging, the bucket is raised in tucked position for emptying its content through the open front end. It offloads its materials to transport in messy way, as swing of its bucket on its suspension cable tends to scatter material. A dragline is versatile excavator with greater digging reach but less digging power than shovel/backhoe. The dragline is especially useful where the area to be excavated is large, too soft for other equipment, or under water.

Even though a dragline is slow in operation and has a smaller rate of output than an equivalent hydraulic backhoe excavator, it can have a long reach when equipped with a long jib/boom. With a 15 m jib, it can throw its bucket 20–25 m out from the equipment because of flexible cable suspension; hence its use for riverbed excavation and bank-side trimming. The dragline can also be operated to cut and grade an embankment slope below its standing level, or for dumping soil or rock on such a slope. A skilled operator would be able to place the bucket accurately to a desired position.

Draglines may be categorized into the following four types:
- Track-mounted (crawler)
- Wheel-mounted (wheeler)
- Truck-mounted
- Walking

Crawler draglines can operate on soft soils where wheeler or truck-mounted draglines are unsuitable. However, crawler types are very slow and need transport for shifting them from site to site.

Walking draglines have several advantages over crawler types such as large sizes, long jibs/booms with corresponding large working ranges, simple structural design, simplified manoeuvrability, and low pressure under the bases. Due to their exceptionally long reach, these machines are especially suitable for the construction of embankments and canals, stripping of overburden and dredging in rivers. The jib/boom is rotated to a position opposite the direction of movement when walking operation is needed. The two shoes, one on each side of the revolving frame, are lowered by an eccentric cam until most of the weight of the machine is supported by the shoes, and the base is lifted off the ground. As the cam keeps on rotating, the machine is dragged backward several feet. This process is continued till the desired location is reached. For operation of the dragline, the shoes should be kept suspended off the ground.

The production rate of a dragline can be worked out in the same way as is done in case of a shovel as follows:

$$q_d = \frac{3600 \times V_b \times (A:D)}{CT_d}$$

where q_d = maximum production, yard³/hour
 V_b = bank measure volume in bucket in yard³
 $A:D$ = combined factor for the dragline's angle of swing and depth of cut
 CT_d = cycle time in seconds for dragline operation of 90° swing and optimum cut

$$V_b = \frac{V_1}{S_w}$$

where V_1 = the nominal capacity of dragline bucket

$$S_w = \text{swell factor} = \left[1 + \frac{\% \text{ swell}}{100}\right]$$

For obtaining bank measure volume, the average loose volume should be divided by 1 plus swell, expressed as a fraction. If swell is 20%, the swell factor would be 1.2.

Clamshell

Clamshell (grab) bucket (Fig. 13.9) is different from that of a dragline bucket. However, the jib/boom is the same as that of a dragline. The boom angle may vary within 40°–70°. It has its own specialties of digging deep, narrow, straight-sided excavations and neat handling of materials through small passages. Its digging cycle of operations is more. General use is made of double-cable two-jaw clamshells—two half shells hinged at the centre. The bucket is supported by a hoist cable and closed by another cable. The grab is fitted with interlocking teeth to help

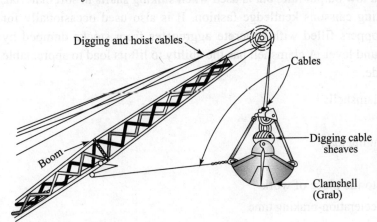

Fig. 13.9 Clamshell

penetration of soil for true application of excavation process, whereas clamshell has no teeth and is used for stockpiling very loose materials such as sand. Buckets are available with or without removable teeth.

Clamshell buckets are available in various sizes and types. Heavy-duty types are used for digging, medium-weight types for general purposes and light-weight types are for re-handling light materials. Heavier buckets are needed for underwater digging. Buckets of varying capacities are available in all the three types. However, the maximum size of a bucket that should be attached to the hoist cable should match the lifting capacity of the crane engaged. Permissible load for the radius of operation should be meticulously calculated from the graphs furnished by the particular manufacturer as lifting capacities of different manufacturers may vary.

Clamshells are used for:
- Handling loose materials like sand, gravel, crushed stone, etc.
- Removing materials from inside cofferdams, pier foundations, sewer manholes, sheet pile-lined trenches etc
- Vertically lifting materials from one location to another
- Dredging in confined areas using grab buckets

To dig, the hoist cable brake is released allowing the bucket to contact soil so that its weight will drive it into soil for a good bite. Bucket with teeth is required for digging harder type of materials. The digging drum clutch is then engaged to pull the bucket jaws together. They first push the dirt inward, then curve and close under it.

The clamshell bucket has a pincer movement, hydraulically operated, and is principally used for the construction of diaphragm walls. The bucket is fixed to a long rod, which is lowered and raised down a frame held vertically (or at an angle) so that it can cut trenches up to 30 m deep in soft material, usually up to 0.6 m width. The machine rotates so that the clamshell can be emptied to a waiting transport.

The clamshell has a low output rate, but is used when sinking shafts in soft material, especially when sinking caissons kentledge fashion. It is also used occasionally for keeping aggregate hoppers filled with concrete aggregates from stocks dumped by delivery trucks at ground level. A clamshell has the ability to lift its load to appreciable heights above the grade.

The cycle time of clamshell:
- Loading bucket
- Lifting and swinging load
- Dumping load
- Swinging back to the place of work
- Acceleration-deceleration-braking time

Grader

A grader (Fig. 13.10) is a kind of equipment specifically developed for accurately trimming, shaping and finishing rather than excavating and hauling and is required where surface has to be smooth and level without undulations and ridges to a given line or contour. A grader performs what a bulldozer can do, but more accurately and smoothly. It is also used extensively for maintaining haul roads to facilitate movement of scrapers.

Graders are used for:
- Shaping or profiling including forming the camber according to the specifications
- Trimming
- Finishing so as to make the surface smooth and level without undulations and ridges
- Shallow ditching
- Terracing/smoothing slopes
- Spreading and mixing earth and construction materials
- Scarifying hard soils
- Clearing roads and construction sites of snow/debris

Graders are deployed for:
- Road building
- Earth-fill dam construction
- Maintaining haul roads

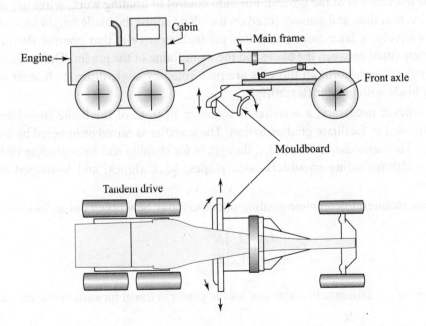

Fig. 13.10 Grader

A grader comprises:
- A main frame, which works as a bridging beam between the front axle and the rear two axles in tandem. The tandem wheels have a fully articulated suspension
- The mouldboard, which is located at the centre of a long wheelbase, is hung from the bridging frame or pivoted to the rear frame to allow a reduced turning circle, increased manoeuvrability and to permit off-set axle grading
- The engine is supported on the rear frame on pneumatic tyres
- The cabin is located adjacent to the engine in modern machines

A grader is self-propelled pneumatic tyre construction equipment supported normally on three axles. Graders are extensively deployed on road construction work involving massive earthwork requiring heavy and robust construction equipment. The three-axle graders have proved their worth to meet such demand. They are used for finishing to fine limits large areas of ground that have been bulldozed or scraped to the required formation level. However, graders' low motive power is not sufficient for their deployment in excavation work.

The three-axle arrangement with the blade hung within the main frame between the front and rear axles and the fully articulated suspension ensures a higher accuracy of grading. The mouldboard on modern graders is operated hydraulically from the driver's cabin. The pitch of the blade is adjusted to suit the work involved. It is tilted backwards when cutting and forwards when spreading. When the blade is more upright, the mixing and rolling that would be given to the material being spread would also be more. In case the blade is left without exercising any control, the blade's operation would be guided by the unevenness of the ground. For auto control of grading work, wires are set along the direction line, and sensors fitted on the blade adjust the blade height automatically. Alternatively, a laser beam activates photoelectric cells that control the hydraulic adjusters fitted between the blade and the main frame of the grader. The mouldboard of the grader can perform all those as are performed by angle dozer, bulldozer or tilting dozer blade with greater flexibility.

Whenever necessary, a scarifier is mounted in front of the blade (mouldboard) to breakup soil to facilitate grading action. The scarifier is raised or lowered by hydraulic power. The basic use of the grader, though, is for shaping and final grading of the total road width including shoulders, side slopes, back slopes, and V-shaped drainage ditches.

Time required to complete grading operation may be worked out as follows:

$$T = \left[\frac{d_f}{V_f} + \frac{d_r}{V_r}\right]\frac{N}{E}$$

where d_f = Distance in linear feet for the grader to travel forward in one direction per cycle

d_r = Distance traveled returning to begin the next grading cycle

V_f = Average speed of forward travel in feet per minute
V_r = Average speed of the return travel in feet per minute
N = Number of forward passes the grader must make past a given point in the length of grading operation
E = Efficiency of operation of the motor grader (70 – 90 %)

It is apparent from the above equation that the productivity of grading operation is worked out on the basis of time to do the work.

Hydraulic excavator

Excavators are used for:
- Excavating
- Trench cutting
- Trimming the base of trench
- Breaking up of stony formation
- Ripping of soft rock
- Loading trucks and scrapers
- Digging out tree stumps and other buried objects
- Laying of pipes

The production of an excavator is a function of the digging cycle. The cycle time of a hydraulic excavator:
- Loading the bucket
- Swinging with a loaded bucket
- Dumping the bucket load
- Swinging with an empty bucket

The above cycle time is dependent on the equipment size and the prevailing job conditions. A small excavator, for example, can usually cycle faster compared to a large one. Nevertheless, it would handle less productive loads per cycle. The excavator would slow down as the load conditions become more severe like the soil getting harder and the trench getting deeper. In such a case, time taken to fill the bucket would be more.

A hydraulic excavator (Fig. 13.11) is versatile construction equipment used as a backhoe. It consists of a hydraulically operated boom to which is attached a one or two part dipper stick and bucket. The bucket cuts back towards the equipment base. The digging arm is also called backhoe.

Two types of excavators are available for different kinds of use—the 360° and the 180° construction equipment. In case of 360° excavators, the entire assembly and operator's cabin are mounted on a rotating base that allows excavation and loading to be carried out in any direction. The base is generally mounted on tracks but can be mounted on tyred wheels also. Bucket capacity of tracked excavator could be from about 0.25 m³ to about 30 m³. In case of wheeled excavator, bucket capacity could be from about 0.135 m³ to about 2 m³. The larger size equipment can dig up to a depth of about 12 m. Special buckets may be attached with hydraulic system for grading and trimming. The hydraulic system gives the equipment high penetration force as well as high degree of precision. 360° excavators are used for heavy-duty work.

In case of 180° excavators, hydraulic system is of smaller capacity compared to 360° excavators as 180° excavators are meant for general excavation and loading jobs. The hydraulic arm is mounted at the back of the construction equipment and can only operate within 180° range only.

The larger size can cut to a depth of 6 or 7 m and excavate a face of the same height, slewing to load to trucks alongside. It can be used for lifting pipes into trenches, and 'bumping down' loose material in the base of a trench with the underside of its bucket. It can usually excavate trenches in all materials except rock; but sometimes has trouble in getting out hard bands of material that are horizontally bedded or which dip away from the machine. It can have a toothed bucket capable of breaking up a stony formation, be fitted with a ripper tooth for soft rock or a hydraulic breaker for hard materials, or have a smooth-edged bucket for trimming the base of a trench. Wide ranges of such materials are available. The smallest size could be used on small building sites. The larger sizes could be used for large trench excavation and general excavation of all kinds.

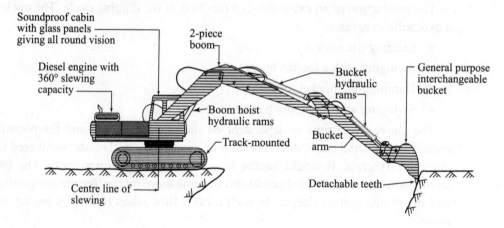

Fig. 13.11 Hydraulic backhoe

Trenching machine

The trenching machines can be used either for excavation of trenches for laying utility pipes and cables or drainage ditches in all kinds of soils except rock. In favourable job and soil conditions, trenching machines may be used for sewers or construction of shallow diaphragm walls. These are fast digging equipment with positive control of depths and widths of trenches. On laying of pipes and cables, the trenches are backfilled. But a ditch is always left open.

Trenching machines are available in various sizes for digging trenches of varying depths and widths. They are track-mounted for increased stability and for distributing load over a greater area.

Wheel-type trencher

These machines (Fig. 13.12) are available with maximum depths exceeding of 2.5 m, with trench widths varying 300 mm or less to 1500 mm or so. Many of them are available with 25 or more digging speeds to permit the selection of the most suitable speed for any job condition.

The excavating part of the machine comprises a power-driven wheel with a number of removable buckets mounted on it. The buckets are equipped with cutter teeth and are available in varying widths. Side cutters may be attached with the buckets to increase the width of a trench. While the machine advances forward slowly, the rotating wheel is lowered to the required depth. The spoil is picked up by the buckets delivering the spoil on to an endless belt conveyor that can be adjusted to the spoil on either side of the trench or can deliver directly to dump trucks.

For hard ground the machine has special cutters cutting a groove at either side of the trench, with a third bucket cutter chain to remove the dumpling of material between.

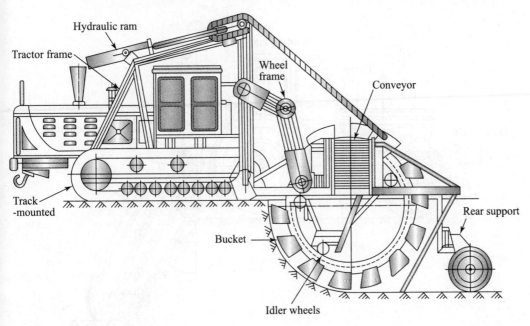

Fig. 13.12 Wheel-type trencher

Ladder-type trencher

By installing extensions to the ladders or booms, and by adding more buckets and chain links, it is possible to dig trenches in excess of 9 m with the large machines. Trench widths in excess of 3.5 m may also be dug. Most of these machines (Fig. 13.13) have booms whose lengths may be varied, thereby permitting a single machine to be used on trenches varying considerably in depth. This eliminates the need of owning a different machine for each depth range. A machine may have 30 or more digging speeds to suit the needs of any given range.

The excavating part of this type comprises two endless chains, which travel along the boom/jib, to which there are cutter buckets fitted with teeth. In addition, shaft-mounted side cutters may be installed on either side of the boom to increase the width of a trench. As the buckets move up the underside of the boom/jib, they bring out earth and deposit it on a belt conveyor, which discharges it along either side of the trench. As a machine moves over uneven ground, it is possible to vary the depth of cut by adjusting the position, but not the length, of the boom.

A modification of the ladder-type machine is one with a vertical boom. This machine is available with seven different boom sizes, which permit trench depths varying from 1.2–2.5 m with trench widths varying from 355–610 mm.

The ladder type machines are not suitable for excavating trenches in rocks or where large quantities of ground water, combined with unstable soil, prevent the walls of a trench from remaining in place.

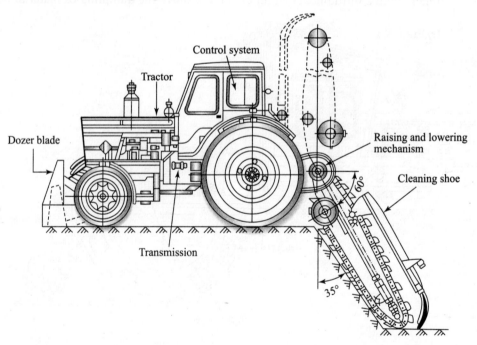

Fig. 13.13 Ladder-type trencher

Production rate of trenching machine depends on:
- Soil class
- Depth and width of trench

- Extent of shoring required
- Topography
- Extent of vegetation like trees, stumps and roots
- Physical obstructions like buried pipes, sidewalks, paved streets, buildings, etc.

A trencher is the most suitable equipment for excavating trenches. The selection depends on:
- Site conditions
- Depth and width of trench
- Soil class
- Condition of groundwater
- Disposal of excavated materials
- Available construction equipment at site

Wheel type trenching machine is most suitable for relatively shallow and narrow trench. A backhoe or a dragline is suitable when excavation is to be carried out after blasting. If the soil is unstable water-saturated material, it may be necessary to deploy a dragline or backhoe or clamshell and let the walls establish a stable slope. If it is necessary to install solid sheeting to hold the walls in place, then a clamshell, which can excavate between the struts/braces, would work satisfactorily.

Rollers

Earth on excavation becomes loose and bulky. Filling earth materials, therefore, needs to be compacted to prevent distortion, settlement and softening. The extent of possible compaction of any filling depends on the type of soil material and moisture content. Soil may be compacted naturally by time and weather. But nature is very slow and unreliable. Heavy rolling would consolidate granular soils significantly and satisfactorily. But compaction of cohesive soils would be relatively slow process under sustained loading. Different types of construction equipment have been developed to obtain satisfactory compaction in different types of soils, and surface finishes such as tar macadam, etc. The different compacting machineries are classified into the following distinct ways:
- Static weight
- Kneading action
- Vibration
- Impact force

A heavy, cylindrical roller is intended primarily to impart dead or static weight. The pneumatic tyres of construction equipment or vehicles gives a kneading action in having their load distributed a bit outward as well as downward on the supporting material. A vibratory action is built into some compaction equipment by the use of eccentrically revolving weights. There are other mechanisms also to cause vibration. An impact force is most obvious when a heavy weight is raised and allowed to fall on to the surface. Consolidating operation by the rollers rely on the deadweights or on vibration in case of lightweight rollers. Deadweights can be as high as 16 tonnes or more.

Static weight rollers

Single-axle smooth wheel rollers are used for:
- Compaction of soil by self-weight/deadweight
- Rendering more even surface

The single-axle roller comprises a frame and smooth steel cylinder loaded with sand or water to increase deadweight. The roller can either be trailed type or self-propelled (powered by internal combustion engines) type. As the roller is pulled ahead, a wave of soil is pushed up in the direction of movement. With successive passes over the ground surface, the roller gradually effects compaction of soil. A self-propelled roller produces more even surfaces. However, wave formation would be larger as the weight of the roller is increased and the diameter decreased. The magnitude of wave formation would depend on the composition of soil. This is true for both types of rollers—trailed or self-propelled. This type of deadweight rollers may be used on most types of granular soils but not as effectively as rollers fitted with vibratory mechanism. The working speed of the compacting roller should not exceed 1.6 to 2 km/hr except during finishing passes when the speed could be 2.5 to 3 km/hr.

Three-wheeled (smooth wheel) loaders are used for:
- Compacting bituminous materials on road surfacing operations
- Negotiation of bends and curves without causing irregular folds or creases

The three-wheeled roller (Fig. 13.14) comprises a wide front roller and two narrower rear rollers in tandem and predominantly used for compacting bituminous materials on road surfacing operations using deadweight. Most three-wheeled and also some tandem rollers are steered by the front drum with the drive provided through the rear rollers, usually set to overlap the track of the front roller by about 150 mm to prevent ridge formations. Differential gearing on the rear axle helps equipment to accommodate bends

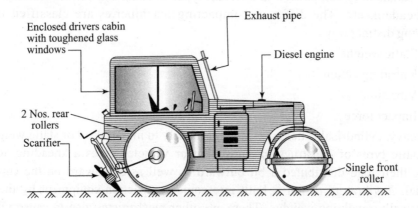

Fig. 13.14 Three-wheeled roller

and curves without causing irregular folds and creases. All smooth static rollers very effectively compact a thin layer on and immediate below the surface, but unfortunately the soil below about 150 mm depth remains virtually undisturbed.

Pneumatic-tyre rollers

Pneumatic-tyre rollers are used for:
- Rolling base courses and blacktops on roads
- Filling for large earthwork involving soil of loamy texture

The pneumatic-tyre rollers are successful in most types. The pneumatic-tyre roller is designed as construction equipment that combines the kneading action with static weight. Tractors provide the power that drives rubber-tyre roller. The deadweights of the rollers are gained by filling the rollers with water, sand or pig iron. These rollers have two axles with odd number of tyres like seven (three front and four rear) or nine (four front and five rear) right up to nineteen (nine front and ten rear) tyres so arranged that the paths of the rear overlap those of the front to prevent ridges. The wheels have independent couplings combined with swivel action to distribute the weight of each tyre evenly on surfaces to negotiate bends and curves without causing irregular folds and creases. Controlling the ground contact pressure is a major criterion on selection of types of rollers to be deployed. In this respect, rubber-tyre roller has a major advantage of ground contact pressure as follows:
- Altering the weight of the equipment
- Increasing/decreasing the number of wheels to be fitted
- Increasing the tyre width
- Changing the contact area of the tyre by varying tyre pressure by inflating/deflating

The towed type pneumatic roller unit has only one axle with only about four tyres to remain within one lane's width. The towed type ballast boxes have pneumatic wheels with larger tyres than those mounted on the self-propelled units.

Moving first forwards and then backwards at speeds up to 20 km/hr produce the best results on compacting earth fill or rolling blacktop. Whereas four to six passes may be sufficient for blacktop, four to eight passes may be necessary for compacting soils.

Sheep's-foot rollers

Sheep's-foot rollers are used for:
- Compacting highly cohesive soil materials, but most effective in sandy soil with clay binder
- Kneading the soil particles together by high pressure transmitted by each sheep's foot

A sheep's-foot roller is different from the above smooth wheel loaders. It comprises a smooth steel cylinder on which projecting feet are mounted. In the earlier models, the projections resembled the shape of a sheep's foot. About 10 to 20 such feet are arranged around the rim at between 100 to 200 mm centres along the axis; several different shapes of foot are available to suit different soil types. Each projecting foot can sink down about 150 mm into loose materials, which is the main advantage of this type of rollers. The small surface area of each foot transmits high pressure, and the feet sink down to knead and tamp the fresh material into the previously compacted layer. The roller's weight exerts pressure on the lift as a whole. With repeated presses, as the lift becomes compacted, the feet gradually rise to ultimately emerge out of the fill.

The sheep's-foot type rollers work more effectively by moving through loose materials rather than rolling over the same because the effectiveness of this equipment's use is its rolling resistance to motion.

Vibratory rollers

Vibration generated in a machine-part that is in contact with the ground produces a rapid series of impacts causing pressure waves that penetrate the soil setting its particles in motion. The combined effect of this motion and the static weight exerted by the machine-part would initiate rearrangement of the soil particles forcing them into a compact structure with minimum of voids.

Vibratory rollers are used for:
- Compacting and consolidating granular soils
- Compacting hardcore or similar bed materials in small areas
- Improving compaction using vibration mechanism

The vibratory action causes deeper penetration of compaction-effort on most granular soils over and above the static weight and kneading action. This would allow more depth of filling before compaction is started. However, vibratory action would not be effective in soils with cohesive materials exceeding 15%.

The vibratory roller comprises a roller drum with one or more eccentric weight/s inside the drum. Vibration is produced by rapid rotation of axle carrying eccentric weights. It is transmitted to the vibratory parts through ordinary bearings and insulated from non-vibrating parts of the equipment by flexible pads or hangers called isolators. The eccentrically rotating weight applies a centrifugal force. The maximum depth up to which filling can be compacted would depend on the total dynamic load comprising the static load plus centrifugal force. On this basic principle, different types of vibratory rollers are manufactured.

A vibratory roller may be hand guided or towed or self-propelled. The weight of self-propelled vibratory rollers varies between 500 kg to 5 tonnes. For achieving required compaction densities, a towed vibratory roller would weigh up to 25 tonnes or so. In this dynamic type of rollers, external weights are hung from the support to increase deadweight instead of loading the drum with water, sand etc.

Impact compactors

In areas where there is space restriction, deployment of rollers would not be possible. Soil in such cases is compacted by tamping.

There are two types of impact compactors:
- Vibratory-plate type
- Impact type

Both types deliver impact blows in rapid succession on the surface of materials to be compacted with short bounces.

In vibratory-plate type compactors, tamping action may be generated by revolving eccentric weights inside a box-like container with a plate for the bottom to act on the material. The impact force in the vibratory-plate compactor may be achieved better by having eccentric weights operate in pairs. If a pair of eccentric weights rotates in the opposite directions (vide Fig. 13.15), the centrifugal forces in combination move both downward and upward directions. If the eccentric weights are large enough, they would lift the plate compactor off the surface; and when they are directed vertically downward their centrifugal forces add to the static and dynamic weights of the falling plate. The vibrating mechanism may be driven by electric, hydraulic, pneumatic or mechanical power. A plate compactor requires fewer passes than a vibratory roller to get the desired compaction.

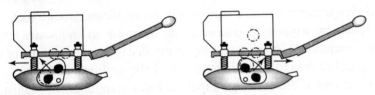

Fig. 13.15 Vibratory-plate compactor

In impact type compaction, mechanical tampers may be used by jump pounding, by leaving the shoe on the ground and hammering it or by weight dropping. The tampers may be operated by diesel, pneumatic, hydraulic or electrical power. The rammer includes a single cylinder, two-cycle engine. The working piston is attached to the foot. This type of machine weighs 50 kg or so and is operated at the rate of 800 to 3000 blows per minute, with jump strokes of 40 to 75 mm off the ground. The hand-operated plate tamper, which has a short stroke, has a petrol engine that operates a reciprocating type vibrator, with frequencies between 4000 to 6000 vibrations per minute. The tamper weighs from 180–680 kg.

A freefall hammer is also used for compaction. Here kinetic energy is used by dropping a heavy flat weight from the jib of a crane. This method allows compaction of layers of soil several metres below the surface.

High-speed rollers

High-speed rollers are used for:
- Compaction at high production rate
- Compacting action combines static weight, impact and kneading
- Compaction is possible in different kinds of soils (cohesive, plastic, silts, gravels and hardcore)

The high-speed roller comprises a pinpointed frame for steering and is driven on four polygonal wheels. Polygonal design ensures compaction of the furrows made by the front segments by the rear segments so as to assure complete coverage during a pass. This method of compaction combines static weights, impact as well as kneading thus providing considerable advantages over other types of roller on large areas involving cohesive and plastic soils, silts, gravel and hardcore.

Tunnel boring machine

Tunnel boring machine (TBM) is an expensive equipment and its deployment would be economical if the length of the proposed tunnel exceeds 2 km. The tunnel boring machine is continuously operating equipment different from such cyclic operation as drilling, blasting and mucking.

The earth's mineral material may be classified simply as rock or soil. The category known as soil is the uppermost material near or on the surface that has been weathered, decomposed, transported, and deposited. It comes from the rocky material that is generally under the surface, though it may be on the surface because of an earthquake or volcanic upheaval followed by centuries of the weathering and reforming process.

A tunnel driven under the earth's surface will generally go through rock of varying consistency. It may be dripping or saturated with water due to high water table and the rock's cracks or variations. The rock may be subdivided into soft, medium, or hard rock. Soft rock includes beach sand, consolidated clay, chalky bedrock, and loose or fractured rock. At the other end of the scale, hard rock includes granite, gneiss, schist, taconite, and hard diorite. The medium rocks are sedimentary rock, sandstone, limestone, chert, and shale.

A tunnel boring machine comprises:
- A full-size cutter head that rotates or oscillates to cut the ground
- A complete or partial tubular casing or shield
- A crowding or propelling mechanism with provision for steering
- A muck (excavated/blasted spoil) disposal system
- Electric and hydraulic power units—electrical power is the primary source of power and hydraulic rams are needed for thrust, torque as well as movement of the machine

The cutter head (Fig. 13.16) rotates on a central structural member. The head is so fabricated as to hold a number of cutters. There should be just enough openings or slots for spoil removal to take place efficiently while the cutter head moves on. The spoil may be collected in buckets or scoops and then may be loaded into a belt conveyor system of at least 750 mm width.

As the cutter head revolves as a single unit, there would be difference in speed between the cutters near the centre and those at the rim. This may be taken care of by varying the spacing as

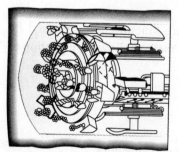

Fig. 13.16 Cutter head of TBM

well as types of cutters. Design in some machines provides arrangement for the central part to be driven separately at higher rpm.

In machines where the cutter head oscillates, each of a number of radial cutter arms has motion like windshield wiper. Their arcs overlap, and distance of oscillating motion can be adjusted. This type is generally used in soft or variable ground.

Cutters may be of various designs—roller or disc-type or planetary cutter. Roller cutters are conical in cross-section and are effective in cutting by crushing or pulverizing action. In case of very hard rock, the tunneling machine has to exert high pressure by its thrust. The roller cutters work well in combination with disc cutters, as roller cutters alone wear out fast when working in hard rock. Use of disc cutters requires less energy, less horsepower and less thrust.

The cutters of planetary type rotate in different directions on their separate axes. These cutters are not dependent on the leading face of the tunnel boring machine. They can operate on a movable arm at the machine's front end. All these make the planetary cutter useful to bore a tunnel face that is not circular.

When the rock bored through is very hard and stable, shield may not be necessary. Otherwise, a boring machine is protected by a tubular shield or casing (Fig. 13.17). Minimum that is done is to shield the roof section. The quality and extent of shielding would depend on the material through which boring is to be done.

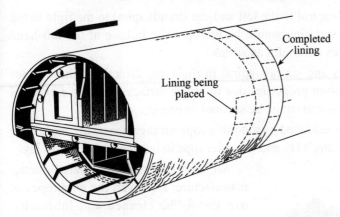

Fig. 13.17 Complete or partial tubular casing or shield

The shield may be arranged in two longitudinal sections. The front section moves with the cutter-head, and the rear one is anchored with the tunnel or its lining. The rear one works as a thrust block for hydraulic cylinders to engage the cutter-head into excavating.

In soft soil susceptible to flowing or caving, cutter-head works inside the shield. The top and occasionally the sides and bottom of the shield penetrate the soil mass before the cutters reach the same.

Hydraulic rams are needed for thrust, torque, and movement of the machine. Hydraulic power controls installed in tunnel boring machine are used to hold the machine in position or move it along the tunnel path. Hydraulic power generates huge axial thrust, and equally high rotational torque. The thrust works to crowd the cutter-head into the rock. The machine's reaction is provided through ribs and radial jacking feet operated against the tunnel sides at right angles to the machine's direction. These retractable feet also help to move the machine along its path.

The power requirement of tunnel boring machines varies considerably depending on the materials to be bored, the method of boring and the auxiliary parts. The cutting-head meant for soft materials may require motor of only 75 HP or so, whereas the total power requirement for boring in hard rock may be as high as 2000 HP. To keep the tunneling work on course, laser guidance system is used.

13.5 PLANTS FOR TRANSPORTATION, MOVEMENT, AND HANDLING

Wire ropes (Cables)

Wire rope is the simplest device that is used for lifting materials or plants and machineries. A rope is normally used for lifting operation in conjunction with a pulley block. Both rope and pulley block invariably form part of construction equipment that is deployed for lifting operation. Even the large crane relies upon the basic principles of lifting tackle—wire rope, pulley blocks, and winch.

Wire ropes (Fig. 13.18) are made of metal wires—individual carbon-steel wires are twisted together to form strands and the strands are twisted together around a steel core to form a rope or cable. The lay of a wire rope is the direction of twist of the wires in the strands and of the strands in the wire ropes. The load carrying capacity of the rope depends on its diameter and the number of wires it is made of. A 6×19 rope means 6 strands, and each strand of 19 wires. Wire rope is manufactured with a safety factor of five. The three common lays of wire ropes are:

- Ordinary lays: the wires spiral to the left and the strands spiral to the right in the right hand lay and the spiraling would be the opposite in case of the left-hand lay—these types of ropes are useful as slings.
- Lang's lay—both wires and strands spiral in the same direction. These ropes usually have better wearing properties due to larger surface area of the external wires. However, both the ends must be secured to prevent spinning.
- Non-rotating lay: this is achieved by a double rope arrangement—the inner rope can be in right-handed Lang's lay and the outer rope in left-handed ordinary lay.

All wire ropes are lubricated during manufacture. However, the wire ropes in use should be cleaned and lubricated when exposed to working environment. Wire ropes must be inspected before first use and given a thorough examination every six months. The record of such inspection and examination should be maintained.

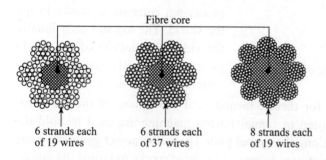

Fig. 13.18 Wire ropes

Lifting attachments

Slings are used extensively for moving materials in bundles. Reinforcing steel bars, structural steel sections, precast concrete units, and similar items may be bundled with slings for lifting and shifting. The lifting capacity of slings depends on the angle formed at the top under the crane's hook. A 19 mm diameter sling can lift about 5.3 tonnes at an angle of 60°. The same sling lifts about 4.3 tonnes at 90° and 3.0 tonnes at 120°. These figures are indicative of the possible variation.

Lifting beams are sometimes used to support slender and fragile materials like precast concrete beams. Lifting brackets and pincers (claws adapted for grasping) are used for lifting and shifting packaged materials. Firebricks, tiles, and electrical insulators are packaged well for transportation and delivery. These packages can easily be handled and moved at sites using brackets and pincers.

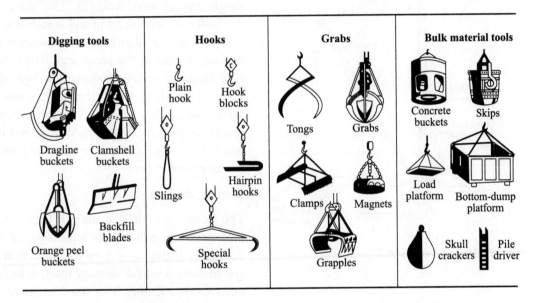

Fig. 13.19 Crane lifting attachments

Concrete skips are used for both lifting and placing concrete. Skips of different shapes and dimensions are used for placing concrete in different situations. Skips can be used for:
- Side opening type skip for placing from the sides as in thin concrete walls
- Bottom opening type skip for mass concrete work
- Tipping type skip for placing concrete in confined places like walls, columns, etc.

Some crane lifting attachments are shown in Fig. 13.19.

Winches

A winch and a drum with wire rope wound around it are required for most lifting/hoisting operations. The winch is powered by compressed air, electricity, hydraulic power or diesel engine.

The speed of lifting of most winches may be controlled to suit the kind of load being lifted. The speed of lifting may vary from 20 m/min to 100 m/min. The modern winches are capable of fine controlling speeds of both lowering and lifting.

Shear legs

The equipment comprises a pin-jointed frame fabricated using steel tubes or steel sections, a winch and pulley block plus lifting tackle. Shear legs are used for vertical lifting or lowering. However, horizontal movement can be effected when the shear legs are mounted on a barge or pontoon. Shear legs are used extensively at construction sites.

Guyed derrick

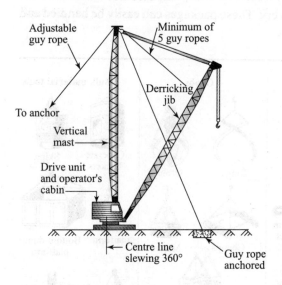

Fig. 13.20 Guyed derrick

This simple and inexpensive non-mobile construction equipment (Fig. 13.20) comprises a single lattice mast and jib. The mast stands vertically on solid bearing and is stabilized by at least five anchored guy ropes in the same way as is done in case of the shear legs. The low pivot type jib is of such length as to allow 360° slewing without fouling with the guy ropes. The arrangement allows also luffing (alteration in its radius of action). Deployment of guyed derrick would be cost-effective if its use is required over a long period of time.

Scotch derrick

The Scotch or stiff-leg derrick (Fig. 13.21) is preferred over either shear legs or guyed derrick for heavy lifting over long and high reaches. It comprises a vertical slewing mast, a luffing jib and two rope drums. Two fixed lattice members called stays support the mast, which is usually made of steel plates. The mast is free to rotate on bearings at its top and bottom supports. Two stays are fixed to the top of the slewing mast at an angle of 45°. Horizontal stays (struts) connect the bottom of the mast with the bases of the 45° inclined stays thus forming triangular frame of the lattice stabilizing members. The mast, therefore, is capable of slewing 270° due to inclined stays. The stays are relatively short compared to the jib length. Heavy kentledge is required at the bases of the mast as well as the inclined stays to resist overturning unless they are bolted solidly to concrete foundations. The weight of kentledge at each base should be about four times the lifting capacity of the stiff-leg type crane.

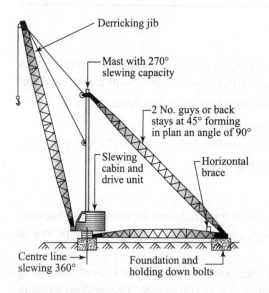

Fig. 13.21 Scotch derrick

Of the two rope drums, one is used for lifting and the other for luffing. The two drums are driven by electricity, hydraulic power, steam or diesel engine. Usually electricity is used to provide power to the drums.

Track-mounted crane

Static cranes have limited reaches. At construction sites, the ground conditions may not be favourable for movement of cranes mounted on wheels. Crawler cranes can move around most unfavourable sites. These cranes do not need firm level prepared surfaces for traversing. The weight of the crane plus load in case of a crawler crane is spread over a large bearing area under wide and long tracks. Crawler cranes are advantageously deployed for lifting and shifting small to medium category loads like concrete skips, reinforcing steel, fabricated steel, formwork, equipment etc at construction sites.

A crawler crane can be deployed for excavation work by converting it into dragline or equipping it with grab.

A crawler crane is built in three sections as follows:
- Base frame
- Superstructure
- Jib

The base frame is fabricated of welded steel sections. The two machine axles form part of the base frame. This fabricated frame supports the weight of the engine, gear system, winches, controls, cabin, jib and counterweight.

The superstructure has a revolving frame sitting on a large turntable capable of slewing 360°. This frame is supported on a base frame. The engine, gears, winches and counterweight are accommodated on and around the turntable.

The two winches comprise two rope drums with independent clutching and braking facilities. The winch with its drum near the jib is for lowering and lifting load using its hook. The rear winch is used for luffing the jib. The winches rely on mechanical or hydraulic system for power transmission.

The jib is of lattice construction with additional sections and fly jibs to obtain the various lengths and capacities. The basic jib is assembled in two sections. Intermediate sections may be added to extend jib lengths. The top section incorporates a head sheave for light loads of up to 10 tonnes and a hammerhead boom for heavy loads. The lifting capacity of a crane depends on the operating radius of its jib. The maximum load lifted at the minimum radius defines the crane's capacity.

Because of constraints, straight jibs often cannot be extended beyond a limit. In such cases, a fly jib can be attached at the end of the straight jib. The fly jib is of similar lattice construction as the main jib. The length of the fly jib is adjusted to suit the job requirement. With a third winch, the fly jib can be used independently for both lifting and lowering.

The track-mounted crane (Fig. 13.22) has a disadvantage in its movement from a construction site to another site. It needs a low-loading truck for transportation.

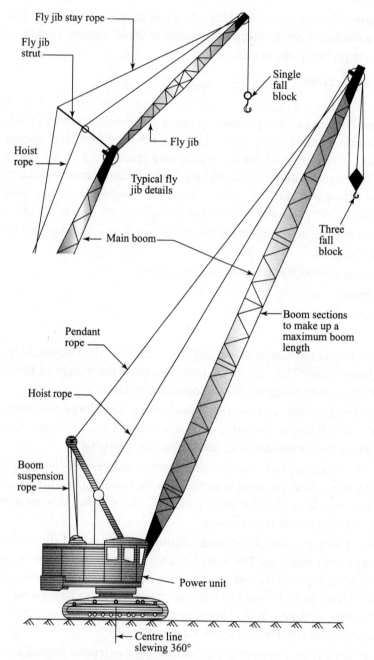

Fig. 13.22 Track-mounted crane

Wheel-mounted cranes

A wheel-mounted self-propelled crane has the clear advantage of mobility. The wheel mounted cranes may be broadly classified as: (i) similar to the track-mounted crane with lattice jib but on wheels instead of track moving at 8–10 km/hr speed (ii) mobile telescopic jib vehicles moving at 30–40 km/hr speed on highways.

Wheel-mounted crane with strut-boom

The wheel-mounted crane with strut-boom (Fig. 13.23) is also built in the three following sections:

- Base frame
- Superstructure
- Boom

The base frame comprises a welded steel chassis. Power is transferred to the wheels via the gear system to the gearbox on the drive axle. The other axle is used for steering. Apart from the usual two-wheel drive chassis, cranes with four-wheel drive chassis are also available.

The boom and the winches including the rope drums are arranged in the same way as are done in case of the track-mounted crane. The wheels are driven by mechanical or hydraulic transmission. As the luffing rope takes the tension, the boom remains in compression and designed like strut.

The wheel-mounted crane is operated on hard grounds for handling and moving relatively light loads. The load is generally transferred to the grounds directly through heavy-duty tyres. However, thick steel discs are sometimes mounted on the tyres one each on each side of the tyres to dampen their bouncing when lifting and moving.

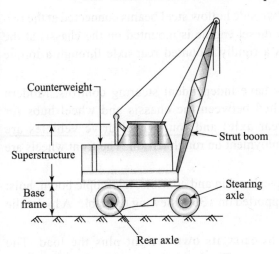

Fig. 13.23 Wheel-mounted crane with strut boom

Jack outriggers are incorporated in larger capacity cranes to increase their operating range. These outriggers are housed in heavy steel compartments and extended or retracted by winding mechanism. The entire crane including the tyres is raised clear of the ground when lifting. The outriggers should be either supported on foundations or on solid grounds in case of heavy lifting operation. With outriggers, the crane would remain static.

The truck is driven from the conventional position in the cabin, but the crane is operated from a different position.

Wheel-mounted crane with cantilever boom

A cantilever boom is pivoted on a specially designed truck (Fig. 13.24) at a much higher position on the superstructure compared to wheel-mounted crane with strut-boom. The cantilever jib crane, therefore, provides greater clearance for handling bulky loads.

A cantilever jib has a wide base tapering to the sheave support and pin-jointed to the superstructure at the underside of its lower end and restrained by the luffing rope running over a sheave along the top side. The top of the boom is thus in tension and the bottom is in compression.

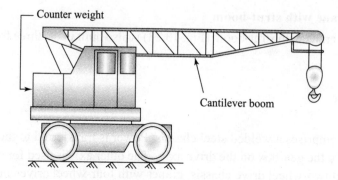

Fig. 13.24 Wheel-mounted crane with cantilever boom

Self-propelled telescopic boom crane

The self-propelled telescopic boom crane (Fig. 13.25) is more versatile and can easily move quickly from site to site at 20–30 km/hr. The telescopic action of the boom is very flexible.

The chassis (base frame) comprises two side hollow steel beams connected at the two ends by similar hollow crossbeams. A diesel engine is mounted on the chassis at the rear of the vehicle. The engine drives a rigidly mounted rear axle through a torque converter and power-shift gearbox.

The front and the rear wheels may have independent steering controls. Modern chassis has hydraulic cylinders attached between the chassis and wheel hubs for steering control. Both two-wheel (rear axle) and four-wheel drive vehicles are available. Four-wheel drives are for deployment on rough terrain. The front wheels are used for steering.

The superstructure comprises the driver's cabin and controls, telescopic boom, hoist and counterweight. The whole unit is supported on a 360° slewing turntable. A hydraulic motor activates the slewing operation.

The cantilever boom is designed to carry its own weight plus the load. The counterweight balances possible overturning. The boom is pin-jointed to the superstructure and luffing is effected by two or more hydraulic rams. The boom itself comprises three or four sections, one sliding within the other. A hydraulic ram and an expanding chain provide the telescoping action.

Because of the self-weight and the weight of the hydraulic telescoping rams, the payload is less compared to the strut-boom. And because of development of high tensile alloy steel, longer telescopic jibs of trapezoidal section made from such steel plates are available now. The reduction of self-weight increases the operating range of the crane.

A hydraulic motor operates the hoist. This motor is in circuit with the main hydraulic pump.

For improving manoeuvrability particularly for warehouse use or similar work, some versions of the crane have four-wheel drive and four-wheel steering that enable the crane to move diagonally.

The lifting capacity of the crane is highly improved when outriggers are used to provide wide and stable base. These outriggers can be hydraulically operated and controlled from the driver's cabin independently.

The main boom may be provided with the operational facility of a fly-jib to extend its operational radius. It may be in alignment with the main boom, or offset up to 25°. On some versions, the fly-jib may be telescoped. In the others, the fly-jibs are swung into positions by means of the hoist and the guy ropes.

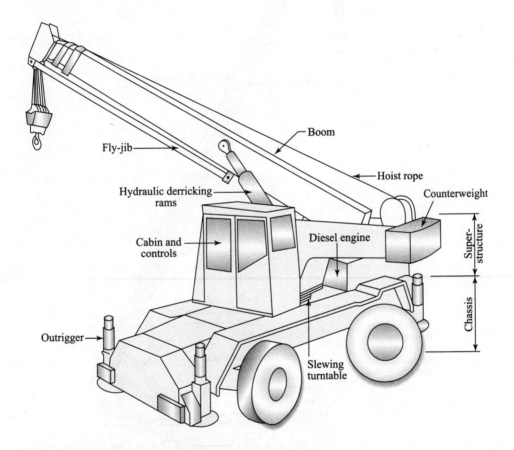

Fig. 13.25 Self-propelled telescopic boom crane

Truck-mounted telescopic boom crane

A truck-mounted telescopic boom crane (Fig. 13.26) moves relatively at high speed on roads/highways and can be deployed at site quickly.

The truck chassis comprises two universal beams with integral outrigger boxes. The chassis supports the power units, transmission system, cabin, boom, counterweight and hoists. The number of axles and wheels of a truck depends upon the travelling weight of the truck.

454 Construction Technology

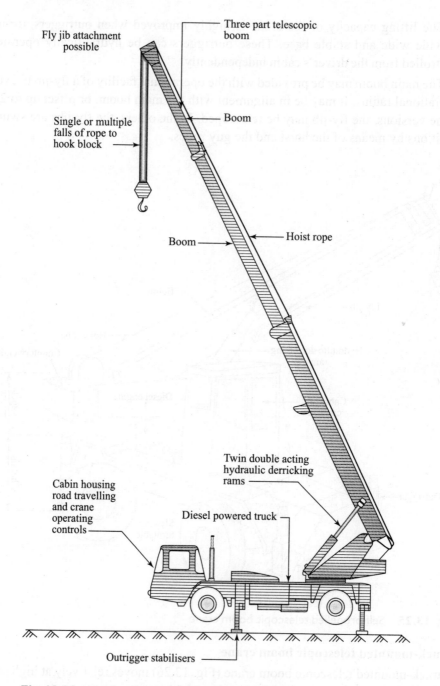

Fig. 13.26 Truck-mounted telescopic boom crane

The superstructure is similar to that of a self-propelled telescopic boom crane except for the counterweight, which is placed at a lower position to improve travelling stability. The whole unit is supported on a 360° slewing turntable. On the smaller cranes, single cabin is provided for crane operation and travelling with a single diesel engine that powers all mechanical parts. However, separate cabins and power sources are provided for travelling and lifting in larger cranes. In the larger cranes, the hydraulic pumps and motors are accommodated on the superstructure.

For travelling on the highways, it is necessary that the boom is retracted and well secured in the horizontal position.

Truck-mounted strut-boom crane

This type of crane (Fig. 13.27) is used for heavy-duty lifting of load. The truck has a stiff chassis mounted on two or more axles depending upon the total travelling weight on roads and highways. The chassis includes a diesel engine, transmission system and outriggers. This engine is used only to supply power for transportation. The superstructure comprises the driver's cabin and controls, strut-boom, hoist, luffing drums, counterweight and an independent diesel engine to supply power for crane's operation. Independent hydraulic motors are used for precision lifting and lowering. The whole unit is supported on a 360° slewing turntable mounted on the chassis.

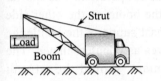

Fig. 13.27 Truck-mounted strut-boom crane

The basic boom length comprising the top and bottom sections only can be packed down neatly for travelling on roads/highways.

The large capacity crane with more sections requires another carrier for transporting the boom and fly-jib. The power for slewing, hoisting and luffing is supplied by electric motors connected to the main diesel engine. Extra outriggers and kentledge are also necessary for handling heavy loads. In calculating loads, weight of wire ropes, lifting tackles and handling devices should be added in case of a crane deployed for heavy-duty lifting work.

Tower cranes

As the name suggests, the jib of the crane is supported at the top of a tall tower-like mast. The boom is set at a height to clear all obstructions as required for medium to high-rise structures. This configuration allows the crane to be positioned very close to the structure under construction. If necessary, a tower crane may even be positioned in the structure under construction also.

The boom of a tower crane may be of two types:

- Luffing
- Horizontal

Luffing boom tower crane

The luffing boom tower crane (Fig. 13.28) comprises a vertical lattice mast and a luffing boom, which can be easily raised to clear nearby obstructions when slewing. The mast is mounted on a sturdy turntable. The luffing wire

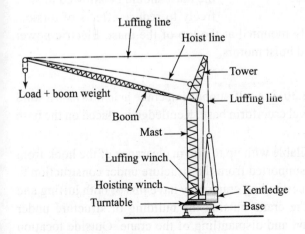

Fig. 13.28 Luffing boom tower crane

rope is connected to the kentledge at the base to counterbalance overturning. In some models of the crane, both the turntable and the kentledge are located at the top of the mast so that all the moving parts are located above the structure under construction. The whole unit is placed on an appropriate foundation with the mast firmly anchored to it with holding down bolts or a mast base section cast into foundation. Electrical motors are used to drive the hoist and luffing winches and also for slewing.

Horizontal boom tower crane

The horizontal boom tower crane (Fig. 13.29) comprises a vertical lattice mast that supports a horizontal boom in two parts. The larger part or section, which carries a trolley or saddle travelling on guides along the length of the boom, is used for lifting. As luffing is not possible in a horizontal boom, moving the trolley can only change the radius of operation. The shorter part of the boom on the other side of the mast supports the kentledge and serves as counterbalance. The tower mast should be strong enough to counter the overturning moment during lifting or wind disturbances and transfer the same to the foundation. Main counterweight is also located at the base. Wind speed exceeding 61 km/hr would require suspension of lifting and lowering operations. And the mast should be allowed to slew freely to reduce effects of torsion.

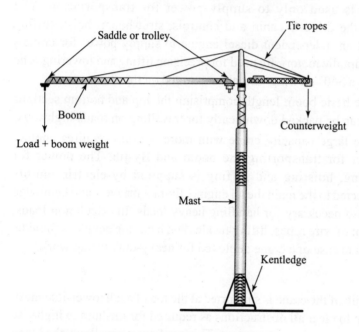

Fig. 13.29 Horizontal boom tower crane

A 360° slewing turntable is generally mounted at the top of the mast. Electric power provides the drive for the slewing and hoist motors.

Freestanding tower crane

The freestanding tower crane (Fig. 13.30) is a self-supporting crane held down on a solid foundation. The mast is bolted to a steel cruciform base. Kentledge is placed on the base to counter overturning of the crane.

Freestanding tower cranes are available with up to 100 m clearance of the hook from the ground level. The crane must be supported from the structure under construction to increase the clearance further. The freestanding crane may have jibs of both luffing and saddle types. It is better to locate the crane outside the building or structure under construction to facilitate both erection and dismantling of the crane. Outside location would also help in avoiding redoing the portions that would have been left out in case the crane was located inside. However, it is to be ensured that the crane's reach does not fall short of the construction requirement.

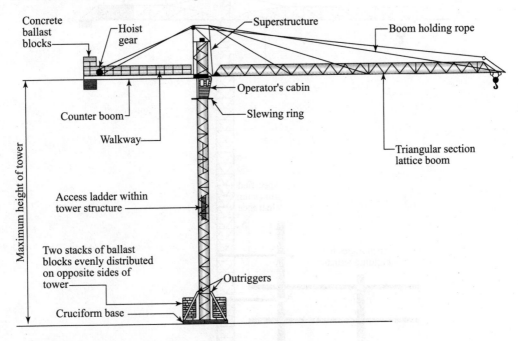

Fig. 13.30 Freestanding tower crane

Supported static tower crane

The supported static tower cranes (Fig. 13.31) are similar in construction to the freestanding tower cranes, but they can be used for lifting to heights in excess of that possible with the freestanding and travelling tower cranes. For this, the crane mast should be tied in to the structure under construction to maintain unsupported heights as per the recommendations of the crane manufacturers. The mast is tied to the structure using adequately designed single or double lattice frame to resist all horizontal forces to provide adequate stability. By judicious selection of supporting positions, hook clearance may be extended from 100 m to 200 m or more. Supported static tower cranes normally have horizontal jibs. As most of the overturning moment is resisted by the structure, the foundation of a mast would be relatively simple. The mast is anchored to the foundation with holding down bolts.

Rail-mounted tower crane

For better site coverage with a tower crane, its base is often rail-mounted. The crane can travel on specially laid railway track with the load on the hook. The track may be straight or curved or may even be in gradient. However, the gradient should not exceed 1 in 200 and the curve should not be less than 11 m in radius depending, however, on mast height. It is essential that the rail tracks are accurately laid, well drained, regularly inspected and properly maintained so as to ensure stability of the crane. This could be a disadvantage, as the track area cannot be used for any other purpose.

458 Construction Technology

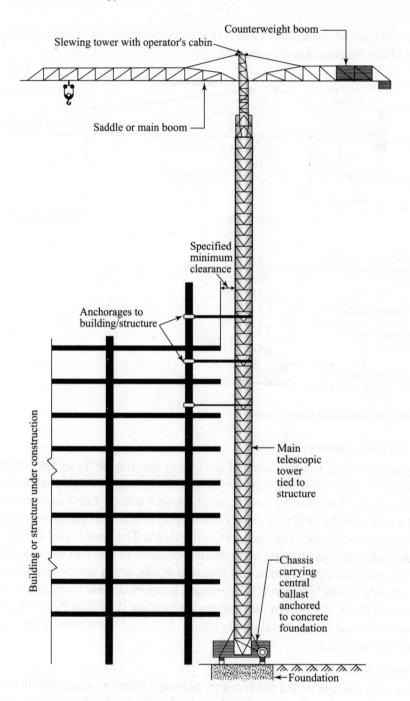

Fig. 13.31 Supported static tower crane

Mechanized Construction 459

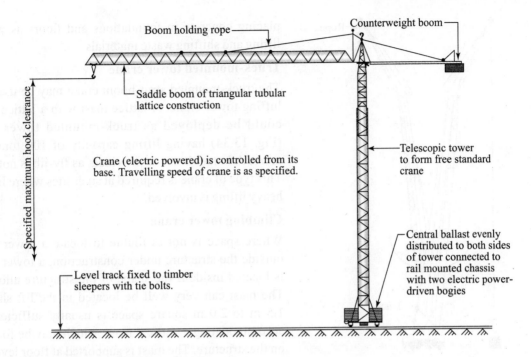

Fig. 13.32 Rail-mounted tower crane

For straight track, the sturdy welded steel base frame mounted on four wheels is of simple construction. For curved track, the base frame is of more complicated construction as required for negotiating curves. The rails are pivoted to the arms welded to the base frame to allow negotiation of curves. The travel wheels are driven through gears to enable the crane to move as fast as safe with load being carried. The crane may also be provided with special braking system to allow fine movement when precise positioning of the load is required.

Rail-mounted cranes (Fig. 13.32) are powered by electricity. Power supply cable should be wound round a spring-loaded drum, which would draw in the cable when the crane would move in the reverse direction thereby reducing the risk of the cable being cut or trapped.

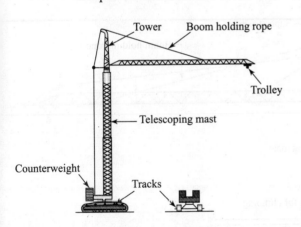

Fig. 13.33 Track-mounted tower crane

Track-mounted tower crane

The track-mounted tower crane (Fig. 13.33) is of relatively small capacity of up to 1500 kg at 40 m radius. The crane is brought to site on a low loader. This crane can be used for

460 Construction Technology

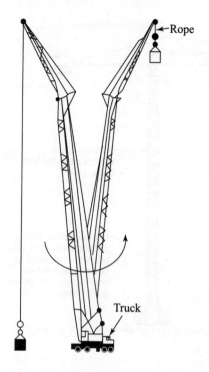

Fig. 13.34 Truck-mounted tower crane

placing concrete in foundations and floors as well as lifting and shifting waste materials.

Truck-mounted tower crane

The truck-mounted strut-boom crane may be used as a luffing tower crane. A lattice mast with a lattice boom could be deployed as truck-mounted tower crane (Fig. 13.34) having lifting capacity of 100 tonnes or more. A longer boom is provided, as fly-jib is not used. This type of crane is required at such sites where limited heavy lifting is involved.

Climbing tower crane

Where space is not available to locate a tower crane outside the structure under construction, a tower crane is located inside if the shape of the structure allows it. The mast can very well be located in the lift shaft as 1.5 m to 2.0 m square space is usually sufficient for accommodating a vertical mast, which may be founded on the structure. The mast is supported at floor levels by steel collars, frames and wedges (Fig. 13.35). The two methods of raising the mast are by using winches or hydraulic jacking. Winch or jack is made an integral part of the system. Specially made mast sections are raised

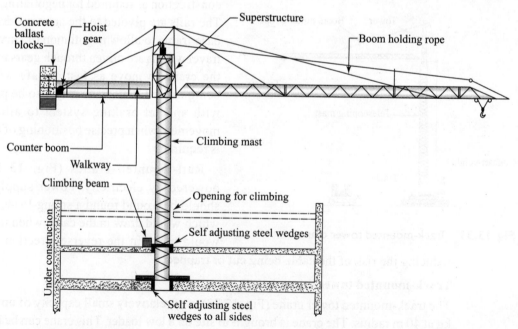

Fig. 13.35 Climbing tower crane

from one floor level to next upper floor level by winching or jacking. The boom is made from small manageable sections to make assembling, handling and dismantling easy. Dismantling requires the boom and the mast to be taken apart in sections and lowered to the ground by winches.

Gantry crane

The gantry cranes are used for:
- Very efficient material handling is possible with this type of crane
- Quite efficient in manufacturing precast concrete members
- Very useful on launching bridge girders and decks

The gantry crane (Fig. 13.36) is a portal frame fabricated from structural steel sections and components. The portal legs are mounted on rail tracks and the bridging beam spans the entire gantry platform, which serve as storage, pre-assembly and manufacturing area. The gantry crane can also be used for erection of equipment and launching of bridge girders and decks. The structure is of lattice frame construction. The mast and horizontal jib crane can be used for heavy handling and erection work. The whole crane structure is designed to resist horizontal and vertical forces. The lifting tackle supported on the portal beam is powered by electricity. A gantry crane can be installed either inside a shop or out in the open yard. Standard models are available having capacities ranging from 5–30 tonnes. However, gantry cranes can be designed, made and installed to meet specific requirements.

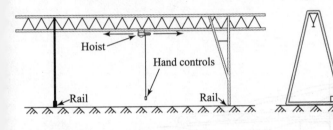

Fig. 13.36 Gantry crane

Hoist

The hoist is used as vertical elevator to transport personnel and materials quickly by means of a moving level platform in the construction of massive structures like dams, power station buildings, cooling towers, chimney stacks etc. A hoist can be used conveniently to supply concrete and other materials in high-rise construction as well. Hoists can be used in combination with tower cranes to speed up construction work at high elevations by reducing lifting time involved in tower crane operation. Hoists can be effectively used both in high-rise and low-rise construction where deployment of tower crane/s may not be cost-efficient. Hoists are designed for lifting either materials or passengers. But the recent trend is to design hoists for transportation of both passengers and materials. A hoist is not a costly device. Its maintenance cost is also low. Hoists are, therefore, widely used lifting devices. Its disadvantage compared to a tower crane is that it can lift materials up to a platform. Further arrangement is needed for lifting or shifting from the platform. However, lifting of materials by hoists to a particular floor/level on completion of the structure is relatively easy compared to lifting by tower or mobile cranes.

The hoists are basically of two forms:
- Mobile hoists
- Static hoists

Mobile hoists

The mobile hoist (Fig. 13.37) comprises a base frame, a winch powered by diesel/petrol engine or electric motor and two mast sections. Four screw jacks are incorporated with the base for levelling the unit when in operational mode. All mobile hoists should be positioned on a firm level base and jacked to ensure stability. The hoists are quite robust and transported from one construction site to another on trucks, or may even be towed.

A mobile hoist may be of 24 m or so height, but generally its lifting height for 250–1000 kg materials is 15 m or so. A mobile hoist does not need any tying with the structure.

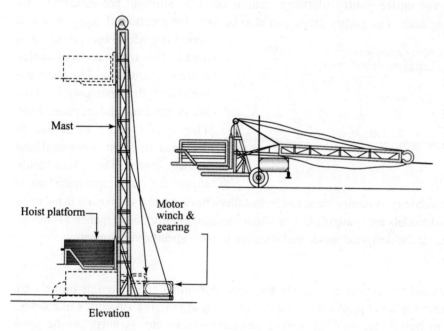

Fig. 13.37 Mobile hoist

Static hoists

The static hoists (Fig. 13.38) like the mobile ones comprise a mast or tower with the lift platform either cantilevered from the small section mast or centrally suspended with guides on either side within an enclosing tower. In both the forms, the mast must be in plumb vertically and tied firmly to the structure or the scaffold at appropriate intervals for stability.

The static hoists are installed for operation at a particular location of the site for a long period of time. Hoists are very useful for lifting materials to different floors/levels on construction of basic structures, as the use of tower cranes for such lifting is not so easy. Hoists in combination with concrete pumps can be better utilized in the construction of tall chimneystacks, lift shafts, cooling towers, etc.

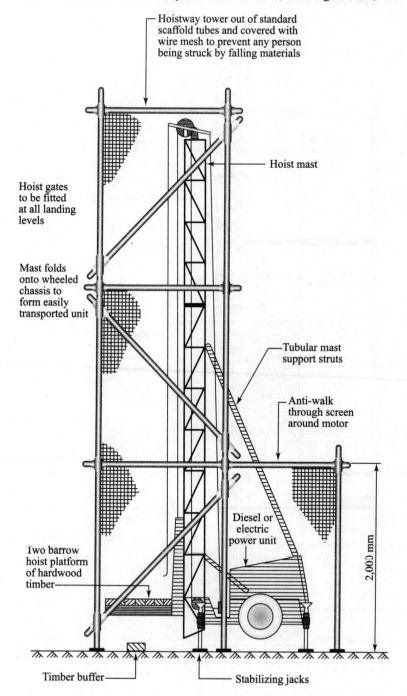

Fig. 13.38 Static hoist

Passenger hoists

The passenger hoist (Fig. 13.39) is a static hoist that can be of cantilever or enclosed variety. The cantilever type may be either a single front-mounted cage or two side-mounted cages.

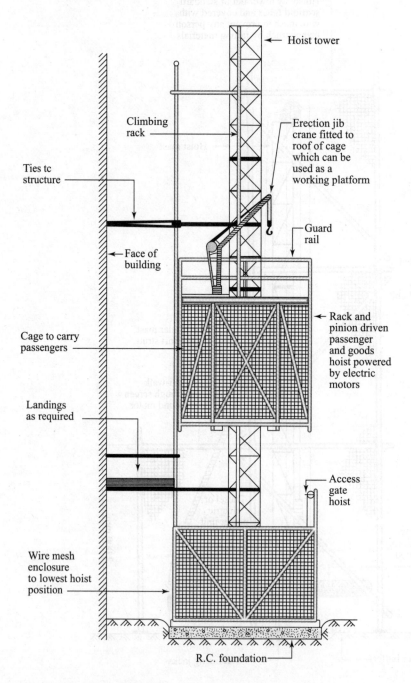

Fig. 13.39 Passenger hoist

hoist with one side-mounted cage is also used. The other form comprises a passenger hoist within an enclosing tower. A 12-passenger hoist of about 100 m height with hoisting speed of 40 m/min is one of the standard hoists used at construction sites.

Concrete hoists

The concrete hoist can easily be used in high-rise structures construction in place of tower crane for concreting work. The hoist comprises a skip, which is fitted inside the mast or outside, lifts concrete to heights to discharge into hoppers or wheelbarrows.

Forklift truck

The forklift truck (Fig. 13.40) is deployed for both transporting and lifting thus obviating the necessity up to a limit of deployment of a separate transport vehicle or lifting device.

These trucks are, however, used for mainly handling palleted materials quickly and efficiently over uneven and rough terrain of construction sites. Many materials are now so packaged as to suit forklift handling. At constructions sites, forklifts are utilized for lifting materials up to 5 tonnes. Some innovated forklifts have provisions for forward reaching.

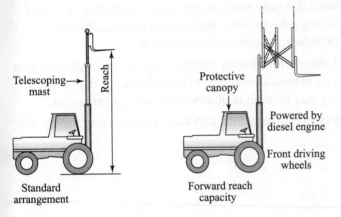

Fig. 13.40 Forklift truck

The forklift comprises a chassis supporting a diesel engine over the rear axle and lifting mast at the front. Forklifts are now available with multiple choices of front-wheel or rear-wheel drive and four-wheel drive with various mast heights and lifting capacities. Both steering and mast operations are controlled hydraulically. For heavy-duty assignment at rough and uneven terrain of construction sites, the equipment is robustly built and fitted with large and wide tyres.

The advantage of this equipment is that it can unload from any side of a delivery transport. The forward extending truck in particular can often reach across the full width of the carrier vehicle.

Materials unloaded from the transport can be transferred directly to the place of work or stacked in systematic stockpiles by deploying forklift. Double handling can thus be avoided.

In a forklift, load nearer the ground level reduces the possibility of overturning. The mast is fitted with hydraulic rams to move the load radially for manoeuvring of the equipment in motion. Modern forklifts are being fitted with telescopic boom to place load inside building structures at heights up to 10 m. This facility may be turned into access platforms for rendering such work as cladding, painting, glazing, street-lighting etc.

Aerial cableway

The aerial ropeway (Fig. 13.41) comprises:
- A track cable firmly anchored at the two ends to robust steel towers
- A wheeled carriage fabricated of light steel sections/plates containing two or more wheels for moving on the track cable
- Two other sheaves to guide the hoist line
- A looped travel rope to move the carriage between the towers
- Hoist rope on the carriage is independently operated from a hoist winch
- Winches for (i) moving the carriage (ii) hoisting as well as lowering the load between the towers

The aerial cableway is feasible for only short spans, as the rope that is used for hoisting tends to sag and gets entangled with other ropes. Nevertheless, this method of lifting and moving is fast and, therefore, efficient and economical.

The aerial cableway is used mainly for placing concrete over a long extended area beyond the capacity of a crane as in the construction of river barrages, dams, bridges, viaducts etc. The aerial cableway may be used in quarries and storage yards also.

The cableways may be of various types to suit the particular project requirements. The most common types are:
- Fixed cableway
- Luffing cableway
- Radial cableway
- Parallel moving cableway

The fixed cableway may be used to service only a straight-line area as both the head and tail towers are held in fixed positions. These types of cableways are, therefore, mainly used for transportation of materials over rivers and ravines.

The luffing cableway has its robust head tower fixed at one end while the tail tower is mounted on a ball pivot base. The hinged base allows the luffing of tail tower by 20° with respect to the vertical on either side. The luffing tower is guyed firmly at the rear and loosely at the sides to allow swaying of the tower. The luffing cableway covers more area than a fixed cableway.

The radial cableway also has its properly designed head tower fixed at one end while the tail tower is mounted on rail tracks for movement radial with respect to the head tower. The radial movement is restricted within 40° with respect to the head tower. The head tower is guyed on both sides for stability against horizontal thrust due to radial movement at the other end. The radial cableway covers more area than luffing cableway and is used in dam and bridge construction.

The parallel moving cableway has its both towers on rail tracks for parallel movement. The parallel movement is not to be considered easy and requires two operators at two ends. The parallel moving cableway is used for placing concrete in the construction of wide dams.

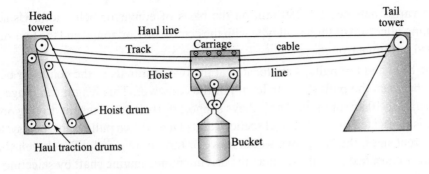

Fig. 13.41 Aerial cableway

Belt conveyor

The belt conveyor (Fig. 13.42) comprises:
- Endless flexible moving belt
- Idler rollers on which the belt is supported
- Two pulleys—a head pulley and a tail pulley
- A motor to drive the head pulley—in the event high driving forces are required, it may be necessary to use more than a pulley—the pulleys are then arranged in tandem to increase the area of contact with the belt
- A tensioning device to take care of slack in the belt
- A steel frame to support the whole system

The belt is made up of cotton or nylon sacking vulcanized with natural or synthetic or blended rubber to provide strength to take tension and absorb impact, as the belt has to be strong and hardwearing to be durable. The top as well as bottom surfaces of the belt need to be covered with a layer of rubber, PVC or any such material to protect the belt from abrasion and impact loading. The thickness of this layer varies between 2–16 mm on the top surfaces and 1–8 mm on the bottom surfaces depending on the materials to be conveyed. The belt is formed into a trough shape at the top to increase the conveying capacity and reduce spillage. Usually the troughing idlers are of three-roll type with the centre roll horizontal and the two side rolls inclined to form a trough. The standard troughing angle for side rolls is 20° but modern belt materials allow idlers to be set at such steep angle as 55°. Belt width may vary from 400 mm to 3000 mm. The conveyors are capable of carrying loads up to considerable height at an inclination ranging from 15° to 25°. A belt changes shape as it goes around a pulley as it is thicker on the slack side than on the loaded side.

The spacing of idler rollers along the conveyor belt depends upon the weight of materials to be carried as well as the belt construction and the angle that forms trough. This spacing is normally 1.0–1.5 m, but is reduced to 300 to 400 mm near the loading point for support against impact for carrier or troughing idlers. The return idlers are so spaced as to reduce sagging of the belt. This spacing is 2.0–3.5 m. The diameter of idler rollers varies between 50–250 mm on the basis of conveyor belt width. Usually the troughing idlers are of three-roll type with the centre roll horizontal and the two side rolls inclined to form a trough.

The head and tail pulleys are located at the two terminals of the conveyor belt. The length between the pulleys is the length of the conveyor. This length can range from a few metres to 10 km or more. The pulleys support the belt and transmit driving power. A pulley is fabricated of welded steel sections and plates. If two pulleys in a belt system are of different sizes, the larger one would move or turn more slowly. The driven shaft can be made to turn faster or slower than the driving motor/engine shaft by selecting pulley sizes judiciously.

A belt drive can be made to provide several speed ratios by mounting several sizes of pulley side by side on the same shaft. The belt is shifted from one pulley to another as required.

A belt conveyor is supported at the two terminals and intermediate sections. At the terminals, structural supports are required for the pulleys, belt tension take-ups, drive, chutes, screens, feeders and other related equipment. The intermediate structures comprise stringers made up of steel sections to support the troughing and return idlers.

All belt conveyors in general are driven by electric motors. The capacity of the drive motor is sized up on the basis of the width and length of the conveyor belt and load to be carried. More than one motor can be provided if so required.

The total power requirement is the sum total of the following power requirements:

- Moving the empty belt over the idlers
- Moving the load on the belt horizontally
- Lifting or lowering the load on belt vertically
- Turning all pulleys
- Making up drive losses
- Operating a tripper, if one is used

Conveyor belts are not commonly used equipment at construction sites. However, these conveyors are used in tunneling work, land reclamation work with dredged

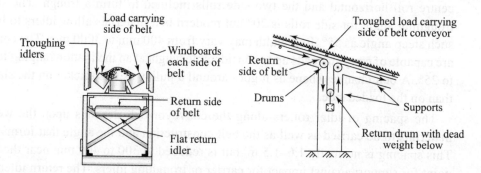

Fig. 13.42 Belt conveyor

materials and in conveying coarse aggregates from quarries for concrete work. But there are portable frames that support conveyor belts, rollers and drive motors. The length of such portable conveyor is normally 3–6 m, but models of lengths up to 20 m are also available. The portable conveyors are inexpensive versatile device that can be used extensively at construction sites because of simplicity, smooth operation and uniform discharge. See also 'belt drives' under Section 13.2 above and 'belt conveyors' under Section 13.6.

Monorail

The monorail comprises:
- Monorail track assembled from short rail sections
- Rail is supported on steel section or fabricated beam that is again supported on trestles, which rest on foundations or bolted/welded to structural steel members
- A self-travelling skip (about 0.5 m^3 capacity), which can be tipped on either side for dumping concrete
- A motor drives the self-travelling skip at inclination up to 1 in 20
- Four idler rollers are spring loaded against the flange of the rail for steadying the skip during travelling

The monorail is used for placing concrete.

13.6 CONCRETE MIXERS AND PUMPS

Mixing of properly proportioned graded fine aggregates, graded coarse aggregates, cement and water results in the production of concrete.

Cement is a bought-out item. Different types of cement and water quality have been discussed in Chapter 7, Sections 7.3 and 7.5.

The sizes of the constituents of the aggregates depend upon the grading required. Sizes from 150 mm up to 3 mm can be obtained by crushing rock. The rock is obtained by blasting (vide Chapter 5). Different types and quality of aggregates have been discussed in Chapter 7, Section 7.4.

Modern crushers are capable of handling large blocks of over 1 m in dimension with the choice of suitable machinery depending upon the costs of reducing the sizes of the fragments at the quarry and the method of transport to the plant. On completion of crushing, washing and sorting into various sizes is generally carried out for proper storage.

The equipment and machineries required for production of aggregates at the required rate of production are:
- Feeders and hoppers
- Primary crushers
- Secondary or reduction crushers
- Conveyors to transport materials between processes

- Screens and washing plant for separating, grading and directing the sizes to the bins
- Storage bins and discharge hoppers

Crushing Plant

Friable rocks suitable for use in concrete production are fed into the crushers for breaking into smaller pieces. The crushed materials will not be of uniform size, but graded. Grading in accordance with specific proportions would depend on setting, type of available rocks and crusher. Further crushing or screening and sorting may be necessary to conform to the requirement. The crushing is done in three types of crushers:

- Primary crushers
- Secondary or reduction crushers
- Tertiary crushers

The stones brought from quarries are fed into crushers to reduce them in sizes. The reduction may be performed by attrition, pressure or compression, impact, shear, or combinations.

- Attrition is wear resisting property of stone—extent of rubbing and grinding down by friction. It is most effective with friable materials. The attrition action is most desirable when the requirement is maximum fine materials.
- Pressure is squeezing action between two surfaces like in jaw crushers. This action is specified when material is hard, tough, and abrasive and minimum of fines is required.
- Impact is the instantaneous, sharp blow delivered by a hammer on the material, which shatters it into many smaller pieces as in hammer mill and impact breakers.
- Shear is a cutting or slicing action as may be seen in gyratory or roll crushers.

The reduction in size is expressed as a ratio of reduction. The ratio of reduction is the ratio of the distance between the fixed and moving faces at the top divided by the distance at the bottom of a crusher. Thus if the distance between the two faces of a jaw crusher at the top is 16" and at the bottom is 4", the ratio of reduction is 4.

The ratio of reduction in the case of a roll crusher is the ratio of the dimension of the largest stone that can be nipped by the rolls divided by the setting of the rolls, which is the smallest distance between the rolls.

Primary crushers

The primary crushers are:
- Jaw crusher
- Gyratory crusher
- Hammer mill
- Impact breaker

Jaw crusher

The jaw crusher that crushes rocks or friable raw materials is relatively of simpler design and smaller size but power consumption and crushing cost is more compared to gyratory crusher. The crusher shape is rectangular from the top and triangular from the side. The usual construction is called swing-jaw. The materials to be crushed are fed from the top into the space between two jaws, one of which is stationary and the other is movable that causes jaw action. The distance between the jaws diminishes as the materials move downward under the effect of gravity and the movable jaw until discharged from the bottom opening. The movable jaw is tough enough to exert sufficient pressure to crush the hardest rock. The jaw plates are made of manganese steel. The jaw crushers are of three types:

- Double toggle
- Single toggle
- Double jaw

The one side of the jaw of the double toggle jaw crusher (Fig. 13.43) is fixed; the other swinging movable jaw is simply suspended from a shaft mounted on bearings on the crusher frame. The bottom hinged point generates crushing motion obtained by a rotating eccentric shaft raising and lowering a rod connected by two toggles to the jaw and a fixed reaction point. A high pressure is exerted near the bottom of the swing jaw, which partially closes the bottoms of the two jaws. The vertical movement is communicated horizontally to the jaw by double toggle plates. This operation is repeated as the eccentric shaft is rotated. The opening and closing movement of the jaw allows discharge of crushed fragments.

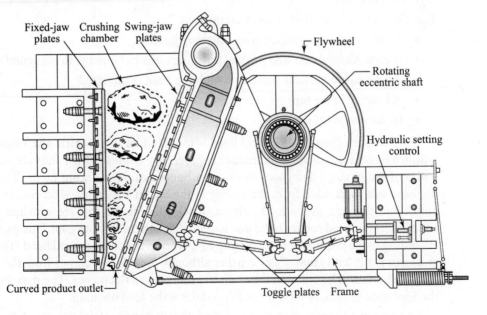

Fig. 13.43 Double toggle jaw crusher

The single toggle jaw crusher (Fig. 13.44) has an eccentrically rotating shaft at the top of the movable jaw for generating motion. Its bottom-hinged point has a single toggle connected to a fixed point. When the eccentric shaft is rotated, it gives the movable jaw a vertical and horizontal motion. This arrangement results in more fines compared to double toggle version. The capacity of a single toggle crusher is usually less than that of a double toggle crusher.

The double jaw crusher has no fixed side. With both the sides moving, the result is not satisfactory. The double jaw crusher is not in much use.

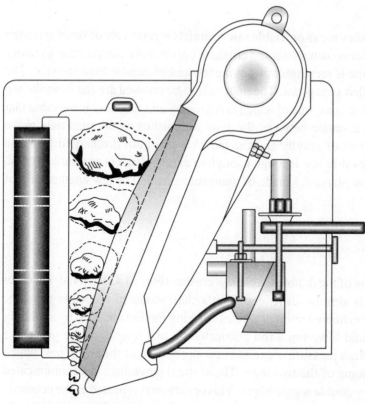

Fig. 13.44 Single toggle jaw crusher

Gyratory crusher

The gyratory crusher (Fig. 13.45) comprises:
- A hardened cone-shaped steel crushing head
- A cone-shaped cast iron or steel crushing chamber lined with hardened steel plates
- An eccentric vertical shaft on which the head is mounted
- A bearing at the top
- Gears and shaft drive

The hardened cone-shaped steel head is mounted on an eccentric shaft that oscillates within a larger wedge-shaped annular chamber. The angles of the cones are such that the width of the passage decreases toward the bottom of the working faces. The shaft and the head are suspended from the spider at the top of the frame. The eccentric shaft is connected through gears to the drive. Rocks fed at the top are crushed inside the chamber as the head rotates to be discharged through the bottom gap. This gap can be set according to the requirement by raising or lowering the cone-shaped head. The action to an extent is similar to that of jaw crusher although the squeeze comes from the side rather than the bottom. The curve of the chamber also breaks flat pieces. The maximum size of the feed rock materials should not exceed $0.8 \times$ the feed opening.

The speed of cone-shaped head and the distance of travel must be carefully synchronized. A wide space allows pieces to fall freely compared to a narrow one. Wide space coupled with slow movement would allow pieces to fall apart before the next impact. Fast gyration and short travel would not allow the pieces to fall far enough, thereby wasting power.

Compared to jaw crushers, gyratory crushers are more expensive as they consume more power and require more vertical space for installation. However, gyratory crushers have greater production and the products are finer and more uniform.

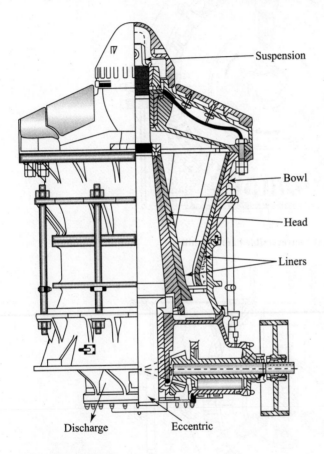

Fig. 13.45 Gyratory crusher

Hammer mill

The hammer mill (Fig. 13.46) comprises:
- Robust housing
- Horizontal shaft rotor/s extending through the housing
- A number of arms and hammers attached to a spool which is mounted on the/each rotor
- One or more manganese steel or other hard steel breaker plates
- A series of grate bars whose spacing(s) may be adjusted to regulate the widths of openings through which the crushed stones flow

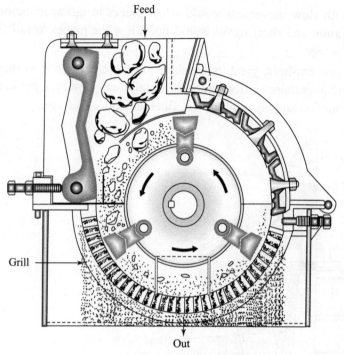

(a) Nonreversible hammer mill

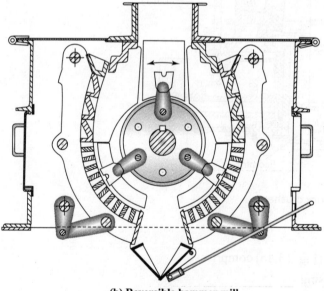

(b) Reversible hammer mill

Fig. 13.46 Hammer crusher/mill

The hammers are linked to one or several rotors. The protruding hammers rotating at high speed strike the fed rock materials breaking them into smaller pieces and driving them against breaker plates, which further reduce their sizes. The chamber base below is lined with grate bars through which crushed stones are discharged. The grates may be adjusted to produce relatively finer materials. Hammer mills have the largest reduction among all types of crushers. They can reduce rocks of large sizes into 25 mm size pieces in single pass. They are used for primary crushing of soft and medium rock. The size of hammers is related to the feed opening.

Impact breaker

The impact breaker (Fig. 13.47) is a variation of the hammer mill. It has protruding breaker plates or bars or grates located above the rotors. The materials struck by the hammers are thrown against the breakers resulting in further breaking into smaller pieces. There is no grate bars at the bottom. The extent of splintering is, therefore, dependent upon the impact velocity. Possible clogging of materials should be avoided by controlling sizes of feed materials as well as the rate of feeding. The impact breakers are suitable for breaking soft materials so as to avoid too much wear and tear.

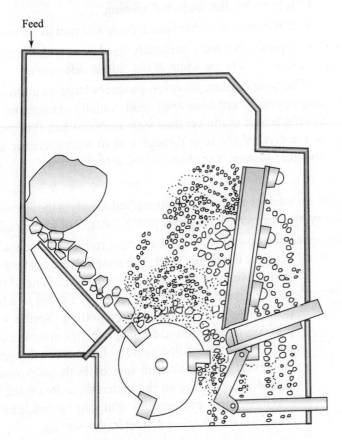

Fig. 13.47 Impact breaker

Secondary and reduction crushers

The stone aggregates obtained by primary crushing need to be further reduced in size for concrete production. Generally screening separates materials larger than a specified size after primary crushing. The materials retained need to be reduced by further crushing in secondary crushers.

The secondary or reduction crushers are:
- Cone crusher
- Roll crusher
- Hammer mill

Cone crusher

A cone crusher (Fig. 13.48) is capable of producing large quantities of uniformly fine-crushed stone. A cone crusher works in the same way as a gyratory crusher works, but differs from it in the following respects:

- It has shorter cone
- It has a smaller receiving opening
- It rotates at a higher speed, from 430 rpm to 580 rpm
- It produces a more uniformly sized stone with the maximum size equal to the width of the closed-side setting

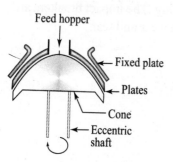

Fig. 13.48 Cone crusher

The cone crusher, however, produces large quantities of fine aggregates. Its crushing head, made usually of manganese steel, is broader but shallower than a gyratory crusher. It is connected to the drive at the base through a shaft with eccentric bushing and gear system. The shaft rotates at high speed.

Roll crusher

Roll crushers (Fig. 13.49) are used for producing additional reductions in the sizes of stone after the output of a quarry has been subjected to one or more stages of prior crushing.

The roll crusher comprises two counter-rotating hard steel rolls, each mounted on a separate horizontal shaft. The materials are fed through the gap between the two rolls. The position of one of the two rolls is fixed. The other roll is spring loaded to adjust its position to provide the desired setting. Spring loading prevents damage due to feeding oversize or non-crushable materials. The rolls usually have smooth surfaces. However, corrugation on one or both sides could be provided depending on the materials to be crushed. The maximum size of materials that can be fed into a roll crusher is 0.85 × roller diameter plus setting.

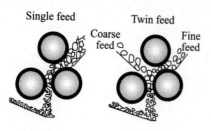

Fig. 13.49 Roll crusher

Hammer mill

See under the primary crushers (Fig. 13.46).

Tertiary crushers

The tertiary crushers are used to produce fine aggregates, such as sand, from stone that has been crushed to suitable sizes by other crushing equipment. It is not uncommon for specifications for concrete to require the use of a homogeneous aggregate, regardless of size. If crushed stone is used for the coarse aggregate, sand manufactured from the same stone would satisfy the specifications.

The tertiary crushers are:
- Roll crushers
- Ball mill
- Rod mill

Roll crushers

See under the secondary crushers (Fig. 13.49).

Ball mill

Ball and rod mills have cylindrical or conical shell, rotate on horizontal axis and are charged with grinding medium such as steel balls or rods. The steel balls in ball mills (Fig. 13.50)

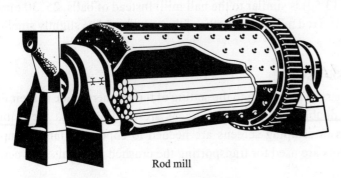

Rod mill

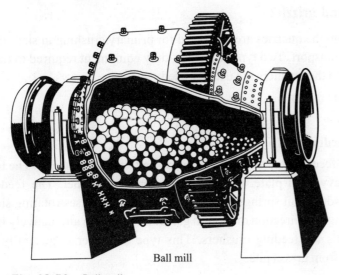

Ball mill

Fig. 13.50 Ball mill

deliver the impact necessary to grind the stone and produce fines with smaller grain sizes compared to those produced by a rod mill.

A ball mill comprises a rotating drum, lined on the inside with a hard mineral wearing surface, equipped with a suitable support or trunnion arrangement at each end, with a driving gear at one end. It is operated with its axis in a horizontal position. Crushed stone, which is fed through the trunnion at one end of the mill, flows to the discharge at the other end. As the mill rotates slowly, the stone is constantly subjected to the impact of the tumbling balls, which produce the desired grinding. The feed materials can be fed into the drum in wet or dry condition. The size of the mill is specified by the diameter and the length of the shell such as 2000×3600.

The rotating drum of a ball mill contains 25–50 mm diameter balls, which occupy about 30% of the drum volume. 15 mm and down stone materials are generally fed into the drum from one end. As the drum rotates, the stone materials are crushed into fine aggregates by the balls and discharged at the other end through a sieve.

Rod mill

The rod mill (Fig. 13.50) is similar to the ball mill. Instead of balls, 25–30 mm diameter steel rods are used in a rod mill. The lengths of the rods should be slightly smaller than the length of the drum.

Screening of Aggregates

Materials are discharged according to the setting of the crushers made at their discharged points. Normally, double or triple deck screens are used except at the primary stage where grizzle bars or scalping screens are needed to cope with large lumps of rock materials. Conveyors are used for transporting the crushed materials between stages or to stockpiles.

Apron feeders and grizzles

Rock materials from the quarries are delivered for primary crushing in sizes depending upon the mode of transport. Two types of heavy-duty equipment required to receive and feed the materials to the crushers are usually used:
- Vibrating plate feeder
- Apron feeder

Vibrating plate feeder

The vibrating plate feeder comprises: (i) feeder pan and (ii) vibrating unit. The feeder pan comprises: (a) heavy steel plate, (b) side liners, and (c) hopper. The feeder pan is mounted on heavy-duty coil springs. The vibrating unit comprises rotating shafts. The shafts' counteracting and direction of motion generates vibration alternatively lifting and conveying materials for feeding crushers. This type of feeder is generally used for materials obtained from gravel pits.

Apron feeder

The apron feeder comprises: (i) continuous chain formed by a series of pans, (ii) steel idler rollers, (iii) a head roller linked through worm reduction gearing, and (iv) an electric or diesel motor. The rate of feeding is controlled by the gear system. The apron feeder is used for feeding quarried materials.

Grizzles (Scalping Unit)

A grizzle comprises a spring-mounted frame with strong parallel bars forming the top deck and robust perforated plate for the lower deck. The unit has a vibrating mechanism. The whole unit is slightly inclined downwards toward the crusher. A grizzle is generally placed at the end of the apron feeder. It is used mainly to screen out materials smaller than the crusher setting. In case of a double tier unit, the top deck is used to divert oversize materials away from the crusher. The proper size materials are fed into the crusher. The undersize materials are fed into the crushers of smaller settings.

Screens

Once primary crushing is done, the discharged rock materials are either conveyed for further crushing or stockpiled according to sizes on screening by using double or triple deck screening. The screens are generally of vibrating type. However, revolving drum types are also used sometimes.

Inclined vibrating screens

Wire meshes of regular sizes are used to sort out different sizes of rock materials for transferring to the storage bins. Screens may be arranged in tiers for screening progressively to smaller sizes. In case of inclined vibrating screens, the whole unit is set on an incline. The materials rolling down each deck are transported for storage in bins. A rotating robust eccentric shaft effects the vibration in screens.

Horizontal vibrating screens

The horizontal vibrating screens are similar to the inclined vibrating screens without the inclination. However, two counter rotating eccentric shafts generate vibration here.

Revolving drum screen

The perforated revolving drum rotates on an inclined axis. The sizes of the perforations increase in stages along the length of the drum with a part of the drum having solid shell without perforations. The smaller size materials are screened first into an outer drum. The solid portion of the drum prolongs sieving time of coarse materials. The oversize materials are discharged into the open at the lower end.

Washing

The crushed rock materials can easily be washed during screening by spraying water over the screens to clean the aggregates of contaminating materials adhering to them. Both coarse and fine aggregates having about 3% contaminants can also be washed separately in washing plants before screening. The method of washing depends on the types of both the materials and contaminants.

The common methods are:

- Simple wash drum—this works like a revolving drum screen. The dirty water is discharged from the end.
- Screw classifier—it comprises a feed box, settling hopper and rotating screw set at an incline. This type of classifier is used for pit sands. Water and sand (fine aggregates) are continuously fed into the feed box. Wastewater is drained out. The rotation speed of the screw classifier must be slow in case of fine materials, as otherwise very fine materials would be washed out.
- Scrubbing drum—it comprises a rotating drum fitted with paddles, spiral blades etc. The aggregates are fed at one end and collected at the other end. The water is forced into the drum from the opposite end. The whole unit can be horizontal but placed usually at an inclination.

For production of concrete of uniform quality, each batch of materials should be like every other batch to be produced and placed for a particular construction work. This would require weight as well as quality and characteristics of all the ingredients not vary from batch to batch.

Aggregates produced by crushing rocks should be segregated in more size-wise ranges and stored separately. More number of sieves should be used during screening of aggregates to have more ranges of aggregate materials. More storage bins in a batching plant would mean more uniformity in continuous batching.

The procurement sources of the ingredients should be the same for production of concrete of uniform quality. This would require procurement from a single source for a particular ingredient. The sole intent of single-source procurement of each individual ingredient of concrete would ultimately be uniform batching that would ensure more consistent strength of concrete when tested. Special attention is necessary to procure cement from a particular source only.

Apart from uniform size-wise grading, moisture content in the aggregates should also be uniform. Stockpiles of aggregates particularly fine aggregates (sands) should be drained for a day or two to assure uniform moisture content.

Where supplied in bulk, cement can be stored in storage bins. For concrete production, cement can be batched by weight and fed into the mixer drum. Manufacturers very often supply cement in sacks. Sacked cement is subject to weight variation, which could jeopardize quality of concrete produced with sacked cement.

Concrete Production

Concrete batching or mixing plants are of several types. The points to be kept in view while selecting the plants for concreting work are:

- Batching
- Mixing
- Transporting/conveying
- Placing

The following points should be taken into consideration while selecting the type of batching plant or mixer required:
- Maximum hourly output in m^3 as required/planned
- Target total output in m^3 as required/planned
- Mode of transporting concrete
- Discharge height of batching plant/mixer compatible with the mode of transport

There are three general types of mixers:
- Plant mixers are generally mounted as a stationary part of a concrete batching plant
- Paving mixers mounted on wheels or tracks are designed to mix the ingredients while in motion along the roadways so that the mixed-product can be placed when ready
- Truck mixers or transit mixers are so designed as to allow the ingredients to mix while the mixers are in transit to the site of placing

Drum mixers

The drum mixers are stationary mixers comprising either tilting or non-tilting drums. The tilting type comprises a slightly conical shaped drum with single opening for both receiving the ingredients as well as discharging the produced concrete. The ingredients are loaded from one side, and the concrete produced is discharged from the opposite side by tilting the drum by 50° to 60°. The drum may be horizontal or tilted slightly upward during loading and mixing. The drum mixers can accommodate up to almost 3 m^3 per batch producing concrete from 4–90 m^3/h. The helical baffles that are fitted inside the drum move the ingredients from the feeding end to the bottom. Reversing the direction of rotation of the drum results in discharging of the produced concrete by counter action of the baffles. Drum rotation at 10 rpm is normal.

The non-tilting type stationary rotating drum is designed with two openings. The ingredients are fed by means of chute from one side and discharged from another side without reversing the direction of rotation of the drum. Concrete is discharged through gates or by chutes that direct the concrete from the rotating drum. However, the discharge time in this type is longer compared to the tilting type.

Pan mixers

The pan mixer is mounted on a vertical shaft with attached blades. Loading of the ingredients is done from the top and discharging the product through a sliding trap in the bottom along a chute down to waiting transport. In some models, there are rotating blades and rotating mixing compartment. Mixing is done with blades rotated within the stationary pan at 14–30 rpm, which on some models is rotated counter to the blades. Pan sizes vary from 0.14–2.8 m^3 producing 4–100 m^3/h. For application on producing low slump concrete, output of concrete is low while power consumption is high. This type of mixers is used in shops for making concrete products.

Batching plants

When a large quantities of concrete need to be produced, a central mixing plant with storage bins of adequate capacities is usually installed. The plant must be accessible for transportation of both ingredients for batching and placing of concrete produced at the desired location.

The batching plant comprises:
- Storage bins
- Weigh hoppers
- A central plant mixer

A plant that has a central plant mixer is referred to as a central mix plant. Where the aggregates, cement and water are batched and then discharged into transit-mix truck for mixing in transit to the site of placing, the plant is called as transit-mix plant. In case water is added in transit from a tank fitted in the truck-mixer or water is mixed at the job site, the plant is known as a dry-batch plant.

The batching of the ingredients, mixing and then placing concrete are issues all related to material handling. The ingredients like aggregates, cement, water and admixture need to be delivered to the batching plant directly or stored for later transfer to the plant's storage bins. The ingredients are delivered to the mixer after batching. The concrete produced is delivered to the job site by truck, transit mixer, concrete bucket, conveyor belt, chute etc. Aggregates are delivered by trucks, rail wagons or on barges. Cement is delivered by trucks or rail wagons. Water is pumped from the nearest available sources. Aggregates may be stored in stockpiles and then transferred to the storage bins by using conveyor belts, bucket elevators or clamshell buckets. Aggregates may also be delivered directly to the storage bins if conveying system is readily available for the purpose. Cement is transferred to watertight silos/bins by pneumatic pumping or by using screw-conveyor or bucket elevator. Water and liquid admixtures may be pumped to storage vessels.

Batching plant storage

The storage bins are available in almost all conceivable size and shape. The bins may be square, rectangular, hexagonal, octagonal or round in cross-sectional shape. There may be as few as a single or as many as sixteen or more aggregate compartments. For cement also, there may a single or as many as six or more compartments for different types of cement. These compartments may be used for storage of pozzolonic materials also. Scales are fitted with each bin or hopper for automatic control of weigh batching using load cells. The weighing hoppers are properly sealed and vented to preclude dust during operation. The batching plant should be equipped with a suitable batch counter that would correctly indicate the number of batches proportioned. The batch counter would correctly indicate the number of batches proportioned as the batch counter cannot be reset.

Paving mixers

Paving mixers are mounted on tracks. Mixing takes place as the mixers move on the tracks. There are generally two or three drums mounted on a single shell. The dry batch of cement and aggregates is discharged into the first drum from a skip, which is loaded with the dry batch from a truck-mixer. The water and admixture are added from containers mounted on the paving mixers. Both water and admixture are batched volumetrically. The ingredients are mixed in the first drum and then transferred to the second drum. The empty first drum is immediately loaded with another dry batch of cement and aggregates. When concrete is produced on proper mixing of the ingredients, the product is discharged into a bucket, which is supported by a boom that forms part of the paving mixer. This bucket is moved out laterally for dumping the concrete on the ground. A mixer timer with a mechanical interlock on the bucket is a part of the mixer. The batch size varies from 0.96–1.16 m^3.

Truck/transit mixers

A truck or transit mixer comprises a drum mounted on a truck or a trailer. The ingredients in the drum are either mixed or agitated while the truck or trailer is moving. Agitation is necessary to avoid segregation of the constituent ingredients of the mixed concrete.

There are three types of truck/transit mixers:

- Inclined-axis revolving drum
- Horizontal-axis revolving drum
- Open-top revolving blade or paddle mixer

The inclined-axis type is the most common. The concrete produced is discharged by reversing the direction of rotation of the drum as is done in the case of tilting drum mixer when the blades/baffles tend to screw the concrete to the drum opening. Truck mixers may be used for total mixing of the ingredients thus producing concrete. This may be done at the yard before the mixer moves to the job-site for concrete placing. The mixing may be carried out at the job-site also in which case water is added before mixing is started. The mixing may also be done en route to the job site. Truck mixers are generally used to complete mixing when a central plant mixer is used to mix the concrete but not totally. When the truck/transit mixers are deployed to transport concrete from central plant mixer, they are generally meant to operate as agitating units against possible segregation. A truck mixer normally hauls about 10 percent more central plant mix concrete over transmit mix concrete.

Mixing speed and agitating speeds of the drum or blades would be different - not less than 4 rpm and not more than 12 rpm in case of mixing, and not less than 2 rpm or not more than 6 rpm in case of agitating. In order to handle the material faster, drums are often operated at higher speeds while charging or discharging. Concrete should be mixed up to 70 to 100 revolutions after loading all the ingredients in the drum including the mixing water. The mixer should be rotated a minimum of 30 revolutions at mixing speed if any water is added after the concrete has been mixed. Any additional rotation of the drum should be at agitating speed. However, the overall total number of revolutions should not exceed 250.

Concrete Pumps

Pumping concrete is easier, quicker, cheaper and better method compared to other methods of placing concrete. A concrete pump can place a lot of concrete in a short period of time both horizontally and vertically. And deployment of concrete pump would allow other lifting and handling construction equipment to be utilized for more arduous work. If concrete is not of specified quality with adequate cement content, it may overstrain the pump, cause extra wear and tear and even create a blockage. If pump operation is all right otherwise, blockages occur mainly due to the following two reasons:

- Water is forced out of the concrete mix resulting in bleeding and consequent blockage by jamming
- Very high frictional resistance due to the nature and quality of the ingredients in the concrete mix

The above two reasons may be attributed to:
- Defective concrete mixes
- Failure of the mixer
- Defective pipelines
- Inadequate or over-wet lubricating layer

To pump concrete, the pressure developed by the pump has to overcome:
- Frictional resistance of the pipe line wall
- Resistance created at bends and tapers
- Pressure head when concrete is to be placed at a level higher than the pump discharge level
- Inertia of concrete in the pump

The pump output depends on several factors:
- Horizontal pipe length
- Vertical pipe length
- Diameter of delivery pipeline
- Length of flexible hose
- Reducers in pipeline in case of changes in pipe diameter
- Workability of concrete
- Cohesiveness of concrete
- Types of aggregates used in concrete

Concrete pumps are available in many sizes and forms:
- Diesel, petrol, or electrically powered
- Stationary or mobile
- Mounted on skids, trailers, trucks or truck-mixers

- Valves of different designs
- Placement booms of different designs and shapes—the reach of boom may be up to 600 m or more
- Hand operated distributors—better way of manual placement of concrete compared to using rubber hose
- Sturdy design of pumps for heavy handling and rough usage at construction sites, and easy maintenance

Pump operation

- A steady supply of pumpable concrete is necessary for effective pumping. Pumpable concrete is produced by consistently batching and mixing properly graded uniform aggregates and other ingredients thoroughly so that there is no variation in mix consistency and workability. In other words, pumpable concrete should be such that it can be pumped without segregation and bleeding. Cement content should be correct. Pumping would not be possible if concrete is unduly harsh, inadequately mixed, non-cohesive and not right in consistency. Natural gravels produce better pumpable concrete mixes as they are generally rounded and a wide range of sizes is available compared to crushed rock aggregates.
- Before commencement of pumping, a grout mixture should be flushed through the pump and hose/pipeline to provide a thin lubricating layer of mortar. Initially inserting a plug into the pipeline and then forcing it out with the grout achieve best results.
- An agitator helps to maintain the concrete in agitated form in between two batches of concrete fed into the hopper. The concrete in the hopper needs to be agitated to keep the concrete moving and the concrete level must be constantly topped up to prevent air pockets forming in the pipeline.
- Pumpable quality of concrete means that its consistency should be such as to allow its pumping without segregation and bleeding of concrete. Pumpable concrete has good cohesion and high workability. Design of pumpable concrete is much specialised. Pumping tends to be more efficient with concrete having a slump of 50–75 mm with a high percentage of fines. Diameter of placing pipeline should be at least 3 to 4 times of the maximum size of aggregates.
- Pipeline for delivery of the concrete occupies very little space and can be easily extended or removed. Provision of a compressed air supply nearby is useful for clearing obstructions in the pipeline.
- For cleaning out the pipeline, a section of the pipeline is removed and the hopper/receiver is pumped clean. A plug made of sponge ball and a cylinder made from old saturated bags is inserted into the rest of the pipeline. This plug is pushed out using water pressure. Compressed air is potentially hazardous and should generally not be used, as the possibility of concrete emerging at high velocity exists. A catch-basket, which should remain properly supported, should be fitted at the end

of the pipeline. Cleaning of the pipeline is very essential for successful pumping operation.

- When pumping downward 15 m or more, an air release valve at the middle of the top bend should be provided to prevent vacuum or air buildup.

There are four types of concrete pumps:
- Piston type
- Pneumatic type
- Squeeze type
- Hydraulic type

Piston pump

A piston pump (Fig. 13.51) comprises:
- A receiving hopper (0.1 to 1.5 m^3) fitted with agitator for remixing the concrete to maintain consistency and uniformity
- A rotary inlet valve—located under the hopper—does not totally open
- A rotary outlet valve—located at the discharge line
- A horizontal cylinder
- A single-action piston—300 mm stroke—about 48 strokes per minute
- A flywheel or crank or chain drives the pistons—basic power is supplied by diesel, petrol or electric/hydraulic motors

As the piston is drawn back, the inlet valve opens and the outlet valve closes. Concrete flows from the hopper into the cylinder by gravity. Because of granular ingredients of concrete, the inlet valves do not close completely. The piston then pushes the concrete from the cylinder into the discharge pipeline for placing. Most pumps comprise two cylinders so as to ensure continuity of concrete flow. Cylinder diameter varies between 160 to 200 mm and more.

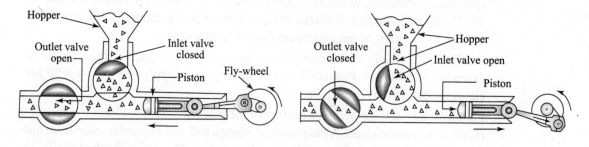

Fig. 13.51 Piston pump

Pneumatic pump

A pneumatic pump (Fig. 13.52) comprises:
- A compressor
- An air receiver tank
- A pressure vessel whose top can be sealed
- A discharge box

A pneumatic pump uses the pushing power of compressed air. Concrete is charged into the pressure vessel (capacity 0.25–1.0 m^3) and the vessel is tightly sealed. Compressed air (pressure 2.5 to 3 bars, 1 bar = 10^5 N/m^2) from air receiver tank is then supplied to the top of the vessel. The resulting pressure developed in the vessel in effect blows the concrete through delivery pipe connected at the bottom into a reblending discharge box to bleed off the air and to prevent segregation and spraying. When the vessel is emptied of concrete, the supply of compressed air is shut off. The vessel is recharged with concrete for repeating the process. For larger output in concrete placing, several vessels are used to maintain a more uniform supply of concrete for placing. Delivery pipes of 150–200 mm diameter are typical and bends must be anchored securely to withstand the pressure of fast moving concrete. Compressors that have minimum capacity of 3.5 m^3/min are used for supply of compressed air. Roughly, a compressor must be capable of supplying compressed air approximately 15 times the volume of concrete to be pumped.

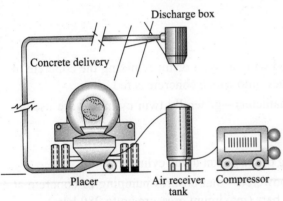

Compressor builds up air pressure in tank which forces concrete in placer through the line

Fig. 13.52 Pneumatic pump

Squeeze pressure pump

A squeeze pressure pump (Fig. 13.53) comprises:
- A receiving hopper with three remixing blades
- A flexible hose is connected at the bottom of the receiving hopper
- A metal drum, which is maintained under high vacuum
- Rotating type rollers

The flexible hose connected at the bottom of the receiving hopper enters the metal drum at the bottom and secured half circle around the inside periphery of the airtight drum and emerges at the top. Hydraulically powered rollers successively flatten the flexible hose and squeeze the concrete out at the top. The operation takes place in a vacuum at 0.8–0.9 bar and thus the hose regains its shape. A steady supply of concrete in the tube from the receiving hopper is thus maintained. The hose size is limited to 75–100 mm in diameter.

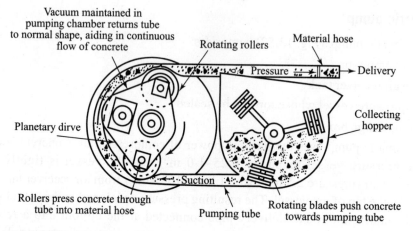

Fig. 13.53 Squeeze pressure pump

Hydraulic concrete pump

A hydraulic concrete pump comprises:

- A receiving hopper, which is fitted with an agitator for remixing the concrete to maintain uniformity and consistency, into which concrete is fed
- A cylinder (150 to 230 mm in diameter)—generally twin cylinders are used—strokes 600–2100 mm
- A piston in each cylinder
- A control valve to control working of the piston in the cylinder
- A drive unit—diesel or electric powered—that ensures pumping of concrete at pressure ranging from 100 to 200 bars (maximum pressure up to 280 bars)
- Delivery pipeline/placer boom

Hydraulic pressure pushes the concrete out along the pipe for placement. During the return stroke on release of the pressure, concrete is sucked from the concrete hopper. Hydraulic pump is powered by diesel or electric motors. Oil is the working fluid that is used to operate one or more pumping cylinders. Twin cylinders are more commonly used. Pumps with output of 200 m^3/hr or more are now available.

Hydraulic pumps have the following advantages:

- The pump design is based on long piston stroke so that fewer strokes are required for concrete placing resulting in reduced frictional resistance in conveying concrete
- The pressure exerted on concrete may be regulated to control vertical lift by pumping
- Acceleration of the piston at the start of the delivery stroke and then sustaining steady and even flow ensures smooth discharge through the pipeline

The twin cylinder hydraulic concrete pumps are capable of 200 m^3/h output through a hose of 150 mm diameter. Pumping so much concrete would cause handling problems during placement. In practice, therefore, 100 mm diameter hose is used for an output of

35–45 m³/h. The control system regulates pumping output by varying oil flow and pressure cutting down output to 20–25 m³/h or even less. Hydraulic pumps can be deployed to place concrete over a distance of 600 m and to heights of 380 m or more.

Pipeline

Most concrete is pumped to the placement areas through rigid pipe or a combination of rigid pipe and heavy-duty flexible hose connected by means of detachable couplings. Rigid pipeline supplying concrete is made from seamless pipes of high quality precision steel, aluminium or plastic and is available in sizes from 80 to 200 mm in diameter. 100 mm diameter pipes are widely used. Aluminium pipeline is fraught with the problem of possible reaction with alkalis in Portland cement and, therefore, is not used. Steel pipeline, which is preferred for pumping concrete, has much better abrasion resisting property during conveying of concrete. Rigid pipe is furnished usually in 3 m lengths so that a person can handle it. The twin cylinders of the concrete pump are connected to the pipeline through a Siamese twin or Y pipe section. A Siamese pipe section at the pump outlet is made up of a taper wedge and a clamping device for connecting it to the pipeline. Selection of pipe diameter depends on the following factors:

- Quality of concrete mix to be pumped with regard to its consistency
- Maximum size of aggregates—pipeline should be at least 3 to 4 times the maximum size of the coarse aggregates
- Size and type of concrete pump
- Horizontal length and vertical height involved in placement of concrete
- Smoothness of inside of the pipeline

Flexible pipeline (hose) is made of rubber or spiral wound flexible metal or plastics. Compared to the rigid pipeline, flexible pipeline presents more resistance to the concrete flow. Flexible pipes are interchangeable with rigid pipe sections. This allows use of flexible pipes in difficult placement areas. In general, 100 mm diameter is the largest flexible pipe size that can be handled by a person.

The total effect of bends, vertical rise and flexible hoses on an installed pipeline is converted in terms of equivalent horizontal pipe length. This co-relation for over 100 mm internal diameter pipeline is as follows:

- A 90° bend   = 10 m horizontal length
- A 45° bend   = 5 m horizontal length
- A 22.5° bend   = 3 m horizontal length
- 1 m vertical rise = 6–8 m horizontal length
- 1 m rubber pipe = 3–6 m steel pipe

Vertical pumping up to 382 m height was carried out in Jin Mao building in China. The longest horizontal distance concrete pumped is 2015 m in Le Refrain tunnel in France.

Available pipeline accessories are:

- Curved sections of rigid pipe

- Rigid and flexible sections in varying lengths
- Pin and gate valves to prevent backflow in the pipeline
- Switch valve to direct the flow into another pipeline
- Swivel joints and rotary distributors
- Connection devices to fill forms from the bottom up
- Splints, rollers and other devices for protection of conduit over rock, concrete, reinforcing steel, and forms, and to provide lifting and lashing points
- Extra strong couplings for vertical runs, and exposed and inaccessible areas
- Transitions for connecting different sizes of pipe
- Air vents for downhill pumping
- Cleanout equipment

Couplings

The couplings that are used to connect both rigid and flexible pipe sections must be strong enough to withstand handling during pipeline installation, misalignment and inadequate supports. Coupling types vary with the size of pipe being used, the larger sizes requiring considerably heavier fittings. Couplings should be designed to allow replacement of any pipe section without moving other pipe sections, and should provide a full internal cross-section with no constrictions or crevices to disrupt the smooth flow of concrete. The couplings in use are of four types:

- Quick action pipe coupling tightened by driving in two wedges
- Quick action pipe coupling (hoop coupling) tightened by pivoting a lever
- Toggle-type pipe coupling (split coupling) which is most commonly used
- Screw coupling

The snap-on type coupling is provided with a rubber gasket to ensure that there is no leakage of slurry.

Placer boom

Placer booms of two types are available:

- Truck-mounted and coupled with the pump on the same truck
- Tower mounted

A placer boom can be easily utilized for placing concrete in columns, deep walls, slabs and beams. The use of a placer boom facilitates better and faster placing of concrete as well as reduces dependence on manual efforts in shifting the flexible hose and adjusting delivery lines time and again.

Using a tower crane (Fig. 13.55), the placer booms can be lifted and placed on a tower mast projecting from the floors. Larger size booms can cover a distance of about 50 m slewing 360° completing a circle. They can reach straight down, straight out and at all points without missing a spot. The operation can be remote controlled from the base.

Mechanized Construction 491

50 m truck mounted placer booms (Fig. 13.54) are presently available with quick detaching arrangement. To disconnect this type of boom from the truck, only thing that needs to be done is to remove the four metal pins. Booms of these types can be mounted, removed and remounted very fast using a crane.

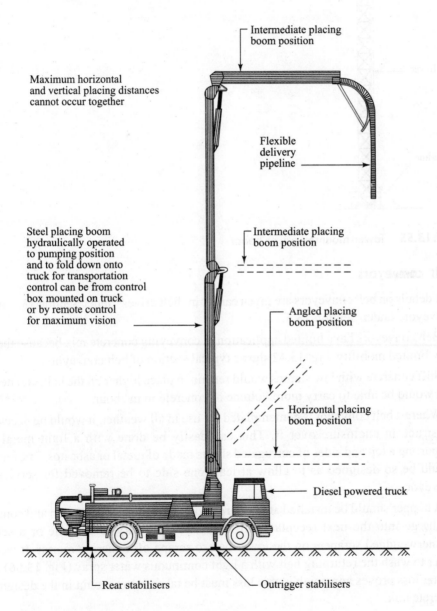

Fig. 13.54 Truck-mounted placer boom

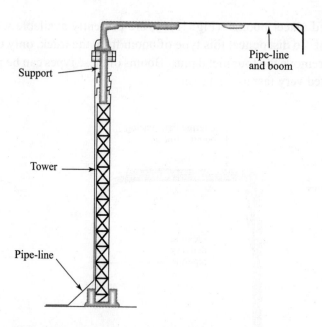

Fig. 13.55 Tower mounted placer boom

Belt conveyors

The details on belt conveyors are given earlier in 'belt drives' under Section 13.2 and 'belt conveyor' under Section 13.5.

Belt conveyors have limited application in conveying concrete mix because they have very limited mobility. Fig. 13.42 shows typical section of belt conveyors.

Stiff concrete with low slumps would remain in place higher on the belt, and hence the belt would be able to carry more volume of concrete in m^3/hour.

Where a belt conveyor is contemplated for use in all weather, it would be necessary to construct an enclosure over it. This can easily be done with a light metal frame supporting a top and sides of corrugated sheets made of metal or asbestos. The enclosure should be so designed as to allow at least one side to be removed for servicing the conveyor.

A hopper should be installed at discharge points to reduce segregation and control the discharge into the next receptacle. It would be necessary to use one or a series of counterweighted scrapers on the lower belt to scrape off adhering mortar. It would be better to wash the returning belt with a light continuous water spray (Fig. 13.56). If the mortar loss proves appreciable, this loss must be taken into account in the design of the concrete mix.

In such cases as construction of heavy structures like dams where large volumes of concrete need to be moved, belt conveyors have proved a good solution on overcoming the problem of moving concrete from the batching/mixing plant to a central distributing plant.

Mechanized Construction

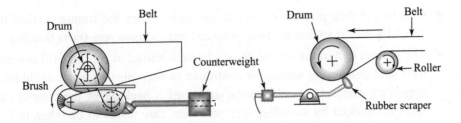

Fig. 13.56 Belt conveyors–cleaning method

Tremie

Placing of concrete under water becomes unavoidable in many cases, although placing of concrete in open air is always preferable. For example, the sealing of the bottom of a cofferdam requires placing of concrete under water. This would pose the problem of loss of cement and saturation of concrete to a minimum. This problem is minimised simply and economically by using tremie (vide Chapter 7, Section 7.17). The tremie is a watertight straight vertical pipe of 150–350 mm diameter (250 mm diameter is a common size) having a funnel at the upper end into which the concrete is fed. It must be watertight, smooth and sufficiently long to reach the place of concreting from above the level of water. Tremie pipes are used in the construction of piled foundations, basements and also diaphragm walls.

The tremie (Fig. 13.57) must be supported to allow it to be raised with the progress of work. It can be hung from a crane or supported from a frame fitted above the water level. Several tremie pipes may be used depending on the volume of work involved. Concrete from the tremie would spread up to about 3 m (approximately 30 m^2) and placing of concrete should be planned accordingly.

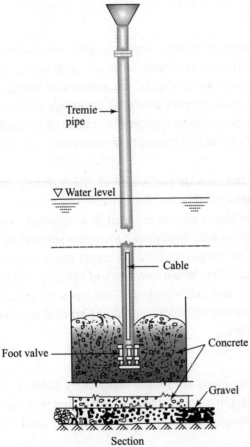

Fig. 13.57 Tremie

To start placing of concrete:

- The bottom of each tremie should be closed with plastic or cement bags.
- The bottom end of the tremie is positioned over the location where concrete is to be placed and lowered till the bottom end rests on the formation level.

- Concrete is then poured on continuous basis. When the tremie is filled up, it is carefully raised so that the plug is forced out and concrete starts flowing.
- The bottom end of the tremie should be kept buried in concrete till concreting is over, and concreting should be continued to keep air and water excluded. Once deposited, concrete should not be disturbed. Checking of the concrete surface level is checked by sounding and sufficient care needs to be taken to keep the concrete surface horizontal so as to avoid entrapping of underwater debris.
- As the depth of placed concrete increases, tremie length is adjusted by removing pipe sections as required.
- All underwater methods result in producing a large amount of laitance on the concrete surface. All laitance should be removed after dewatering before more concrete is deposited.

13.7 SCAFFOLDING

Scaffolding (or scaffold) is temporary erection of timber or steel structures to facilitate construction, alteration or demolition of structures or to allow hoisting, lowering or standing/supporting of materials, men etc at heights. In construction work, scaffolding may be deemed as 'falsework' as such support would be necessary till the permanent structure becomes self-supporting. Scaffolding is generally erected of steel tubes clipped or coupled together and adequately laced and braced to make strong, stable and safe for working on the platforms.

Scaffolding is erected along the outside perimeter of the building or structure preferably on level ground. Pressure on soil is distributed evenly by placing each upright tube/pole on base plates over sole plates so that the load is distributed evenly on the ground. The top of the upright is fitted with head or cap plates for supporting formwork for slab or platform. A single row of upright scaffold is erected partly supported by the structure under construction for stability. When two rows of uprights are erected, the scaffolding becomes independent and self-standing. Two rows of uprights can be supported on rails for mobility. How the scaffolding would be erected would depend on the planning of construction activities.

Single row of uprights

A single row of upright scaffolding pipes are joined together by ledgers. Putlogs are connected by the ledgers and masonry walls under construction for stability. The platform thus made for raising masonry walls would need a ladder to reach the platform (vide Fig. 13.58).

Double rows of uprights

In independent scaffolds with double rows of uprights (Fig. 13.59), ledgers connect the two rows of uprights. Transoms connect the ledgers in the transverse direction. The scaffolding structure under construction is further strengthened by cross-bracings as

Mechanized Construction 495

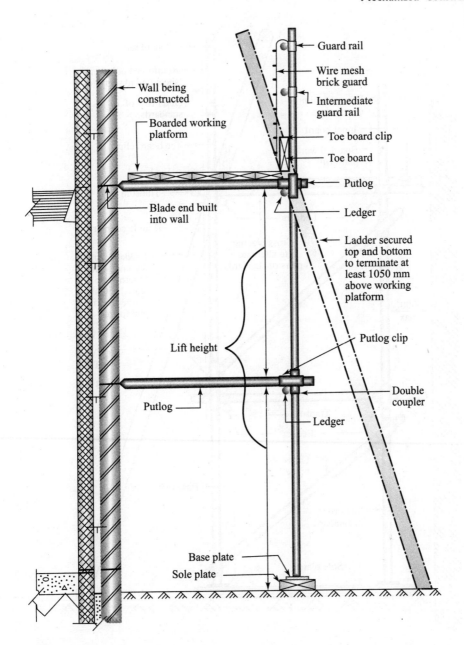

Fig. 13.58 A single row of uprights

required. The working platform is made accessible using a portable ladder, which is secured very carefully both at the base as well as at the top.

Truss-out scaffold

In truss-out scaffolds (Fig. 13.60), the platform frame is made independent of the building or structure under construction, although supported entirely on the completed

496 Construction Technology

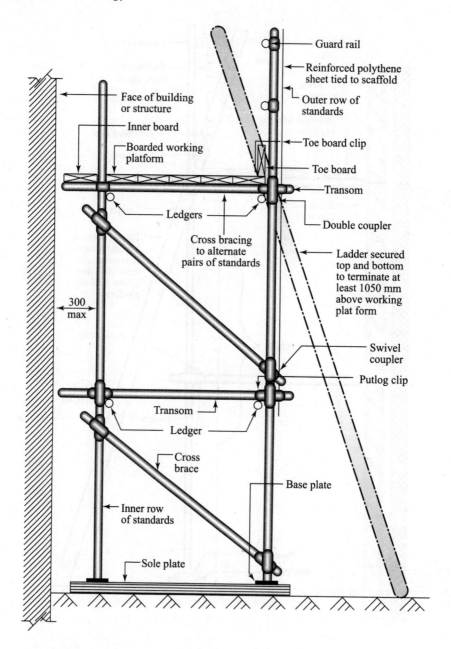

Fig. 13.59 Double rows of uprights

building or structure. This type of scaffold is used where erection of conventional scaffold from the ground level is not desirable. This type of scaffold should be erected only when highly skilled personnel required for the job are available at the site.

Mobile scaffold

In mobile scaffolds (Fig. 13.61), the scaffold is fabricated in the form of tower that is mounted on wheels. The working platform (minimum size 1.2 m × 1.2 m) can be easily

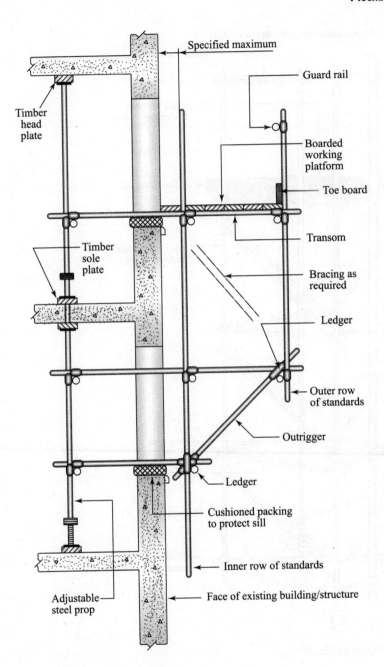

Fig. 13.60 Truss-out scaffolds

moved on wheels to different working positions mainly for work on the ceilings or maintenance work. A ladder is fitted on one side of the tower for making the platform accessible. A mobile tower should be accurately designed and properly braced and its maximum height should be less than 10 m.

498 Construction Technology

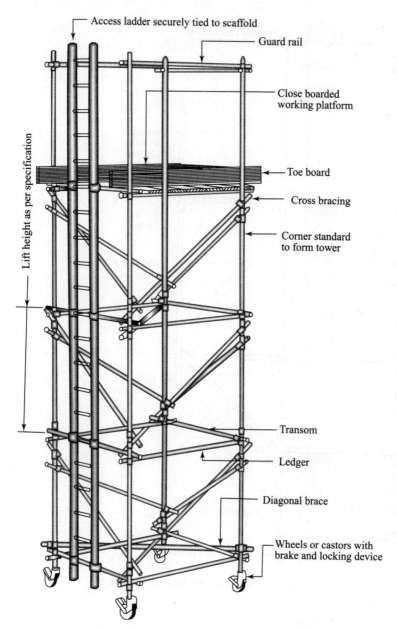

Fig. 13.61 Mobile scaffolds

Suspended scaffolds

In slung scaffolds, a working platform is fabricated in conventional manner. The platform comprises ledgers, transoms, timber boards, guard rails and toe boards. The size of the platform should be less than 2.5 m × 2.5 m. This type of scaffold is suspended from the main structure by means of wire ropes or steel chains. The platform should not be held in position by using less than six evenly spaced wire ropes or steel chains securely anchored at both ends.

In suspended scaffold, a cradle like working platform is suspended from cantilever beam or outrigger for temporary access to the face of a building for carrying out cleaning or maintenance work at heights. A suspended scaffold should be made of platform boards, guard rails and toe boards in conformity with the safety standards. The cradle may have manual or electrical power control.

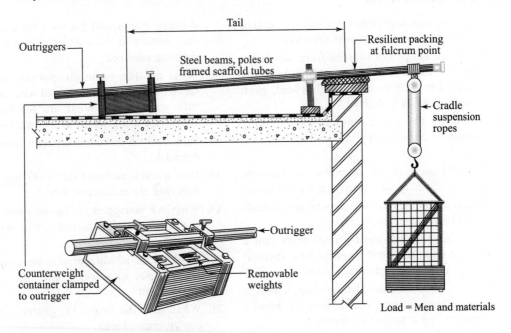

Fig. 13.62 Suspended scaffolds

There are other types of scaffolds to suit different purposes. Erection of timber or bamboo pole scaffolds is common in India especially for masonry and painting work.

SUMMARY

Innovative technology has now made onetime arduous construction activities not only easily manageable but also in achieving improved quality products in less time. Construction equipment is being modernized by incorporating precision instruments and software. Construction materials are producing high performance products by using continuously improved additives that are the results of worldwide research. There is no alternative to adopting mechanized construction methods to achieve more in less time by planning larger outputs.

Earthmoving is bulky work and is increasingly being mechanized in India. The range of earthmoving equipment available worldwide is vast and such equipment can be deployed suitably in all kinds of excavation and filling, both on-shore and off-shore. Details of earthmoving equipment generally deployed in construction projects are given in this Chapter.

Concrete is the most widely used construction material. In a construction project, if aggregates are brought to the site using belt conveyors and concrete is placed by pumping from below the central batching plant, it could be a continuous process like assembly like production in industries. Details of all sorts of equipment, starting from crushing aggregates to concrete production and placing are included in this chapter. In addition, plants for transportation, movement, and handling, as well as falsework such as scaffolding, form a part of this chapter.

REVIEW QUESTIONS

1. Why is mechanization based on rented construction equipment cost effective?
2. What are the power sources from which the power transmitted by a rotating machine component such as a shaft, flywheel, gear, pulley, or clutch is derived?
3. There are two basic types of earthmoving equipment. Name them giving examples of each type.
4. Tyres of construction equipment are required for specific purposes. What are the purposes?
5. Describe how the track of a crawler unit is made up and how the track is driven.
6. What are the five basic components that a tractor comprises of? Explain how thermal energy is utilized in driving a tractor.
7. What is the definition of bank volume? What is the difference between a scraper bowl's capacity and a scraper's capacity?
8. What is the difference between a tractor and a bulldozer?
9. A bulldozer can be used for ripping soft rock. What are the requirements of ripping action?
10. What are the advantages of self-propelled scraper over towed scraper (towed by tractor)?
11. Cycle time of shovel comprises four operations. What are these operations? A backhoe is a variation of a face shovel. What is the variation?
12. "A grader performs what a bulldozer can do, but more accurately and smoothly." Elaborate this statement.
13. How is a ladder type trenching machine different from a wheel type one?
14. Explain briefly how different types of compacting equipment are classified into four distinct ways.
15. As the cutter head of a tunnel boring machine revolves as a single unit, there would be difference in speed between the cutters near the centre and those at the rim. How is this problem resolved?
16. What are the basic components of lifting tackle? Discuss how the metal wires are turned into wire ropes? What are the sources of energy for using winch as a power unit?
17. How a scotch derrick is different from a guyed derrick?
18. How a track-mounted (crawler) crane can be deployed for excavation work?
19. Why jack outriggers are incorporated in larger capacity wheel-mounted crane with strut boom?
20. What are the advantages of tower cranes? How many types of tower cranes are there for deployment?
21. What are the uses of gantry cranes in construction work?
22. What are the advantages of static and mobile hoists?
23. What kind of materials can be quickly or efficiently handled by forklift trucks?
24. Discuss the types of construction work in which cableway can be useful. What are the most common types of cableways?
25. There are many types of crushers used for primary, secondary, and tertiary crushing. Name the processes used for each type of crushing.
26. What does a concrete batching plant comprise of? Discuss how a truck or transit mixer is used for mixing and transporting concrete at the same time.
27. What are the salient points in concrete pumping operation?
28. Discuss how concrete is placed under water.
29. Why is scaffolding (or scaffold) deemed as falsework?

14
Quality Control and Assurance

INTRODUCTION

Customers would feel assured about the quality of work undertaken and services rendered if quality remains integrated into the system. Projects involve a number of agencies which execute the construction work and/or render services. If different agencies pursue different quality management systems, it would be difficult for them to satisfy the customer about the quality of work executed and services rendered.

A formal set of standards, that is internationally recognized, is always preferable to customers as well as executing agencies compared to bespoke quality systems. ISO 9000:2000 provides a unique set of standards; it does not deal with any particular product but assesses a quality system as a whole stressing on continual improvement. The standard recognizes that a company cannot function effectively and prove that it has been doing so unless there are clearly kept records of what has been done and the procedure followed for doing that.

14.1 DEFINITIONS

Previously, quality was defined by the engineer. At present, quality is what the client wants and depends on the knowledge base of the workers right up to the lowest level which would require change of the mindset of both workers and managers. Quality, therefore, should be integrated into the system and not enforced from the outside. That makes ISO 9000 quality system very relevant. Quality can be sacrificed neither stage-wise nor totally.

Process

Process is a set of inter-related or interacting activities that transform inputs into outputs. Inputs to a process are generally outputs of other processes. The processes in an organization are generally planned and carried out under controlled conditions to add value. A process where the conformity of the resulting product cannot be readily or economically verified is frequently referred to as a 'special process'.

Product

Product is the result of a system of activities which uses resources to transform inputs into outputs. Resources may include personnel, finance, facilities, equipment, techniques, and methods. There are four generic product categories: (i) services such as transport, (ii) software such as computer programme, (iii) hardware such as engine mechanical part, and (iv) processed materials such as lubricant. Many products comprise elements belonging to different generic product categories. The product would then be known on the basis of the dominant element.

Procedure

Procedure is specified as the means of executing an activity or process. The procedure can be documented or not documented. When a procedure is documented, it is referred to as 'documented procedure' or 'written procedure'. The document that contains a procedure may be called a 'procedure document'.

Quality

Every product, system, or process has a set of inherent characteristics. These characteristics need to fulfil the requirements of the customers and other interested parties. The degree to which such inherent characteristics fulfil the requirements determines the quality—poor, good, or excellent.. 'Inherent' implies existing as a permanent characteristic. The requirements in project construction include conformance to specifications, standards, codes, and drawings.

Grade

Grade is the category or rank given to different quality requirements for products, processes, or systems having the same functional use. 'Regrade' is related to 'nonconformity'.

Customer

Customer is an organization or person that receives a product. The terms consumer, client, end-user, retailer, beneficiary, and purchaser stands for customer. A customer can be internal or external to the organization.

Customer satisfaction

Customer's perceptions of the degree of fulfilment of the requirements and expectations—the requirements are specified, implied, or obligatory. A transaction is time- and event-specific. Customer complaints are a common indicator of low customer satisfaction but their absence does not necessarily imply high customer satisfaction. Even when the customer requirements have been agreed with the customer and fulfilled, this does not necessarily ensure high customer satisfaction.

Management

Management means a person or a group of people with authority and responsibility for the conduct and control of an organization.

Top management

Top management comprises a person or a group of people who direct and control an organization at the highest level. Through leadership and actions, the top management can create an environment where people are fully involved and in which a QMS can operate effectively.

Quality management

Quality management stands for coordinated activities to direct and control an organization with regard to quality. The sense in which 'management' is used becomes clear when some form of qualifier is prefixed before it, for example, 'quality' in this case.

Quality management system

Quality Management System (QMS) is responsible for directing and controlling an organization with regard to quality and is that part of the organization's management system that focuses on the achievement of results, in relation to the quality objectives, to satisfy the needs, expectations, and requirements of interested parties, as appropriate.

Interested party

A person or group having an interest in the performance or success of an organization—customers, owners, people in an organization, suppliers, bankers, unions, partners, or society. A group can comprise an organization, a part thereof, or more than one organization.

Quality policy

Quality policy presents the overall intentions and direction of an organization related to quality as formally expressed by the top management. Quality management principles in the ISO 9000 standard can form a basis for the establishment of quality policy. Quality policy provides a framework for the setting of quality objectives.

Quality objective

Quality objectives are generally specified for relevant functions and levels in the organization related to quality based on the organization's quality policy and the commitment to continual improvement, and their achievement needs to be measurable.

Quality planning

Quality planning forms part of quality management focused on setting quality objectives and specifying necessary operational processes and related resources to fulfil the quality objectives. Establishing quality plans can be part of quality planning.

Quality plan

Quality plan is a document specifying which procedures and associated resources should be applied by whom and when to a specific project, product, process, or contract. These procedures generally include those referring to quality management processes and product realization processes. A quality plan often makes reference to parts of the quality manual or to procedure documents. A quality plan is generally one of the results of quality planning.

Quality control

Quality control is part of quality management and is focused on fulfilling quality requirements to attain predetermined qualitative characteristics related to materials, processes, and services.

Quality assurance

Quality assurance is part of quality management focused on providing confidence in fulfilling the quality requirements.

Quality improvement

Quality improvement is part of quality management focused on increasing the ability to fulfil quality requirements related to any aspect such as effectiveness, efficiency, or traceability.

Continual improvement

Continual improvement is recurring activity to increase the ability to fulfil the requirements. The process of establishing objectives and finding opportunities for improvement is a continual process through the use of audit findings and audit conclusions, analysis of data, management reviews, or other means and generally leads to corrective or preventive action.

Actions for improvement of a Quality Management System in increasing the probability of boosting the satisfaction of customers and other interested parties include the following:

(i) Understand current and future customer needs, meet customer requirements, and strive to exceed customer expectations
(ii) Leadership should be such as to create and maintain the internal environment congenial so that people become completely involved in achieving the organization's objectives
(iii) People at all levels form the core of an organization and their total involvement enables their abilities to be used for the organization's benefit
(iv) Desired result is achieved more efficiently when activities and related resources are managed as a process
(v) Identifying, understanding, and managing inter-related processes as a system contributes to the organization's effectiveness and efficiency in achieving its objectives
(vi) Continual improvement of the organization's overall performance needs to be a permanent objective of the organization
(vii) Effective decisions are based on the analysis of data and information
(viii) Identifying, understanding, and managing interrelated processes as a system contributes to the organization's effectiveness and efficiency in achieving its objectives

Effectiveness
The extent to which planned activities are realized and planned results achieved signifies effectiveness.

Efficiency
Efficiency is the relationship between the result achieved and the resources used.

Organization
An organization is a group of people and facilities with an orderly arrangement of responsibilities, authorities, and relationships—this definition is valid for the purposes of Quality Management System standards. Company, corporation, firm, enterprise, etc., are examples of an organization, which can either be public or private or even a joint sector enterprise.

Organizational structure
Organizational structure is orderly arrangement of responsibilities, authorities, and relationships between people. The organizational structure is often provided in a quality manual or a quality plan. For a project, the scope of an organizational structure can include relevant interfaces to external organizations.

Infrastructure
Infrastructure system denotes permanent facilities, equipment, and services needed for the operation of an organization.

Work environment
A set of conditions under which work is performed—conditions include physical, social, psychological, and environmental factors like temperature, recognition schemes, atmospheric composition, and ergonomics (vide Chapter 15, Section 15.1).

Supplier
Supplier is an organization or person that provides a product or service or information. A supplier can be internal or external to the organization and can be producer, distributor, retailer, or vendor of the product. Contractually, a supplier is sometimes called 'contractor'.

Project
A project is a unique process comprising a set of coordinated and controlled activities with start and finish dates undertaken to achieve an objective conforming to specific requirements including the constraints of time, cost, and resources. An individual project can form part of a larger project structure. The outcome of a project may be one or several units of product.

Quality manual
Quality manual is the document specifying the Quality Management System of an organization. These manuals can vary in detail and format to suit the size and complexity of an individual organization.

Quality audit

Audits are systematic, independent, and documented process for determining the extent to which the Quality Management System requirements are fulfilled and for identifying opportunities for improvement.

Record

Record is a document stating the results achieved or providing evidence of the activities performed. Records can be used, for example, to document traceability and provide evidence of verification, preventive action, and corrective action. Records are not to be under revision control.

Objective evidence

Objective evidence is data supporting the existence or verity of something and it can be obtained through observation, measurement, test, or other means.

Inspection

Inspection is conformity evaluation by observation and judgement accompanied as appropriate by measurement, testing, or gauging.

Test

Test is the process of determining one or more characteristics according to a procedure.

Verification

Verification is confirmation, through the provision of objective evidence, that specified requirements have been fulfilled. Confirmation could comprise activities such as performing alternative calculations, comparing a new design specification with a similar proven design specification, undertaking tests and demonstrations, and reviewing documents prior to issue.

Validation

Validation is confirmation, through the provision of objective evidence, that the requirements for a specific intended use or application have been fulfilled. The used conditions for validation can be real or simulated.

Qualification process

Qualification process is to demonstrate the ability to fulfil the specified requirements. This can involve persons, products, processes, or systems like auditor qualification process, material qualification process, and so on.

Review

Review is activity undertaken to determine the suitability, adequacy, and effectiveness of the subject matter to achieve established objectives. The review can also include the determination of efficiency.

Audit

Audit is systematic, independent, and documented process for obtaining audit evidence and evaluating it objectively to determine the extent to which audit criteria are fulfilled. Internal audits, sometimes called first party audits, are conducted by, or on behalf of, the organization itself for internal purposes and can form the basis for an organization's self-declaration of conformity. External audits include what are generally termed 'second or third party audits'. Second party audits are conducted by parties having an interest in the organization, such as customers, or by other persons on their behalf. Third party audits are conducted by external independent organizations. When two or more auditing organizations cooperate to audit a single auditee jointly, this is termed 'joint audit'.

Auditor

Auditor is a person with the competence to conduct an audit.

Audit team

An audit team comprises one or more auditor/s, auditors-in-training, and technical experts when required for conducting an audit. One auditor in the audit team is generally appointed as audit team leader. Observers can accompany the audit team but do not act as part of it.

Technical expert

A knowledgeable person in the audit team is a technical expert. This person provides specific knowledge of or expertise on the subject to be audited.

Auditee

Auditee is an organization being audited.

Competence

Competence is the demonstrated ability to apply knowledge and skills.

Traceability

The ability to trace the history, application, or location of that which is under consideration by means of recorded identifications. In case of a product, traceability can be related to: (i) the origin of materials and parts, (ii) the processing history, (iii) the distribution and location of the product.

Conformity

Conformity is the fulfilment of the specified requirement.

Nonconformity

Nonconformity is the non-fulfilment of the specified requirement.

Defect

Non-fulfilment of a requirement is related to an intended or specified use. The term 'defect', compared to non-conformity, has legal connotation and should be used with extreme caution.

Preventive action

Preventive action is exercised to eliminate the cause of a potential nonconformity or other undesirable potential situation in order to prevent occurrence.

Corrective action

Corrective action is taken to eliminate the cause of a detected nonconformity or other undesirable situation to prevent recurrence. There can be more than one cause for nonconformity. There is a distinction between correction and corrective action.

Correction

Correction is action taken to eliminate a detected nonconformity. A correction can be made in conjunction with a corrective action. A correction can be, for example, rework or regrade.

Rework

Action is taken on a nonconforming product to make it conform to the requirements. Unlike rework, repair can affect or change parts of the nonconforming product to make it acceptable for the intended use.

Regrade

Alteration of the grade of a nonconforming product is made in order to make it conform to the requirements that differ from the initial ones.

Repair

Repair is action taken on a nonconforming product to make it acceptable for the intended use. Repair includes remedial action taken on a previously conforming product to restore it for use, for example, as part of maintenance. Unlike rework, repair can affect or change parts of the nonconforming product.

Scrap

Scrap means the action taken on a nonconforming product to preclude its originally intended use. In a nonconforming service situation, use is precluded by discontinuing the service.

Information

Information is meaningful data.

Document

Document is information and its supporting medium. Record, specification, procedure document, drawing, report, and standard are documents. The medium can be paper, magnetic, electronic or optical computer disc, photograph or master sample, or a combination thereof. A set of documents, for example, specifications and records, is frequently called 'documentation'. Some requirements like the requirement to be readable relate to all types of documents; however, there can be different requirements for specifications—the requirement to be revision-controlled and for records – the requirement to be retrievable.

Specification

Specification is document stating requirements. A specification can be related to activities like procedure document, process specification, and test specification, or products like product specification, performance specification, and drawing.

14.2 ISO 9000 QUALITY SYSTEM

ISO 9001:2000—a formal set of standards that are internationally recognized is always more preferable to standards of customers or contractors. ISO 9001 standards have been devised by an international committee, and are a formal set of Quality Management System. 'A Quality Management System embraces all areas of the organization: marketing, contract acceptance, product design, production, delivery, service, finance, and administration'. The objective of a Quality Management System is to ensure that all conforming products and services are delivered to the customers for which an optimum level of quality is to be maintained through all the stages of the quality cycle of the production process. The entire purpose is to ensure customer's satisfaction as the phrase "the customer is king" is truer today than ever before.

ISO 9001 does not deal with any particular product but rather assesses the system as a whole. It is focused on an organization's effectiveness in delivering a quality product or service to the customer from the contractual and design phase to production/installation, storage, and maintenance. The standard recognizes that an organization cannot function smoothly/efficiently and prove that it has been doing so unless there are clearly maintained records of what has been done and what procedures have been followed. The employees would not be able to perform at the optimum level if they are not briefed precisely about their responsibilities and if there are no clear guidelines about their specific duties.

ISO 9000 standards are not certified by the ISO itself, but by the local bodies like Bureau Veritas, TUV, Det Norsk, and other such organizations accredited by the National Organizations for Standardizations like the Bureau of Indian Standards. In terms of self-assessment or assessment by the customers, third-party assessment is more objective.

Investment in the Quality Management System, apart from responding to customer expectations, has resulted in benefits to the efficiency of the organization, its operations, and economic performance, as well as to the quality of its products and services. The revised standards would be of great help for such organizations as are striving to go beyond compliance with the QMS requirements for the certification purposes.

An organization accredited by ISO used to be audited on the basis of 20 clauses contained in ISO 9001:1994. The 20 clauses were as follows:

Clause 4.1: Management responsibility
- Quality policy—to ensure the organization's commitment to quality
- Organization—to establish who is responsibile for what
- Management review—to establish that the quality system is working

Clause 4.2: Quality system
- General—to have a 'quality manual' that will act as a road map of the quality system
- Quality system procedures—to set procedures and to ensure that they are followed
- Quality planning—to establish plans of the organization in order to achieve quality

Clause 4.3: Contract review
- Purpose—to ensure that the firm understands and meets its customers' requirements

Clause 4.4: Design control
- Purpose—to translate customers' needs into specifications—design and development planning, identification and allocation of resources, organizational and technical interfaces, definition and control of design inputs/outputs and interfaces, design verification, design validation, review/approval/recording, and control of design changes

Clause 4.5: Document control
- Purpose—to provide precisely the document or information needed-review and approval of documents by authorized persons, making available the updated versions of necessary documents at appropriate locations, removal of obsolete documents, authorization, and recording of changes to documents

Clause 4.6: Purchasing
- Purpose—to avoid problems caused by purchased materials

Clause 4.7: Control of customer-supplied product
- Purpose—to ensure that the customer-supplied products remain fit for use

Clause 4.8: Product identification and traceability
- Purpose—to enable the organization in keeping track of supplies, components, and finished products

Clause 4.9: Process control
- Core requirement—identify and plan processes, use suitable equipments and set up an adequate processing environment, prepare work instructions so that quality would not be affected because of non-availability of such instructions, comply

with reference standards and quality plans, monitor key characteristics and features during production, approve processes and equipments as required, establish criteria for workmanship in the form of samples or illustrations, establish suitable equipment maintenance procedures, maintain records on qualified processes and personnel

Clause 4.10: Inspection and testing
- Purpose—to check that incoming goods, processes, etc., meet the established requirements

Clause 4.11: Control of inspection, measuring, and test equipment
- Purpose—to ensure that the correct measuring/testing equipment is used to check work and that the equipment operates effectively and gives reliable results

Clause 4.12: Inspection and test status
- Purpose—to identify products that are ready to proceed to the subsequent stages of processing or dispatch—use of markings, tags, labels, or similar identifiers – use of routing cards, inspection records, job cards, and similar items – use of physical location

Clause 4.13: Control of non-conforming product
- Purpose—to establish ways of identifying and dealing with non-conformity

Clause 4.14: Corrective and preventive action
- Purpose—to ensure that problems are solved and to prevent their recurrence – disposition of non-conformity by corrective or preventive action

Clause 4.15: Handling, storage, packaging, preservation, and delivery
- Purpose—to ensure that goods are appropriately handled

Clause 4.16: Control of quality records
- Purpose—to establish proofs (records) of actions taken

Clause 4.17: Internal quality audit
- Purpose—to establish that the organization is performing what it had promised

Clause 4.18: Training
- Purpose—to ensure that all concerned personnel are adequately trained

Clause 4.19: Servicing
- Purpose—to establish procedures for after-sales service

Clause 4.20: Statistical techniques
- Purpose—to establish whether the organization requires sampling and statistical techniques

These 20 clauses of the ISO 9001:1994 have been superseded by the clauses of ISO 9001:2000 as enumerated in the following for the understanding of the requirements of the QMS. The readers must use the original documents issued by the Bureau of Indian Standards for proper understanding and actual implementation of the system. Detailed information on the ISO 9000 family of standards may be found on the following website: www.iso.org.

ISO 9000:1994 version was management system-focused so that the compliance could be obtained by an organization on the basis of its management system, regardless of the quality of the actual product and services they provided. The ISO 9000:2000 version links product and service quality to system quality through a focus on customer satisfaction and builds into the process the demonstration of continual improvement. The new standard is a far more solid basis for the QMS.

An individual cannot document a QMS. All personnel who are responsible for any part in the system need to be involved. This amounts to that the quality system, by definition, has to be built from the operational level up and cannot be imposed by external consultants/experts. The quality system must be a practical working system so that it ensures that consistency of operation is maintained and it may be used as a training aid.

The QMS is built primarily through the audit and review mechanism. The requirement to audit that the system is functioning effectively as planned and to review possible system improvements, utilizing audit results, should ensure that the improvement cycle is engaged through the corrective action procedures. The overriding requirement is that the systems must reflect the established practices of the organization, improved where necessary to bring them into line with current and future requirements.

ISO 9001:2000 – Quality Management Systems – Requirements

ISO 9001:1994 has been revised into five main sections as follows:

- Quality Management Systems
- Management responsibility
- Resource management
- Product realization
- Measurement, analysis, and improvement

All these five sections have been further divided into sub-sections. What is worth noting is that none of the requirements of the ISO 9001:1994 have been removed.

The main clauses of ISO 9001:2000 are discussed here. The first three sections are general in nature—Scope, Normative Reference, and Terms and Conditions.

4 Quality Management Systems (QMS)

4.1 General requirements

The organization is required to establish, document, implement, and maintain QMS and continually improve its effectiveness in conformity with the ISO requirements. The key issue here is continual improvement.

The organization's obligations under this clause are as follows:
- Identification of the processes needed for the QMS ensuring their application all through the organization determining the sequence and interaction of the processes

- Determination of the criteria and methods needed for ensuring effective operation and control of these processes
- Mobilization of resources and gathering relevant information for supporting the operation and monitoring of these processes – the processes should be monitored, measured, and analysed
- Implementation of such actions as required for achieving results as planned and continual improvement of the processes

The organization must ensure control over processes such as outsourcing as this could affect product conformity with the requirements. The control of such outsourced processes shall be identified within the QMS which should be regarded as sacrosanct.

4.2 Documentation requirements

4.2.1 General

An organization's documentation depends on its size, type, complexity, and competence. It is, therefore, possible that documentation can differ from one organization to another to control processes. A distinction is made between the documentation required by the ISO standard and that required by the organization. The documented procedure under the ISO means the procedure is established, documented, implemented, and maintained. The preparation, issue, and change of documents that specify quality requirements or prescribe activities affecting quality shall be controlled to ensure that the correct documents are being followed. Such documents, including changes thereto, shall be reviewed and approved for release by authorized personnel.

The QMS documentation ought to include:
- A quality policy and quality objectives—stated and documented
- A quality manual
- Documented procedures required by the ISO
- Documents that are required by the organization to ensure effective planning, operation, and control of the processes
- Records that are required by the ISO

4.2.2 Quality manual

The organization must establish and maintain a quality manual that has to be like a roadmap of the QMS and controlled. The contents of this manual are:
- The scope of the QMS as well as the details of and justification for any exclusions
- Documented procedures or reference to them
- Description of the interaction of the processes of the QMS

4.2.3 Control of documents

The documents required by the QMS should be totally controlled. The records are special types of documents and are controlled according to the QMS requirements.

The defined and needed controls are contained in the established documented procedure.

- Documents are approved for adequacy and accuracy prior to issue for use
- Documents are reviewed and updated as and when required, and re-approved
- Current revision status and changes effected should be identified for compliance
- Relevant versions of the documents as are applicable are readily available at the points of use—such documents need to be legible, readily identifiable, and retrievable
- Documents of external origin are identified for controlled distribution
- Drevention of unintended use of obsolete/superseded documents for the purpose of which such documents are suitably identified if they are retained for any purpose instead of being destroyed

4.2.4 Control of records

Records that constitute documentary evidence of quality conforming to the ISO requirements and effective operation of the QMS should be specified, prepared, and maintained. Records should be legible, readily identifiable, and retrievable and should also be protected against damage, deterioration, or loss. A documented procedure should be established to define the controls needed for the identification, storage, retrieval, protection, retention time, maintenance, and disposition of quality records.

5 Management responsibilities

5.1 Management commitment

This clause emphasizes the higher management's commitment to the implementation of the QMS and meeting the customer needs as well as regulatory and legal needs so as to demonstrate their commitment to the organization and other concerned personnel. The focus should be on the customer's needs. The higher management should remain committed to continual improvement of the effectiveness of the QMS by:

- Communicating the importance of meeting the customer needs as well as statutory, regulatory, and legal requirements to the organization
- Establishing the quality policy
- Establishing the quality objectives
- Conducting management reviews
- Ensuring the availability of resources

5.2 Customer focus

This clause emphasizes that all efforts on the QMS should be focused on the customer needs and expectations. It is related directly to the higher management's commitment of ensuring that the customer requirements are determined and fulfilled with the aim of enhancing customer satisfaction thereby gaining customer confidence.

5.3 Quality policy

The top management will spell out its quality policy with commitment to continual improvement. The policy would be defined and documented. The quality policy is in conformity with the organization's goals and customer's expectations and needs and will form one element of the corporate policy. The policy statement continues to be reviewed thus becoming a controlled document. The top management will ensure the following:

- Quality policy is appropriate to the purpose of the organization and is suitable for the customer needs
- Quality policy includes commitment on compliance with the requirements and continual improvement of the effectiveness of the QMS at all levels of the organization
- Quality policy provides a framework for establishing and reviewing the quality objectives
- Quality policy is communicated and understood within the organization
- Quality policy is regularly reviewed for its suitability and objectiveness

5.4 Planning

5.4.1 Quality objectives

There has to be greater emphasis on the quality objectives consistent with the quality policy and the responsibility for establishing them clearly lies with the top management. The quality objectives must be established across the entire spectrum of the organization's activities including those needed to meet the requirements for the ultimate product. These objectives must be measurable and the responsibilities of each function and level in the organization must be defined.

5.4.2 QMS planning

The top management is responsible for making plans to ensure that the objectives of the organization are met. The management has to ensure the following:

- QMS is planned to meet the requirements of the quality objectives of the organization as well as the general requirements of the QMS
- The integrity of the QMS is maintained when changes to the QMS are planned and executed

5.5 Responsibility, authority, and communication

5.5.1 Responsibility and authority

The top management should ensure that the responsibilities and authorities are defined and communicated within the organization.

5.5.2 Management representative (MR)

The top management would have to appoint a member of the management who, irrespective of other responsibilities, should have defined responsibility and authority for the following:

- Ensuring that processes needed for the QMS are established, implemented, and maintained
- Reporting to the top management on the performance of the QMS and need for improvement, if any

- Ensuring the promotion of awareness of customer requirements throughout the organization

5.5.3 Internal communication

The top management should ensure that:

- Appropriate communication processes are established within the organization
- Communication takes place regarding the effectiveness of the QMS

5.6 Management review

5.6.1 General

Reviews the QMS should be carried out by top management at defined intervals with records to indicate actions decided upon to ensure its continuing suitability, adequacy, and effectiveness. This review should include assessing opportunities for improvement and the need for changes to the QMS, including the policy and quality objectives. The effectiveness of these actions should be considered in subsequent reviews. The records generated on account of these reviews should be maintained.

5.6.2 Review Input

The input to the management review should include information on current performance and improvement opportunity:

- Results of quality audits
- Customer feedback
- Process performance and product conformity analysis
- Status of preventive and corrective actions
- Follow-up actions from previous management reviews
- Changes that could affect the QMS
- Recommendations for improvement

5.6.3 Review output

The output from the management review should include decisions and actions concerning:

- Improvement of the effectiveness of the QMS and its processes
- Improvement of product related to customer requirements
- Resource needs

6 Resource management

6.1 Provision of resources

The organization should determine and provide the necessary resources for:

- Implementing and maintaining the QMS and continually improving its effectiveness—continual improvement is the key word
- Enhancing customer satisfaction by addressing customer requirements

Information is ever increasingly a vital resource and an organization needs to define and maintain the current/dated/documented information for achieving conformity of products and/or services.

6.2 Human resources

6.2.1 General

The organization should select and assign such personnel as are competent on the basis of appropriate education, training, skills, and experience to those activities that impact the conformity of product and/or services. The key word here is 'competent'. Competence, including achieved qualification levels, needs to be demonstrated and documented. This should be applicable also for the subcontracted work at construction sites.

6.2.2 Competence, awareness, and training

The organization's responsibilities are as follows:
- To determine the necessary competency needs for personnel who perform work to achieve conformity of product and/or service
- To address these needs by providing training or executing alternative actions
- To evaluate the effectiveness of the action taken on training or the alternatives on a continual basis
- To ensure that the personnel deployed are aware of the relevance and importance of the activities they are involved in and how they contribute to the achievement of the quality objectives
- To maintain appropriate records of education, training, skills, and experience of all concerned personnel

6.3 Infrastructure

This is a basic requirement. The organization should identify, provide, and maintain the infrastructure needed for achieving conformity to product requirements. Infrastructure includes, as applicable:
- Buildings, workplace, and associated utilities
- Process equipment—both hardware and software
- Supporting services such as transport, communication, power, security, etc

6.4 Work environment

The organization should determine and manage a conducive work environment for achieving conformity to product requirements.

7 Product realization

7.1 Planning of product realization

The essential requirement of the organization is to plan and develop the processes needed for product realization—to convert customer requirements into customer satisfaction by providing the required product and/or services. Product realization can also be planned so as to remain consistent with the requirements of the other processes of the QMS.

The organization should determine the following appropriately in planning product realization under controlled conditions:

- Quality objectives and requirements for the product
- The necessity of establishing processes, documents, and providing resources specific to the product
- Verification, validation, monitoring, inspection, and test activities specific to the product as required and the criteria for acceptability
- Records needed to be generated for providing evidence that the realization processes and resulting product meet the requirements

The output of this planning should conform to the organization's operating practices including the documented quality plans.

7.2 Customer-related processes

One of the first processes to be established is that for identifying customer requirements.

7.2.1 Determination of the requirements related to the product

The aim is to establish the needs and requirements of the customer and may be prior to receiving an order or contract. Emphasis is given to the implied needs of the customer and end users. Accordingly, the organization shall determine the following:

- Requirements as specified by the customer covering delivery and post-delivery activities
- Requirements that are not specified by the customer but are necessary for specified and intended use, if known
- The obligations related to the product including regulatory and statutory requirements
- Any additional requirements as deemed necessary by the organization

7.2.2 Review of requirements related to the product

On determination of the requirements related to the product, the organization should review the same. This review should be carried out prior to the organization's commitment regarding the supply of a product to the customer following submission of tenders, acceptance of contracts or orders, and acceptance of changes to contracts or orders. The organization should ensure the following:

- Product requirements are clearly defined
- In the absence of the customer's documented statement of requirements, acceptance of the offered requirements has to be confirmed by the customer
- Ambiguities in contract or order requirements differing from those previously expressed are definitely resolved
- Organization has the ability to meet the defined requirements
- Results of the review and actions that arise from the review are recorded
- In case the product requirements are changed, the relevant documents are amended and the concerned personnel are made aware of the changed requirements

7.2.3 Customer communication

This is a new requirement for a successful organization to establish and implement effective arrangements for communication and liaison with customers in relation to:

- Product information
- Enquiries, contract or order processing, including amendments
- Customer feedback, including customer complaints and other reports relating to nonconformities

7.3 Design and development

7.3.1 Design and development planning

The organization shall prepare plans for each design and development activity and control the design and development of product and/or services. The design shall be defined, controlled, and verified.

During the design and development planning, the organization shall determine/define/ensure:

- The stages of design and development process
- Appropriate review, verification, and validation of each design and development stage
- The responsibilities and authorities for design and development activities
- The interfaces between different groups involved in design and development should be managed so as to ensure effective communication and clear assignments of responsibilities
- During the progress of the design and development, planning output should be updated

7.3.2 Design and development inputs

Design inputs relating to the product requirements should be appropriately specified on the basis of time schedule and correctly translated into design documents. The design interfaces should be identified and controlled.

The design inputs include:

- Functional and performance requirements
- Applicable statutory and regulatory requirements
- Information derived from previous similar designs
- Other requirements essential for design and development

These inputs should be reviewed for adequacy by persons other than those involved in actual design. Requirements should be complete, unambiguous, and non-conflicting with each other. Design changes, including field changes, should be governed by control measures commensurate with those applied to the original design.

7.3.3 Design and development outputs

The outputs of design and development process should be provided in a form that enables verification against the design and development input requirements and should be approved prior to release for execution.

The design and development outputs should include:
- Conformance to the input requirements for design and development
- Appropriate information on purchasing, production, and service provision
- Product acceptance criteria
- Delineation of the characteristics of the product that are essential for its safe and proper use

7.3.4 Design and development review

Systematic reviews of the design results at appropriate stages of design should be planned and conducted and such reviews should be documented. The participants at each design review should include representatives of functions concerned with the design or development stages being reviewed. The purpose of the systematic reviews is to:

- Evaluate the ability of the results of design and development to fulfil the requirements
- Identify the problems and propose necessary remedial actions

Records of the results of such reviews and necessary actions, if any, should be maintained.

7.3.5 Design and development verification

At appropriate stages of design, verification should be performed in accordance with planned arrangements to ensure that the design and development outputs have met the design and development input requirements. Records of the results of the verification and necessary actions, if any, should be maintained.

7.3.6 Design and development validation

Design and development validation should be performed as per the design and development planning to ensure that the resulting product meets the requirements for the specified application or known intended use. Wherever practicable, validation should be completed prior to the implementation or delivery of the product. Records of the results of validation and necessary actions, if any, should be maintained.

7.3.7 Control of design and development changes

All design and development changes or modifications need to be identified as early as possible, and records thereof maintained. The changes should be reviewed, verified, and validated appropriately and approved by the authorized personnel before implementation. The review of design and development changes should include evaluation of the effect of the changes on constituent parts or product already delivered. Records of the results of the review of changes and necessary actions, if any, should be maintained.

7.4 Purchasing

7.4.1 Purchasing process

The purchase requirements should be specified fully and accurately and the organization should ensure that the purchased item or product conforms to the specified requirements. The type and extent of control applied to the supplier and purchased product should depend on the effect of subsequent product realization or the final product. The ability of the suppliers to supply item or product conforming to the organization's requirements should be evaluated and the selection made on merit.

For the kind of control delineated above, it is essential to establish criteria for selection, evaluation, and re-evaluation. Records of the results of evaluations and necessary actions, if any, should be maintained.

7.4.2 Purchasing information

The purchasing document should contain information describing the product to be purchased, including where appropriate:

- Requirements for the approval or qualification of the product, procedures, processes, facilities, and equipment
- Requirements for qualification of the personnel involved
- QMS requirements

The organization should ensure the adequacy of the specified requirements contained in the purchasing documents prior to their release for action.

7.4.3 Verification of purchased product

It is absolutely necessary for the organization to establish and implement the inspection or other activities necessary for ensuring that the purchased products conform to the specified purchase requirements. The organization should specify the intended verification arrangements and method of product release in the purchasing information in case the organization intends to perform verification at the supplier's premises.

7.5 Production and service provision

7.5.1 Control of production and service provision

Production and service provision are carried out by the organization under controlled conditions. ISO 9001:2000 version defines six key elements of controlling production:

- Information on product characteristics
- Work instructions, as required
- Suitable equipment for production, installation and service provision
- Measuring and monitoring devices
- Measuring and monitoring activities
- Process of release, delivery, and after sales services

7.5.2 Validation of processes for production and service provision

The organization should identify and plan the production and service provision processes that are directly related to quality and ensure that these processes are performed in controlled conditions. The organization should validate any processes for production and service provision where the resulting output cannot be verified by subsequent monitoring or measurement. This includes any process where deficiencies become apparent only after the product is in use or the service has been provided. The organization should define arrangements for validation that includes, where applicable:

- Criteria for review and approval of the processes
- Approval of equipment and qualification of personnel
- Use of defined methods and procedures
- Requirement of records
- Revalidation

Evidence of the validated processes, equipment, and personnel needs to be recorded and maintained.

7.5.3 Identification and traceability

The onus is on the organization to decide whether it is necessary to identify the product's time of production, receipt/delivery, installation, and servicing. The organization should identify:

- The product by suitable means, where appropriate, during the product realization
- The product status corresponding to monitoring and measurement requirements

Where traceability is required, the organization should control and record the unique identification of the product.

7.5.4 Customer property

Customer property is an item or material that the customer wants the organization to incorporate into the product or use it for service being provided by the organization. The management may decide that a procedure is required for the control of customer property by identifying, verifying, protecting, and safeguarding the same. In case of any customer property being lost, damaged, or otherwise found to be unsuitable for use, this should be reported to the customer and records maintained. The customer property may include such property as covered by intellectual property rights that would demand stringent control.

7.5.5 Preservation of product

At all stages of internal processing and delivery to the destination, the organization must make provisions for safeguarding the conformity of the product by proper identification and appropriate handling, packaging, storage, and protection. Preservation should also be applicable to the constituent parts of a product.

7.6 Control of measuring and monitoring devices

The organization is required to determine the monitoring and measurement that need to be carried out and also to identify the monitoring and measurement devices needed to provide evidence of conformity to the product to the requirements as determined. It is also necessary that the organization establish processes to ensure that monitoring and measurement be carried out in conformity with the monitoring and measurement requirements. Measuring, inspection, and test equipment should be used in a way which ensures that any measurement uncertainty, including accuracy and precision, is known and is consistent with the required measurement capability. Any test equipment software should meet the applicable requirements for the design and development of the product. When used in monitoring and measurement of specified requirements, the ability of computer software to satisfy the intended application should be confirmed. This should be undertaken prior to initial use and reconfirmed as necessary.

Where it is required to ensure valid results, measuring and monitoring devices should be:
- Calibrated or verified at specified intervals or prior to use, against measurement standards traceable to national/international standards – in case of such standard/s not existing, the basis used for calibration or verification should be recorded
- Adjusted or re-adjusted as required
- Identified to enable the calibration status to be determined
- Safeguarded from adjustments that would invalidate the measurement result
- Protected from damage and deterioration during handling, maintenance, and storage

Furthermore, the organization should assess and record the previous measuring results when the equipment is found not to conform to the requirements. The organization should take appropriate action on the equipment and any product that is affected. The records of the results of calibration and verification should be maintained.

8 Measurement, analysis, and improvement

8.1 General

The organization should plan and implement the monitoring, measurement, analysis, and improvement processes needed for:
- Demonstrating conformity of the product
- Ensuring conformity of the QMS
- Continual improvement of the QMS

For this, determination of applicable methods, including statistical techniques, and the extent of their use would be necessary.

8.2 Monitoring and measurement

8.2.1 Customer satisfaction

The methods of obtaining and monitoring customer requirements for measurements of the performance of the QMS and the customer's perception as to whether the organization has met such requirements should be determined. As the measurements of the performance of the QMS make progress, the organization should progressively monitor information relating to customer perception and use this information effectively. Customer satisfaction must be a primary measure of the system output.

8.2.2 Internal audit

The organization should conduct internal audits at planned intervals/schedules for the purpose of determining whether the QMS:

- Conforms to the planned arrangements of the ISO and to the QMS requirements as established by the organization
- Is effectively implemented and maintained

An audit programme should be planned taking into account the following:

- Status and importance of the processes
- Areas to be audited
- Results of the previous audits
- Defined audit criteria, scope, frequency, and methods
- Objectivity and impartiality of the audit process—this should be the basis of the selection of auditors and conduct of the audits
- Auditors should not audit their own work

The procedures defining the responsibilities and requirements for planning and conducting the audits and reporting the results as well as maintaining the records should be documented. The management responsible for the area being audited should ensure that:

- Action is taken without undue delay to eliminate detected non-conformities and their causes
- Follow-up activities include the verification of the actions taken and reporting of the verification results

8.2.3 Monitoring and measurement of processes

The requirements of monitoring and measurement processes are as given under:

- The organization should apply suitable methods for monitoring and, where applicable, measurement of the QMS processes—these methods should demonstrate the ability of the processes to achieve the planned results
- If results are not achieved as planned, appropriate correction and corrective action should be taken so as to ensure conformity of the product

8.2.4 Monitoring and measurement of product

The organization should ensure the following:

- Monitoring and measurement of the characteristics of the product are carried out to verify that product requirements have been fulfilled
- Monitoring and measurement as mentioned above should be executed at appropriate stages of product realization process in accordance with the planned arrangements
- Maintaining evidence of conformity with the acceptance criteria should be recorded—such records should indicate the assigned person/s authorizing release of the product
- Release of the product and delivery of the service should be effected only on completion of planned arrangements satisfactorily

8.3 Control of nonconforming product

The organization should define and maintain documented procedures to ensure that product that does not conform to the product requirements is identified and prevented from unintended use or delivery. The organization should deal with nonconforming products by:

- Eliminating the detected nonconformity by taking the required action
- Authorizing its use, release, or acceptance under concession by a relevant authority and, where applicable, by the customer
- Taking action to preclude its original intended use or application
- Maintaining records of the nature of nonconformities and any subsequent actions taken including concessions obtained

A nonconforming product that is corrected should be subjected to re-verification so as to demonstrate its conformity to the requirements. As and when a product's nonconformity is detected after delivery or use for a while, the organization should take action appropriate to the effects, or potential effects, of the nonconformity.

8.4 Analysis of data

The suitability and effectiveness of the QMS should be demonstrated by the organization by determining, collecting, and analysing appropriate data. The organization should also evaluate where continual improvement of the effectiveness of the QMS can be made. The data should include that generated as a result of monitoring and measurement and from other relevant sources. The data analysed should provide information on:

- Customer satisfaction
- Conformity to the product requirements
- Characteristics and trends of processes and products including opportunities for preventive action
- Suppliers

8.5 Improvement

8.5.1 Continual improvement

Continual improvement of the QMS can be effected by the organization through:

- Quality policy
- Quality objectives
- Audit results
- Data analysis
- Corrective and preventive actions
- Management review

8.5.2 Corrective action

The organization should establish a process for eliminating the causes of nonconformities so as to prevent recurrence. Nonconformity reports (NCR), customer complaints, and other suitable QMS records are useful as inputs to the appropriate corrective action process. A documented procedure should be established together with the procedures for the corrective action process to define requirements for:

- Reviewing nonconformities including customer complaints, if any
- Investigating and determining the causes of nonconformities
- Evaluating the need for action to eliminate causes of nonconformities to preclude recurrence
- Determination and implementation of corrective action
- Recording of the results of action taken
- Following up to review the corrective action taken

8.5.3 Preventive action

The organization should establish a process for eliminating the causes of potential nonconformities to preclude their occurrence. QMS records and results from the analysis of data should be used as inputs for this and responsibilities of preventive action established. The preventive actions should be appropriate to the effect of the potential problems. A documented procedure should be established together with the procedures for the preventive action process to define requirements for:

- Identifying potential product, service, and process nonconformities
- Investigating the causes of potential nonconformities and recording the results
- Determining and implementing preventing action needed to eliminate causes of potential nonconformities
- Recording of results of action taken
- Reviewing the preventive action taken

SUMMARY

In a highly competitive global scenario, customers must be convinced about the quality plan of producers and service providers. They are becoming increasingly quality conscious. They know that quality is far more important in the long run compared to cost consideration. To hold on to customers, producers of products and service providers would require such quality plan as would directly define their efforts towards customer satisfaction and continual improvement.

ISO 9000:2000 standard recognizes that an organization cannot function smoothly/efficiently and prove that it has been doing so unless they have clearly kept records of what has been done and what procedures have been followed. The employees would not be able to perform at the optimum level if they are not briefed precisely about their responsibilities and if there are no clear guidelines about their specific duties. The employees should be trained on the basis of systematic planning.

ISO 9000 standards are not certified by the ISO itself, but by the local organisations accredited by the National Organisations for Standardizations like the Bureau of Indian Standards. In terms of self-assessment or assessment by the customers, third party assessment is more objective.

REVIEW QUESTIONS

1. Define quality, quality policy, and quality management systems (QMS).
2. What should be the ultimate intent of the QMS?
3. What is quality improvement? What is meant by continual improvement?
4. What is audit? What are first party, and second- or third-party audits?
5. The 20 clauses of the earlier ISO 9001:1994 have been revised into five main sections in ISO 9001:2000. Name the revised five sections.
6. What is quality manual? What does it contain?
7. What is quality policy (QP)? What is the top management required to ensure through the QP?
8. What are the responsibilities of the management representative (MR)?
9. What kind of communication should be established with the customer?
10. What is the purpose of internal audit?
11. How should a nonconforming product be dealt with?
12. How can continual improvement of the QMS be affected?

15
Safety

INTRODUCTION

Safety means achieving proper operating conditions, prevention of accidents, or mitigation of the consequences of accidents. A considerable number of personnel get injured every year, seriously or fatally, while engaged in construction work. Problems arise at the construction sites only when the safety measures are bypassed. Therefore, it becomes the responsibility of the management to safeguard the safety and welfare of everyone assigned to construction activities.

Effective health and safety management is founded on the provision of a safe and healthy working environment with safe system of work at its core. The key to success is to ensure that health and safety aspects are carefully planned, organized, monitored, controlled, and reviewed. This chapter deals with the salient features of safety, starting from housekeeping to environment, accidents, and the consequences of accidents.

15.1 BASIC PRINCIPLES ON SAFETY

Safety is inherent in right attitude. Individual safety depends on one's attitude and sense of overriding priority attached to safety. One with the proper attitude would always return home safely on completion of a day's work in the same way as one arrived without suffering from injury or health hazard during the course of the day's work (Section 1.11 of Chapter 1).

The management should follow 'cradle to grave' approach on the safety issue by:

- Ensuring that safety is inherent in design—the basis of design work is 'safety first'
- Selecting proper people
- Motivating them by imparting training
- Ensuring that no one is overworked

Training is learning that changes attitude. The knowledge of what constitutes a safe approach in different situations and at different phases of construction work at the construction sites must be acquired by professional training. An agency engaged in construction work, therefore, should arrange for comprehensive training. The various requirements for a successful training are as follows:

- Active commitment, support, and interest of the top management—an environment must exist where a trained person would be able to apply his acquired knowledge and developed skill through training
- Adequate fund and organizational structure to provide opportunity for training
- Availability of pertinent expertise for imparting training by knowledgeable and experienced trainers

In training the employees, the management imparts knowledge and skill. What the employees gain by training, apart from knowledge and skill, is motivation. They have to be given information and knowledge that accidents are not inevitable but are caused. They need training to develop skills and recognize the need to comply with and develop safe system of work, and to report unsafe conditions and practices. Their attitude and awareness related to safety needs continual improvement and the social environment of the workplace must foster good safety and health practices.

The employee must be trained on what he is required to learn by the management. This is known as pertinent knowledge. The employee must know how to carry out the assigned responsibilities skilfully and safely. The development of skill should, therefore, be part of the training schedule.

'If in doubt, ask' is an excellent motto; but 'if in doubt, don't do it' is probably a better piece of advice. The project work is executed by a number of agencies involved in the process. Such execution exposes men and materials at sites to safety hazards. All agencies at the sites, therefore, should state their safety policies and execute their assigned work in accordance with the stated policies. The safety policy should be formulated out of a sense of priority for safety for the entire period of implementation and should originate from the top most level of management of each executing agency. The safety policy of each agency must comply with the prevailing statutory laws and rules.

As a number of agencies are involved at sites carrying out inter-dependent activities, it would not be possible for each agency to pursue its safety policy independently. The owners/customers should, therefore, take the initiative to form safety committees at sites comprising designated representatives of all involved agencies. An owner's representative should be the head of the safety committee stressing the importance of achieving proper operating conditions so as to prevent accidents or to mitigate the consequences of accidents. This would result in protection of site personnel, public, property, environment, and achieving planned progress. Accident is not only costly but also delays progress of the project work. If any agency deviates from any safety procedure or statutory rule, the owner's representative would have the authority to stop work.

The members of the safety committee at a project site must meet regularly at short intervals to review the status of safety at the site. The purposes of such meetings are listed here:

- To stress repeatedly on inculcating safety awareness among the involved agencies engaged at the construction site by spearheading campaign
- The flow of information between different agencies

- To apportion the responsibilities of different agencies in inter-related and overlapping work—overlapping work is susceptible to hazards
- To review the accidents to determine follow-up actions needed and to take preventive measures—minutes of meetings should, therefore, be sent to all agencies promptly

Each executing/construction agency (contractor) in turn should have a safety committee of its own comprising members who are involved in actual construction level right up to the rank of workers. Safety awareness should be stressed to such an extent that safety should be ingrained in their behaviour and change their attitude positively.

Regarding subcontracting, a contractor must hire only competent and resourceful subcontractors. The contractor's key responsibilities are:

- Framing and implementing health and safety plan
- Co-ordination of work and co-operation with subcontractors to eliminate unsafe practices
- Providing information on project risks
- Ascertaining and approving a subcontractor's plan of execution of high risk operations
- Ensuring that a subcontractor complies with the safety practices followed at the site and also the statutory rules and regulations
- Monitoring a subcontractor's performance on health and safety issues on a continuous basis
- Permitting entry of only authorized persons for execution of work
- Spearheading campaign on health and safety
- Updating files on health and safety

The health and safety plan should include:
- General description of the activities involved
- Time schedules for completion of each activity
- Known health and safety risks
- Information needed by a subcontractor to demonstrate its resourcefulness and competence
- Information provided by the subcontractor/s on health and safety

By its own nature, the construction work presents a multitude of hazards for workers in handling heavy weights, placing large quantities of concrete, working at heights, and operating at all weather conditions. The agencies working at the sites would have to ensure safety measures against all kinds of hazards. As accidents resulting from hazards may delay project implementation, owners or their designated representatives should ensure that executing agencies fulfil their contractual commitments safely and timely with assured quality. As evident from this, the safety features need to include precautionary measures against gravitational forces, mechanical work injuries, burn injuries, pressure hazards, electrical hazards, fire and fire-related hazards, explosion hazards, toxic hazards, radiation hazards, noise pollution hazards, and so on. Each agency must ensure that: (i) safety system, (ii) mitigating system, (iii) emergency alarm, and (iv) evacuation plan work properly.

Ergonomics is the study of engineering aspects of the relation between human workers and their working environment. The working environment, among other things, includes objects like tools and equipment, chairs, tables, and steps. Ergonomics promotes well-being at work by addressing all aspects of the work environment. It is concerned with the application of scientific data on human capabilities and performance to the design of workplaces, tools, equipments, and systems. Other aspects of the wide scope of ergonomics include organizational arrangements that aim to limit the potentially harmful effects of physically demanding jobs on individuals. They include selection and training, matching personal skills to job demands, job rotation, and the setting of appropriate work breaks. Examples of the positive use of ergonomics include the design of tools and the limiting of weights in sacks and bags. In line with ergonomic consideration, an ideal location of a fire extinguisher should be near exit and stair landings that can offer immediate access.

The 'cradle to grave' system approach as mentioned before requires safety to be taken into consideration right from the design stage covering ergonomics as well. The following list gives the definitions of all basic safety-related terms:

- Danger—a risk to health or bodily injury
- Accident—an incident plus the resulting undesired consequences like physical harm, property damage, interruption in progress of work
- Incident—the sequence of events or actions resulting in the undesired consequences
- Event—an internal or external occurrence involving equipment performance or human action that causes a system upset
- Failure—the inability of a system, subsystem, or component to perform its required function
- Injury—consequence of an incident
- Hazard—an intrinsic property or condition that has the *potential* to cause an accident
- Initiating event—an event that will result in an accident unless systems or operators intervene
- Intermediate event—an event in a sequence that helps propagate the accident or helps prevent the accident or mitigate the consequences
- Reliability—the probability that a system, subsystem, or component will perform its intended function for a specified period of time under normal conditions
- Risk—the combination of the *likelihood* of occurrence of an abnormal event or failure and the consequences of that event or failure to persons, materials, or environment

15.2 HOUSEKEEPING

A construction company with good housekeeping practices will always have a good foundation for higher productivity, better quality, reduced costs, greater safety, and higher employee morale so as to excel in performance. Housekeeping has, therefore, bearing on safety and reflects the management's general attitude towards safe execution practices. This apparently unimportant issue makes a big difference in employee morale and ultimately his attitude, which is related to safety.

The following '5S' housekeeping as practiced in Japan is worth emulating:

- Seiri (Sort)—identify unnecessary items in the workplace and get rid of them
- Seiton (Systematize)—arrange necessary items in good order to facilitate easy picking for use; a place for everything and everything in its place
- Seiso (Sweep)—clean the workplace completely so that there is no dust on floor, machine, and equipment
- Seiketsu (Sanitize)—maintain high standard of housekeeping and workplace organization at all times
- Shitsuke (Self-discipline)—train people to follow good housekeeping disciplines autonomously

Based on the '5S', the following guidelines may be followed:

- IS:4082 for recommendation on stacking and storage of construction materials at site and IS:7969 safety code for handling and storage of building materials are safety-related codes issued by the Bureau of Indian Standards (BIS)
- All kinds of materials, trash, construction equipment, tools, and tackles should be safely stored away from accessible areas so that the same does not fall from heights or accidents may not happen due to tripping, slipping, or falling over materials and equipment lying around
- Waste materials should be removed at regular intervals and always at the end of working day
- Excavated spoil/debris should be dumped in areas earmarked for the purpose and waste materials should be disposed in rubbish bins
- Materials should be piled at a safe distance from approaches and roads
- Portable equipment should be restored to storage place after use. Tools should be kept in tool boxes
- Spilled oil and grease in workplaces and accesses should be cleaned and covered with materials like saw dust, sand, ash, etc
- Metal containers with lids should be used for storing flammable materials
- Protruding nails from timber should be removed or hammered down
- While storing materials, it is to be ensured that access to fire extinguishers is not blocked
- A routine inspection is needed at the end of each working day to confirm that everything is left in a safe condition
- Good housekeeping should be everyone's responsibility

15.3 PERSONAL SAFETY

A useful safety motto for those involved in piling work is the '5W':

- Wear hard helmets as protection against falling objects from heights
- Wear ear plugs as noise level may exceed 90 dB
- Wear goggles or safety glasses to protect eyes
- Wear safety shoes or boots as protection against falling objects from heights
- Wear hand gloves (a dolly can catch fire during driving because it becomes very hot)

Apart from piling, all these safety gears and safety belts are required for personal safety at the construction sites for different kinds of construction activities. The safety belts are essential for working at heights – each worker must wear safety belt and tie it with something already secured. The hand gloves must be worn for working on material-handling jobs, welding, gas-cutting, pneumatic breaking, etc. Wearing of protective goggles or using face shields is necessary for performing welding, gas cutting, grinding, pneumatic breaking, brick dressing, etc. Ear plugs must be worn in noisy areas having a noise level above 90 dB as already mentioned.

General guidelines of personal safety are as follows:

- IS:1179 – Equipment for eye and face protection during welding; IS:2925 – Specification for industrial safety helmets; IS:6994 – Specification for safety gloves: Part I – leather and cotton gloves; IS:7293 – Safety code for working with construction machinery; IS:8519 – Guide for industrial safety equipment for body protection; IS:8520 – Guide for selection of industrial safety equipment for eye, face, and ear protection; IS:8521 – Industrial safety face shields: Part I – With plastic visor; IS:8522 – Respirators, chemical cartridge; IS:8523 – Respirators, canister type (gas masks); IS:8940 – ode of practice for maintenance and care of industrial safety equipment for eyes and face protection; IS:10245 (Parts I to III) – Specification for breathing apparatus; IS:10667 – Guide for selection of industrial safety equipment for protection of foot and leg; and SP 53 – Hand-operated tools: safety code for the use, care, and protection are all safety-related codes issued by the BIS
- Even with safety gears provided, one should continue to remain alert and vigilant about personal safety
- Concentration of hazardous dust and fume at sites should be within permissible limits
- Access to and exit from the work place should be hazard-free at all levels
- Work permit is needed for entry or working in danger-prone zones like blasting site, stacks, silos, wells, cofferdams, etc (IS:5216: Part I – latest revision issued by the BIS)
- Safety guards and safety devices of equipment and machinery should be maintained properly

- Dark areas should be illuminated—irrespective of this, one should carry a torch while moving in dark areas
- Personnel engaged in radiography should undergo medical examination periodically
- Safety nets should be provided maximum 6 m below the place where workers are engaged in execution work at heights
- Overexertion of individuals should be avoided

15.4 FIRE PROTECTION

Fire safety can be ensured if flammable materials are kept away from heat and oxygen. The three ingredients of fire are fuel, oxygen, and a source of ignition (heat). There would be no fire if any of them is removed. Precautionary measures, therefore, should be planned before initiating any construction work:

- Use of less flammable materials like timber, bamboos, coal, paints, etc.
- Quantities of flammable materials should be kept to the minimum, for example, for a shift at any location for uninterrupted work – the remainder should be kept in fireproof stores
- Flammable solids, liquids (in containers), and gas (in cylinders) should be separately stored away from oxygen and oxidizing materials
- Cigarette butt or match stick could be the cause of ignition; smoking, therefore, should be prohibited

Precautions should be taken in transporting, storing, handling, and using flammable materials. Most electrical fire can be attributed to defects such as insulation failure, overloaded conductors, and poor connections. Fire extinguishers should be located at strategically convenient points and the personnel trained to use them. A fire tender should remain parked at a site round-the-clock. Adequate water should always be available for fire-fighting. Buckets filled with sand should be kept all over the site for smothering fire. Smouldering fire, if any, should be extinguished.

The fire detection cum alarm system for the protection of life and property detects a fire at the earliest and raises an alarm for immediate fire-fighting. A fire detector is designed to detect one or more of the three characteristics of fire: heat, smoke, and radiation (flame). The type of detector to be installed would depend on the nature of the construction work and circumstances involved:

- Heat detectors—(i) fixed heat detectors respond when the ambient temperature reaches a constant high value ($58°C$ or $88°C$); (ii) rate of rise heat detector measures the rate of increase in the ambient air temperature over a given period of time and responds when the increase is greater than normal or when the temperature reaches $58°C$, whichever is the earlier
- Smoke detectors—(i) ionization smoke detector responds when smoke (or combustion products) enters the detector and causes changes in ionization currents within the detector; (ii) optical smoke detector containing a light source

and photoelectric cell responds when light is absorbed or scattered by smoke particles
- Flame detector is a detector that looks at a given area and responds when it detects infrared or ultraviolet radiation. Because of its inability to detect smouldering fire, a flame detector would be used in specialized applications or as a supplement to heat or smoke detectors

In a *Background Paper*, the Confederation of Indian Industry, Eastern Region, concluded the chapter on *Fire Safety* by issuing 'Ten Commandments' as follows. Loss Prevention Association of India Limited advocates the same commandments on fire safety:

1. Have a written down Fire Prevention Plan for your Company and ensure that it is sincerely implemented.
2. Identify and eliminate fire risks or reduce them to the maximum extent possible.
3. Train and retrain your employees in fire prevention and fire fighting.
4. Install suitable fire protection equipment and make sure that your employees know how to use it in case of fire.
5. Regularly inspect your fire safety equipment so that it does not fail in an emergency. Have adequate water supply.
6. Establish an Emergency Plan in close coordination with the public fire department.
7. Take the utmost care while handling flammable materials. Provide special protection for major fire risks.
8. Follow good housekeeping practices, because a clean house is a safe house.
9. Protect the plant against hazards within and outside by having suitable construction.
10. Never violate fire safety laws—they are meant for your protection.

National Building Code, Part IV issued by the BIS is a comprehensive document covering all aspects of fire safety. BIS plays a major role in formulating the general standards and policies related to fire safety and fire fighting. National Building Code contains a list of the relevant BIS codes in its Appendix A. Some of the safety-related codes are: IS:933 – Specification for portable chemical foam fire extinguisher; IS:934 – Specification for portable water type fire extinguisher (soda acid); IS:2878 – Specification for fire extinguisher, carbon dioxide type (portable and trolley-mounted); IS:5507 – Specification for 50 litre capacity chemical fire engine, foam type; IS:6070 – Code of practice for selection, operation, and maintenance of trailer fire pumps, portable fire pumps, water tenders, and motor fire engines; IS:8758 – Recommendations for fire precautionary measures in the construction of temporary structures and pandals; and IS:10204 – Specification for portable fire extinguisher mechanical foam type.

15.5 ELECTRICAL SAFETY

Electrical safety assumes high importance as the ratio of fatalities to injuries is higher for electrical accidents compared to most other categories of injury. Electricity constitutes a hidden danger as it cannot be seen. It can light up a devastating fire or cause a fatal accident by electrocution. However, if properly used, electricity is the most versatile form of energy for performing work. Electrical hazards are mainly of two types:

- Persons or livestock accidentally or due to negligence come in contact with live parts of an installation and get shocks or get electrocuted
- Heat is generated due to loose contacts, improper installation, or improper use of electrical equipment resulting in breaking out of fire

Preventive measures against these hazards lie in proper design of the electrical installation wherein safety would be in-built. The design must provide:

- Protection against both direct and indirect contact
- Protection against thermal effects in normal service
- Protection against both overcurrents or fault currents
- Protection against overvoltage

Protection can be effected by means of proper insulation, proper clearances, effective earthing, sensitive relaying, and judicious operation and maintenance. All cabling and installation on the load side of the supply point must comply with the relevant statutory requirements as indicated in the following documents, which cover mainly the safety aspects, to assure electrical safety:

- Indian Electricity Act, 1910
- Electricity (Supply) Act, 1948
- Indian Electricity Rules, 1956
- National Electric Code, 1985
- Other relevant rules of the Generating and Supply Authorities

The BIS brought out the National Electrical Code that describes the procedures and provides guidelines for selection of equipment to meet certain requirements. The Indian Electricity Rules stipulate that all installation work should be carried out in accordance with the National Electrical Code and various codes of practices issued by the BIS. This stipulation assures safety.

In a *Background Paper*, the Confederation of Indian Industry, Eastern Region, concluded the chapter on *Electrical Safety* by issuing 'Ten Commandments' as follows:

1. Study the network carefully and coolly, and then chalk out your line of action, step by step.
2. Check that your diagram book is up-to-date.
3. Do not assume anything. Preconceived notions or biased ideas are our greatest enemy.
4. Before carrying out an operation, check with the Control Room for any last minute alteration.
5. Remember, live and dead bars look exactly alike. So,

6. Do not touch a conductor unless it is earthed. And
7. Do not earth a conductor unless it is proved to be dead by live-line detector or any other suitable gadgets.
8. Have your own check points. Never depend on others'.
9. Plug all your back-feed points; while working on bus keep it shortened and earthed with shorting clips.
10. Check, recheck, and check again. Proceed only when you are fully confident; if not, do not hesitate to cancel the job.

Temporary electrical cables should be either laid overhead or alternatively protected from damage by effectively covering the same. Reinforcing steel rods or any metallic part of structure should not be used for supporting wires and cables, fixtures, equipment, earthing, etc. Buried cables should be protected and de-energized before taking up any excavation or blasting work. Power supply circuits other than that of illumination should be switched off beyond working hours. While working near energized circuits, line voltage should be ascertained, and safe distance should be maintained accordingly. Before any maintenance work is taken up on electrical installation/equipment, the circuits should be de-energized and ascertained to be dead by positive test with an approved voltage-testing device. Switches or breakers should be provided with locking means in the 'off' position. Protection against direct and indirect contact with electricity should be provided by suitable means such as:

- Enclosure
- Insulation
- Extra low voltage wherever possible
- Safety isolation
- Earthing of neutral
- Earthing of normally dead parts
- Current-operated earth leakage circuit breakers
- Insulation control devices
- Isolated neutral

Employees working on electrical installation/equipment should be trained in fire-fighting, first aid, and artificial resuscitation techniques. IS:1416 – Safety Transformers; IS:5216 – Recommendation on Safety Procedures and Practices in Electrical Work, Part I: General, Part II: Life Saving Techniques; and SP 31 – Treatment for electrical shock are all safety-related codes issued by the BIS.

15.6 MECHANICAL HANDLING

Cranes and hoists, and powered trucks and forklifts are the primary construction equipment for mechanical handling at construction sites. Mechanical handling techniques in deployment of such equipment has considerably improved efficiency and safety, but has at the same time added other sources of potential injury. Safety related to operating conditions and site hazards, operator's skill, efficiency, and attitude can be jeopardized. All construction equipment should be of good construction, made from sound materials of adequate strength, free from defects, and suitable for the purpose in terms of capacity, size. and type. All equipment should be tested and regularly examined to ensure their integrity. Mechanical handling should be carried out generally in daylight hours in fair weather. Otherwise, appropriate precautionary measures need to be taken to ensure safe working environment. All equipment should always be used properly with utmost care and top most priority to safety.

The problems that can be foreseen on the performance of mechanical handling equipment are as follows:

- Overturning – Due to weak support, operating beyond the rated capacity, striking obstructions, manoeuvring with load elevated in case of powered truck, sudden braking, hitting obstructions, driving downward with load in front, turning or crossing ramps at an angle, shifting loads, and unsuitable road or support conditions
- Overloading – By exceeding the rated capacity or operating radii or failure of safety devices
- Jib of crane breaking or falling (other than overturning or support failure)
- Collision – With overhead cables or structures, with other cranes/vehicles, stacks, pipes, etc.
- Failure of support– Not so solid ground/base, outriggers not extended
- Floor failure – If floor capacity is less than load plus equipment/truck weight
- Fall from platforms – Using the unprotected hoist platform as a working platform or means of access, equipment failure
- Operator errors – From impaired or restricted visibility, poor eyesight, poor training – Miscalculation in manoeuvring the jib or load
- Loss of load – From failure or bumping of the hoist platform
- Explosion and fire – May be caused by electrical shorting, leaking fuel pipes, and from hydrogen generation during battery charging – A truck itself can be the source of ignition if operated in a flammable environment

The several safety guidelines are enumerated in the following:

- Load indicators of two types should be incorporated in the cranes: (i) load/radius indicator and (ii) automatic safe load indicator providing audible and visual alarm. Visually, a coloured flashing light is signalled. Overloading for mechanical handling should not be permitted. All controls should be clearly identified and tested to ensure that they are in order and would function correctly as and when required.
- Limit switches would prevent hook or sheave block to be wound up to the cable drum.

- Safe access should be provided for the operator and for use during inspection and maintenance/emergency.
- From the operator's position, both the hook and load should be clearly visible and the controls should be within easy reach. An alternative to this is remote control.
- Passenger/s should not be carried without written authorization and never on lifting tackle and goods hoists.
- Testing/examining of lifting tackle like chains, slings, wire ropes, eyebolts, and shackles is essential to ensure that each item be free from damage and knots as appropriate, be clearly marked for identification and safe working load, and be properly used. Wire ropes and slings should be kept well lubricated. The load should be free of obstruction and properly slung before it is lifted.
- Crane operators and slingers should be fit and strong enough for the assignment apart from being competent and need to be trained and re-trained.
- For lifting, lowering, or shifting, normal signalling system should be displayed in visual posters and strictly followed. Hand signals should be adopted. Verbal signals should be totally avoided. The operators should take signals only from one person.
- Operating area should be free of overhead cables and bared power supply conductors. Solid supporting base should be available for the equipment. Possibility of hitting other cranes or structures should be examined and avoided.
- Jerking of load should be avoided during both lifting and lowering.
- When jacks are used for lifting, the load should be centrally placed.
- None should be permitted to stand or move under lifted or hoisted load.
- For hoists, overtravel switches need to be incorporated or physical barriers provided to prevent continued operation of hoist platform at the top and bottom of travel.

15.7 TRANSPORTATION

Management of site traffic should be well planned keeping in view that accidents involving site transport result often in fatalities. Total safety plan of any sizable project should cover traffic management also. Planning of transportation system should include the entire site starting from the entry point ending at the point of exit. There should be separate entry and exit points for roads and railways. On site, roads should be suitable for movement of vehicles/mobile construction equipment with unobstructed visibility and space for manoeuvring especially at the road crossings. It would be ideal if separate entry and exit points for roadways and walkways are provided. With separate entry and exit points, it would be still better if one-way roads are made. 'Free-for-all' approach causes jam and untangling such jam delays progress of work. Disciplined traffic management can be enforced if the same forms part of the on-site training programme.

On-site roads should be so constructed as to be suitable for heavy-duty vehicular movement maintaining gradients, geometry, surface widths, and curves conforming to the provisions of the *Indian Road Congress Handbook*. Roads that are not suitable for vehicular transportation should be fenced off or blocked. Slippery roads should be covered with sawdust, sand, or ash. Potholes should be repaired or filled up temporarily with graded aggregates and clearance should be adequate for the movement of vehicles or operation of construction equipment.

While roads should be suitable for vehicular movement, vehicles also should be certified as roadworthy by the competent government authority. In addition, only properly qualified and licensed (with valid license) personnel should operate/drive construction equipment/transport vehicles, if weather permits, after checking brakes and other safety devices.

For safe handling and transportation of materials and equipment, overloading or loading oversize loads or jerking should be avoided. The loads must be properly secured by lashing and they should not project beyond decks on any side except for a maximum of 1.5 m at the rear. As for overdimensional loading, special precautions should be taken including provision of red lamps along periphery at night. A loaded stationary vehicle or a vehicle under loading/unloading should be effectively put on brakes or blocked.

A driver of a vehicle should not move in the reverse direction unless assisted by a signal man, who should be in a position to have a clear view of the driver and the space behind the vehicle during the reversing operation. Speed limits for different types of vehicles should be specified and displayed at strategic locations at the site. Overspeeding should not be allowed in any case.

No one should try to ride a vehicle unless there is a seat for the rider. Riding on top of loads should not be allowed as there would always be possibility of shifting/slipping/rolling of load due to application of brakes or turning on bends. No passenger should be allowed to get on a truck or tractor hauling flammable materials. No one should get off a vehicle till it grinds to a stationary position and there is a safe place for landing. No one inside or attending a vehicle conveying flammable materials should smoke or use matches or lighters.

15.8 WELDING AND FLAME CUTTING

Hazards related to welding and flame cutting are:

- Heat – direct/radiated
- Sparks
- Fumes/harmful rays
- Electrical arc/shock

The various safety guidelines can be listed as follows:

- IS:817—Code of practice for training and testing of metal arc welders; IS:818—Code of practice for safety and health requirements in electric and gas welding and cutting operations; IS:1179—Specification for equipment for eye and face protection during welding; IS:1393—Code of practice for training and testing of oxy-acetylene welders; IS:2598—Safety code for industrial radiographic practice; and IS:3016—Code of practice for fire precautions in welding and cutting operations are all safety-related codes issued by the BIS on welding and gas cutting.

- As evident from the above, welding and gas cutting work demand precautionary measures before these operations are permitted near areas where combustible materials are stored, or near materials or plant where explosives or flammable dust, gases, or vapour are likely to be present or given off.
- Adequate ventilation in confined spaces should be arranged by means of exhaust fans or forced draught as the conditions require thereby assuring safe welding and gas cutting operations.
- Welding and gas cutting should not be done on drums, barrels, tanks, etc. containing explosives or flammable materials unless the containers have been emptied, cleaned thoroughly by steam, or other effective means making sure that no flammable material is present.
- Welding and gas cutting operations should be carried out only by qualified and authorized persons who should have 'permit to work' issued in their names (IS:5216, Part I).
- A welder should wear goggles with suitable filter lenses and such clothing/overall as is free from grease, oil, or other flammable materials. He should wear safety footwear, gloves, and use safety shield for personal protection from heat and hot metal splashes. Respiratory mask or breathing apparatus should be kept within easy reach for exigency.
- The cables from welding equipment should be placed in such a way that they are not run over by vehicular traffic. Double earthing should be provided.
- Welding and gas cutting torches should be lit by means of friction lighters, pilot flames, or such safe methods, but match sticks should not be used.
- Area around the place of welding should be barricaded to protect others from the harmful rays emanating from the work. When welding or gas cutting is carried out at heights, safety measures should be taken to prevent sparks or hot metal falling on persons or on flammable materials. Fire extinguishers should be available near the location of welding operations.
- Welders should not tamper with safety devices and valves on gas cyliners.
- When acetylene cylinders are coupled, flash arresters should be inserted between the cylinder and coupler block or between the coupler block and the regulator. Only acetylene cylinders of approximately equal pressure should be coupled. Gas should not be taken from a cylinder unless a pressure-reducing regulator has been attached to the valve.
- All cylinders must be used and stored while in upright position only. They must be stored away from open flames and other sources of heat.
- Oxygen cylinders must not be stored near other cylinders containing gas or oil, grease or other flammable materials.

- Oxygen pressure for welding and gas cutting operations should always be high enough to prevent acetylene flowing into oxygen cylinder.
- Welding and gas cutting equipment including hoses and cables should be well maintained.
- Gas lines and compressed air lines should be identified by suitable colour codes for easy identification to prevent fire and explosion hazards. Hoses especially designed only for welding and flame cutting operations should be used. The hoses for acetylene and oxygen should be of different colours.
- Safety precautions for electric arc welding include: (i) when welding is carried out near pipelines carrying flammable materials, a separate earth conductor should be connected to the job directly from the site instead of using the pipelines as earth conductor; (ii) body contact with the electrode or other live parts of electric welding equipment should be avoided; (iii) utmost care should be exercised to prevent accidental contact of electrodes with ground; (iv) the welding cables should not be allowed to get entangled with power cables. In addition, it is to be ensured that the cables are not damaged by movement of materials.
- Radiography should be performed in accordance with all applicable safety requirements. Official regulations should be laid down on site storage, handling, and use of radioactive materials or apparatus generating radiation for non-destructive radiographic tests.
- Workers involved in radiography should undergo medical examination periodically.
- Those engaged in radiography should be allowed to work or stay in danger zone. The radiation zone should be cordoned off with proper signals to stop entry of others.

The regulation of possession, filling, delivery, dispatch, handling, examination, and testing of gas cylinders falls under the Gas Cylinder Rules, 1981. A gas cylinder is a metal container intended for storage and transport of compressed gas and having a volume exceeding 500 ml but not exceeding 1000 litres.

15.9 SCAFFOLDS AND LADDERS

Scaffolds and ladders are meant for safe access for executing short duration and emergency work from the ground level or from a part of any permanent structure or from other available means of support. Suitable means of access should be chosen based upon an evaluation of the work to be done, duration of the work involved, working environment, and capability of personnel engaged for the work. The support base of scaffolds and ladders should be stable.

The scaffolds and ladders should be designed accurately considering the load to be sustained and considering a factor of safety of six. These enabling facilities should be securely anchored or fixed with adjacent structures, if any and should be periodically inspected for necessary repair/replacement.

Sufficient safety guidelines have been incorporated in IS:3696 – Safety code for scaffolds and ladders (Part I deals with scaffolds and Part II with ladders) and IS:4014 (Parts I and 2) – Code of practice for steel tubular scaffolding issued by the BIS.

A ladder should be of rigid construction having sufficient strength for the intended loads and should either be made of good quality wood or metal. All ladders should be maintained well to keep them in safe working conditions. Other safety guidelines for ladders are listed here:

- A ladder must be securely fixed at the top and bottom and must be propped if it is longer than 4.5 m to prevent sagging
- A ladder must extend a minimum distance of 1 m beyond the landing it would serve.
- The correct angle for a ladder should be 1:4 (H:V)
- A ladder with missing rung/s or rungs soiled with slippery substances should not be used
- A ladder must have a firm and level footing—loose materials should be removed from the bottom of a ladder
- A single portable ladder should not be over 6 m in length
- A metal ladder must not be used for electrical work or located near electric circuit or equipment
- A ladder should not be used for carrying materials in hand when climbing as both the hands should be free

Rigidly constructed scaffolds of adequate load-bearing capacity should be used as safe means of access to places of work that cannot be done from the ground level or from other means of support. Scaffolds, if possible, should be load tested or otherwise designed with a factor of safety of six as mentioned before. All scaffolds should be securely supported or suspended and properly braced as and when required to ensure stability. Other safety guidelines for scaffolds are as follows:

- A scaffold should be erected by qualified scaffold erectors under the constant supervision by a qualified scaffold supervisor
- During erection or dismantling of scaffolds, a red tag 'Not Safe for Use' must be placed at the access points to warn workers/supervisors against the use of such scaffolds
- Chains, wire-ropes, or other lifting materials used for the suspension of scaffolds must be of adequate strength and of tested quality – all such chains and ropes should be properly fastened to safe anchorage points
- Scaffolds more than 3.5 m above the ground level or floor level with stationary support or suspended from any overhead support should be protected by at least 1-m high guardrail properly attached, bolted, braced, or otherwise secured with only such opening as required for the delivery of materials
- Scaffold platforms must not be overloaded

- Working platforms should have at least 150 mm high toe boards to prevent tools and materials from falling down
- Every scaffold platform should be kept free from unnecessary obstruction, materials, rubbish, and from any protruding nails
- Working platforms, gangways, stairways, etc., should be so constructed as not to sag unduly or unequally
- All scaffolds should be maintained to assure continued safety

Ladders or scaffolds should not be removed while in use.

15.10 FABRICATION AND ERECTION

Major safety guidelines on fabrication and erection of structural steelwork are discussed under different sections in this chapter:

- Quality of materials – covered by BIS codes
- Housekeeping (Section 15.2)
- Transportation (Section 15.7)
- Mechanical handling (Section 15.6)
- Welding and flame cutting (Section 15.8)
- Personal safety (Section 15.3)
- Fire safety (Section 15.4)
- Electrical safety (Section 15.5)
- Scaffolds and ladders (Section 15.9)
- Stability during erection
- Rigidity of connections

Other safety guidelines include the following:

- IS:7205—Safety code of erection of structural steel work; IS:8989—Safety code for erection of concrete framed structures; and IS:11057—Industrial safety nets are all safety-related codes issued by the BIS on fabrication and erection work.
- Materials received at the site for fabrication must conform to the relevant specifications. This may require testing at the site or at any recognized laboratory.
- Raw or fabricated materials received at the site should be checked for physical or other defects, if any, for possible rectification/replacement as the case may be.
- Raw materials for fabrication should be stacked properly and in an orderly manner so that safety is not jeopardized.
- Steel sections/shapes or fabricated structures or structures under fabrication should be placed/stacked/supported in a safe and orderly manner considering possible rolling, sliding, or toppling.
- For steel fabrication and erection work, qualified welders and gas-cutters should be deployed.
- Drilling machine, grinding machine, etc should be provided with safety guards.

- Lifting and hauling tackles, slings, ropes, etc should be fit for use. Slinging should be done properly so as to avoid slipping.
- Care should be taken to secure erected members in positions using erection bolts, slings, and guys as required. Bracings and ties should be erected on priority for stability.
- Care should be taken against falling/dropping of tools, tackles, etc from heights. Loose tools, bolts, nuts, etc should be kept on trays/boxes, which should be properly anchored.
- To ensure stability of the structure during progressive erection, proper sequence of erection should be followed.
- For safe erection, only one person should be assigned for giving signals.
- Erection should not be done in adverse weather conditions or at night without making appropriate arrangement to ensure safety. Outdoor erection work should normally be carried out in fair weather during the day when visibility is clear.
- Structural members should be preassembled and fastened to the extent possible before erection.
- Before lifting for erection, steel trusses should be sufficiently shored, braced, or guyed till permanently secured in position.
- Guide rope must be used for guiding load that is being lifted.
- Safety nets with suitable and sufficient anchorage should be provided not more than 6 m below temporary platforms on which workers are engaged in erection work at heights.
- The area where erection work is in progress should be cordoned off by barricades to prevent unauthorized entry into the area.

15.11 EXCAVATION

The key to safety in excavation work is planning. All excavations, however small, should be subjected to risk assessment. The main hazards to be controlled include those listed here under:

- IS:3764—Safety code for excavation work issued by the BIS—This code is for reference.
- Collapse of sides—the sides of excavations must be sloped to a safe angle, not steeper than the angle of repose of any particular soil. If it is not possible to provide the required slope, the sides of the excavation should be securely supported by timber or any other type of shoring or sheet piling where there is a danger of collapsing or dislodgement of earth or any material. Regular inspection of the work is also a safety requirement.
- Falling materials from sides—spoil heaps should be kept at least 1.5 m or half of the depth of excavation, whichever is more, away from excavation edges, and vehicles like trucks and dumpers should be kept well away from the edges. Those within an excavation pit should wear safety helmets.

- Falls into the excavation—every accessible part of an excavation, pit or opening in the ground into which there is any danger of persons falling should be fenced off with a barrier up to a height of at least 1 m as close to the edge of excavation as possible. Movement of construction equipment and vehicles should be so controlled as to prevent overloading pressures to the soil in the area.
- Undermining the structures in the area—excavation or earthwork below the foundation level of any adjoining building or structure should not be taken up unless adequate preventive measures are taken against damage to the existing structure or fall of any part. The investigation should be made in advance to determine whether structural support is needed.
- Discovery of underground services—a safe system of work is essential to avoid sudden discovery of services which, if interrupted, could be dangerous to the excavation worker and to those supplied by the service. Safe digging practices include manual digging where necessary. Cable locators should always be used to add to information from plans, which is often relatively inaccurate.
- Access and egress to the excavation—ladder or other fixed access to excavation pits is essential for emergency escape. Ladders should be placed at every 25 m and be fixed securely at the top. There should not be any problem on step-off from the top rung of a ladder. The ladder should be extended from the bottom of the pit to at least 1 m above the surface of the ground.
- Fumes and lack of oxygen—in deep excavations, oxygen supply may be limited and fumes may accumulate. This may be hazardous in confined space. How bad the hazard is could be checked by experienced and competent persons from time to time depending on the gravity of the prevailing conditions.
- Occupational health considerations—the workers having bodily cuts and bruises and engaged in excavation work in contaminated soil must wear protective clothing including safety boots and gloves. They must maintain good standard of personal hygiene. Arrangements should be made for their regular medical check-up.

The safety guidelines for mechanized excavation and earthwork are outlined here:

- One should keep safe distance from construction equipment.
- One should never walk close behind construction equipment or under its boom.
- One should always look both ways and listen before crossing a haul road; scrapers can move very fast. The haul roads are meant for disposal of excavated spoil at designated dumping area using scrapers or scrapers pushed by bulldozers.
- Pedestrians and light vehicles should not travel on the haul road.
- One should not park anything on a haul road.
- Generally, haul roads should be maintained in good condition by regular grading and compaction as this would reduce braking distances. In the summer, dust should be kept to a minimum by regular watering.

- Haul road gradients should be kept to a minimum and large cross-falls should be avoided.
- Overhead cables and height restrictions should be clearly marked with signs and tapes.

15.12 BLASTING

Blasting in effect is the conversion of a chemical substance in the explosive into a gas thereby producing enormous pressure in the confined place of a blast hole thus breaking rocks by the huge compressive forces generated. Blasting is also, apart from excavation, effective in demolition.

The Explosives Act, 1884 and the Explosives Rules, 1983 govern the storage, transport, and use of explosives. A contractor must hold a valid license under the Explosives Rules, 1983 for possession and/or use of explosives.

Excavation in rocks is performed in daylight by blasting at such time as notified to all. Muffling is done to prevent splinters from flying. Personnel engaged for blasting should be trained, experienced, and licensed. Detonators and explosives are stored separately in magazines and also transported to sites separately. The condition of blast holes should be checked before inserting cartridges. Guards with red flags and whistles should be posted at the periphery of hazardous area to keep people away at a safe distance during blasting. The excavated pits should be fenced to prevent people from falling in the same.

IS:4081, the safety code for blasting and related drilling operations, issued by the BIS is relevant on rock excavation by blasting.

The safety guidelines for transportation of explosives include the following:

- The word 'Explosives' should be marked on the front, rear, and both sides in contrasting colour with the background in any vehicle transporting explosives or a red flag with the word 'Explosives' or 'Danger' be displayed in a conspicuous place of the vehicle
- Vehicles transporting explosives should not be overloaded and in no case should the explosive containers be piled higher than the closed sides of the body. Any vehicle with an open body should have a tarpaulin to cover the explosives containers
- All vehicles when transporting explosives should be in safe working conditions
- The floor of a vehicle should be without any exposed metal on the inside of the body.
- No trailer should be attached to a vehicle hauling explosives
- Passengers or other unauthorized persons should not be allowed to ride a vehicle transporting explosives

Following are the safety guidelines for handling explosives:

- Packages or containers of explosives should not be thrown or dropped while being loaded and unloaded or otherwise handled, but they should be carefully deposited and stored or placed in such a manner as to prevent the packages or containers from sliding or falling or being otherwise displaced

- Containers of explosives should not be opened inside a magazine or within 15 m of a magazine
- Tools made of woods or other non-metallic material should be used for opening containers of explosives
- Explosives/detonators issued to individual workers should be placed in separate insulated carriers or containers equipped with lids for safe handling
- None except an attendant should be permitted to move with explosives/detonators when the same are being transported in a shaft, slope, or other underground working

Following are the safety guidelines for storing explosives:

- A magazine should be constructed after obtaining 'No Objection Certificate' from the licensing authority under the Explosives Rules, 1983.
- Only trained and competent operators should handle and store explosives.
- Explosives and detonators should be stored separately in detached, dry, ventilated, bulletproof, and fire-resistant magazines, away from other buildings, railways, and roads/highways.
- A magazine for the storage of dynamite should be constructed in such a manner that it would prevent freezing of the dynamite during extended periods of cold weather. If the dynamite does freeze, it should be thawed before it is handled or used, as the danger of premature firing is much greater when it is frozen.

15.13 FORMWORK

Risks involved in erection and stripping of formwork are related to the following:

- Form materials—must be strong quality-wise (with the correct moisture content in case of timber)
- Form supports—strengthening and supporting of forms should be adequate and should be based on: (i) loading of wet concrete, self-weight of forms plus props/shores/ties/bracings and superimposed loads to be borne, and also of wind/flowing water; (ii) spans; (iii) setting temperature of concrete; and (iv) rate of placing concrete
- Form stripping—danger of materials falling during removal of forms and supports/props/stiffeners
- Premature form stripping—forms should not be stripped before the concrete has attained sufficient strength for self-support

The safety guidelines involved in formwork are as listed here:

- The materials of formwork should be carefully examined before being used to ensure soundness and strength of the same.
- Formwork should be examined, erected, and dismantled under the supervision of qualified and experienced personnel, and as far as practicable by workers familiar with such work.

- For concrete work, supports for formwork should be properly designed by taking into consideration the loads, spans, setting temperature, and rate of placing concrete.
- Forms as well as supporting props should be adequately braced horizontally and diagonally in both longitudinal and transverse directions to develop total load-carrying capacity.
- Props or shores should be made of steel or straight-grained timber strong enough to support vertical loads and lateral thrusts. Both props and shores should rest on firm and unyielding bases.
- Heavy building materials should not be dumped on forms.
- To prevent danger from falling parts when forms are being stripped, forms should be as far as practicable be moved down as a whole. Otherwise, the remaining parts should be supported.
- Forms and form panels should be provided with adequate number of U-bolts or other lifting attachments.

15.14 CONCRETING

As the concreting operation is started, forms become working containers, loads and thrusts develop, noise disturb communication, equipment is set working, and a considerable tonnage of materials begin to be moved. This is when all the careful attention to the planning of working method attains a peak where training and experience means the difference between safe working and unsafe working. Hazards of concrete work are related to:

- Construction equipment—safe operation and use
- Shuttering, scaffolding, walkways, and access ladders—safety is to be ensured
- Concrete placing—safe use of buckets, slings, U-bolts, etc., should be ensured
- Hardened concrete—formwork should not be stripped before concrete has attained sufficient strength to bear self-weight and possible superimposed load

The safety guidelines relating to concreting are enumerated as under:

- Concrete buckets for use with cranes and aerial cableways should be free from any projections from which accumulated concrete could fall. Utmost care is needed to ensure that a bucket would not fall due to snapping of sling or dislodging of journal bearing or any other reason.
- Loaded concrete buckets should be guided into position by appropriate means. During tipping of concrete bucket, workers should be moved away from the range of any kickback due to concrete that sticks to the bucket.
- A winch should be located in a strategic position so that its operator would be able to see filling, hoisting, unloading, and lowering of bucket.
- In case of pumped concrete, scaffolding carrying the concrete delivery pipe should be strong enough to support the pipe conveying concrete with all the concerned workers on the scaffolding at the same time.

- Concrete pipes should be properly secured at the ends and curves, and should be provided with air release valves near the top. A pipe should be securely attached to the pump nozzle by a bolted collar or equivalent means.
- When concrete pipes are cleaned by compressed air, all workers not involved in the process should be kept away at a safe distance.
- Pressure gauges on the pumps should be checked at the beginning of every shift.
- Only physically fit workers should operate vibrators. Despite this, all possible measures need to be taken to reduce vibration transmitted back to such workers.
- Regarding electrical vibrators, safety measures should be ensured as per the provisions of Section 15.5.
- Reinforcing steel rods projecting from floors or walls should be provided with protection as a safety measure. The rods should also not be stored on scaffolding or formwork in such a way as to endanger stability.
- For a manually operated mixer, no one should stand under its hopper. For safe operation of such mixer, condition of wire-rope and hoisting/anchoring brake/s need to be checked daily.
- More dust is generated in manual batching and mixing of concrete. For 'Personal safety', refer to Section 15.3.

15.15 FLOORS

The hazards to be considered during construction of floors along with connecting staircases include fall from heights because of unprotected edges and openings. Keeping such hazards in view, the safety guidelines to be followed are described hereunder:

- Fall prevention should be kept in view right from the design stage and sequence of construction or erection should be planned accordingly. This way, safety can be assured more effectively.
- All floor openings and floor edges should be protected with temporary hand railings. Small floor openings should be covered with steel plates/gratings or timber planks duly reinforced/fixed.
- Similarly, edges of staircases should also be protected by temporary hand railings or temporary masonry work.
- The floor edges should be protected with temporary hand railings till such time the side cladding is taken up. The edges of unfinished slabs also should be protected with temporary barriers.
- Similar safety measures should be followed in case of steel structures also. Temporary railings or coverings may be tack-welded in case of steel structures
- Proper housekeeping would eliminate hazards in movement on floor. For guidelines on housekeeping, refer to Section 15.2.
- Illumination level should be sufficient to ensure movement or working on any floor hazard-free. For the required illumination level, refer to Section 2.2 of Chapter 2.

- Areas on floor where work is in progress should be barricaded. Unauthorized persons should be cautioned against hazard of trespassing. Appropriate warning signs must be provided on the barricade/s cautioning unauthorized persons to stay away.

15.16 ENVIRONMENT AT SITE

The basic approach should be to suppress/control pollution at the source rather than its abatement once generated/caused. Environmental protection benefits include reduced damage to the environment by controlling emissions and adverse impacts on the ecosystems, and reduced demand on natural resources. The short-term objectives for the reduction of environmental impact of any process are:

- Reduced consumption of resources
- Reduction of emissions and other by-products of the process to air, water, and land
- Reduction of production (of waste) and increased recycling of waste

A contractor, in the longer-term, should examine their activities to minimize the consumption of energy resources and water onsite or offsite including offices by assessing actual consumption as accurately as possible and setting targets to be reached.

Initial Ground Contamination

Initial ground contamination, if any, and its extent should be established during soil investigation and necessary action should be enforced to restore a polluted site to a condition suitable for execution of construction work safely.

Waste Management and Pollution Control

- Waste should be prevented or reduced at the source by proper planning and designing whenever feasible
- Waste that cannot be prevented or reduced should be reused whenever feasible
- Waste that cannot be prevented, reduced, or reused should be recycled in a safe and environment-friendly manner whenever feasible—recycling is treating a material to recover a usable product
- Waste that cannot be prevented, reduced, reused, or recycled should be treated in a safe and environment-friendly manner and the energy should be derived from it whenever feasible
- Disposal or release into the environment as a last resort in a safe and environment-friendly manner whenever feasible
- Hazardous waste materials are: asbestos and asbestos-containing products, flammable materials, lead and lead compounds, organic halogen compounds, acids and alkalis, inorganic metallic and non-metallic compounds

Air Pollution Control

- Control of air pollution requires limiting the emission of air pollutants into the atmosphere
- Six major air pollutants are: (i) carbon monoxide, (ii) ozone, (iii) sulphur dioxide, (iv) nitrogen dioxide, (v) particulate matter like dust, smoke, mist, fog, smog, and sprays, and (vi) lead (automobile emissions)
- There are also air emission standards for friable asbestos during a number of activities and disposal of asbestos or asbestos-containing materials
- Pollution from the use and maintenance of motor vehicles and construction equipment should be controlled
- Releases from stationary sources such as boilers, furnaces, and the burning of waste should be controlled

Water Pollution Control

- Water should be physically, biologically, and chemically free of pollution
- Toxic or conventional pollutants include: (i) pollutants affecting pH value; (ii) pollutants affecting temperature; (iii) flammable pollutants; and (iv) pollutants that require pre-treatment
- Spillage from oil storage tanks and surrounding dyke area on account of: (i) loading/unloading, (ii) ruptured hydraulic or fuel line or tank, (iii) release from petroleum/gas/oil tanks, (iv) leaking underground storage tanks, and (v) manhole contamination should be controlled
- Each contractor must furnish all information including the means of disposal and containment of spillages, if any

Noise Control

Noise can be defined as unwanted sound. Awareness of this problem has increased in recent years and actions are being taken to solve them. A noise problem comprises three inter-related problems: (i) source, (ii) receiver, and (iii) transmission path between the source and the receiver. The transmission path is usually the atmosphere. Any comprehensive discussion of the environment must now include possible noise pollution. The safety guidelines to control noise pollution are:

- Procuring equipment with in-built noise control
- Moving the noise source away from the work area
- Building sound-absorbing total enclosure surrounding the noise source
- Using silencers whenever feasible
- Using personal safety gear (as discussed in Section 15.3)

The protection of environment is governed by the Environment (Protection) Rules under the Environment (Protection) Act, 1986.

15.17 FIRST AID

First aid is the first help received by an injured person before proper medical help is made available or till the arrival of a doctor. Medical help should be available at the shortest possible time.

First aid facilities should be available with all agencies engaged in construction work. However, it would be better if the customer provides a room, which should be easily accessible, for rendering first aid during working hours. Well-equipped ambulance/s should also remain parked at a site for emergency use. The medical services that should be available in the work place would depend on the nature of the work and the related hazards, the number of employees, and the location of the workplace relative to the availability of external medical assistance. Shift working should be taken into account in counting the total number of workers. In addition, the ratio of persons trained for rendering first aid to the total number of workers should be kept in view.

The first-aid room should have a first-aid box containing at least: (i) roll of 25 or 50 mm bandages for bandaging sterile dressing over wounds; (ii) triangular bandages; (iii) assorted band-aids; (iv) antiseptic cream/lotion/wipes; (v) antihistamine cream; (vi) 25 or 50 mm adhesive tapes; (vii) roll of absorbent cotton; (viii) bottle of calamine lotion; (ix) safety pins; (x) snub-nosed scissors; (xi) pair of round-ended tweezers/forceps; (xii) two wooden splints; (xiii) eye bath/eye pad; (xiv) pen torch; (xv) soluble painkiller; and (xvi) instruction book on first aid. The contents of the first-aid box should be inspected regularly by qualified personnel and the items used up or those not in usable condition should be replenished or replaced.

One rendering the first aid should, if possible, note down the following about the person to be given the first aid:

- Identity and address
- Blood group
- Condition of heart
- Diabetic patient or not
- Epileptic patient or not

There should be prompt action in rendering first-aid services in the cases of severe bleeding and suspended breathing. All injuries, major or minor, should be treated, reported, and recorded. The contractor must submit to the owner fortnightly a report on the accidents that occurred during the preceding fortnight describing the circumstances leading to the accidents and the extent of damages and injury caused by them.

Cell phones or walkie-talkie should be used for better communication between different organizations and different locations (horizontal or vertical) in case of accidents causing injuries or resulting in fatalities (Section 2.2 of Chapter 2).

15.18 ACCIDENTS

Majority of the accidents are the direct results of unsafe activities and conditions – they are the immediate or primary causes of accidents. Relatively, a few are caused by unsafe conditions. Accidents are also caused by unpreventable secondary reasons (external to the component or system), but very rarely – they are usually harder to seek out and identify. Accidents can be avoided by taking corrective actions on unsafe activities and conditions.

The unsafe acts include:

- Negligence on using personal safety gears
- Working without authority
- Making safety devices inoperative
- Operating and working at unsafe or wrong speed
- Using unsafe/defective equipment, using hands, etc.
- Using equipment in the wrong way or for the wrong tasks
- Disconnecting safety devices such as guards
- Unauthorized servicing and maintaining of moving equipment
- Leaving equipment in dangerous condition
- Unsafe loading, placing, mixing, etc.
- Taking position disregarding safety
- Smoking in hazardous areas or drinking alcohol
- Distracting, teasing, horseplay, etc.

The unsafe conditions include:

- Ineffective housekeeping
- Unsafe or inadequate illumination
- Improper ventilation
- Excessive noise
- Inadequate fire warning systems
- Unguarded or inadequately guarded moving machine parts
- Missing platform guardrails
- Defective condition—rough, sharp, slippery, etc.
- Unsafe design or construction
- Hazardous arrangement or processes like piling, storing, layout, etc.
- Hazardous atmospheric conditions

Both the unsafe conditions and acts can be eliminated by effective management control and strict supervisory intervention. Accident prevention is a management function. Accidents are costly and the cost of an accident can be worked out. The management should investigate the causes of accidents and take corrective and preventive measures promptly.

The indirect costs of an accident include:
- Cost of lost time of the injured employee
- Cost of time lost by other employees who stop work because of an accident
- Cost of time lost by supervisory staff to: (i) assist the injured employee; (ii) investigate the cause of the accident; (iii) arrange for a substitute to replace the injured employee; (iv) prepare an accident report
- Cost related to damaged equipment or other property
- Cost related to spoilt or damaged materials
- Cost due to delayed progress on the project
- Cost of paying wages to the injured employee during the period of injury
- Cost of lost production resulting from the disruption of work of other employees for a while following an accident

Accidents at the construction sites can be classified as follows:
- Uncontrollable contact between men and equipment or between men and materials such as cranes, trucks, and material storage
- Failure of temporary structures such as forms, scaffolds, ramps, ladders, cofferdams, cut sheets, etc.
- Inherent engineering hazards such as the use of explosives, presence of injurious gases, toxic dusts, etc.
- Unsafe practices of individual workers or personal hazards resulting from the carelessness of workers

An analysis of the causes given here indicates that most of the accidents could be avoided through the application of an effective safety programme. The likelihood of a sequence of events that could cause an accident should first be assessed:
- Tabulate different stages or phases of the system's mission; system's operation should be diagrammed; systems/sub-systems' logical relation should quantify the failure probabilities
- List the risk sensitiveness in each phase; identify weakness in a system; identify those which contribute most heavily to risk; suggest ways to minimize
- Select proper people, train them, and do not overwork them

The seven principles to be observed in formulating strategies for control and management of safety and health at construction sites by accident prevention are as follows:
- Avoid a risk altogether, if possible, by eliminating the hazard
- Tackle risks at source like designing floors without floor opening/s
- Adapt work to the individual when designing work areas and selecting methods of work; ergonomics aims in improving the interface between people and their workplaces
- Use technology to improve conditions – quieter equipment would not create noise problem

- Give priority to protection for the entire workplace rather than to individuals; personal safety gear should not be the only safety measure
- Ensure that everybody understands the means to remain safe and healthy at work
- Make sure health and safety management is accepted by every individual and that it applies to all aspects of the organization's activities

There are no standard or formal requirements for calculating frequency and severity rates of accidents that are the indices used in computing injury rates.

$$\text{Frequency rate} = \frac{\text{Number of injuries} \times 1{,}000{,}000}{\text{Total number of hours worked}}$$

$$\text{Severity rate} = \frac{\text{Total days lost plus notional days charged} \times 1{,}000{,}000}{\text{Employee hours of exposure}}$$

The frequency rate as shown is essentially a weighted frequency rate that permits the days lost due to temporary total disability to be recorded for fatalities and permanently disabled cases. The notional days often used are 6,000 (20 working years at 300 days per year) for death or permanent disability and 1,800 for the loss of an eye. There are scheduled charges for permanent partial disabilities based on the Workman's Compensation Act (India). The frequency rate can be improved by artificial means like providing temporary jobs to those who are convalescing.

SUMMARY

Safety is jeopardized by some physical failure and individual error. Based on this, a good and effective safety programme has to be planned, implemented and followed consistently. This requires total commitment from the topmost management. Not only that, safety should be the concern of also design engineers, manufacturers, importers and suppliers' engineers. 'A permit to work' procedure is required as part of a safe system of work when, because of potentially hazardous circumstances, there is a need to strictly control access into areas, rooms, confined spaces and/or control specific work to be carried out on plant or equipment where there is inherent danger. Safety programme should also aim at prevention of adverse environmental effects like release of toxic gases, water pollution, noise etc on the neighborhood. While a 'Caution' notice signals warning against interference with work/equipment, a 'Danger' notice signals warning against risk to health or bodily injury.

The modern concept of safety relies more on involving the employees directly in the health and safety programme apart from prescribing their specific assignments. A judicious combination of application of behavioural sciences and engineering, and approach based on technology is a must in preventing accidents. Behavioural science is relevant as safety is inherent in right attitude and sense of priority. Training is essential as knowledge is prerequisite for any successful activity. In training the employees, the management imparts knowledge and skill. What the employees gain on training, apart from knowledge and skill, is motivation. Accident is not only costly but also delays progress of the project work.

REVIEW QUESTIONS

1. What do the personnel engaged in construction work acquire when training is imparted?
2. Why is it necessary to form a safety committee at site? Who should comprise the safety committee? Who should be at the head of the safety committee and why?
3. What is ergonomics? How is ergonomics relevant in construction work?
4. How is hazard different from risk?
5. Describe 5S housekeeping as practised in Japan.
6. For personal safety, what are the 5Ws followed by those involved in piling work?
7. Fuel, heat, and oxygen are the three ingredients required to cause fire. What are the 'Ten Commandments' on fire safety?
8. What are the 'Ten Commandments' on electrical safety? How temporary electrical cables should be laid at construction sites?
9. What may go wrong in the performance of mechanical handling equipment?
10. What should be done for safe handling and transportation of materials and equipment?
11. What are the safety precautions that should be taken for electric arc welding?
12. What are the important safety guidelines for design, erection, and use of scaffolds?
13. What kind of care should be taken to secure erected structural steel members in positions?
14. What are the main hazards involved in excavation in soil?
15. What are the important safety guidelines for transportation, handling, and storing of explosives used for excavation by blasting?
16. What are the risks involved in erection and stripping of formwork?
17. What kinds of hazards are related to concrete work?
18. What is the main safety hazard involved in floor construction?
19. What are the short-term objectives for the reduction of environmental impact of any process?
20. What kinds of facilities should be available at the site for 'first aid'?
21. Name ten unsafe acts that may result in accidents.

Bibliography

Aitcin, P. C. and A. M. Neville, *High-performance Concrete Demystified*, Concrete International, 1993.

American Welding Society (AWS), Structural Welding Code, 1st ed., (superseding earlier edition), 1972.

American Welding Society and Welding Institute of Canada, Module 6 *Electrodes and Consumables for Welding*.

Antia, K. F., *Railway Track*, The New Book Company Private Limited, Mumbai, 5th ed., 1960.

Ataev, S. S., *Construction Technology*, Mir Publishers, Moscow, (reprinted) 1985.

Blodgett, Omer W., *Design of Welded Structures*, published by the James F. Lincoln Arc Welding Foundation, 1966.

Bonnett, C. F., *Practical Railway Engineering*, Imperial College Press, London, 2nd ed., 2005.

Borshchov, T., R. Mansurov, and V. Sergeev, *Land Reclamation Machinery*, Mir Pub'ishers, Moscow, English translation, 1988.

Bowles, J. E., *Foundation Analysis and Design*, The McGraw-Hill Companies Inc., 5th International ed., 1997.

Bray, R. N., A. D. Bates, and J. M. Land, *Dredging*, Butterworth-Heinemann, 1996.

Bungey, J. H., *The Testing of Concrete in Structures*, *General Catalogue*, Surrey University Press, New York, USA, 1st ed., 1982.

Capper, L. P. and W. F. Cassie, *The Mechanics of Engineering Soils*, Asia Publishing House, Mumbai, 1st Indian ed. (reprinted), 1962.

Cement Manufacturers' Association, *Handbook on Cement Concrete Roads*, New Delhi, 1993.

Chen, W. F. (Editor-in-chief), *The Civil Engineering Handbook*, CRC Press, USA, 2nd ed., 2003.

Chudley, R., *Construction Technology* (Vol. 4), Longman Group Limited, London, 1st ed., 1977.

Chudley, R., revised by R. Greeno, *Building Construction Handbook*, Addison Wesley, Longman Group Limited, England, 3rd ed., 1999.

Clayton, C. R. I., M. C. Mathews, and N. E. Simons, *Site Investigation*, Blackwell Science Ltd, Oxford, UK, 2nd ed., 1995.

Collocott, T. C. and A. B. Dobson (editors), *Dictionary of Science and Technology*, Allied Publishers Private Limited, (reprinted) 1984.

Construction EQUIPMENT & MACHINERY in India – A compilation of articles published in Civil Engineering & Construction Review (1988 – 1990).

Cooke, T. H., *Concrete Pumping and Spraying; A Practical Guide*, Thomas Telford Ltd, 1990.

Day, D. A., *Construction Equipment Guide*, John Wiley & Sons, 1973.

Deshmukh, R. T. and D. J. Deshmukh, *Winning Coal & Iron Ore* (Vol. 1), published by D. J. Deshmukh, Amaravati, Maharashtra, June 1967.

Dorf, R. C. (Editor-in-Chief), The Civil Engineering Handbook (Part A & B), CRC Press, USA, Jaico Publishing House, 1st ed., 1998.

Emmitt, S. and C. A. Gorse, *Barry's Advanced Construction of Buildings*, Blackwell Publishing, 1st ed., 2006.

Fintel M. (Editor), *Handbook of Concrete Engineering*, 2nd ed., CBS Publishers and Distributors, Delhi, 1st Indian ed., 1986.

Fleming, W. G. K., A. J. Weltman, M. F. Randolph, and W. K. Elson, *Piling Engineering*, John Wiley & Sons Inc., 2nd ed., 1992.

French, C. W. and A. Moktharzadeh, *High Strength Concrete: Effects of Materials, Curing and Test Procedures on Short-term Compressive Strength*, PCI Journal.

Garg, S. P., *Groundwater and Tube Wells*, Oxford & IBH Publishing Co., 2nd ed. (reprinted), 1985.

Gaylord, E. J. (Jr) and C. N. Gaylord, *Structural Engineering Handbook*, McGraw-Hill Book Co., USA, 1968.

Gourd, L. M., *Principles of Welding Technology*, E. Arnold, London, 2nd ed., 1986.

Griffith, A. and P. Watson, *Construction Management Principles and Practice*, Palgrave Macmillan, 1st published 2004.

Grundy, J. T., *Construction Technology* (Vol. 1 & 2), Viva Books Pvt. Ltd, 1st South Asian ed., (published by arrangement with Edward Arnold Publishers Ltd., London), 1998.

Gustafsson, R., *Swedish Blasting Technique*, SPI, Gothenburg, Sweden, 1973.

Harris, F., *Modern Construction and Ground Engineering Equipment and Methods*, Longman Scientific & Technical, Longman Group Limited, UK, 2nd ed., 1994.

Havers, J. A. and F. W. Stubbs (Jr), *Handbook of Heavy Construction*, McGraw-Hill, New York, 1971.

Hewlett, P. C., *Lea's Chemistry of Cement and Concrete*, Butterworth-Heinemann, 4th ed., 2004.

Hofler, J. and J. Schlumpf, *Shotcrete in Tunnel Construction*, Reinhardt + Reichenecker GmbH, Edition 09/2004.

Holt, A. S. J., *Principles of Construction Safety*, Blackwell Publishing (Wiley Group), 2001.

Indian Roads Congress, *Specifications for Road and Bridge Works*, 4th revision, 2001.

John, R. D., S. Ramalingam, and V. R. Rengasamy, *Durability of Concrete in Aggressive Environment in Space Projects* – An Article.

Kalissky, V., A. Manzon, and G. Nagula, *Automobile Truck Driver's Manual*, Mir Publishers, Moscow, 2nd printing, 1988.

Kendrick, P., M. Copson, S. Beresford, and P. McCormick, *Roadwork Theory and Practice*, Butterworth-Heinemann, 2004.

Khanna, P. N., *Indian Practical Civil Engineers' Handbook*, Engineers Publishers, New Delhi, 2001.

La Londe Jr., W. S. and M. F. Janes, *Concrete Engineering Handbook*, McGraw-Hill Book Company, 1st ed., 1961.

Langefors, U. and B. Kihlstrom, *Rock Blasting*, John Wiley & Sons, 3rd ed., 1979.

Lea, F. M., *The Chemistry of Cement and Concrete*, Edward Arnold Publishers Limited, 3rd ed. (reprinted), 1980.

Lewitt, E. H., *Hydraulics and Fluid Mechanics*, Sir Isaac Pitman & Sons Ltd, London, 10th ed. (reprinted), 1959.

McGraw-Hill's Construction Contracting, USA 'Handbook of Construction Techniques'.

Mehta, P. K., and J. M. Monteiro, *Concrete Microstructure, Properties, and Materials*, Indian Concrete Institute, supported by ACC, (reprinted) 1999.

Merritt, F. S. and J. T. Ricketts, *Building Design and Construction Handbook*, McGraw-Hill Book Co., 1968.

Montgomery, R. L. and J. W. Leach, *Dredging and Dredged Material Disposal* (Vol. 1 & 2), American Society of Civil Engineers, New York, 1984.

Moskvin, V., F. Ivanov, S. Alekseyev, and E. Guzeyev, *Concrete and Reinforced Concrete Deterioration and Protection*, Mir Publishers, Moscow, English translation, 1983.

Neville, A. M. and J. J. Brooks, *Concrete Technology*, ELBS with Longman Group Ltd, England, ELBS ed., (reprinted) 1994.

Neville, A. M., *Maintenance and Durability of Concrete Structures*, ICI Bulletin No. 59, April-June 1997.

Neville, A. M., *Properties of Concrete*, Longman Group Ltd, UK, ELBS 3rd ed. (reprinted), 1988.

Newman, J. and Choo Ban Seng, *Advanced Concrete Technology* (4 volumes), Elsevier Butterworth-Heinemann, Oxford, UK, 1st ed., 2003.

Nichols Jr., H. L. and D. A. Day, *Moving the Earth*, McGraw-Hill (a division of The McGraw-Hill Companies), 4th ed., 1998.

O'Brien, J. J., J. A. Havers, and F. W. Stubbs Jr., *Standard Handbook of Heavy Construction*, McGraw-Hill Book Co., 2nd ed., 1971.

Perry's Chemical Engineers' Handbook, McGraw-Hill International, 6th ed., 1984.

Peurifoy, R. L., *Construction Planning, Equipment, and Methods*, McGraw-Hill Book Co. Inc., International student edition, 1956.

Raina, V. K., *Concrete Bridge Practice Analysis, Design and Economics*, Tata McGraw Hill, 1st ed., 1990.

Rakshit, K. S., *Design and Construction of Highway Bridges*, New Central Book Agency, 1st ed., 1992.

Ramchandran, V. S. and J. J. Beaudoin (Editors), *Handbook of Analytical Techniques in Concrete Science and Technology, (Principles, Techniques and Applications)*, William Andrew Publishing/Noyes Publications, USA, 2001.

Ray, I. and A. K. Chakraborty, *Speciality Admixtures: A State-of-the-art Report* – An article published in the souvenir of All India Seminar on Construction Chemicals: present status and scope for improvements on July 30, 1998.

Richardson, J. G., *Supervision of Concrete Construction* (Vol. 1 & 2), Palladian Publications Ltd., London, 1st published 1986 (Vol. 1) & 1987 (Vol 2).

Robinson, R. and B. Thagesen, *Road Engineering for Development*, Taylor & Francis, 2nd ed., 2004.

Rodichev, V. and G. Rodicheva, *Tractors and Automobiles*, Mir Publishers, Moscow, 2nd printing, 1989.

Roninson, E.G., Motor Graders, Mir Publishers, Moscow, English Translation, 1985.

Rzhevsky, V. V., *Opencast Mining Unit Operations*, Mir Publishers, Moscow, English translation, 1985.

Santhakumar, A. R., *Concrete Technology*, Oxford University Press, New Delhi, 1st ed., 2007.

Sarkar, S. K., *Building a Thermal Power Project – A Case Study*, published in the Seminar Papers dated 2–4 January 1998 of the Institution of Engineers (India), Ranchi.

Sarkar, S. K., *Concreting in Cold Weather*, published in the Indian Concrete Journal, January 1977.

Sarkar, S. K., *How to Achieve Maximum Benefits from Slipforms*, published in the Indian Concrete Journal, April 1977.

Sarkar, S. K., *Mechanised Construction in India*, published in the Seminar Papers dated 14 June 1991 of the Institution of Engineers (India), Kolkata.

Sarkar, S. K., *Role of Management on Safe Implementation of Project Maintaining Quality,* published in the Seminar Papers dated 19–20 March 2004 of the Institution of Engineers (I), Kolkata.

Second International Symposium on Sprayed Concrete, Norway, 23–26 September 1996 – Proceedings.

Sen, B., *Chemical Composition and Physical Structure of Concrete* – An article by the former Director of B. E. College (deemed University) published in connection with the Workshop on 'High Performance Concrete' on 8 March 1997 in Kolkata.

Sen, B., *Construction Chemicals* – 'Keynote Address' by the former Director of B. E. College (deemed University) at the All India Seminar in Kolkata on Construction Chemicals: Present Status and Scope for Improvements, 30 July 1998.

Sen, B., *Lightweight Concrete* – An article by the former Director of B. E. College (deemed University) published in connection with the Workshop on 'High Performance Concrete' on March 8, 1997 in Kolkata.

Sengupta, P. C., *How to Map the Earth, True to Size and Shape*, Barman & Co (Technical Publications), 1st ed., 1954.

Sharma, P. C. and D. K. Aggarwal, *Machine Design*, Catson Publishing House, Ludhiana, 3rd ed. (reprinted), 1983.

Singh, J., *Heavy Construction Planning, Equipment and Methods*, Oxford & IBH Publishing Co. Pvt Ltd, New Delhi, 1993.

Singh, V., *Wells and Caissons*, Nemchand & Bros, Roorkee, UP, 2nd ed., 1981.

Smith, N. J., *Engineering Project Management*, Blackwell Science Ltd, Oxford, UK, 2nd ed., 2002.

Steel Designers' Manual, Constructional Steel Research and Development Organization, Crosby Lockwood, London, 4th ed. (2nd ELBS edition), 1972.

Steel Designers' Manual, The Steel Construction Institute, Graham W. Owens and Peter R. Knowles (Editors), Blackwell Science Ltd, Oxford, 5th ed. (reprinted), 1998.

Supplement to IEI News, Institution of Engineers (India), December 1999.

Supplement to IEI News, Institution of Engineers (India), March 2003.

Terzaghi, K. and R.B. Peck, *Soil Mechanics in Engineering Practice*, John Wiley & Sons, 2nd ed., 1967.

Testing Equipment for the Construction Industry, General Catalogue, Ultra Technologies Pvt. Ltd, New Delhi, 6th ed., 2005.

The British Steel Piling Co. Ltd, *The B. S. P. Pocket Book Piling Edition, Technical Information and Tables on pile driving operations,* 8th ed., 1963.

The Indian Roads Congress, *Handbook of Quality Control for Construction of Roads and Runways*, 1st revision, New Delhi, 1977.

The Indian Roads Congress, *Highway Research Bulletin No.41,* 1990.

The Institution of Structural Engineers issued in conjunction with the Cement and Concrete Association, UK, *Report on concrete practice, PART ONE: Materials and Workmanship* issued in May 1963.

The US Department of the Interior Bureau of Reclamation, *Concrete Manual*, 7th ed. (revised reprint), 1966.

Tomlinson, M. J., *Foundation Design and Construction*, Prentice-Hall (an imprint of Pearson Education), Essex, England, 7th ed., 2001.

Tomlinson, M. J., *Pile Design and Construction Practice*, E & FN Spon (an imprint of Chapman & Hall), London, 4th ed., 1994.

Twort, A. C. and J. G. Rees, *Civil Engineering: Supervision and Management*, Elsevier Butterworth-Heinemann, Oxford, UK, 4th ed., 2004.

Ugural, A. C., *Mechanical Design: An Integrated Approach*, McGraw-Hill (Higher Education), 2004.

Unwalla, B. T., *Concrete Technology-An Overview* – An invited 'State-of-the-art' article published in the Institution of Engineers (I) Journal in November 1980.

Urquhart, L. C., *Civil Engineering Handbook*, McGraw-Hill Book Co. Inc., USA, 4th ed., 1959.

Waddell, J. J. and J. A. Dobrowolski, *Concrete Construction Handbook*, McGraw-Hill Inc., USA, 3rd ed., 1993.

Warren, D. R., *Civil Engineering Construction: Design and Management*, Macmillan, Hampshire, 1996.

Zaitsev, Yu. V., LrK. L. Ovsyannikov, and V. F. Promyslov, *Design and Erection of Reinforced Concrete Structures*, Mir Publishers, Moscow, English translation, 1986.

Index

A

Abrasion
   drilling 115
   resistance 223
Absolute weight strength 120
Abutment 312
Accident 531
   causes of 12
Acid 273
   attack 182
   pickling 304
Acoustic
   emission method 256
   pressure 192
Active earth pressure 49
Additives 348
Admixture 173, 231
   air-entraining 203
   anti-washout 204
   blended 205
   extended set control 202
   gas-forming 205
   retarding 203
   self-curing concrete 204
   set accelerating 202
   use of 200
Aerial cableway 466
Aggregates 172, 220, 251
   alkali 204
   coarse 198
   graded 198
   normal 198
   screening of 478
   type and grading of 222

Air pollution control 552
Alkali silica reaction 192
Aluminium profiled sheets 325
ANFO 124
Angle of repose 48
Anti-corrosive painting 302
Apron feeder 479
Arris 312
Asbestos cement sheets 325
Asbestos-free profiled
  sheets 325
    functional requirements
     of 325
Ashlar masonry 312, 315
Atterberg limits 49
Audit 507
Auditee 507
Auditor 507

B

Backhoe 427
Backing 312
Balanced cantilever
   bridge 298
Ball
   mill 477
   penetration test 176
Bar 344
Barrel vault 330, 332
Basic mechanics of
  breakage 104
Batching 480
   and mixing 373
   plant storage 482
   plants 482

Bats 312
Batter 312
Beam/girder
   construction 322, 323
Bed 312
   leveller 89
Belt
   conveyor 467, 492
   drives 395
Benching 106
Bentonite 71
Binder 347
BIS 533
Bit 111
Bitumen 328
   grouting 76
   putty 344
Bituminous
   concrete 365
   macadam 363
Blast
   cleaning 304
   pressure 192
Blastability 106
Blasthole 111
   drill 115
   pressure 121
Blasting 105, 547
   conventional smooth 130,
    131
   delay 127, 129
   instantaneous 129
   patterns and firing
    sequence 128
   theory 105

Blasting caps   126
   electric detonators   126
   fuse/non-electric   126
Blending of
   three admixtures   206
   two admixtures   205
Block-in-course masonry   316
Bolting   262
Bond   312
   English   316
   flemish   316
   heading   316
   Herringbone   317
   raking   317
   stretching   316
Bonded terrazzo   346
Bonding agent   247
Bottom charge   106, 109
Brakes   398
Brick bonding   311
Bricklaying   317
Bricks   311, 320
   first class   311
   second class   311
   third class bricks   311
Brickwork   312
Brucite   184
Bulk
   density   48
   excavation   55
Bulldozer   414
Burden   105, 106
Bush chains   395
Bushings and bearings   392
Buttress   312

C

Cable-stayed bridge   299
Caissons   76
   box   79
   monolithic   78
   open   77
   pneumatic   79
Capillary fringe   375
Capillary pores   179, 201, 327
Capping layer   361

Carbonation   180
Cartridge hammers   404
Cast-in-situ reinforced concrete roofs   322
Cavitation   190
Cellulose   273
Cement   172, 251, 320
   content   222
   grout   74
   concrete pavement   368
   paste   173
   plant recycling   366
Ceramic   342
   veneer   312
Chain drives   394
Chalk putties   344
Chase   312
Chemical
   attack   184
   grout   75
   stability   223
Chipboard   208
Chloride induced corrosion   185
Churn drill   113
Circular saws   404
Cladding   309
Clamp vibrators   405
Clamshell   431
   buckets   432
   dredger   94
Clay   46
Climbing tower crane   460
Closer   312
Clutches   397
Coal tar putty   344
Cofferdam   59
   cellular   62
   double skin   62
   single skin   61
Collision   538
Coloured plaster finish   340
Column
   charge   107
   forms   210
Communication facilities   26
Compacting factor test   175
Compaction   48, 359

Competence   507
Compressibility   48
Compression   104
   couplings   391
Compressive strength   222
Compressors   401
   reciprocating type   401
   rotary screw type   402
   rotary vane type   402
Concrete
   reinforced   368
   dowel-jointed   368
   foam   327
   hardened   177
   heavyweight   224
   high performance   227
   lightweight   219
   plastic   174
   prestressed   239
   ready-mixed   226
   self-compacting   229
   semi-dense bituminous   364
   un-reinforced   368
Concrete   172, 250, 252
   block   313
   cracking   180
   hoists   465
   production   232, 480
   propertie   173
   pumps   484
   shell roofs   329
Concreting   549
   extreme weather   232
   hot weather   232
   train   371, 372
   underwater   241
Concussion   136
Conformity   507
Connections   262
Conoid shells   332
Consolidation   48
Construction activities   2
   auxiliary   3
   general   2
   specialized   2

Construction
    documents   9
    equipment   29
    estimating   5
    joints   369
    materials   27
    methods   26
    process   3
    records   10
    schedule   5
Construction workers   4
    semiskilled   4
    skilled   4
    unskilled   4
Container's efficiency factor (CEF)   414
Contiguous piling   71
Control of
    documents   513
    records   514
Cooling measures   233
Coping   312
Corbel   313
Correction   508
Corrective action   508, 525
Corrosion   156
    chloride induced   185
    inhibitor   247
    -inhibiting admixtures   203
Coupling link   394
Couplings   390, 490
    clamped   391
    rigid   390
Course   313
CPM   6
Cramp   312
Crashing   7
Crawler
    dozer   415
    draglines   430
Cross-ply rubber tyre   406
Crushers
    cone   476
    jaw   471
    gyratory   472
    primary   469, 470

reduction   469, 476
roll   476, 477
tertiary   477
Crushing plant   470
Culverts   380
Curing   232
    of concrete   248
    methods of   248
Cushion   162
    blasting   135
Customer   502
    communication   519
    focus   514
    satisfaction   502
    -related processes   518
Cutter head   444
Cuttings   420
Cycle time   88
    of a shovel   426
    of clamshell   432

D

Damp-proof membrane   345
Danger   531
Decorative plastering   339
Deep bored wells   64
Defect   508
Deflagration   123
Dense graded bituminous macadam   364
Density   222
Depth per bit   112
Derrick   41
    guyed   448
    scotch   448
Detonation   123
    pressure   121
    velocity of   120
Diaphragm
    pump   403
    walls   69
Diffusion   180
Discharge hoppers   470
Document   508
Dolly   162
Domes   335

Double curvature shells   329
    rows of upright   495
    toggle jaw crusher   471
    acting hammer   160
    plate clutch   397
Dowel bar   313, 371
Down-the-hole pneumatic drill   117
Dragline   429
Drainage   375
Drawbar   412
    horsepower   412
Dredged materials   101
Dredger   86, 87
    amphibious   100
    bucket   99
    cutting suction   92
    dipper   98
    dustpan   94
    grab hopper   95
    scraper   100
    suction   91
    trailing suction hopper   88
Drifter   113
    drills   116
Drillability   110
Drilling
    bits   112
    efficiency   110
    equipment   111
    of blastholes   110
    pattern   111, 128
    rate   110
    square   128
    staggered   128
Drills   403
Drip   313
Drop hammer   158
Dry
    density   48
    drill   113
    rubble   314, 315
Durability   179, 229
Dye-penetration test   288
Dynamite   123

E

Earthwork 355
Ease of
    handling 207
    tripping 207
Edge preparation 271
Effectiveness 505
Effects of sea water 184
Efficiency 505
Efflorescence 183
Electric
    drive 399
    tools 403
Electrical
    power 19
    safety 535
Electro-osmosis 66
Electrodes 272
EDM equipment 352
Electromigration 180
Embankment 56, 357
Engineering 1
Equipment selection 386
Erection 291
    by pushing 297
    of bridges 296
    using crane 298
    using gantry girder 298
Ergonomics 531
Ettringite 183
Event 531
Excavation 56, 355, 545
    mechanized 53
    rock 103
    sloped 55
    surface 103
    confined 54
Expanded polystyrene
  blocks 327
Explosion and fire 538
Exploratory boreholes 46
Explosive 104, 105, 118
    sensitivity and
        sensitiveness 119

F

Fabrication 215
    and erection 544
Face 106, 310, 313, 341
    height 105
    shovel 424
Failure 531
    of support 538
Falsework 31
Fastening 292
Feeders and hoppers 469
Fibre
    base felt 328
    -reinforced concrete 237
Filling 420
    ability 231
Final
    location survey 352
    set 196
Finishing 232, 235
Finishing coats 349
Fire
    detector 534
    exposure 190
    fighting 25, 534
    protection 534
    resistance 223
First aid 25, 553
Fixed
    cableway 466
    -form paver 374
Flame
    cleaning 304
    detector 535
Flanged couplings 391
Flash set 196
Flashover tendency 119
Flat
    belts 396
    roof 322
    -slab construction 322
Flexible couplings 391
Floating coats 338
Floor failure 538
Floors 345, 550
Flow test 176

Fluid coupling 392
Fluid torque converter 392
Fly-rock 137
Footing forms 209
Forklift truck 465
Formwork 206, 250, 300, 548
Fragmentation 108, 109
Frame 344
Free face 105
Freestanding tower crane 456
Freeze-thaw action 187
    resistance against 224
Frequency rate 556
Friction grips 295
    piles 140
    welding 264
Frog 313
Full trailer 40
Fumes 121

G

Gabions 360
Gantry crane 461
Gauges of tracks 37
Gears 389
Gel
    pores 201
    water 201
Geotextiles 360
Girder 323
Glass 343
    bullet-resistant 343
    clear window 343
    corrugated 343
    fibre reinforced cement 323
    fibre reinforced plastic 323
    heat-absorbing 343
    laminated 343
    obscure 343
    obscure wired 343
    plate and float 343
    polished wired 343
    tempered 343
    tinted and coated 343
    transparent mirror 344
Glazing 343, 344

Global positioning systems   352
Grade   502
Grader   86, 433
Granite   342
Gravels and crushed stones   198
Greases   394
Grizzles   479
Ground freezing   66
Groundwater   375
    control   56
Grout   247, 302, 302, 313
    masonry   313
Gullies   377

H

H-piles   147
Hammer mill   473, 476
Hammers   404
Handling and transportation of units   289
Hauling and handling   38
Hanging leader   163
Hazard   531
Header   313
Heaped capacity   40
Heat
    detectors   534
    input   275
    insulation   326
    of hydration   196
Heat-affected zone   275
Heating measures   236
High explosive   123
High-speed rollers   443
Hoist   41, 461
Horizontal boom tower crane   456
Hot deep galvanized corrugated sheets   325
Hot dip galvanizing   306
Housekeeping   532
Human resources   517
Hydrated cement paste   201
Hydration   195
Hydraulic   399
    backhoe dredger   97

    concrete pump   488
    excavator   435
    gear pump   400
    impact hammer   161
    motors   400
    pile driver   167
    vane pumps   400
Hydrostatic piston pump   400
Hyperbolic paraboloid shells   332

I

Illumination   22
Impact
    breaker   475
    compactors   359, 442
    hammers   158
    pile driving   157
In-situ recycling   366
Incident   531
Inclined
    planes   396
    vibrating screens   479
Information   508
Infrastructure   505, 517
    development   17
Initial
    ground contamination   551
    set   196
Initiating
    event   531
    explosives   125
Injury   531
Inspection   506
Insulation   324, 326
Integrity testing   170
Interested party   503
Intermediate event   531
Internal
    audit   524
    communication   516
Intumescent coatings   349
ISO
    9000 quality system   509
    9001:1994   510
    9001:2000   509, 512

J

Jackhammer   113, 116
Jacking pit   81
Jamb   313, 344
Jaw
    clutch   397
    crusher   471
Jigsaws   404
Job-built forms   209

K

Kieselguhr   123
Kinds of drilling   111

L

Laboratory   25
Ladder   543
Laitance   190
Leveling   292
Levers   388
Limestone   342
Line drilling   134
Lining   292
Linking roads   21
Lintels   313
Liquid limit   49
Load test   167, 168
Load-bearing wall   310, 313
Long-span barrel vaults   331
Longitudinal joints   369
Loss of load   538
Low explosive   122
Lubricants   393, 394
    synthetic   394
Luffing
    boom tower crane   455
    cableway   466

M

Magnetic particle test   288
Management   502
    commitment   514
    representative   515
    responsibilities   514
    review   516
Mandrel   163

Index

Manhole 377
Manual preparation 304
Masonry 309, 311, 313
   bonding 311
Materials 207, 215, 225, 227, 230, 238
   for embankment formation 358
Mean sea level 52
Mechanical handling 537
Mechanical preparation 304
Mechanization 9
Metal 209, 348
   inert gas 267
Micro
   piles 155
   pores 327
Minimum putty 344
Mix 225, 231
   design 221, 480
Mixer
   central plant 482
   drum 481
   pan 481
   paving 483
   truck/transit 483
Mobile
   hoists 462
   scaffold 497
Moisture content 47
Monitoring 523
Monolithic terrazzo 346
Monorail 469
Mortar 311, 313, 320, 339
Mullions 313

N

National
   building code 535
   electrical code 536
Neat cement finish 340
Negative skin friction 142
Nitroglycerin 123
Noise pollution 135
   air-blast 136
   dust 135
   vibration 136

Non-destructive testing 253
Nonconformity 507

O

Objective evidence 506
Operator errors 538
Ordinary plastering 338
Organization 505
Overloading 538
Overturning 538

P

Paint coatings 305
Painting 347
Paints 349
   flame-retardant 349
   fungicide 349
   heat-resisting 349
   masonry 349
   water-repellent 349
   waterproofing 349
Parallel moving cableway 466
Particle board 208
Partitions 313
Passenger hoists 464
Passive earth pressure 49
Pebble-dash finish 341
Pendentive domes 335
Penetration 269
Percussion drill 111, 112
Percussive boring 84
Permanent
   exclusion 67
   forms 323
Permeability 179, 180, 223
Personal safety 533
PERT 6
Pier 313
Pigments 347
Pilaster 313
Pile/s
   bored 152
   bored cast-in-situ 152
   box 147
   continuous flight auger 153
   displacement 143

   driven cast-in-situ displacement 149
   end-bearing 140
   helical cast-in-situ displacement 151
   hollow precast tubular 145
   mini 155
   preformed steel 146
   precast concrete 144
   replacement 151
   tubular section 146
Pile
   driving equipment 157
   helmet 139, 162
Piling rig 164
Pipe
   bursting 84
   jacking 32, 81
Pipeline 489
Pipes 379
Piston
   drill 117
   pump 486
Placer boom 490
Placing 216, 480
Placing and curing 234, 237
Planning 515, 539
   of product realization 517
Plaster 348
   finish 341
Plastering 338, 340
Plastic
   glazing 344
   limit 49
Plasticity
   index 50
   of soil 49
Plumbing 292
Plywood 208
Pneumatic
   dredgers 100
   motor 398
Poker vibrators 405
Polymer
   concrete 245, 328
   -impregnated concrete 243

Index

-modified cement
    concrete 244
Porosity 48
Portland cement 196
Ports 38
Position of test welds 285
Post-tensioning 241
Power sources 401
Pozzolona 183
Pre-splitting 133, 357
Pre-tensioning 240
Precast reinforced concrete
    roofs 324
    structures 293
Prefabricated forms 209
Prefabrication 30
Preheating 277
Preliminary
    coat 338
    survey 352
Pressure grouting 74
Preventive action 508, 526
Primers 349
Procedure 502
Process 501
Product 501
Product realization 517
Production 233, 235, 236
    measurement 413
Productivity 9
Project 15, 505
    linear 384
    fast-tracked 384
    reports 9
    site development 50
Pseudo-random noise code 353
Pulse echo method 257
Pump operation 485
Pumping aid 205
Pumps 402
    centrifugal type 402
    displacement type 402
    submersible type 402
Purchasing
    information 521
    process 521

Putty 344

Q

QMS planning 515
Qualification
    of welders 284
    process 506
Quality 11, 502
    assurance 11, 504
    audit 506
    control 11, 231, 504
    improvement 504
    management system 503, 512
    manual 505, 513
    objective 503, 515
    plan 503
    planning 503
    policy 503, 514
Quoin 313

R

Rabbet 344
Radial
    cableway 466
    cracks 104
    -ply rubber tyres 406
Radiography 288
Rail-mounted tower
    crane 457
Rails 344
Rebound test 253
Receiving and unloading 291
Receiving pit 82
Reconnaissance survey 352
Record 506
Refractory masonry 318
Regrade 508
Reinforcing
    steel 214, 252
    steel bars 249
Relative
    bulk strength 120
    weight strength 120

Reliability 11, 531
Remote positioning unit 353
Repair 508
Residual stresses 281
Resistivity 193
Resource management 516
Review 506
    input 516
    output 516
Revolving drum screen 479
Rework 508
Rigid concrete pavement 366
Rimpull 412
Ripper 417
Ripping 110, 417
Risk 531
Riveting 262
Roads 351
Rocks
    drillability of 109
    igneous 109
    metamorphic 109
    sedimentary 109
Rod mill 478
Roller 359, 439
    chains 394
    link 394
Rolling resistance 409
Roofing 324
Roof 322
    covered with sheets 325
    covering 325
Rope-suspended leader 164
Rotary
    drilling 113
    drilling rig 114
Rotary-percussion drilling 115
Rough-coat finish 341
Rubble
    random coursed 314, 315
    square coursed
    un-coursed 314, 315
Rubble masonry 314
Rutile 273

S

Safety 11, 528
    committee 529
    fuse 125
    guidelines 538, 544, 546, 548, 549, 550
    hazard 326
    weathering 185
Sand drains 67, 376
Sands 197
    calcareous 197
    siliceous 197
    bulking of 48
Sandstone 342
Sash 344
Scabbing 105
Scaffold 542, 543
    slung 498
    suspended 498
    truss-out 495
Scaffolding 31, 318, 319, 494
Scattering 108
Science 1
Scrap 508
Scraped finish 341
Scraper 419
    types of 422
Screens 479
Screw
    classifier 480
    drivers 404
Scrubbing drum 480
Secant piled walls 72
Seepage pressure 48
Segregation 177
Self-propelled
    scraper 422
    telescopic boom crane 452
Setting
    out 51
    time 177
Severity rate 556
Shafts 390
Shanks 417

Shear 104
Shear legs 448
Sheep's-foot rollers 441
Sheet piling 58, 68
Shell
    roofs 329
    structures 329
Shoring 32
Short-span barrel vaults 331
Shot drill 115
Shotcrete 217
Shoulders, islands, and median 362
Shovel's production 427
Shrinkage and distortion 278
Shrinkage
    factor 413
    limit 49
    percent 413
Silent chain 395
Sill 344
Simple wash drum 480
Single
    acting hammers 159
    curvature shell structures 329
    plate (disc) clutch 397
    row of uprights 494
    shot grouting 75
    toggle jaw crusher 472
Skin
    construction 329
    friction 140, 142
Slab and beam forms 210
Slings 294
Slipform 213
Slump test 175
Slurry 124
    seals 365
    trench cut-off wall 73
Smoke detectors 534
Soils 44
Soil investigation 46
Specific gravity 47

Standard penetration test 47
Solid
    brickwork 316
    masonry wall 314
Sound insulation 326
Spacing of holes 108
Special
    dredgers 100
    forms 213
Specification 509
Sponge finish 341
Spreader bar 295
Springs 388
Spuds 87, 163
Squeeze pressure pump 487
Stabilization 360
Static hoists 462
Static weight rollers 440
Stiles 344
Stone masonry 312, 314
    hard 314
    soft 314
Stones 320
Stoper 113
Storage after sorting 291
    bins 470, 482
    facilities 24
    preservation 41
Store life 122
Strength 177
Stress corrosion 193
Stress-relieving 281
Stretcher 314
Stripping of forms 247
Struck
    capacity 39
    volume 413
Structural steel 261
Sub-bases 361
Sub-grade 360
Subcontractors 530
Submersible pump 403
Subsurface water drainage 376

Index

Sulphate attack 182
Sump pumping 57, 58
Superplasticizer 231
Supplier 505
Supported static tower crane 457
Surface 103
   cutting 104
   dressing 365
   preparation 303, 348
   quarrying 104
   stripping 104
   water drainage 376
Surfactants 201
Survey 352
Suspension bridge 301
Swell factor 413
Sympathetic detonation 119

T

Tamping vibrators 405
Tangents 353
Technical expert 507
Technology 2
Temporary exclusion 56, 57
Ten commandments 535, 536
Tensile strength 223
Tension 104
Terrazzo 345
Test 506
Textured finish 341
Theodolites 352
Thermal
   influences 191
   insulation over roofs 326
Thin grouted membrane 73
Throw 108
Thrust boring 82
Timber 208, 348
   domes 336
   piles 148
TNT 124
Topsoil 46
Torque wrench 292

Total charge 107
Tower cranes 455
Traceability 507
Track-mounted crane 449, 459
Tracks 407
Tractor 39, 40, 410
Training 4, 12, 528
Transit 232, 234, 235, 237
Transportation 539
   airways 39
   road 36
   railway 37
   waterway 38
Transverse joint 314, 369
Tremie 493
Trenching machine 436
Trencher
   ladder-type 437
   wheel-type 437
Trenchless technology 81
Trial pits 47
Truck-mounted
   strut-boom crane 455
   telescopic boom crane 453
   tower crane 460
Trucks 39
Tungsten inert gas (TIG) 267
Tunnel boring machine 444
Two shot grouting 75

U

Ultrasonic
   inspection 289
   pulse method 255
Under-reaming of pile 154
Underbed 324
Undercoats 349
Unsafe
   acts 554
   conditions 555

V

V–belts 396
Vacuum lifter 296

Validation 506
Vane pumps 400
Veebee consistometer test 176
Veneer 314
Verification 506
Vertical-down technique 270
Vertical-up technique 270
Vibrating plate feeder 478
Vibratory
   pile drivers 164
   rollers 442
Viscosity agent 231
Void ratio 48

W

Wagon drills 116
Walking draglines 430
Wall 314
Wall forms 212
Washing 479
Water 172, 253, 320
   bound macadam 361
   : cement ratio 196
   content 222
   injection dredger 90
   logging 21
   pollution control 552
   reducing admixtures 201
   requirement 20
   resistance 121
   table 47, 375
   tightness 178
Waterproofing 324, 328
   over roofs 327
Weathering 314
Weigh hoppers 482
Weld cracks 277
Weldability 274
Welders 541
Welding 262
   and flame cutting 540
   cold-pressure 264
   fusion 264
   hot-pressure 264

manual metal arc 267
mechanizing arc 282
positions 269
submerged arc 267
technology 263
Well point system 63
Wet mix macadam 361

Wheel tyres 405
Wheel-mounted cranes 450
  with cantilever boom 451
  with strut-boom 451
Wheeler dozer 415
Winches 447
Window 344

Windsor probe test 254
Wire ropes 446
Work environment 505, 517
Workability 174
Working drawings 261
Workmanship 320
Wythe 314